LIBRARY SERVICES

Oligonucleotide Synthesis

METHODS IN MOLECULAR BIOLOGY™

John M. Walker, SERIES EDITOR

300. **Protein Nanotechnology:** *Protocols, Instrumentation, and Applications*, edited by *Tuan Vo-Dinh*, 2005
299. **Amyloid Proteins:** *Methods and Protocols*, edited by *Einar M. Sigurdsson*, 2005
298. **Peptide Synthesis and Application**, edited by *John Howl*, 2005
297. **Forensic DNA Typing Protocols**, edited by *Angel Carracedo*, 2005
296. **Cell Cycle Protocols**, edited by *Tim Humphrey and Gavin Brooks*, 2005
295. **Immunochemical Protocols, Third Edition**, edited by *Robert Burns*, 2005
294. **Cell Migration:** *Developmental Methods and Protocols*, edited by *Jun-Lin Guan*, 2005
293. **Laser Capture Microdissection:** *Methods and Protocols*, edited by *Graeme I. Murray and Stephanie Curran*, 2005
292. **DNA Viruses:** *Methods and Protocols*, edited by *Paul M. Lieberman*, 2005
291. **Molecular Toxicology Protocols**, edited by *Phouthone Keohavong and Stephen G. Grant*, 2005
290. **Basic Cell Culture**, *Third Edition*, edited by *Cheryl D. Helgason and Cindy Miller*, 2005
289. **Epidermal Cells**, *Methods and Applications*, edited by *Kursad Turksen*, 2004
288. **Oligonucleotide Synthesis**, *Methods and Applications*, edited by *Piet Herdewijn*, 2005
287. **Epigenetics Protocols**, edited by *Trygve O. Tollefsbol*, 2004
286. **Transgenic Plants:** *Methods and Protocols*, edited by *Leandro Peña*, 2005
285. **Cell Cycle Control and Dysregulation Protocols:** *Cyclins, Cyclin-Dependent Kinases, and Other Factors*, edited by *Antonio Giordano and Gaetano Romano*, 2004
284. **Signal Transduction Protocols**, *Second Edition*, edited by *Robert C. Dickson and Michael D. Mendenhall*, 2004
283. **Bioconjugation Protocols**, edited by *Christof M. Niemeyer*, 2004
282. **Apoptosis Methods and Protocols**, edited by *Hugh J. M. Brady*, 2004
281. **Checkpoint Controls and Cancer, Volume 2:** *Activation and Regulation Protocols*, edited by *Axel H. Schönthal*, 2004
280. **Checkpoint Controls and Cancer, Volume 1:** *Reviews and Model Systems*, edited by *Axel H. Schönthal*, 2004
279. **Nitric Oxide Protocols**, *Second Edition*, edited by *Aviv Hassid*, 2004
278. **Protein NMR Techniques**, *Second Edition*, edited by *A. Kristina Downing*, 2004
277. **Trinucleotide Repeat Protocols**, edited by *Yoshinori Kohwi*, 2004
276. **Capillary Electrophoresis of Proteins and Peptides**, edited by *Mark A. Strege and Avinash L. Lagu*, 2004
275. **Chemoinformatics**, edited by *Jürgen Bajorath*, 2004
274. **Photosynthesis Research Protocols**, edited by *Robert Carpentier*, 2004
273. **Platelets and Megakaryocytes, Volume 2:** *Perspectives and Techniques*, edited by *Jonathan M. Gibbins and Martyn P. Mahaut-Smith*, 2004
272. **Platelets and Megakaryocytes, Volume 1:** *Functional Assays*, edited by *Jonathan M. Gibbins and Martyn P. Mahaut-Smith*, 2004
271. **B Cell Protocols**, edited by *Hua Gu and Klaus Rajewsky*, 2004
270. **Parasite Genomics Protocols**, edited by *Sara E. Melville*, 2004
269. **Vaccina Virus and Poxvirology:** *Methods and Protocols*, edited by *Stuart N. Isaacs*, 2004
268. **Public Health Microbiology:** *Methods and Protocols*, edited by *John F. T. Spencer and Alicia L. Ragout de Spencer*, 2004
267. **Recombinant Gene Expression:** *Reviews and Protocols, Second Edition*, edited by *Paulina Balbas and Argelia Johnson*, 2004
266. **Genomics, Proteomics, and Clinical Bacteriology:** *Methods and Reviews*, edited by *Neil Woodford and Alan Johnson*, 2004
265. **RNA Interference, Editing, and Modification:** *Methods and Protocols*, edited by *Jonatha M. Gott*, 2004
264. **Protein Arrays:** *Methods and Protocols*, edited by *Eric Fung*, 2004
263. **Flow Cytometry**, *Second Edition*, edited by *Teresa S. Hawley and Robert G. Hawley*, 2004
262. **Genetic Recombination Protocols**, edited by *Alan S. Waldman*, 2004
261. **Protein–Protein Interactions:** *Methods and Applications*, edited by *Haian Fu*, 2004
260. **Mobile Genetic Elements:** *Protocols and Genomic Applications*, edited by *Wolfgang J. Miller and Pierre Capy*, 2004
259. **Receptor Signal Transduction Protocols**, *Second Edition*, edited by *Gary B. Willars and R. A. John Challiss*, 2004
258. **Gene Expression Profiling:** *Methods and Protocols*, edited by *Richard A. Shimkets*, 2004
257. **mRNA Processing and Metabolism:** *Methods and Protocols*, edited by *Daniel R. Schoenberg*, 2004
256. **Bacterial Artifical Chromosomes, Volume 2:** *Functional Studies*, edited by *Shaying Zhao and Marvin Stodolsky*, 2004
255. **Bacterial Artifical Chromosomes, Volume 1:** *Library Construction, Physical Mapping, and Sequencing*, edited by *Shaying Zhao and Marvin Stodolsky*, 2004

METHODS IN MOLECULAR BIOLOGY™

Oligonucleotide Synthesis

Methods and Applications

Edited by

Piet Herdewijn

*Rega Institute, Medicinal Chemistry Lab,
Katholieke Universiteit Leuven,
Leuven, Belgium*

HUMANA PRESS ✻ TOTOWA, NEW JERSEY

© 2005 Humana Press Inc.
999 Riverview Drive, Suite 208
Totowa, New Jersey 07512

www.humanapress.com

All rights reserved. No part of this book may be reproduced, stored in a retrieval system, or transmitted in any form or by any means, electronic, mechanical, photocopying, microfilming, recording, or otherwise without written permission from the Publisher. Methods in Molecular Biology™ is a trademark of The Humana Press Inc.

All comments, opinions, conclusions, or recommendations are those of the author(s), and do not necessarily reflect the views of the publisher.

This publication is printed on acid-free paper. ∞
ANSI Z39.48-1984 (American Standards Institute)
Permanence of Paper for Printed Library Materials.
Cover illustration: Figure 1 from Chapter 18 "High-Throughput Production of Optimized Primers (Fimers) for Whole-Genome Direct Sequencing," by Nikolai Polushin, Andrei Malykh, A. Michael Morocho, Alexei Slesarev, and Sergei Kozyavkin.

Production Editor: Wendy S. Kopf.
Cover design by Patricia F. Cleary.

For additional copies, pricing for bulk purchases, and/or information about other Humana titles, contact Humana at the above address or at any of the following numbers: Tel.: 973-256-1699; Fax: 973-256-8341; E-mail: humana@humanapr.com; or visit our Website: www.humanapress.com

Photocopy Authorization Policy:
Authorization to photocopy items for internal or personal use, or the internal or personal use of specific clients, is granted by Humana Press Inc., provided that the base fee of US $25.00 per copy is paid directly to the Copyright Clearance Center at 222 Rosewood Drive, Danvers, MA 01923. For those organizations that have been granted a photocopy license from the CCC, a separate system of payment has been arranged and is acceptable to Humana Press Inc. The fee code for users of the Transactional Reporting Service is: [1-58829-233-9/05 $25.00].

Printed in the United States of America. 10 9 8 7 6 5 4 3 2 1

Library of Congress Cataloging in Publication Data

e-ISBN: 1-59259-823-4
ISSN: 1064-3745

Oligonucleotide synthesis : methods and applications / edited by Piet Herdewijn.
    p. ; cm. -- (Methods in molecular biology ; 288)
  Includes bibliographical references and index.
  ISBN 1-58829-233-9 (alk. paper)
  1. Oligonucleotides--Synthesis--Laboratory manuals.
  [DNLM: 1. Oligonucleotides--chemical synthesis--Laboratory Manuals. 2. Oligonucleotides--chemistry--Laboratory Manuals. QU 25 O465 2004] I. Herdewijn, Piet. II. Series: Methods in molecular biology (Totowa, N.J.) ; v. 288.
  QP625.O47O54 2004
  572.8'5--dc22
                                    2004002114

Preface

Nucleic acids chemistry has been fundamental to molecular biology and recently it has moved into therapeutic areas. The evolution of this chemistry has been tremendous during the last ten years and new techniques are regularly introduced and existing protocols are ameliorated. It is not possible to give an overview of the evolution of the different aspects of nucleic acids chemistry and their applications within the context of one book. The aim of *Oligonucleotide Synthesis: Methods and Applications* is to give the readers an insight into new key developments and to deliver protocols and critical comments for the practical execution of the experiments. Inside details from the inventors of protocols has often proven to be of the utmost importance for the successful application of a new technique.

Oligonucleotide Synthesis: Methods and Applications covers new developments in the fundamental chemistry of nucleic acids as well as new applications of nucleic acids with tremendous potential, such as RNA interference.

High-throughput DNA synthesis is normally done by the phosphoramidite four-step process. Now, a novel two-step cycle, developed by Marvin Caruthers, enables a higher purity DNA to be obtained in a less costly way. New discoveries, such as siRNA, have driven the scientific community to come up with synthetic methods to obtain RNA as efficiently as DNA. Thanks to the development of new reagents, the classical RNA synthesis using a 2'-*O*-tert-butyldimethylsilyl protecting group has reached the level of 99% coupling yields. The synthesis and purification of RNAs by this method is described by Brian Sproat. New methods have also been developed in the field of RNA synthesis. One of these methods uses an inverse protection scheme (acid labile group in the 2'-*O*-position and fluoride labile group in the 5'-*O*-position), and this protocol is covered by William Marshall. Using this method, the oligonucleotide is stored in its 2'-*O*-protected form and final deprotection is carried out before use.

The most widely used oligonucleotides in antisense technology are phosphorothioates, mainly because of the susceptibility of the target RNA to be cleaved by RNase H (when hybridized with phosphorothioate oligonucleotide). The synthesis of phosphorothioates has been improved considerably by the introduction of new sulfur transfer reagents, as described by Yogesh Sanghvi. The guidelines to monitor the RNA cleavage reaction using ribonucleases H have been written by Masad Damha, in which illustrative examples on selecting particularly potent, enzyme eliciting AON are provided by the 2'-fluoroarabinonucleic acids (2'F-ANA) and their analogs.

Although the phosphoramidite approach has become the method of choice for oligonucleotide synthesis, the H-phosphonate alternative, as described by Jurek Stawinski and Roger Strömberg, remains a useful alternative, especially for solution-phase synthesis and when only a small excess of building blocks are used. The original phosphotriester approach to oligonucleotide synthesis has no longer a routine application, but is still very useful, for example for the ring closure of circular oligonucleotides. The synthesis of circular oligonucleotides is described by Enrique Pedroso, and he points to the many potential applications of circular DNA.

The antisense technology has stimulated research on modified oligonucleotides that hybridize very strongly with complementary RNA, mainly as a result of the preorganization of the carbohydrate moiety of the building units of these oligonucleotides into a "northern-type" conformation. The most strongly binding oligonucleotides are LNA, whose synthesis is summarized by Jesper Wengel. The HypNA–pPNA chimers of Vladimir Efimov are water soluble PNA-like hybrids with DNA/RNA binding properties very close to that of PNA itself. Besides sugar modification, duplex stability may also be increased by base modifications. The 7-substituted 8-aza-7-deazapurine base, as described by Frank Seela, can bring the stability of a dA–dT base pair to the same level as for dG–dC pair. The synthesis of hypermodified nucleotides and their incorporation into oligonucleotides is a recent field of research that is of particular importance for investigation on the local structure and the specificity of tRNA. Darell Davis describes here several examples of hypermodified nucleoside phosphoramidites and their incorporation into RNA.

The topic of conjugate chemistry is covered by several important examples. Peptide conjugation, as described by Michael Gait, is an extremely important method to improve cellular delivery, as well as cell-specific targeting of oligonucleotides. Biotin-labeled oligonucleotides are widely used for oligonucleotide detection and isolation. The synthesis of biotin phosphoramidites with extraordinarily long tether and biotin-labeled oligonucleotide is covered by Nikolai Polushin. Ali Laayoun describes the synthesis of reactive reporter groups for labeling of nucleic acids on their phosphates and the results obtained with high-density DNA chip analysis.

Triple helix formation is a unique approach that uses oligonucleotides to target double-stranded DNA, this allows interaction at an early stage of gene expression. Jian-Sheng Sun describes the synthesis of conjugates between triple helix-forming oligonucleotides and camptothecin, which is a way to direct the cleavage activity of topoisomerases.

Before moving to the application section, a chapter is introduced where Michael Göbel describes fluorescence-based on-line detection in RNA electrophoresis. The importance of developing fluorescence technology is that, sooner or later, all techniques in molecular biology and biochemistry that use radioisotope-based protocols, will be replaced by less hazardous (radiation protection) and more user friendly methods (waste disposal, faster analysis).

One of the most widely used techniques for the quantification of nucleic acids is the real-time polymerase chain reaction assay, which uses molecular beacons to monitor the amplification process. Jacqueline Vet describes this procedure in a stepwise fashion. An important factor for the quality of an amplification reaction, is the selection of optimized primers (choice of sequences and adjustment of reaction conditions). Nikolai Polushin and his group have developed modified primers or fimers to direct sequencing of genomic DNA.

The "RNA world" hypothesis has stimulated research on the nonenzymatic template-directed synthesis of RNA. The contribution of Michael Göbel is aimed at providing reliable protocols for carrying out these experiments, using nonradioactive labeling procedures for monitoring the process. The nuclease footprinting technique, initially developed to study protein–DNA interactions, is now widely used to investigate the sequence-selective binding of small molecules to DNA. This technique is described by Christian Bailly, using actinomycin D as a model compound. The search for strong and selective DNA binding ligands is a continuing process, given the need for molecules that may control gene expression. One of the most unique tools for interfering with the function and metabolism of nucleic acids is PNA. The most important properties of PNA and some of their analogs are reviewed by Peter Nielsen.

It is becoming impossible to cover modern techniques in science and overlook library approaches. Depending on the application, the desired library may vary from very low sequence variation, to very high diversity. Seven techniques to construct nucleic acids libraries of different diversity are described by Peter Unrau, whereas Andres Jäschke is concentrating on protocols describing combinatorial nucleic acid libraries applied in the field of ribozymes accelerating a Diels-Alder reaction. The SELEX approach for aptamer design against RNA and protein targets is covered by Jean-Jacques Toulmé.

The last chapter, written by Jean-Remi Bertrand, deals with the most exiting new area of oligonucleotide applications, i.e., the use of small interference RNA (siRNA) as an inhibitor of gene expression, Although the mode of action of siRNA is not fully understood, the use of siRNA as an inhibitor of gene expression (both in vitro and in vivo), has changed the strategy for the analysis of gene functions and the control of their expression.

We hope that *Oligonucleotide Synthesis: Methods and Applications* will give the reader a flavor of several important new directions in nucleic acids chemistry and their applications, as well as provide detailed protocols for carrying out these experiments. Some topics are more specific, other are more general. Rather than presenting a laboratory manual dealing with one specific topic of nucleic acids chemistry, we have decided to cover a broad field of research to make the manual useful for a large audience and attract the interest of scientists in this expanding field of research. The book starts with a new protocol for DNA synthesis and finishes with details of RNA interference experiments.

Most of all, I would like to cordially thank all our contributors, appreciating the time they have taken to write the chapters and knowing that none of them are looking for additional paper work that only keeps them away from their experiments.

Piet Herdewijn

Contents

Preface v
Contributors xiii

1. Synthesis of DNA Using a New Two-Step Cycle
   **Douglas J. Dellinger, Jason R. Betley,
   Tadeusz K. Wyrzykiewicz, and Marvin H. Caruthers** 1
2. RNA Synthesis Using 2'-O-(Tert-Butyldimethylsilyl) Protection
   Brian S. Sproat 17
3. RNA Oligonucleotide Synthesis
   Via 5'-Silyl-2'-Orthoester Chemistry
   **Stephanie A. Hartsel, David E. Kitchen, Stephen A. Scaringe,
   and William S. Marshall** 33
4. Dimethylthiarum Disulfide: *New Sulfur Transfer Reagent
   in Phosphorothioates Oligonucleotide Synthesis*
   Zhiwei Wang, Quanlai Song, and Yogesh S. Sanghvi 51
5. Assay for Evaluating Ribonuclease H-Mediated Degradation
   of RNA–Antisense Oligonucleotide Duplexes
   **Annie Galarneau, Kyung-Lyum Min, Maria M. Mangos,
   and Masad J. Damha** 65
6. Di- and Oligonucleotide Synthesis
   Using H-Phosphonate Chemistry
   Jacek Stawinski and Roger Strömberg 81
7. Solid-Phase Synthesis of Circular Oligonucleotides
   **Enrique Pedroso, Nuria Escaja, Miriam Frieden,
   and Anna Grandas** 101
8. Locked Nucleic Acid Synthesis
   **Henrik M. Pfundheller, Anders M. Sørensen, Christian Lomholt,
   Anders M. Johansen, Troels Koch, and Jesper Wengel** 127
9. Synthesis of DNA Mimics Representing
   HypNA-pPNA Hetero-oligomers
   Vladimir A. Efimov and Oksana G. Chakhmakhcheva 147
10. Base-Modified Oligonucleotides With Increased Duplex Stability:
    Pyrazolo[3,4-d]pyrimidines Replacing Purines
    **Frank Seela, Yang He, Junlin He, Georg Becher, Rita Kröschel,
    Matthias Zulauf, and Peter Leonard** 165

11. Introduction of Hypermodified Nucleotides in RNA
    Darrell R. Davis and Ashok C. Bajji 187
12. Chemical Methods for Peptide–Oligonucleotide
    Conjugate Synthesis
    Dmitry A. Stetsenko and Michael J. Gait 205
13. Biotin-Labeled Oligonucleotides
    With Extraordinarily Long Tethering Arms
    *A. Michael Morocho, Valeri Karamyshev, Olga Shcherbinina,
    and Nikolai Polushin* 225
14. Universal Labeling Chemistry for Nucleic Acid Detection
    on DNA Chips
    *Ali Laayoun, Eloy Bernal-Méndez, Isabelle Sothier,
    Mitsuharu Kotera, and Alain Troesch* 241
15. Postsynthetic Functionalization
    of Triple Helix-Forming Oligonucleotides
    Alexandre S. Boutorine, and Jian-Sheng Sun 251
16. Fluorescence-Based On-Line Detection as an Analytical Tool
    in RNA Electrophoresis
    Ute Scheffer and Michael Göbel 261
17. Design and Optimization of Molecular Beacon
    Real-Time Polymerase Chain Reaction Assays
    Jacqueline A. M. Vet and Salvatore A. E. Marras 273
18. High-Throughput Production of Optimized Primers
    (Fimers) for Whole-Genome Direct Sequencing
    *Nikolai Polushin, Andrei Malykh, A. Michael Morocho,
    Alexei Slesarev, and Sergei Kozyavkin* 291
19. Nonenzymatic Template-Directed RNA Synthesis
    Marcus Hey and Michael Göbel 305
20. DNAse I Footprinting of Small Molecule Binding Sites on DNA
    *Christian Bailly, Jérôme Kluza, Christopher Martin,
    Thomas Ellis, and Michael J. Waring* 319
21. Gene Targeting Using Peptide Nucleic Acid
    Peter E. Nielsen 343
22. Nucleic Acid Library Construction
    Using Synthetic DNA Constructs
    Hani S. Zaher and Peter J. Unrau 359
23. In Vitro Selection From Combinatorial Nucleic Acid Libraries
    Andres Jäschke 379

24. In Vitro Selection Procedures for Identifying DNA
    and RNA Aptamers Targeted to Nucleic Acids and Proteins
    **Eric Dausse, Christian Cazenave, Bernard Rayner,
    and Jean-Jacques Toulmé** *391*
25. Short Double-Stranded Ribonucleic Acid as Inhibitor
    of Gene Expression by the Interference Mechanism
    Jean-Rémi Bertrand, Andrei Maksimenko, and Claude Malvy *411*

Index *431*

Contributors

CHRISTIAN BAILLY • *INSERM U-524, Institut de Recherches sur le Cancer de Lille, Lille, France*
ASHOK C. BAJII • *Myriad Genetics, Salt Lake City, UT*
GEORG BECHER • *Laboratorium für Organische und Bioorganische Chemie, Institut für Chemie, Universität Osnabrück, Osnabrück, Germany*
JASON R. BETLEY • *Emmbrook, Suffolk, UK*
ELOY BERNAL-MÉNDEZ • *Advanced Technology/Molecular Diagnostics, BioMérieux, Marcy L'Etoile, France*
JEAN-RÉMI BERTRAND • *Institut Gustave Roussy, CNRS UNR 8121, Vecterologie et Transfert de Gènes, Villejuif Cedex, France*
ALEXANDRE S. BOUTORINE • *Laboratoire de Biophysique, Muséum Nationale d'Histoire Naturelle, Paris cedex 05, France*
MARVIN H. CARUTHERS • *Department of Chemistry and Biochemistry, University of Colorado at Boulder, Boulder, CO and Agilent Laboratories, Palo Alto, CA*
CHRISTIAN CAZENAVE • *INSERM U386, Modulation Artificielle des Gènes Eucaryotes, Université Victor Segalen Bordeaux 2, Bordeaux, France*
OKSANA G. CHAKHMAKHCHEVA • *Institute of Bio-organic Chemistry, Russian Academy of Sciences, Moscow, Russia*
MASAD J. DAMHA • *Department of Chemistry, McGill University, Montreal QC, Canada*
ERIC DAUSSE • *INSERM U386, Modulation Artificielle des Gènes Eucaryotes, Université Victor Segalen Bordeaux 2, Bordeaux, France and Institut Européen de Chimie et Biologie, Pessac, France*
DARRELL R. DAVIS • *Department of Medicinal Chemistry, University of Utah, Salt Lake City, UT*
DOUGLAS J. DELLINGER • *Agilent Laboratories, Fort Collins, CO*
VLADIMIR A. EFIMOV • *Institute of Bio-organic Chemistry, Russian Academy of Sciences, Moscow, Russia*
THOMAS ELLIS • *Department of Pharmacology, University of Cambridge, Cambridge, UK*
NURIA ESCAJA • *Department de Química Orgánica, Universitat de Barcelona, Barcelona, Spain*
MIRIAM FRIEDEN • *Santaris Pharma, Copenhagen, Denmark*
MICHAEL J. GAIT • *Laboratory of Molecular Biology, MRC, Cambridge, UK*

ANNIE GALARNEAU • *Department of Chemistry, McGill University, Montreal QC, Canada*
MICHAEL GÖBEL • *Department of Organic Chemistry and Chemical Biology, Johann Wolfgang Goethe-University Frankfurt, Frankfurt am Main, Germany*
ANNA GRANDAS • *Department de Química Orgánica, Universitat de Barcelona, Barcelona, Spain*
STEPHANIE A. HARTSEL • *Research, Development and Production, Dharmacon, Lafayette, CO*
YANG HE • *Laboratorium für Organische und Bioorganische Chemie, Institut für Chemie, Universität Osnabrück, Osnabrück, Germany*
JUNLIN HE • *Laboratorium für Organische und Bioorganische Chemie, Institut für Chemie, Universität Osnabrück, Osnabrück, Germany*
PIET HERDEWIJN • *Medicinal Chemistry Lab, Katholieke Universiteit Leuven, Leuven, Belgium*
MARCUS HEY • *Department of Organic Chemistry and Chemical Biology, Johann Wolfgang Goethe-University Frankfurt, Frankfurt am Main, Germany*
ANDRES JÄSCHKE • *Institute of Pharmacy and Molecular Biology, Heidelberg University, Heidelberg, Germany*
ANDERS M. JOHANSEN • *Department of Chemistry, Exiqon A/S, Vedbæk, Denmark*
VALERI KARAMYSHEV • *Fidelity Systems Inc., Gaithersburg, MD*
DAVID E. KITCHEN • *Research, Development and Production, Dharmacon, Lafayette, CO*
JÉROME KLUZA • *INSERM U-524, Génétique Moléculaire et Approches Thérapeutiques de Hémopathies Malignes, Lille, France*
TROELS KOCH • *Department of Chemistry, Santaris A/S, Copenhagen, Denmark*
MITSUHARU KOTERA • *CNRS UMR 5616, Laboratoire d'Etudes Dynamiques et Structurales de la Sélectivité (LEDSS), Université Joseph Fourier, Grenoble, France*
SERGEI KOZYAVKIN • *Fidelity Systems Inc., Gaithersburg, MD*
RITA KRÖSCHEL • *Laboratorium für Organische und Bioorganische Chemie, Institut für Chemie, Universität Osnabrück, Osnabrück, Germany*
ALI LAAYOUN • *Advanced Technology/Molecular Diagnostics, BioMérieux, Marcy L'Etoile, France*
PETER LEONARD • *Laboratorium für Organische und Bioorganische Chemie, Institut für Chemie, Universität Osnabrück, Osnabrück, Germany*
CHRISTIAN LOMHOLT • *Department of Chemistry, Exiqon A/S, Vedbæk, Denmark*
ANDREI MAKSIMENKO • *BioAlliance Pharma SA, Paris, France*
CLAUDE MALVY • *Institut Gustave Roussy, CNRS UNR 8121, Villejuif cedex, France*
ANDREI MALYKH • *Fidelity Systems Inc., Gaithersburg, MD*

Contributors

MARIA M. MANGOS • *Department of Chemistry, McGill University, Montreal QC, Canada*
SALVATORE A. E. MARRAS • *Department of Molecular Genetics, Public Health Research Institute, Newark, NJ*
WILLIAM S. MARSHALL • *Research, Development and Production, Dharmacon, Lafayette, CO*
CHRISTOPHER MARTIN • *Department of Pharmacology, University of Cambridge, Cambridge, UK*
KYUNG-LYUM MIN • *Department of Chemistry, McGill University, Montreal QC, Canada*
A. MICHAEL MOROCHO • *Fidelity Systems Inc., Gaithersburg, MD*
PETER E. NIELSEN • *Department of Medical Biochemistry and Genetics, The Panum Institute, University of Copenhagen, Copenhagen, Denmark*
ENRIQUE PEDROSO • *Department de Química Orgánica, Universitat de Barcelona, Barcelona, Spain*
HENRIK M. PFUNDHELLER • *Department of Chemistry, Exiqon A/S, Vedbæk, Denmark*
NIKOLAI POLUSHIN • *Fidelity Systems Inc., Gaithersburg, MD*
BERNARD RAYNER • *INSERM U386, Modulation Artificielle des Gènes Eucaryotes, Université Victor Segalen Bordeaux 2, Bordeaux, France*
YOGESH S. SANGVHI • *Rasayan Inc., Encinitas, CA*
STEPHEN A. SCARINGE • *Research, Development and Production, Dharmacon, Lafayette, CO*
UTE SCHEFFER • *Department of Organic Chemistry and Chemical Biology, Johann Wolfgang Goethe-University Frankfurt, Frankfurt am Main, Germany*
FRANK SEELA • *Laboratorium für Organische und Bioorganische Chemie, Institut für Chemie, Universität Osnabrück, Osnabrück, Germany*
OLGA SHCHERBININA • *Fidelity Systems Inc., Gaithersburg, MD*
ALEXEI SLESAREV • *Fidelity Systems Inc., Gaithersburg, MD*
QUANLAI SONG • *Medicinal Chemistry Department, ISIS Pharmaceuticals, CA, Carlsbad*
ANDERS M. SØRENSEN • *Department of Chemistry, Cureon A/S, Copenhagen, Denmark*
ISABELLE SOTHIER • *Advanced Technology/Molecular Diagnostics, BioMérieux, Marcy L'Etoile, France*
BRIAN S. SPROAT • *RNA-TEC, Leuven, Belgium*
JACEK STAWINSKI • *Arrhenius Laboratory, Department of Chemistry, Stockholm University, Stockholm, Sweden and Institute of Bioorganic Chemistry, Polish Academy of Sciences, Poznan, Poland*
DMITRY A. STETSENKO • *Laboratory of Molecular Biology, MRC, Cambridge, UK*

ROGER STRÖMBERG • *Scheele Laboratory, Division of Organic and Bioorganic Chemistry, Karolinska Institutet, Stockholm, Sweden*

JIAN-SHENG SUN • *Laboratoire de Biophysique, Muséum Nationale d'Histoire Naturelle, Paris Cedex 05, France*

JEAN-JACQUES TOULMÉ • *INSERM U386, Modulation Artificielle de Gènes Eucaryotes, Université Victor Segalen Bordeaux 2, Bordeaux, France and Institut Européen de Chimie et Biologie, Pessac, France*

ALAIN TROESCH • *Advanced Technology/Molecular Diagnostics, BioMérieux, Marcy L'Etoile, France*

PETER J. UNRAU • *Department of Molecular Biology and Biochemistry, Simon Fraser University, Burnaby, B.C., Canada*

JACQUELINE A. M. VET • *Food Science Division, Bio-Rad, Marnes la Coquette, France*

ZHIWEI WANG • *Medicinal Chemistry Department, ISIS Pharmaceuticals, Carlsbad, CA*

MICHAEL J. WARING • *Department of Pharmacology, University of Cambridge, Cambridge, UK*

JESPER WENGEL • *Nucleic Acid Center, Department of Chemistry, University of Southern Denmark, Odense M, Denmark*

TADEUSZ K. WYRZYKIEWICZ • *Isis Pharmaceuticals, Carlsbad, CA*

HANI S. ZAHER • *Department of Molecular Biology and Biochemistry, Simon Fraser University, Burnaby, BC, Canada*

MATTHIAS ZULAUF • *Laboratorium für Organische und Bioorganische Chemie, Institut für Chemie, Universität Osnabrück, Osnabrück, Germany*

1

Synthesis of DNA Using a New Two-Step Cycle

Douglas J. Dellinger, Jason R. Betley, Tadeusz K. Wyrzykiewicz, and Marvin H. Caruthers

Summary

For the first time a new, two-step method is described for synthesizing deoxyribonucleic acid. The approach uses 5'-carbonate protected 2'-deoxynucleoside-3'-phosphoramidites as synthons and a peroxy anion buffer that removes the carbonate protecting group and oxidizes the internucleotide linkage. Following synthesis via this two-step cycle, oligomers are isolated by standard procedures.

Key Words: Two-step synthesis cycle; 5'-carbonate protection; peroxy anion deprotection and oxidation.

1. Introduction

For sometime now, the method of choice for the chemical synthesis of oligodeoxynucleotides has been the phosphoramidite four-step process that uses the reaction of acid-activated deoxynucleoside phosphoramidites with solid-phase tethered deoxynucleoside (*1,2*). Initially, in **Fig. 1**, *step 1*, the 5'-*O*-dimethoxytrityl (DMT) group is removed from a deoxynucleoside linked to the polymer support (**Fig. 1**). In *step 2*, elongation of the growing oligodeoxynucleotide occurs via the formation of a phosphite triester internucleotide bond. This reaction product is first treated with a capping agent *step 3* to esterify failure sequences and cleave phosphite reaction products on the heterocyclic bases. The nascent phosphite internucleotide linkage is then oxidized to the corresponding phosphotriester in *step 4*. In the first step of each cycle, the DMT group is removed from the growing oligodeoxynucleotide using a large excess of a weak acid, trichloroacetic acid (TCA), in an organic solvent. Further repetitions of this four-step process generate the oligodeoxynucleotide of desired length and sequence. The final product is cleaved from the solid

From: *Methods in Molecular Biology, vol. 288: Oligonucleotide Synthesis: Methods and Applications*
Edited by: P. Herdewijn © Humana Press Inc., Totowa, NJ

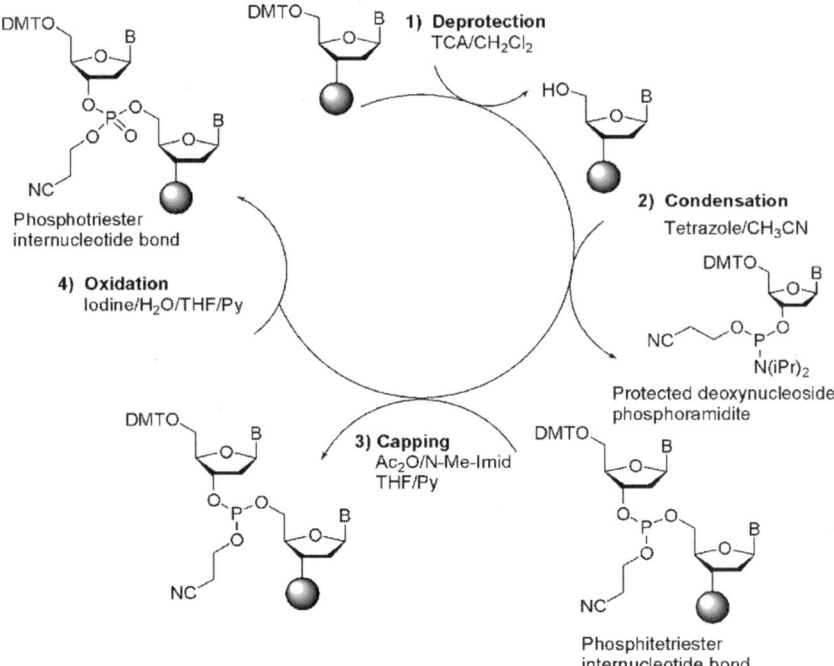

Fig. 1. Solid-phase, four-step phosphoramidite oligodeoxynucleotide synthesis cycle. TCA, trichloroacetic acid; B, appropriately protected deoxynucleoside base; N-Me-Imid, *N*-methylimidazole; ●, controlled pore glass support.

phase and obtained free of base and phosphate protecting groups by treatment of the support with concentrated ammonium hydroxide.

Although oligodeoxynucleotides synthesized via this chemistry are of satisfactory quality for most biological uses (deoxyribonucleic acid [DNA] sequencing, polymerase chain reaction [PCR] applications, and site-specific mutagenesis), many recently discovered applications can be only marginally carried out using this approach. These uses include large-scale synthesis, the preparation of DNA microarrays on planar glass surfaces, and the synthesis of genes or large gene fragments. The problem is side-reactions, which are known to occur at detectable but acceptable levels during routine synthesis. These can rise to unacceptable levels under the conditions required for new, expanded applications. Two of the most important series of side reactions can be traced to the acid detritylation procedure (**Fig. 1**, *step 1*). This step leads to depurination of protected bases and, because the detritylation step is reversible, a series of oligomers lacking one or more of the correct nucleotides. As a consequence, mutations accumulate at levels that seriously limit certain applications of synthetic DNA.

DNA Synthesis Using a New Two-Step Cycle

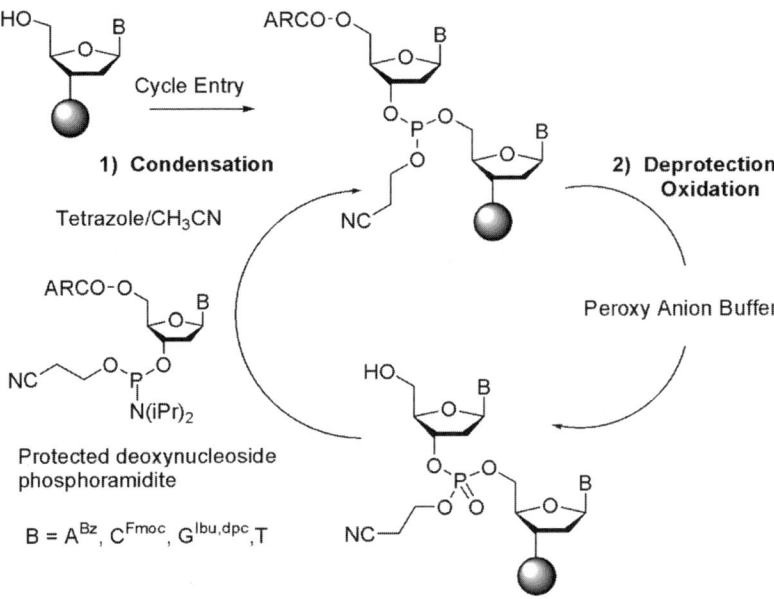

Fig. 2. Solid-phase two-step phosphoramidite oligodeoxynucleotide synthesis cycle.

Because of these limitations, a new two-step synthesis cycle has been developed (**Fig. 2**). The key to this approach is the use of aqueous peroxy anions buffered under mildly basic conditions, pH 9.6, to remove an aryloxycarbonyl (ARCO) group, which is substituted for DMT in the four-step cycle outlined in **Fig. 1**. This peroxy anion solution is both strongly nucleophilic and mildly oxidizing. The strong nucleophilicity of this reagent permits quantitative and irreversible removal of the 5'-O-carbonate protecting group, whereas its mild oxidative potential quantitatively oxidizes the internucleotide phosphite triester without detectable oxidation of the heterocyclic bases. This combination of desirable chemical properties into one reagent (peroxy anion deprotection and oxidation) generates a new, two-step cycle that uses fewer solvents and eliminates acid catalyzed detritylation from the synthesis procedures. As a consequence depurination and reversibility of the detritylation step cease to exist as problems in oligomer synthesis. Because cyclical yields are so high, capping with acetic anhydride can also be eliminated. DNA prepared by this two-step procedure will be useful not only for new, extended applications but also for current applications. Moreover, because the cycle is considerably abbreviated and many steps eliminated (i.e., disposal of chlorinated reagents), the synthesis of DNA is now less expensive. Protocols needed to carry out this two-step cycle are described.

2. Materials

2.1. Instruments

1. Solid-phase DNA synthesis is performed on an ABI 394 automated DNA synthesizer from ABI/PE Biosystems.
2. Nuclear magnetic resonance (NMR) data are recorded on a Varian VXR-300 or a Bruker 400MHz spectrophotometer. Tetramethylsilane is used as an internal reference for ^{1}H and ^{13}C NMR. An external capillary containing 85% H_3PO_4 is used as a reference for ^{31}P NMR.
3. Desorption/ionization time-of-flight (MALDI-TOF) mass data is obtained on a Voyager DE-STR instrument from Perseptive Biosystems by comparison to an internal oligodeoxynucleotide standard.
4. Electrospray (ES) mass spectra are recorded on a Hewlett Packard 5989B mass spectrometer with an HP 9987A ES LC/MS interface coupled to a series II HP 1090 liquid chromatograph.
5. High-pressure (performance) liquid chromatography (HPLC) is performed using an Agilent Technologies 1100 instrument.
6. Autoradiography and fluorescent imaging is performed on a Molecular Dynamics Typhoon Model 8600 (Amersham Biosciences).
7. DNAPac column (25 cm × 4 mm ID) is from Dionex Corp.
8. ODS-Hypersil column (25 cm × 4 mm ID) is from Agilent Technologies.
9. Resource-Q column (30 cm × 6.4 mm ID) is from Amersham Biosciences.

2.2. Reagents

1. Dichloromethane and pyridine are distilled over calcium hydride. Dichloromethane is used immediately after distillation.
2. Anhydrous acetonitrile (Fisher Scientific).
3. Dioxane, pyridine and N,N'-dimethylformamide are purist grade (Fluka).
4. Dichloromethane, pyridine, toluene, chloroform, ethanol, hexanes, ethylacetate, and triethylamine are reagent grade and supplied by Fisher Scientific, Mallenkrodt, or Fluka.
5. Anhydrous sodium sulfate is obtained from Mallenkrodt.
6. 4-Chlorophenyl chloroformate is purchased from Alfa Chemicals.
7. 2-Amino-2-methyl-1-propanol is obtained as a $1.5M$ aqueous solution, pH 10.3, from Sigma Diagnostics.
8. 2-Cyanoethyl-N,N,N',N'-tetraisopropylphosphorodiamidite is obtained from Chemgenes Corporation and purified by distillation under high vacuum.
9. 1H-tetrazole is purchased from Fluka.
10. Snake venom phosphodiesterase I from *Crotalus adamanteus* and alkaline phosphatase are obtained from US Biosciences.
11. Thin-layer chromatography is performed on aluminum-backed silica F_{254} plates from EM Sciences.
12. Medium-pressure, preparative column chromatography is performed using 230–400 mesh silica gel from EM Science.

13. Meta-chloroperbenzoic acid was purchased from Aldrich at 55–83% pure and used without further purification.
14. Control pore glass (CPG) columns having a long chain alkylamino linker, one µmol 2'-deoxynucleoside, and 500-Å glass pore size were purchased from Glen Research Corporation: dA (cat. no. 20-2100-41), AcdC (cat. no. 20-2113-41), dG (cat. no. 20-2120-41), dT (cat. no. 20-2130-41).
15. 2'-Deoxythymidine, 4-N-(9-fluorenylmethoxycarbamoyl)-2'-deoxycytidine, 2-N-isobutyryl-2'-deoxyguanosine, and 6-N-benzoyl-2'-deoxyadenosine were purchased from Glen Research Corporation.
16. 4-Chlorophenyl chloroformate was obtained from Alfa.
17. Anhydrous argon gas was supplied by Airgas, Radnor, PA.
18. Drierite was purchased from W. A. Hammond Drierite Company.

2.3. Solutions and Buffers

1. Saturated aqueous solution of sodium bicarbonate.
2. Saturated aqueous solution of sodium chloride (brine).
3. Solution of ethanol (10%) in water (v/v).
4. 0.45M Tetrazole in anhydrous acetonitrile.
5. Trichloroacetic acid (3%) in anhydrous dichloromethane (w/v).
6. Lithium hydroxide (3%) in water (w/w).
7. Dioxane (32.5 mL) and 30% hydrogen peroxide (10 mL) containing 1.78 g meta-chloroperbenzoic acid.
8. HPLC-grade water from Fisher Scientific.
9. A solution of dioxane/water (1/1, v/v).
10. 50 mM Triethylammonium bicarbonate (pH 7.0 in water).
11. 25 mM Tris-HCl buffer (pH 8.0) containing 5% acetonitrile.
12. An aqueous solution of 1M sodium chloride.
13. An aqueous solution of sodium perchlorate (98%, 10 mM) containing acetonitrile (2%) (v/v).
14. An aqueous solution of sodium perchlorate (98%, 300 mM) containing acetonitrile (2%) (v/v).
15. An aqueous solution of 10 mM sodium hydroxide and 80 mM sodium bromide.
16. An aqueous solution of 10 mM sodium hydroxide and 1.5M sodium bromide.
17. An aqueous solution of 50 mM triethylammonium bicarbonate (TEAB), pH 7.0.
18. 45 mg 2',4',6'-Trihydroxyacetophenone and 4 ammonium citrate crystals in acetonitrile/HPLC-grade water (1/1, v/v).
19. An aqueous buffer of Tris-HCl (32 mM) and magnesium chloride (12 mM), pH 7.5.
20. TBE buffer: an aqueous buffer containing 7M urea, 90 mM Tris-borate, pH 8.3, and 2 mM ethylenediaminetetraacetic acid (EDTA).
21. Solvent A: dichloromethane/ethanol (19/1, v/v).
22. Concentrated ammonium hydroxide was obtained from Mallenkrodt.

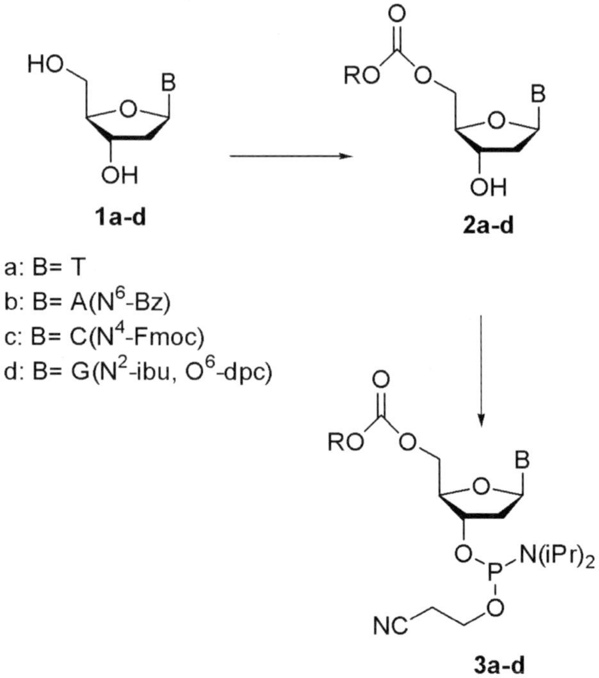
HO
B
O
OH
1a-d
RO
O
B
OH
2a-d
a: B= T
b: B= A(N6-Bz)
c: B= C(N4-Fmoc)
d: B= G(N2-ibu, O6-dpc)
RO
O
B
O
P
N(iPr)2
NC
3a-d

Fig. 3. Synthesis of 5'-*O*-carbonate protected 2'-deoxynucleoside phosphoramidites. Bz, benzoyl; Fmoc, 9-fluorenylmethoxycarbamoyl; ibu, isobutyryl; dpc, *N*,*N*-diphenylcarbamoyl; R, 4-chlorophenyl.

3. Methods

The following methods outline (1) the preparation of synthons useful in this new two-step cycle, (2) the synthesis of DNA, and (3) purification of oligomers.

3.1. Synthons

The preparation of the synthons is described in **Subheadings 3.1.1.–3.1.5.** The chemistry presented in this section is summarized in **Fig. 3**.

3.1.1. 5'-O-(4-Chlorophenyloxycarbonyl)-2'-Deoxythymidine (see ***Fig. 3, 2a***)

2'-Deoxythymidine (4.84 g, 20.0 mmol) is desiccated by coevaporation of entrapped water with anhydrous pyridine (3 × 100 mL) and then redissolved in anhydrous pyridine (800 mL). 4-Chlorophenyl chloroformate (3.08 mL, 22 mmol) is added dropwise, and vigorous stirring is used to dissolve any precipitate formed during addition. After 2 h solvent is removed *in vacuo* and residual pyridine eliminated from the oily residue by coevaporation with 100 mL toluene. The resulting oil is dissolved in 500 mL dichloromethane. This solution is extracted with 240 mL saturated $NaHCO_3$ followed by 250 mL brine, and the aqueous layers are back-extracted with 100 mL dichloromethane. The dichloromethane layer is separated and dried over anhydrous Na_2SO_4. The solvent is decanted, the Na_2SO_4 washed with dichloromethane, and the combined solutions evaporated to yield a viscous oil. Trituration with 300 mL chloroform and filtration affords compound **2a** as a white, microcrystalline solid (4.93 g, 62.2%); mp 178–180°C; R_f (solvent A) 0.22. ^{1}H NMR (dimethyl sulfoxide [DMSO]-d_6, 400 MHz): δ 7.40 (s, 1H), 7.25 (d, 2h, $J = 8.4$), 7.00 (d, 2H, $J = 8.4$), 6.25–6.20 (m, 1H), 4.40–3.95 (m, 4H), 2.30–2.05 (m, 2H), 1.80 (s, 3H). ^{13}C NMR (DMSO-d_6, 100.5 MHz): δ 164.4, 152.6, 150.5, 149.0, 135.5, 131.5, 129.4, 122.0, 110.9, 84.7, 83.5, 70.2, 67.5, 39.8, 12.0. Analysis calculated for $C_{17}H_{17}ClN_2O_7$: C, 51.5; H, 4.3; N, 7.1. Found: C, 51.3; H, 4.5; N, 7.0. electrospray ionization mass spectrometry (ESI-MS): m/z 397 (M + H$^+$, 100).

3.1.2. 5'-O-(4-Chlorophenyloxycarbonyl)-6-N-Benzoyl-2'-Deoxyadenosine (see **Fig. 3**, **2b**)

This compound is prepared analogously to compound **2a** using 6-*N*-benzoyl-2'-deoxyadenosine (7.10 g, 20.0 mmol) to yield a white powder (5.41 g, 53.1%); R_f (solvent A) 0.25. ^{1}H NMR (DMSO-d_6, 400 MHz): δ 9.40 (s, 1H), 8.70 (s, 1H), 8.20 (s, 1H), 7.95 (d, $J = 7.2$), 7.50 (t, 1H, $J = 7.6$), 7.40 (m, 2H), 7.20 (d, 2H, $J = 8.8$), 7.00 (d, 2H, $J = 8.8$), 6.45–6.40 (t, 1H, $J = 6.2$), 5.05–4.30 (m, 5H), 2.90–2.50 (m, 2H). ^{13}C NMR (DMSO-d_6, 100.5 MHz): 165.1, 153.0, 152.4, 151.3, 149.4, 141.7, 133.3, 132.8, 131.4, 129.4, 128.7, 123.2, 122.1, 84.7, 84.3, 71.3, 67.8, 39.9. ESI-MS: m/z 510 (M + H$^+$, 100).

3.1.3. 5'-O-(4-Chlorophenyloxycarbonyl)-4-N-(9-Fluorenylmethoxycarbamoyl)-2'-Deoxycytidine (see **Fig. 3**, **2c**)

4-*N*-(9-fluorenylmethoxycarbamoyl)-2'-deoxycytidine is prepared according to the method of Ti et al. (*3*). Compound **2c** is then prepared analogously to compound **2a** using 4-*N*-(9-fluorenylmethoxycarbamoyl)-2'-deoxycytidine (8.98 g, 20.0 mmol) to yield a white powder (7.57 g, 62.7%); R_f (solvent A) 0.38. ^{1}H NMR (DMSO-d_6, 400 MHz): δ 8.75 (br s, 1H), 8.05 (d, 1H, $J = 7.5$), 7.75 (d, 2H, $J = 7.5$), 7.57 (d, 2H, $J = 7.5$), 7.38 (t, 2H, $J = 7.5$), 7.35–7.25 (m, 5H), 7.10 (d, 2H, $J = 8.8$), 6.30 (t, 1H, $J = 6.4$), 4.55–4.20 (m, 6H), 2.90–2.10 (m, 4H). ^{13}C NMR (DMSO-d_6, 100.5 MHz): δ 162.7, 155.3, 153.1, 152.4, 149.2, 143.8, 143.1, 141.2, 140.9, 131.6, 129.5, 127.9, 127.1, 124.9, 122.2, 120.0, 95.4, 87.4, 70.7, 67.8, 46.5, 41.3. ESI-MS: m/z 604 (M + H$^+$, 100).

3.1.4. 5'-O-(4-Chlorophenyloxycarbonyl)-2-N-Isobutyryl-6-O-N,N-Diphenylcarbamoyl)-2'-Deoxyguanosine (see **Fig. 3**, **2d**)

Compound **2d** is synthesized by a two-step procedure from 2-*N*-isobutyryl-2'-deoxyguanosine (6.75 g, 20.0 mmol). Trace amounts of water are removed from the *N*-protected deoxynucleoside by coevaporation with anhydrous pyridine (3 × 100 mL). The residual material is redissolved in 800 mL anhydrous pyridine and 4-chlorophenyl chloroformate (3.08 mL, 22.0 mmol) added dropwise with vigorous stirring. After 2 h solvent is removed under reduced pressure and residual pyridine eliminated by coevaporation with 100 mL toluene. The resulting oil is dissolved in 500 mL dichloromethane, extracted with 250 mL saturated $NaHCO_3$ and 250 mL brine, dried over Na_2SO_4, and solvent-evaporated to yield a viscous yellow oil. The crude product is then purified by silica gel column chromatography (0–10% ethanol in dichloromethane) to give 5'-O-(4-chlorophenyloxycarbonyl)-2-N,-isobutyryl-2'-deoxyguanosine (6.03 g, 61.3%); R_f (solvent A) 0.12. ^{1}H NMR (DMSO-d_6, 400 MHz): δ 9.25 (br s, 1H), 8.85 (br s, 1H), 8.20 (s, 1H), 7.60 (d, 2H, J = 8.8), 7.35 (d, 2H, J = 8.8), 6.40 (t, 1H, J = 6.4), 4.60–4.20 (m, 4H), 2.75–2.50 (m, 3H), 1.20–1.10 (m, 6H). ^{13}C NMR (DMSO-d_6, 100.5 MHz): δ 180.1, 154.8, 152.6, 149.4, 148.4, 137.5, 130.3, 129.5, 123.1, 120.4, 83.8, 83.0, 70.4, 68.4, 34.8, 13.9. ESI-MS: m/z 492 (M + H$^+$). 5'-O-(4-chlorophenyloxycarbonyl)-2-N-isobutyryl-2'-deoxyguanosine (3.84 g, 7.8 mmol) is dissolved in a mixture of 250 mL dry pyridine and 10 mL dichloromethane. Diphenylcarbamoyl chloride (1.80 g, 7.80 mmol) is added dropwise and the reaction mixture stirred overnight. Water (5 mL) is added and the solvent evaporated under reduced pressure. The resultant oil is purified by silica gel column chromatography and the product eluted using an ethanol gradient (0–10%) in dichloromethane (*see* **Note 1**). Fractions containing ultraviolet-absorbing material are collected and analyzed by thin-layer chromatography. Product fractions are evaporated and the oil triturated with dichloromethane/hexanes (1/1) to yield compound **2d** as a white solid (3.57 g, 66.5%); R_f (solvent A) 0.38; ^{1}H NMR (DMSO-d_6, 400 MHz): δ 10.22 (s, 1H), 8.62 (s, 1H), 7.54–7.16 (m, 14H), 6.46 (t, 1H, J = 6.2), 5.62 (m, 1H), 4.72–4.17 (m, 4H), 2.80 (septuplet, 1H, J = 6.7), 3.00–2.40 (m, 2H), 1.10–1.05 (m, 6H). ^{13}C NMR (DMSO-d_6, 100.5 MHz): δ 175.0, 155.2, 154.3, 152.7, 152.3, 150.3, 149.5, 144.5, 141.6, 130.4, 129.5, 129.4, 127.3, 126.9, 123.1, 120.9, 84.3, 70.5, 68.6, 38.4, 34.7, 19.3, 19.2. ESI-MS: m/z 687 (M + H$^+$, 100).

3.1.5. Synthesis of 2'-Deoxynucleoside-3'-Phosphoramidites (see **Fig. 3**, 3a–d)

Protected 2'-deoxynucleosides *(see* **Fig. 3**, **2a–d**) (5.00 mmol each) and tetrazole (175 mg, 2.50 mmol per compound) are dried separately under vacuum for 24 h. The protected 2'-deoxynucleoside is dissolved in anhydrous dichloromethane (100 mL). 2-Cyanoethyl-N,N,N',N'-tetraisopropyl-phosphorodiamidite (2.06 mL, 6.50 mmol) and tetrazole are then added. The reaction mixture is allowed to stir overnight at room temperature (*see* **Note 2**). A sample of the crude reaction mixture is analyzed by ^{31}P NMR using either an external lock or unlocked. On completion of the synthesis, the reaction mixture is extracted with saturated 150 mL NaHCO$_3$ and 150 mL brine, and dried over Na$_2$SO$_4$. The combined dichloromethane solutions (including backwashes from the saturated NaHCO$_3$ and brine) are concentrated and the crude reaction mixture transferred to a silica gel column preequilibrated with hexanes. The dichloromethane is removed from the column with hexanes and the product eluted using a hexanes/ethyl acetate gradient (50–100% ethyl acetate). After evaporation of solvents under reduced pressure, products are isolated as white, glassy solids in yields varying between 85–95% (*see* **Note 3**). ^{31}P NMR (CDCl$_3$, 162.0 MHz): δ (compound *3a*) 149.01, 149.10; (compound *3b*) 148.86, 149.24; (compound *3c*) 148.60, 148.81; (compound *3d*) 148.99, 149.04.

3.2. Reagents for DNA Synthesis

The next procedures in this process involve the preparation of the reagents needed to synthesize DNA via this two-step approach on solid supports.

3.2.1. Amidites

The appropriately protected 2'-deoxynucleoside-3'-phosphoramidites *(see* **Fig. 3**, *3a–d)* are diluted to a standard concentration of 0.1*M* in anhydrous acetonitrile. These solutions are filtered through a 0.2 µ*M* filter before use. Because these compounds are water sensitive, anhydrous conditions are recommended when handling these reagents.

3.2.2. Coupling Activator

A 0.45*M* solution of tetrazole in dry acetonitrile is used to catalyze the coupling reaction. Prior to attachment to the DNA synthesizer, the solution is filtered through a 0.2 µ*M* filter. Anhydrous conditions are recommended while making and handling this reagent (*see* **Note 4**).

3.2.3. Acid Deprotection

Commercial CPG columns (1 µ*M* scale) containing 5'-dimethoxytrityl protected 2'-deoxynucleosides are currently being used with this procedure. Removal of the 5'-dimethoxytrityl group with 3% trichloroacetic acid in anhydrous dichloromethane is carried out manually. After washing the column with anhydrous dichloromethane, the resulting CPG column, having a 2'-deoxynucleoside with a free 5'-hydroxyl group, is mounted on the DNA synthesizer.

3.2.4. Cyclical 5'-Deprotection With Peroxy Anion Buffer

The peroxy anion buffer is prepared as a two-component system. Solution A is composed of 3% (w/v) lithium hydroxide (10 mL), 1.5*M* 2-amino-2-methyl-1-propanol (10 mL, pH 10.3), and 17.5 mL dioxane. Solution B contains 1.78 g meta-chloroperbenzoic acid, 10 mL 30% hydrogen peroxide, and 32.5 mL dioxane. These solutions are mixed in equal volumes to generate the peroxy anion deprotection buffer, pH 9.6.

Configuring this buffer as a two-component system is necessary because meta-chloroperbenzoic acid decomposes in the presence of lithium hydroxide. This decomposition not only reduces the effective peroxy anion concentration but it also lowers the pH, which decreases the nucleophilicity of the remaining reagent. By combining these solutions daily, a peroxy anion buffer is generated that remains active for 24 h. Meta-chloroperbenzoic acid is purified by extraction with an aqueous 250 mM phosphate buffer, pH 7.2, before use.

3.2.5. Solvents for DNA Synthesis

1. Anhydrous-grade acetonitrile is purchased from Fisher Scientific.
2. Dioxane is purist grade and supplied by Fluka.
3. Water is HPLC grade.

3.3. Instrument Modification

An important aspect of using the peroxy anion reagent on an automated DNA synthesizer is its viscosity. On our older ABI 380B automated synthesizer it is possible to separate the reagent into two parts that are placed in the capping ports. Although these two reagents have different viscosities, it is possible to deliver approximately equal volumes of these reagents to the column. With our newer ABI/PE Model 394 automated synthesizer, this differential viscosity makes it difficult to deliver equivalent volumes of the two components through the capping ports. On this instrument, we premix the two-component reagent and place the mixture in the trichloroacetic acid port. The viscosity effects and column fill times for these reagents vary greatly among commercial instruments and require testing. A consideration that is directly related to the column fill time is washing of the buffered aqueous reagent from the support. Wash times after peroxy anion deprotection are determined empirically and vary with the make and model of the DNA synthesizer. An effective wash cycle involves (1) aqueous dioxane (1:1) to remove residual salts, (2) reagent-grade dioxane, which eliminates water, and (3) anhydrous acetonitrile to remove the final traces of water. Incomplete removal of residual buffering agent can inhibit the acid-catalyzed phosphoramidite coupling reaction.

3.4. The DNA Synthesis Cycle

Synthesis as described in this section is carried out on an ABI/PE Model 394 DNA synthesizer. All syntheses as described are performed on a 1-μM scale using CPG columns linked to 5'-dimethoxytrityl protected 2'-deoxynucleo-

Table 1
Oligodeoxynucleotide Synthesis Cycle

Step	Function	Reagent	Cycle time (S)
1	Wash	Acetonitrile	25
2	Coupling	Amidite (0.1M, 20 Eq) Tetrazole (0.45M, 10 Eq) Anhydrous Acetonitrile	30
3	Wash	Acetonitrile	5
4	Wash	Dioxane	25
5	Deblock	1:1 Mixture of Peroxy Anion Solutions A and B	15
6	Wait		15
7–14	Repeat steps 5 and 6 four times		120
15	Wash	Dioxane:Water (1:1)	45
16	Wash	Dioxane	30

sides. The 5'-dimethoxytrityl protecting group is removed using 3% trichloroacetic acid in dichloromethane prior to automated two-step synthesis. 5'-ARCO modified 2'-deoxynucleoside-3'-phosphoramidites (*see* **Fig. 3**, **3a–d**, 0.1M each) in anhydrous acetonitrile are installed at ports 1–4. Other solutions unique to this two-step cycle are installed at ports 10 (peroxy anion buffer), 15 (dioxane:water, 1:1, v/v) and 19 (dioxane). Solutions unique to capping and oxidation in the four-step cycle are discarded. Removal of the 5'-carbonate following each 2'-deoxynucleotide addition is completed by treatment with peroxy anion buffer (15 s) and a wait step (15 s) repeated a total of five times. After this 5'-carbonate deprotection step, the column is washed with dioxane:water (1:1, v/v), dioxane, and anhydrous acetonitrile. The total cycle time is approx 5.5 min. A summary of the complete synthesis cycle is included in **Table 1** (*see* **Note 5**).

3.5. Work-Up and Analysis of Synthetic DNA

The synthetic DNA is freed of protecting groups and characterized using the following procedures. Analyses of oligomers synthesized using this two-step cycle are presented in **Fig. 4**.

3.5.1. Removal of Protecting Groups

Following completion of all synthesis steps (including removal of the 5'-carbonate from the final synthon added to the growing oligodeoxynucleotide), the product mixture is removed from CPG by treatment with concentrated

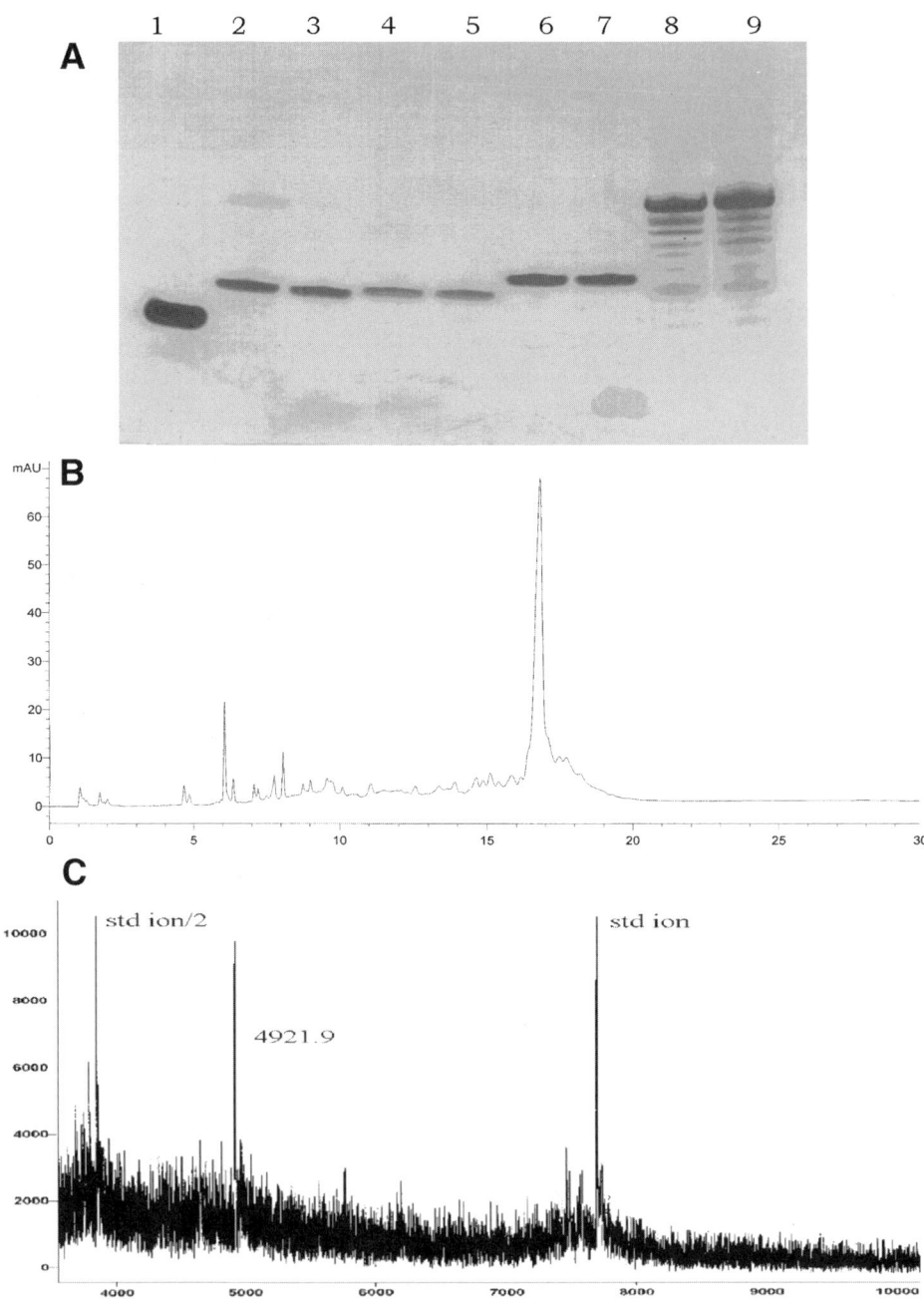

Fig. 4. Oligodeoxynucleotide analysis by polyacrylamide gel electrophoresis, ion-exchange HPLC, and MALDI-TOF mass spectrometry. (**A**) Polyacrylamide gel electrophoresis analysis of unpurified, completely deprotected reaction mixtures.

ammonium hydroxide for 60 min at room temperature. The ammonium reaction mixture free of CPG is cooled in an ice bath, transferred to a screw-cap reacti-vial fitted with a Teflon seal, and the solution is heated at 55°C overnight to remove the heterocyclic base protecting groups. The vial is then cooled in an ice bath, the reaction mixture transferred to an Eppendorf tube, and the solution evaporated to dryness (*see* **Note 6**).

3.5.2. Purification of Synthetic DNA

The dry mixture of reaction products is dissolved in water and purified by HPLC using an ODS-Hypersil (5 µ) column eluted at 1.5 mL/min. The eluant is 0–20% acetonitrile in 50 mM triethylammonium bicarbonate, pH 7.0 (linear gradient, 40 min). Product fractions are concentrated to dryness to remove the volatile buffer, resuspended in an appropriate buffer for biochemical experiments, and frozen (*see* **Note 7**).

3.5.3. Analysis of Synthetic DNA

Several analytical procedures are used to routinely assess the quality of synthetic DNA. These include ion-exchange HPLC, gel electrophoresis, MALDI-TOF mass spectrometry, and enzymic degradation of oligomers. NMR (^{31}P, ^{13}C, and ^{1}H) is indispensable as a research tool for developing new synthetic procedures and analyzing the quality of synthons. However, for analysis of oligomer products it is of marginal value.

1. Ion-exchange HPLC: Over the years we have used numerous ion-exchange columns with various salt gradients. Two sets of conditions we have used recently for oligomers 10–20 nucleotides in length are as follows.
    a. The HPLC is performed on a DNAPac column (25 cm × 4 mm ID) using (A) 25 mM Tris buffer (pH 8.0)/0.05% acetonitrile and (B) 1M NaCl as eluants at a flow rate of 1.5 mL/min. An alternative gradient is (A) 98% 10 mM sodium perchlorate/2% acetonitrile and (B) 98% 300 mM sodium perchlorate/2% acetonitrile at a flow rate of 1.5 mL/min.
    b. The HPLC is performed using a Resource-Q column (30 cm × 6.4 mm ID). Gradient eluants are (A) 10 mM NaOH/80 mM NaBr and (B) 10 mM NaOH/1.5M NaBr at a flow rate of 1.5 mL/min.

Fig. 4. (*continued*) Oligomers were synthesized using the two-step cycle *(lanes 3, 5, 7, and 9)* and the four-step cycle *(lanes 2, 4, 6, and 8)*. Lanes 2, 3: d(T$_2$A$_4$T$_2$); lanes 4, 5: d(T$_2$CT$_2$CT$_2$); lanes 6, 7: d(T$_2$GT$_2$GT$_2$); lanes 8, 9: d(CAGTTGTAAACGAGTT). BPB, bromophenol blue dye marker *(lane 1)*. **(B)** Ion-exchange HPLC of reaction products from the synthesis shown in *lane 9*, in **A**. **(C)** MALDI-TOF mass spectrum of the major product peak observed in the ion-exchange HPLC profile shown in **(B)** (17 min).

2. Gel electrophoresis: Although many different combinations of crosslinked polyacrylamide gels have been tried, for oligomers 10–20 in length, usually 15 or 20% crosslinked gels are used in $7M$ urea containing TBE buffer. Electrophoresis is usually carried out at 50–60°C over 3–6 h using 42-cm plates. A Molecular Dynamics Phosphorimager is used to visualize products on gels.
3. MALDI-TOF analysis of oligodeoxynucleotides: Oligodeoxynucleotides are purified by HPLC using 50 mM TEAB, pH 7.0 (*see* **Subheading 2.3.**), with a 0–20% acetonitrile gradient and, after concentrating to dryness, dissolved in acetonitrile/ultrapure water (1:1) to give a final concentration of 0.005 A_{260} U/µL. The matrix solution is prepared by mixing 45 mg of 2',4',6'-trihydroxyacetophenone and 4 ammonium citrate crystals in acetonitrile/ultrapure water (1:1, 500 µL). The suspension is vortexed, centrifuged, and the clear matrix solution (0.3 µL) is deposited on the probe. The oligodeoxynucleotide solution (0.3 µL) is then deposited on the same spot and the plate left to dry at room temperature for 5 min. All mass spectra are acquired in both negative and positive ion modes with delayed extraction and reflector operation.
4. Enzymatic degradation of oligomers: In a typical experiment the reaction mixture (30 µL) containing 0.3 µg of snake venom phosphodiesterase, 4.02 µg of alkaline phosphatase, buffer (32 mM Tris-HCl and 12 mM $MgCl_2$, pH 7.5), and an oligomer (0.5 A_{260} U) is incubated at 37°C for 4 h. Enzymes are then heat-denatured (95°C, 3 min), and the reaction mixture is diluted to 200 µL with water, centrifuged, and analyzed by HPLC (ODS-Hypersil [5 µ] column 4 × 250, flow 1.5 mL/min, 0–20% acetonitrile in 50 mM triethylammonium bicarbonate, pH 7.0 [linear gradient] in 40 min).

4. Notes

1. Following silica gel column chromatography, products such as compounds **2a–d** and **3a–d** (*see* **Fig. 3**) should be immediately stabilized as solids by removal of solvents. Otherwise trace amounts of water in solvents will cause hydrolysis of the carbonate (**2a–d**, **3a–d**) and the phosphoramidite (**3a–d**).
2. During the preparation of these synthons, diisopropyl ammonium tetrazolide is generated. This salt must be completely removed during flash chromatography. Otherwise it will depress the overall DNA synthesis rate and yield by buffering the condensation reaction. Excessive amounts of this salt can be detected in the ^{1}H NMR of the purified synthons. If excess salt is suspected as a source for low yields, a second silica gel column chromatography should solve the problem.
3. Purification of the synthons **3a–d** (*see* **Fig. 3**) is carried out by flash chromatography on silica gel. It is important to complete this chromatography rapidly. This is because silica gel is somewhat acidic. It will therefore catalyze considerable hydrolysis of the phosphoramidite if the synthon remains in contact with silica gel for some time. Usually we add 0.5–1% triethylamine to the eluting solvent. This base neutralizes the silica gel and therefore reduces hydrolysis of the phosphoramidite product.

4. The quality of tetrazole used to activate the synthons can affect the repetitive yields. Commercially available material, if prepared specifically for oligodeoxynucleotide synthesis ($0.45M$ acetonitrile solution), usually is sufficiently pure for direct use. However to ensure high-quality tetrazole, it can be purified by sublimation before use in DNA synthesis.
5. All nucleic acid chemists know that water (preferably the lack thereof) is the most critical parameter for synthesizing DNA. Certainly this rule applies to the two-step synthesis of DNA as outlined here. Among the various sources of water contamination, the most significant one is acetonitrile as it must be anhydrous for several key steps. These include its use as a solvent for the synthons and also tetrazole. Acetonitrile used to wash the CPG just prior to each condensation step should also be anhydrous. Commercial, synthesis-grade acetonitrile is usually sufficiently anhydrous for these steps. It should, however, be checked with a Karl Fischer titrator before use. To be absolutely sure that you are using anhydrous acetonitrile, distillation from calcium hydride immediately before use is recommended. Another source of water is its adsorption on the synthons during storage. This can be minimized by storage at $-20°C$ under dry argon in a desiccator equipped with Drierite. Prior to opening the desiccator, it should be allowed to attain room temperature. Following transfer of the synthons to vials for use on the synthesizer, these samples should be dried under vacuum overnight to remove final traces of water.
6. A common mistake that occurs during removal of protecting groups relates to the quality of concentrated ammonium hydroxide. Often aliquots of ammonium hydroxide are removed from the same bottle (usually a large bottle) over a prolonged period of time for many different DNA preparations. This is a mistake as ammonium hydroxide is volatile and loses much ammonia after several repetitive openings of a bottle. This problem can be overcome by purchasing small bottles, storing at $-5°C$, and discarding the contents after 3–4 DNA preparations. Alternatively a large bottle can be divided into small samples, stored at $-5°C$, and each aliquot used only once.
7. Triethylammonium bicarbonate is a volatile buffer. It should be stored at $4°C$. Just prior to use, carbon dioxide should be allowed to bubble through the buffer. This leads to reconversion of triethylamine to the bicarbonate salt. The pH should be checked prior to usage to assure that sufficient carbon dioxide has been used to reestablish the correct pH.

Acknowledgments

The authors thank Geraldine Fulcrand, Steve Lefkowitz, Michel Perbost, David Sheehan, Rachel Philosof-Oppenheimer, and Agnieszka Sierzchala for helpful scientific discussions. This work was supported by NIH (GM25680) and Agilent Technologies, Inc.

References

1. Caruthers, M. H., Barone, A. D., Beaucage, S. L., et al. (1987) Chemical synthesis of deoxyoligonucleotides by the phosphoramidite method. *Methods Enzymol.* **154,** 287–313.
2. Beaucage, S. L. and Caruthers, M. H. (2000) Synthetic strategies and parameters involved in the synthesis of oligodeoxyribonucleotides according to the phosphoramidite method. In *Current Protocols in Nucleic Acids Chemistry* (Beaucage, S. L., Bergstrom, D. E., Glick, G. D., and Jones, R. A., eds.), Wiley, New York, Unit 3.3, pp. 1–20.
3. Ti, G. S., Gaffney, B. L., and Jones, R. A. (1982) Transient protection: efficient one-flask syntheses of protected deoxynucleosides *J. Am. Chem. Soc.* **104,** 1316–1319.

2

RNA Synthesis Using 2'-O-(Tert-Butyldimethylsilyl) Protection

Brian S. Sproat

Summary

This chapter enables the reader to carry out the solid-phase synthesis of ribonucleic acid (RNA) using β-cyanoethyl phosphoramidite chemistry combined with *tert*-butyldimethylsilyl protection of the ribose 2'-hydroxyl group. Phosphoramidite monomers are activated with 5-benzylmercapto-1*H*-tetrazole enabling fast and highly efficient coupling to the 5'-hydroxyl group of the support-bound oligonucleotide. On completion of the synthesis, the stepwise deprotection of the nucleobase, phosphate, and ribose protecting groups is carried out using optimized protocols. Subsequently the various high-pressure (performance) liquid chromatography (HPLC) procedures are described enabling the purification and analysis of the RNA. For this purpose anion-exchange and reversed-phase HPLC are used singly or in combination according to the final purity requirement of the RNA.

Key Words: Oligoribonucleotide synthesis; TBDMS; 5-benzylmercapto-1*H*-tetrazole; solid phase; triethylamine tris(hydrofluoride); anion exchange; HPLC; β-cyanoethyl phosphoramidite; chaotropes.

1. Introduction

Until a few years ago the chemical synthesis of ribonucleic acid (RNA) was still very much in its infancy, lagging way behind deoxyribonucleic acid (DNA) technology, with only a few groups worldwide able to synthesize and purify it properly. However, with the very recent intense interest in synthetic RNA, largely triggered by the discovery and utility of small interfering RNAs (siRNAs) *(1,2)* but also owing to the development of ribozymes and aptamers for therapeutic and/or diagnostic applications, RNA synthesis has become routine and highly reliable. This boom in the use of synthetic RNA has not only caused specialty reagent manufacturers to improve the quality of their products but has also created competition between suppliers which has subsequently led to reduced consumer cost.

From: *Methods in Molecular Biology, vol. 288: Oligonucleotide Synthesis: Methods and Applications*
Edited by: P. Herdewijn © Humana Press Inc., Totowa, NJ

Described here are reliable procedures to enable the novice to produce good-quality synthetic RNA either manually or machine assisted. The synthesis is performed on solid-phase using standard β-cyanoethyl phosphoramidite chemistry *(3)* with *tert*-butyldimethylsilyl (TBDMS) protection of the ribose 2'-hydroxyl group *(4)*, meaning that the 3'-terminal nucleoside is anchored via a succinyl linkage to an insoluble matrix, generally aminopropyl functionalized controlled pore glass (CPG) or polystyrene, contained in an appropriate reaction vessel. The nucleobases of the phosphoramidites and functionalized supports are protected with the *N-tert*-butylphenoxyacetyl group to enable mild deprotection of the RNA at the end of the synthesis *(5)*. Reagents are introduced into the vessel for removing protecting groups and enabling chain extension of the RNA, one nucleotide at a time, and excess reagent is simply flushed away with a suitable solvent. The process is cyclical and repeated until the desired length of RNA is obtained. Because there is no intermediate purification, all reactions should be as close to 100% yield as possible. In practice the chain extension reactions have a yield of 98.5–99%. Each cycle comprises a detritylation step that unmasks the 5'-hydroxyl group for chain extension, washing with acetonitrile, a coupling step in which the desired nucleotide as a phosphoramidite building block activated with 5-(benzylmercapto)-1*H*-tetrazole *(6)* is added, a capping step that acylates any unreacted 5'-hydroxyl group, an oxidation step that converts the phosphite triester to a phosphate triester, a further capping step that removes any occluded iodine, and finally a washing step with acetonitrile. The use of 5-(benzylmercapto)-1*H*-tetrazole for activation of the sterically hindered 2'-*O*-TBDMS protected phosphoramidites is strongly preferred over conventional 1*H*-tetrazole regarding both speed and coupling efficiency *(6)*.

Subsequently the fully protected support-bound RNA is deprotected in a stepwise fashion, comprising in the first step cleavage of the linkage to the solid-phase and cleavage of the nucleobase and phosphate protecting groups. In the second step the 2'-*O*-TBDMS groups are cleaved using a special fluoride reagent, namely, triethylamine tris(hydrofluoride) *(7,8)*. For RNAs longer than about 25 nucleotides it is best to leave the 5'-terminal dimethoxytrityl group attached as it is lipophilic and can be used as a purification aid. Purification is achieved by anion-exchange and/or reversed-phase high-performance liquid chromatography (HPLC) according to the length of the RNA and the purity required.

2. Materials
1. 5'-*O*-Dimethoxytrityl-*N*(pac or tac)-2'-*O*-TBDMS-3'-*O*-(β-cyanoethylphosphoramidites) of A, U, C, and G (Pierce, Milwaukee, WI, or Proligo, Hamburg, Germany). Store dry at –20°C.

2. Solid-phase supports, either CPG (Proligo or Pierce) or polystyrene (Amersham Biosciences) functionalized with A, U, C, and G.
3. 5-Benzylmercapto-1*H*-tetrazole (BMT) (emp Biotech, Berlin, Germany).
4. Capping solutions A (fast deprotection) and B (Proligo). *Caution:* Hazardous.
5. Oxidation solution containing iodine (Proligo). *Caution:* Hazardous.
6. Deblock solution composed of 3% trichloroacetic acid in dichloromethane (Proligo). *Caution:* Toxic and corrosive.
7. DNA synthesis-grade acetonitrile containing less than 30 ppm H_2O. *Caution:* Toxic.
8. Assorted 1000 series gas-tight syringes with volumes of 0.5, 1, and 2.5 mL (Hamilton Company; Reno, NV).
9. DNA/RNA synthesizer (Applied Biosystems or Amersham Biosciences) or a glass reaction vessel fitted with a ground glass joint at the top and a fine porosity glass frit and a tap at the bottom.
10. High purity concentrated aqueous ammonium hydroxide. Irritating to eyes and respiratory system. Use in a well-ventilated fume cupboard.
11. Anhydrous 8*M* methylamine in ethanol. Irritating to eyes and respiratory system. Use in a well-ventilated fume cupboard.
12. Anhydrous dimethyl sulfoxide (DMSO) (Fluka, Biotech. grade). Wear gloves.
13. Triethylamine tris(hydrofluoride) (Aldrich). *Hazardous:* wear full protection and use only in a well-ventilated fume cupboard.
14. Anhydrous triethylamine. Irritant; use in a fume cupboard.
15. *N*-Methylpyrrolidone, peptide synthesis grade.
16. Prop-2-yl trimethylsilyl ether prepared according to Jones *(9)*.
17. Diethyl ether. *Caution:* Fire hazard.
18. Biocompatible HPLC equipment (Amersham Biosciences).
19. Anion-exchange HPLC columns: MonoQ 5/5, Source 15Q 16/10 and/or FineLINE 35 pilot column packed with Source 15Q (Amersham Biosciences).
20. Sodium perchlorate. *Caution:* Toxic and corrosive, wear gloves.
21. Disodium ethylenediaminetetraacetic acid (EDTA).
22. Sterile 1*M* Tris-HCl buffer, pH 7.4.
23. Hi-Prep 26/10 desalting column (Amersham Biosciences).
24. Reversed-phase HPLC columns: Hamilton PRP-1, 7 × 305 mm and XTerra™ RP_8, 4.6 × 250 mm (Waters).
25. HPLC-grade acetonitrile. Toxic.
26. Ammonium bicarbonate.
27. Glacial acetic acid. *Caution:* Corrosive.

3. Methods

The methods outlined below describe: (1) the solid-phase synthesis of oligoribonucleotides, manually or machine aided, (2) the deprotection of the oligoribonucleotides, (3) the purification of oligoribonucleotides, and (4) the analysis of the purified product.

3.1. Solid-Phase Synthesis

The solid-phase synthesis of oligoribonucleotides is described in **Subheadings 3.1.1.** and **3.1.2.** For those people without access to a solid-phase synthesizer, the RNA can be synthesized manually with a minimum of equipment. The various steps involved in each cycle of the synthesis are illustrated in **Fig. 1**.

3.1.1. Manual RNA Synthesis

1. Weigh out the requisite amounts of the 4 monomers required in small vials (*see* **Note 1**) that can be closed with a septum and dry them overnight *in vacuo* over separate containers of phosphorus pentaoxide and potassium hydroxide pellets (*see* **Note 2**). For syntheses in the scale range of 1–3 µmol, it is recommended to use 8–10 equivalents of monomer per coupling relative to the amount of support used. For synthesis scales above 5 µmol a monomer excess of five fold is sufficient.
2. Release the vacuum with dry argon, and fit the monomer bottles with tight-fitting rubber septa.
3. Using a gas-tight syringe, dissolve each monomer in the requisite volume of dry acetonitrile to give a 0.1M solution and seal the top with Parafilm. It is not recommended to store the monomer solutions for more than 2–3 d at room temperature.
4. Transfer the requisite amount of CPG carrying the desired 3'-terminal ribonucleoside into the glass reaction vessel (*see* **Note 3**). For a 1-µmol scale synthesis the vessel should have a volume of about 5 mL, whereas for a 10-µmol scale a volume of 20 mL is more appropriate to allow good washing.
5. Add 3% trichloroacetic acid in dichloromethane (deblock solution) to the support using a Pasteur pipet, and let it percolate through. A deep orange color is produced owing to the released dimethoxytrityl cation. Continue to add acid until the effluent is no longer orange.
6. Drain the support using a slight pressure of argon.
7. Wash the CPG batchwise 8–10 times with acetonitrile (best done using a Teflon wash bottle), removing the supernatant each time with argon pressure.
8. Wash the CPG once with very dry acetonitrile, < 30 ppm H_2O, flush away with argon pressure, close the tap, and stopper the vessel.
9. Using two dry gas-tight syringes add the desired monomer as a 0.1M solution in acetonitrile and an equal volume of 0.3M BMT stock solution in acetonitrile to the CPG, stopper the vessel, and agitate several times during a period of 5 min.
10. During the coupling reaction, **step 9**, clean both syringes thoroughly with acetonitrile and store in a desiccator.
11. Drain the CPG, wash once with acetonitrile, and flush away with argon pressure.
12. Add a few milliliters of capping mixture comprising 1 vol of fast deprotection capping solution A and 1.1 vol of capping solution B (*see* **Note 4**), stopper the vessel, and agitate for 1 min, then drain.

2′-O-(Tert-Butyldimethylsilyl) Protection

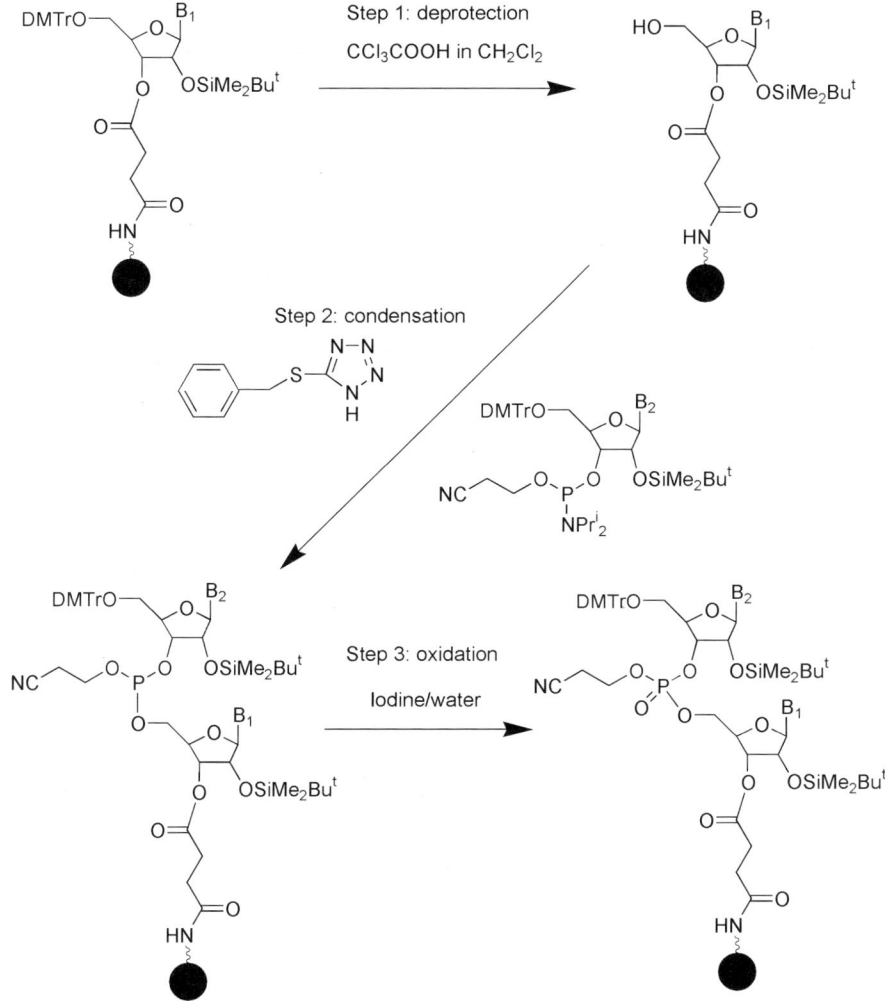

Fig. 1. Scheme illustrating a single cycle of solid-phase RNA synthesis via the phosphoramidite method. The filled-in *black circle* represents the CPG support. B_1 and B_2 represent protected nucleobases, that is, uracil-1-yl, N^4-(4-t-butylphenoxyacetyl)cytosine-1-yl, N^2-(4-t-butylphenoxyacetyl)guanin-9-yl or N^6-(4-t-butylphenoxyacetyl)adenin-9-yl.

13. Wash the CPG once with acetonitrile and flush away with argon pressure.
14. Add a few milliliters of oxidation mixture and allow it to slowly percolate through the CPG for 2 min (*see* **Note 5**). This oxidizes the phosphite triester to a phosphate triester.

15. Drain the CPG and wash once with acetonitrile and drain with argon pressure.
16. Add a few milliliters of fresh capping mixture, agitate for 30 s, and drain using argon pressure.
17. Wash the CPG thoroughly with acetonitrile six times, draining each time in between using argon pressure.
18. Repeat **steps 5–17** as many times as necessary until the desired sequence is reached.
19. For RNAs longer than about 25 nucleotides the final trityl group should be left on as a purification aid. For shorter RNAs remove the final trityl group as in **step 5**, and wash the CPG thoroughly with acetonitrile.
20. Dry the CPG using a stream of argon (*see* **Note 6**).

3.1.2. Automated RNA Synthesis

Follow the instructions for the particular instrument plus the program for the RNA synthesis scale you intend to use. The CPG or polystyrene support is now placed inside a small plastic cartridge. The reagents are available in the correct bottles to fit the various instruments on the market.

3.2. Deprotection

In the first deprotection step the succinate linkage connecting the 3'-terminal nucleoside to the solid support is cleaved, the β-cyanoethyl protecting groups on the internucleotide linkages are removed by β-elimination, and the nucleobase exocyclic amine protecting groups are cleaved. This step is best performed with a 1:1 mixture of concentrated aqueous ammonia and 8M ethanolic methylamine, which prevents premature loss of the TBDMS groups, which would otherwise lead to degradation of the RNA under basic conditions. In the second deprotection step the TBDMS groups are removed using triethylamine tris(hydrofluoride) plus an appropriate solvent.

3.2.1. Deprotection of Base Labile Protecting Groups

1. Transfer the argon-dried support to a screw-top vial or small Duran bottle equipped with a tight-fitting screw top.
2. Add a 1:1 mixture of concentrated aqueous ammonia and 8M ethanolic methylamine. For a 0.2–1-μmol scale synthesis, use a volume of 2 mL, otherwise use 2 mL per μmol of support (*see* **Note 7**). Close the vial or bottle tightly and seal further with Parafilm.
3. Place the vial or bottle in a preheated oven at 65°C for 30 min for small vials or 40 min for larger bottles, which take longer to reach the desired temperature.
4. Allow the vial/bottle to cool completely before opening. This can be speeded up by cooling it in an ice bath.
5. Remove the supernatant and wash the support several times with a few milliliters of ethanol/sterile water (1:1 v/v).

6. Dry the supernatant and combined washings in a Falcon tube in a SpeedVac or, for larger volumes, evaporate to dryness on a rotary evaporator. Do not use water bath temperatures above 30°C for trityl-on material.
7. Dry the residue once by evaporation of absolute ethanol.

3.2.2. Desilylation of Trityl-Off RNA

1. In a Falcon tube dissolve the trityl-off residue obtained as described in **Subheading 3.2.1.** in a 1:1 mixture of dry DMSO and triethylamine tris(hydrofluoride) *(10)*, using 600 µL per µmol and sonicate briefly. If the oligoribonucleotide has been dried down in a glass flask, dissolve it in the required volume of dry DMSO by gently warming the flask with a hair dryer, then transfer the solution to a Falcon tube and add an equal volume of the fluoride reagent in a well-ventilated fume cupboard.
2. Close the tube and seal with Parafilm and place in a preheated oven at 65°C for 2.5 h.
3. Cool the tube to room temperature.
4. Quench the reaction by addition of 2 vol of isopropyl trimethylsilyl ether *(9)*, close, and shake vigorously at intervals during 10 min. A precipitate appears.
5. Carefully open the tube and add 5 vol of diethyl ether; close and agitate.
6. Collect the precipitate by centrifugation at 2200g at 4°C for 5 min.
7. Remove the supernatant by careful decantation.
8. Resuspend the pellet in ether, close the tube, agitate, and collect the precipitate by centrifugation.
9. Repeat **steps 7** and **8**.
10. Dry the pellet carefully *in vacuo*.

3.2.3. Desilylation of Trityl-On RNA

1. In a Falcon tube dissolve the trityl-on residue, as obtained in **Subheading 3.2.1.**, in 600 µL per µmol of a freshly prepared solution of *N*-methylpyrrolidone/triethylamine/triethylamine tris(hydrofluoride) (6:3:4 by vol) *(11)*. For material that has been dried down in a glass flask, dissolve the residue in the minimum volume of dry DMSO, transfer the solution to a Falcon tube, and add the freshly prepared desilylation solution.
2. Perform **steps 2–8** as described in **Subheading 3.2.2.**
3. Dry the RNA pellet very briefly with an argon stream and then dissolve it immediately in sterile 0.1*M* aqueous ammonium bicarbonate ready for immediate purification by reversed-phase HPLC.

3.3. Purification

This section is devoted to the anion-exchange HPLC purification of fully deprotected RNA using gradients of sodium perchlorate *(11)* or lithium perchlorate *(12)* (*see* **Note 8**) as chaotropes, the reversed-phase HPLC purification of trityl-on RNA, detritylation, and desalting.

3.3.1. Anion-Exchange HPLC Purification

Anion-exchange HPLC is recommended as the first or only purification step for oligoribonucleotides shorter than about 25 nucleotides. It generally results in a purity of 95–98%. It is recommended to use a gradient of sodium perchlorate in sterile water/acetonitrile (9:1 v/v) containing 50 mM Tris-HCl buffer, pH 7.6, and 50 µM EDTA. The purpose of the EDTA is to complex traces of heavy metals that could otherwise lead to cleavage and degradation of the RNA. Source 15Q 16/10 columns are recommended for purification of 1-µmol scale syntheses with a flow rate of 5 mL per min. For syntheses in the 10- to 100-µmol scale, these are best purified using a FineLINE Pilot 35 column packed with Source 15Q and eluted at 20 mL per min (*see* **Note 9**). The low-salt or A buffer contains 10 mM sodium perchlorate and the high-salt or B buffer contains 600 mM sodium perchlorate. A gradient from 0–60% B during 40 min gives good resolution. When not in use store the columns in 20% ethanol in sterile water to prevent microbial growth. Prior to use flush the column with several column volumes of sterile water before equilibrating the column with buffer B followed by buffer A. The desired product peak is the late-eluting major component. This material is then desalted as described in **Subheading 3.3.4.** A typical trace of an anion-exchange HPLC purification is shown in **Fig. 2**. In this example the oligomer is a 21-mer synthesized manually on a 20-µmol scale and purified on Source 15Q packed in a FineLINE Pilot 35 column. As can be seen, the failure peaks are very small compared to the product peak, which elutes at 24–27 min.

3.3.2. Reversed-Phase HPLC Purification of Trityl-On RNA

The effect of the highly lipophilic dimethoxytrityl group is to profoundly retard the full-length product when purified on a reversed-phase HPLC column. Although the best separation of failure peaks from the desired trityl-on product peak is obtained using aqueous triethylammonium acetate/acetonitrile buffers, for ease of salt removal and minimal damage to the RNA, the use of ammonium bicarbonate is preferred.

Recommended columns for trityl-on purification are the Hamilton PRP-1, 7 × 305 mm for a few µmol scale syntheses or a 21.5- × 250-mm column for 10–20-µmol scale purifications (*see* **Note 10**). Buffer A is 0.1M ammonium bicarbonate prepared in sterile water and buffer B is 0.1 M aqueous ammonium bicarbonate/acetonitrile (1:1 v/v). Recommended gradient is 0–90% B during 40 min. The failure peaks elute early and are well separated from the desired trityl-on product peak, which elutes last. Collect the product fraction in a polypropylene Falcon tube and dry down on a SpeedVac. Residual ammonium bicarbonate is then removed by lyophilization of the product, which is now ready for detritylation. A typical trace of a trityl-on RNA purification by

2'-O-(Tert-Butyldimethylsilyl) Protection

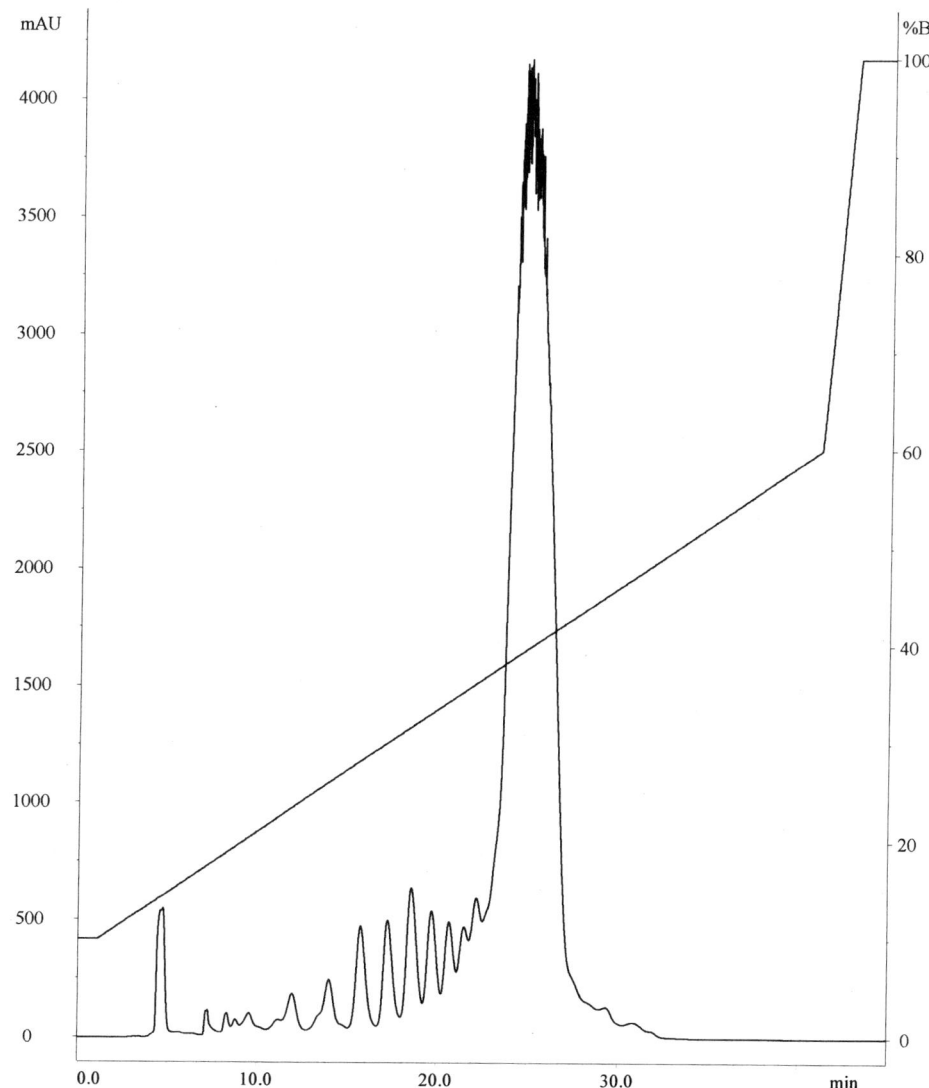

Fig. 2. A preparative anion-exchange HPLC trace of a 21-mer oligoribonucleotide synthesized manually on a 20-µmol scale and purified on Source 15Q packed in a FineLINE Pilot 35 column. Absorbance was monitored at 280 nm.

reversed-phase HPLC is shown in **Fig. 3**. The example shows a trityl-on 58-mer oligoribonucleotide synthesized by machine on a 1-µmol scale and purified on a 7- × 305-mm Hamilton PRP-1 column. The desired product peak elutes at 20–22 min well separated from the trityl-off failure sequences (at 7–12 min).

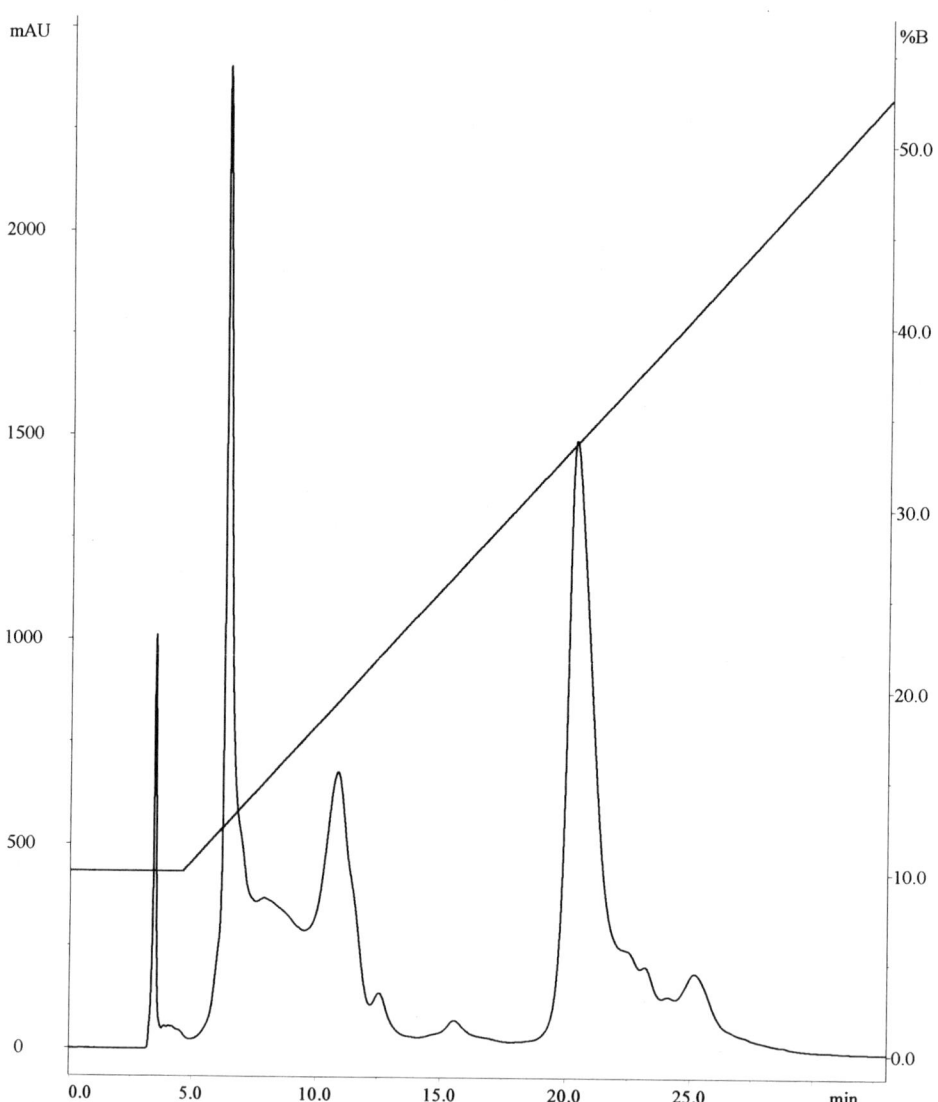

Fig. 3. Reversed-phase HPLC trace of a trityl-on 58-mer oligoribonucleotide purified on a Hamilton PRP-1 column. Absorbance was monitored at 295 nm.

3.3.3. Detritylation of Trityl-On RNA

1. Dissolve the purified trityl-on RNA as obtained from **Subheading 3.3.2.** in 3% sterile aqueous acetic acid (200 µL per µmol) and keep 45 min at room temperature. The pH should be approx 3.5.

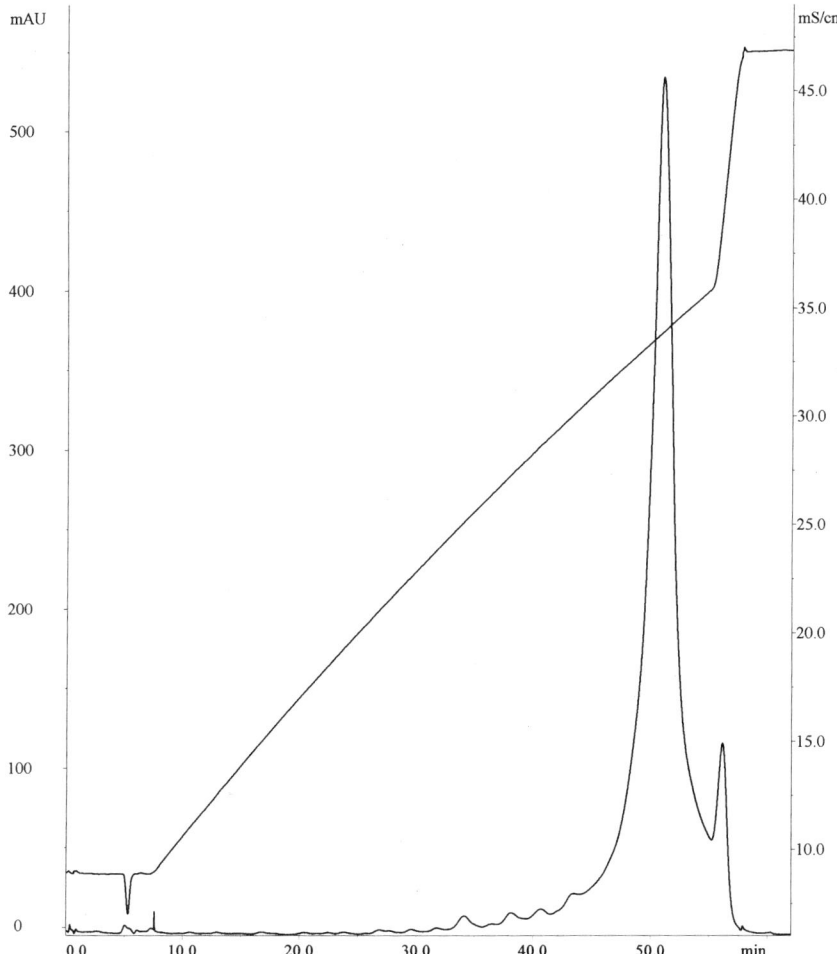

Fig. 4. Anion-exchange HPLC trace showing the purification of a detritylated 58-mer oligoribonucleotide on a Source 15Q 16/10 column. The compound was initially purified trityl-on by reversed-phase HPLC. Absorbance was monitored at 295 nm.

2. Neutralize by careful addition of solid ammonium bicarbonate until evolution of carbon dioxide ceases. The pH will now be approx 7.8.
3. Repurify the product by anion-exchange HPLC as outlined in **Subheading 3.3.1.**, which in addition converts the product into the sodium form.
4. Prior to use the product must be desalted (see **Subheading 3.3.4.**).
   Fig. 4 shows the trace of the anion-exchange HPLC purification of the 58-mer oligoribonucleotide that was initially purified trityl-on by reversed-phase HPLC (see **Fig. 3**). The peak was collected from 49 to 52 min.

3.3.4. Desalting by HPLC

For oligonucleotides purified by anion-exchange HPLC it is necessary to remove the excess salt, buffer, and EDTA from the RNA regardless of the intended application. This is readily achieved using a desalting column such as the Hi-Prep 26/10, which is filled with Sephadex G-25. The sample should be loaded in a volume not greater than 15 mL, but preferably less than 10 mL, to obtain a complete separation between the first eluting RNA and the salt peak that comes after it. The column is stored in and eluted with 20% ethanol in sterile water; this prevents microbial growth. Monitoring of the column effluent by ultraviolet light and conductivity avoids contamination of the RNA caused by incompletely removed salt.

The salt-free RNA in aqueous ethanol is first concentrated in a SpeedVac and finally lyophilized to obtain the pure RNA in its sodium form as a fluffy white solid.

3.4. Analysis of the Purified RNA

Generally the purity of the final product is checked by analytical anion-exchange HPLC using a MonoQ 5/5 column and buffers according to **Subheading 3.3.1.** Alternatives are capillary gel electrophoresis and polyacrylamide gel electrophoresis, both of which operate under denaturing conditions. These methods are not described here.

RNA destined for structural studies, such as nuclear magnetic resonance spectroscopy (NMR) or X-ray crystallography, is also best checked in addition by analytical reversed-phase HPLC on a high-resolution column such as the 5-μM XTerra RP$_8$, 4.6 × 250 mm. In this case a gradient from 0–25% acetonitrile in aqueous ammonium bicarbonate will suffice, combined with a flow rate of 1 mL/min^{-1}. As an example an analytical reversed-phase HPLC trace of the double HPLC-purified 58-mer oligoribonucleotide is illustrated in **Fig. 5**.

As an absolute check on product authenticity, the molecular weight of the RNA should be determined by mass spectroscopy, either electrospray ionization or matrix-assisted laser desorption ionization *(13)* (*see* **Note 11**). This is a must when modified nucleotides are incorporated. In addition, the RNA can also be sequenced by standard methods to check the absolute order of the monomeric units within the sequence. Here the reader is advised to refer to the appropriate literature as these methods are outside the scope of this chapter.

4. Notes

1. The most suitable vials for this purpose are those amber glass bottles that are used by suppliers of DNA and RNA phosphoramidites and fit directly on ABI synthesizers.

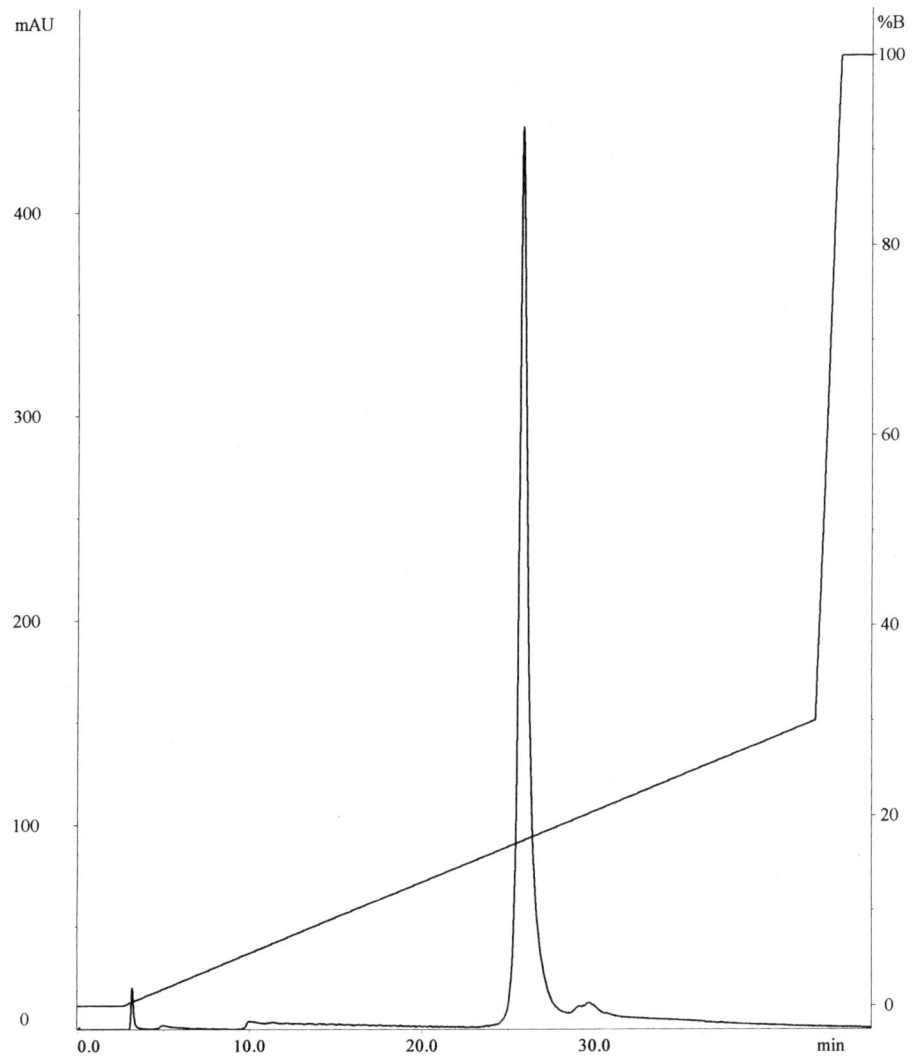

Fig. 5. Analytical reversed-phase HPLC trace of double HPLC purified 58-mer oligoribonucleotide run on a 5-μM XTerra RP$_8$, 4.6 × 250-mm column. Absorbance was monitored at 260 nm.

2. It is essential to allow monomers to reach room temperature before opening and weighing out material; otherwise condensation will occur leading to eventual degradation. The bottle contents should be put back under argon before sealing and storing again at –20°C. The drying step ensures the removal of any residual moisture that could otherwise impair the coupling reaction.

3. Suitable glass reaction vessels in a variety of sizes can be made by any laboratory glassblower following the drawing in the literature *(14)*. The vessel is basically a cylindrical tube, fitted with a porous glass frit at the bottom that is connected to a narrow internal-diameter glass tube closed off by a tap. The top end of the vessel is equipped with a B14 ground glass joint that can be closed with a simple B14 stopper.
4. The capping mixture is best made up in a dry-stoppered flask in a volume sufficient for one day's usage.
5. Because the standard oxidation mixture is only 16 mM in iodine, that is, 16 μmol per mL, use enough mixture for larger scale syntheses to ensure that there is an ample excess of reagent. For a 20-μmol scale synthesis, use 5 mL of the oxidation mixture. Incomplete oxidation will result in cleavage at the phosphite triester bond during the subsequent detritylation step, hence leading to a reduced yield of full-length product.
6. CPG bearing trityl-off protected RNA can be stored cold and dry ready for deprotection at an appropriate time; however, CPG-bearing trityl-on RNA must be deprotected and purified immediately on completion of the synthesis.
7. Always work in a well-ventilated fume cupboard when working with ammonia and methylamine. Avoid too great an air space in the vial or bottle, otherwise most of the ammonia and methylamine will end up in the vapor phase. In the worst case this could lead to an incomplete deprotection.
8. Oligoribonucleotides that contain 4 or more consecutive Gs are notoriously difficult to purify by anion-exchange HPLC because they form tetraplexes and higher aggregates in solution. Such RNAs are best purified using a lithium perchlorate gradient because these structures do not form if lithium ions are present instead of sodium or potassium ions.
9. To avoid product shoot through owing to the ionic strength of the applied sample solution being too high, it is advisable to apply the crude RNA sample to the column dissolved in a volume of 10–50 mL of 50 mM Tris-HCl buffer, pH 7.4, using a 10- or 50-mL super loop. As an alternative, desalt the sample prior to purification.
10. As an alternative for larger scale separations, a FineLINE Pilot 35 column packed with Source 15RPC can be used.
11. RNA samples as sodium salts are not suitable for mass spectrometry. Mass spectrometry samples are best prepared as ammonium salts. This can be done in several ways. One way is to do anion-exchange HPLC using ammonium sulfate for elution, followed by desalting on a small NAP cartridge. A second way is to take a small aliquot of the RNA in its sodium form and exchange the sodium ions with ammonium ions by using ammonium form Dowex 50 resin. A third way is to purify a small sample of RNA by reversed-phase HPLC using the aqueous ammonium bicarbonate/acetonitrile system followed by lyophilization of the residual salt. Once in the ammonium form the RNA should not in contact with glass surfaces again, otherwise sodium and potassium ions will be picked up that will degrade the quality of the mass spectra.

References

1. Ramaswamy, G. and Slack, F. J. (2002) siRNA: a guide for RNA silencing. *Chem. Biol.* **9,** 1053–1055.
2. Couzin, J. (2002) Small RNAs make big splash. *Science* **298,** 2296–2297.
3. Sinha, N. D., Biernat, J., McManus, J., and Köster, H. (1984) Polymer support oligonucleotide synthesis XVIII: use of β-cyanoethyl-N,N-dialkylamino-/N-morpholino phosphoramidite of deoxynucleosides for the synthesis of DNA fragments simplifying deprotection and isolation of the final product. *Nucleic Acids Res.* **12,** 4539–4557.
4. Usman, N., Ogilvie, K. K., Jiang, M.-Y., and Cedergren, R. J. (1987) Automated chemical synthesis of long oligoribonucleotides using 2'-O-silylated ribonucleoside 3'-O-phosphoramidites on a CPG support: synthesis of a 43-nucleotide sequence similar to the 3'-half molecule of an *Escherichia coli* formylmethionine tRNA. *J. Am. Chem. Soc.* **109,** 7845–7854.
5. Sinha, N. D., Davis, P., Usman, N., et al. (1993) Labile exocyclic amine protection of nucleosides in DNA, RNA and oligonucleotide analog synthesis facilitating N-deacylation, minimizing depurination and chain degradation. *Biochimie* **75,** 13–23.
6. Welz, R. and Müller, S. (2002) 5-(Benzylmercapto)-1*H*-tetrazole as activator for 2'-O-TBDMS phosphoramidite building blocks in RNA synthesis. *Tetrahedron Lett.* **43,** 795–797.
7. Gasparutto, D., Livache, T., Bazin, H., et al. (1992) Chemical synthesis of a biologically active natural tRNA with its minor bases. *Nucleic Acids Res.* **20,** 5159–5166.
8. Westman, E. and Strömberg, R. (1994) Removal of t-butyldimethylsilyl protection in RNA-synthesis: triethylamine trihydrofluoride (TEA, 3HF) is a more reliable alternative to tetrabutylammonium fluoride (TBAF). *Nucleic Acids Res.* **22,** 2430–2431.
9. Song, Q. and Jones, R. A. (1999) Use of silyl ethers as fluoride scavengers in RNA synthesis. *Tetrahedron Lett.* **40,** 4653–4654.
10. Vinayak, R., Andrus, A., and Hampel, A. (1995) Rapid desilylation of oligoribonucleotides at elevated temperatures: cleavage activity in ribozyme-substrate assays. *Biomed. Pept. Proteins Nucleic Acids* **1,** 227–230.
11. Wincott, F., DiRenzo, A., Shaffer, C., et al. (1995) Synthesis, deprotection, analysis and purification of RNA and ribozymes. *Nucleic Acids Res.* **23,** 2677–2684.
12. Sproat, B., Colonna, F., Mullah, B., et al. (1995) An efficient method for the isolation and purification of oligoribonucleotides. *Nucleosides Nucleotides* **14,** 255–273.
13. Pieles, U., Zürcher, W., Schär, M., and Moser, H. E. (1993) Matrix-assisted laser desorption ionization time-of flight mass spectrometry: a powerful tool for the mass and sequence analysis of natural and modified oligonucleotides. *Nucleic Acids Res.* **21,** 3191–3196.
14. Sproat, B. S. and Gait, M. J. (1984) Solid-phase synthesis of oligodeoxyribonucleotides by the phosphotriester method. In *Oligonucleotide Synthesis: A Practical Approach* (Gait, M. J., ed.), IRL Press, Oxford, UK, p. 92.

3

RNA Oligonucleotide Synthesis Via 5'-Silyl-2'-Orthoester Chemistry

Stephanie A. Hartsel, David E. Kitchen, Stephen A. Scaringe, and William S. Marshall

Summary

Rapid, reliable, and cost-efficient methods of ribonucleic acid (RNA) oligonucleotide synthesis are in demand owing to an increasing awareness of critical structural, functional, and regulatory roles of RNA throughout biology. The most promising area of growth and development is in RNA interference as an emerging technology for facilitating research in drug discovery and therapeutic intervention. Traditional methods of RNA synthesis, which are based on 2'-silyl protection strategies derived from deoxyribonucleic acid (DNA) synthesis strategies, are limited in their ability to produce oligos of sufficient purity and length for high-throughput applications. The more recently developed 5'-silyl-2'-acetoxy ethyl orthoester chemistry (2'-ACE™), circumvents several limitations of the 2'-silyl approaches. A clear improvement in RNA synthesis technology, 2'-ACE results in faster coupling rates, higher yields, greater purity, and superior ease of handling. Another advantage of the 2'-ACE protecting group strategy is that the molecules can be produced in an intermediately protected form that is soluble in aqueous solutions but resistant to nuclease attack. The chemistry can be scaled up or down and is flexible enough to allow for the incorporation of modifying groups if desired. A detailed description of the 2'-ACE protocol and procedures for end product analysis are presented.

Key Words: RNA oligonucleotide synthesis; oligonucleotide synthesis; RNA synthesis; ribonucleic acid synthesis; 2'-ACE.

1. Introduction

Developments in ribonucleic acid (RNA) biochemistry over the past two decades have clearly demonstrated a wide variety of RNA-mediated events that are important for cellular function. The discovery of small interfering RNAs (siRNAs) and their ability to effectively suppress gene expression in mammalian cells *(1)* has sparked tremendous interest in developing siRNA-based screens to identify and validate disease-relevant targets and to enable advancements in functional genomics. As such, the ability to readily synthesize RNA oligonucleotides in high yield with requisite purity and sequence integrity is becoming increasingly important for discovery biology. Considerable efforts are also being made to design siRNAs for therapeutic intervention.

In the 1980s, RNA synthesis was successfully accomplished by Ogilvie and Usman's application of the 2'-*O*-*t*-butyldimethylsilyl (tBDMS) 2'-hydroxyl protection strategy *(2)* to classical phosphoramidite-based deoxyribonucleic acid (DNA) synthesis *(3)*. The more recent introduction of 2'-*O*-triisopropylsilyloxymethyl (TOM) 2'-hydroxyl protection chemistry provides a more favorable application of 2'-silyl chemistry relative to the tBDMS protecting group, resulting in improved synthesis efficiencies *(4)*. However, although TOM chemistry provides greater synthesis efficiencies relative to tBDMS chemistry, RNA oligonucleotide synthesis using TOM chemistry remains limited in areas such as nucleotide length, purity, and scalability.

Considerable advancements in RNA synthesis have been achieved with 5'-silyl-2'-acetoxy ethyl orthoester (ACE™) chemistry *(5–6)*. 2'-ACE chemistry provides several significant advantages over 2'-silyl-based methods such as 2'-tBDMS and 2'-TOM. The methodology yields a water-soluble intermediate with increased nuclease resistance, superior ease-of-handling properties, and greater capacity for long-term storage. The 2'-ACE intermediate protecting group can be removed in 10 min using mild aqueous acid. Because of the rapid coupling kinetics and high average stepwise coupling yields, 2'-ACE yields a crude product of high enough purity for most biological applications. Crude 21-mers are reliably and routinely produced at greater than 80% purity. Of equal importance, 2'-ACE enables the synthesis of RNA in excess of 70 bases in length, and it is highly scalable in both numbers of sequences produced and scale of synthesis.

2. Materials

2.1. Automated Equipment

A 2'-ACE RNA oligoribonucleotide synthesis is performed on a modified Applied Biosystems 394 DNA/RNA synthesizer. The trityl monitor and its associated valve block are removed. Owing to incompatibility with triethylammonium trihydrogenfluoride, a reagent used in the synthesis process, the glass flow restrictors are removed, and the lines to each column position are replaced with 7/1000 diameter Tefzel™ tubing to restrict the delivery of reagent to the columns. The modified 394 instrument is operated via ABI Oligonet™ software, version 1.1.

The viscosity of the ancillary reagents used with 5'-silyl-2'-ACE chemistry requires a two-stage regulator that is connected to the instrument via 6.35-mm

RNA Oligonucleotide Synthesis

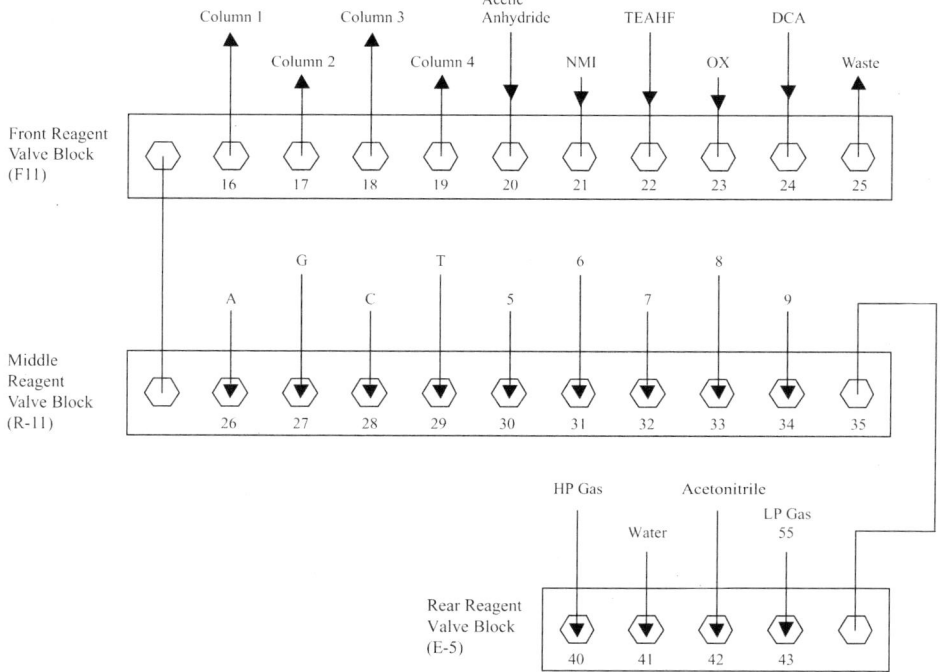

Fig. 1. Schematic diagram of a modified 394 RNA synthesizer.

(1/4-in. od) tubing. The pressure at the second stage is increased to 60 psi. Reagents are placed and arranged on the instrument according to **Fig. 1**.

Matrix-assisted laser desorption/ionization time-of-flight (MALDI-TOF) mass spectroscopy analysis is achieved with the Perseptive Biosystems Voyager™-DE BioSpectrometry Workstation, using a 3-hydroxypicolinic acid (3-HPA) matrix *(7)*. Strong ion-exchange high-pressure (performance) liquid chromatography (HPLC) is conducted on a Waters 2793 Alliance HT system with a Dionex DNA-Pac–PA-100 column (4 × 250 mm).

2.2. Reagents and Solvents

1. Acetonitrile (anhydrous).
2. Dichloromethane (DCM).
3. 0.5M 5-ethylthio-1H-tetrazole (American International Chemical–ETTXCC).
4. Toluene.
5. t-butyl hydroperoxide 70% (v/v) in water (Aldrich-45, 813-9).
6. Acetic anhydride (Aldrich-11, 004-3).
7. N-methylimidazole (Aldrich-33, 609-2).
8. 48% Hydrogen fluoride (HF) in water.

9. Dichloroacetic acid (DCA) (Aldrich-D5, 470-2).
10. 0.1M 2'-ACE-Me-phosphoramidite (Dharmacon).
11. 1M Disodium 2-carbamoyl-2-cyanoethylene-1,1-dithiolate in N,N'-dimethyl formamide 1 (Dharmacon) *(8)*.
12. 40% 1-methylamine in water (Aldrich-42, 646-6).
13. 3'-*O*-linked ribonucleoside on polystyrene support (Dharmacon).
14. 100 mM acetic acid, pH 3.8 (pH with TEMED).
15. N,N,N',N'-tetramethylethylenediamine (TEMED).
16. HPLC buffer A: 5 mM sodium perchlorate, 10 mM Tris (hydroxymethyl)-aminomethane, pH 8.0.
17. HPLC buffer B: 300 mM sodium perchloroate, 10 mM Tris (hydroxymethyl) aminomethane, pH 8.0.
18. 5X Tris-borate ethylenediaminetetraacetic acid (TBE): 445 mM Tris-borate, pH 8.0, 10 mM ethylene diamine tetraacetic acid (EDTA).
19. 8 and 15% polyacrylamide 7M urea stock solutions.
20. Gel loading buffer: 7M urea, 1X TBE (Tris-borate, pH 8.0, 2 mM EDTA).
21. 10% ammonium persulfate (APS).
22. Gel tracking dye.
23. Gel elution buffer: 0.3M sodium acetate, pH 7.5.

2.3. Reagent Preparation

All reagents are prepared and transferred to either 250- or 500-mL amber-colored bottles designed to fit the ABI 394 synthesizer (VWR 15900-138 and VWR 15900-140), except for triethylammonium trihydrogenfluoride (TEAHF), which must be prepared and stored in polyethylene. The Nalgene PassPort™ (VWR 16057-168) is compatible with the TEAHF reagent and is threaded to fit ABI 394.

1. Add 20 mL of 1-methylimidiazole, 10:90 v/v, to 180 mL anhydrous acetonitrile.
2. Add 20 mL of acetic anhydride, 10:90 v/v, to 180 mL anhydrous acetonitrile.
3. Add 3 mL of 3% dichloroacetic acid, 3.97 v/v, to 97 mL of dichloromethane.
4. In a 2-L separatory funnel, shake 360 mL of 70% *t*-butyl hydroperoxide with 1.5 L of toluene. Allow layers to separate for 30 min and discard the bottom aqueous layer. Decant the top layer into a bottle, and place in a freezer with a sintered glass funnel. Allow the water to freeze, then using vacuum filtration quickly separate the reagent from the ice crystals using the chilled fritted filter. Store reagent at −20°C.
5. Dissolve 32.54 g of 0.5M 5-(ethylthio)-tetrazole in 500 mL of acetonitrile. Add a Molecular Trap™ sieve bag at least 48 h before placing the reagent on the instrument.
6. In a Nalgene cylinder, add 160 mL of triethylamine to 240 mL of N,N'-dimethyl formamide while stirring. Slowly add 15.6 mL of 48% hydrogen fluoride (HF). *Caution:* Use protective gloves and apron when handling HF. Pour contents into a 500-mL Nalgene bottle (Nalgene 2099-0016). Store at −20°C.

7. Dissolve 0.5 g of 1M disodium-2-carbamoyl-2-cyanoethylene-1,1-dithiolate trihydrate (S_2Na_2) in 9 mL of dimethyl formamide and 1 mL of HPLC-grade water. Store solution at –20°C. Fresh reagent should be prepared weekly.
8. HPLC buffer A: Dissolve 0.7 g $NaClO_4$, 0.61 g Tris, 20 mL MeCN, and 0.18 mL perchloric acid in HPLC-grade water, and bring the volume up to 1 L using HPLC-grade water.
9. HPLC buffer B:. Dissolve 42.14 g $NaClO_4$, 0.61 g Tris, 20 mL MeCN, and 0.18 mL perchloric acid in HPLC-grade water, and bring the volume up to 1 L using HPLC-grade water.
10. 5X TBE electrophoresis buffer: Dissolve 216 g Tris, 110 g boric acid, and 7.44 g EDTA in molecular-grade water to a final volume of 4 L.
11. 1X TBE electrophoresis buffer: Bring 200 mL of 5X TBE up to 1 L with molecular-grade water.
12. 8% polyacrylamide 7M urea stock solution: Mix 400 mL of 40% acrylamide solution (29:1 acrylamide:*bis*) with 400 mL 5X TBE buffer, dissolve 841 g urea into the solution, and dilute to 2 L with molecular-grade water.
13. 15% polyacrylamide 7M urea stock solution: Mix 750 mL of 40% acrylamide solution (29:1 acrylamide:*bis*) with 400 mL 5X TBE buffer, dissolve 841 g urea into the solution, and dilute to 2 L with molecular-grade water.
14. Gel loading buffer: Dissolve 0.6 g Tris and 21.0 g urea in molecular-grade water, and bring to a final volume of 50 mL.
15. 10% APS: Dissolve 5.0 g ammonium persulfate in molecular-grade water, and bring to a final volume of 50 mL.
16. Gel tracking dye: Dissolve 0.6 g Tris, 5.4 g bromophenol blue, 5.4 g xylene cyanol, and 21.0 g urea in molecular-grade water, and dilute to a final volume of 50 mL.
17. Gel elution buffer (0.3M sodium acetate): Dissolve 12.3 g NaOAc in molecular-grade water to a final volume of 500 mL.

3. Methods
3.1. Protected Ribonucleoside Monomers

5'-*O*-silyl-2'-*O*-ACE protected phosphoramidites (**Fig. 2**) are prepared and purified according to previously published procedures *(5)*.

Briefly, monomer synthesis begins from standard base-protected ribonucleosides [rA(ibu), rC(acetyl), rG(ibu) and U]. Orthogonal, 5'-silyl-2'-ACE protection and amidite preparation is then accomplished in five general steps (**Fig. 3**):

1. Simultaneous transient protection of the 5'- and 3'-hydroxyl groups with 1,1,3,3tetraispropyldisiloxane (TIPS).
2. Regiospecific conversion of the 2'-hydroxyl to the 2'-*O*-orthoester using tris(acetoxyethyl)orthoformate (ACE orthoformate).
3. Removal of the 5',3'-TIPS protection.
4. Introduction of the 5'-*O*-silyl ether protecting group using benzhydryloxy*bis*-(trimethylsilyloxy)-chlorosilane (BzH-Cl).

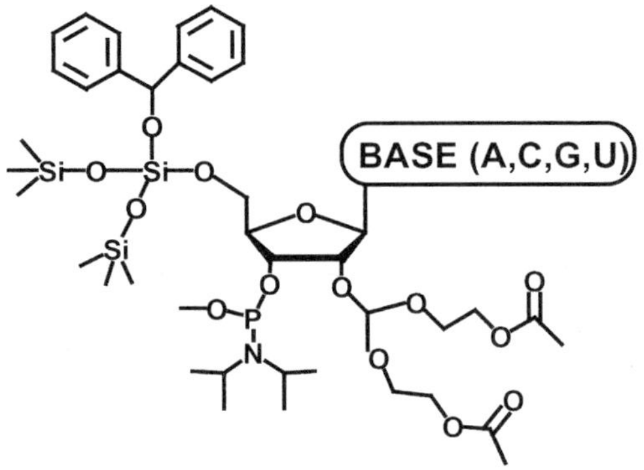

Fig. 2. Structure of phosphoramidite monomers used in 2'-ACE RNA synthesis.

5. Phosphitylation of the 3'-OH with *bis*(*N*,*N*'-diisopropylamino)methoxyphosphine.

The fully protected, phosphitylated monomer is an oil. For ease of handling and dissolution, the phosphoramidite solution is evaporated to dryness in a tared flask to enable quantitation of yields. The phosphoramidite oil is then dissolved in anhydrous acetonitrile, distributed into synthesis vials in 1.0-mmol aliquots, and evaporated to dryness under vacuum in the presence of potassium hydroxide (KOH) and P_2O_5.

3.2. Synthesis of Oligoribonucleosides

5'-silyl-2'-ACE oligoribonucleotide synthesis begins with the appropriately modified 3'-terminal nucleoside attached through the 3'-hydroxyl to a polystyrene support (*see* **Note 1**). The solid support contained in an appropriate reaction cartridge is then placed on the appropriate column position on the instrument. A synthesis cycle is created using the delivery times and wait steps outlined in **Table 1**. The recommended cycle for each base addition is illustrated in **Fig. 4**.

Initial detritylation: The first step in the synthesis cycle is the removal of the 5' *O*-DMT from the nucleoside-bound polystyrene support using 3% DCA in DCM (*see* **Note 2**).

1. *Coupling:* The 5-ethylthio-1*H*-tetrazole solution is delivered to the solid support, followed by simultaneous delivery of an equal quantity of activator and phosphoramidite solution. Depending on the desired sequence and synthesis scale, excess activator and activator plus amidite are alternately delivered repeatedly to increase coupling efficiency, which is typically in excess of 99% per coupling reaction.

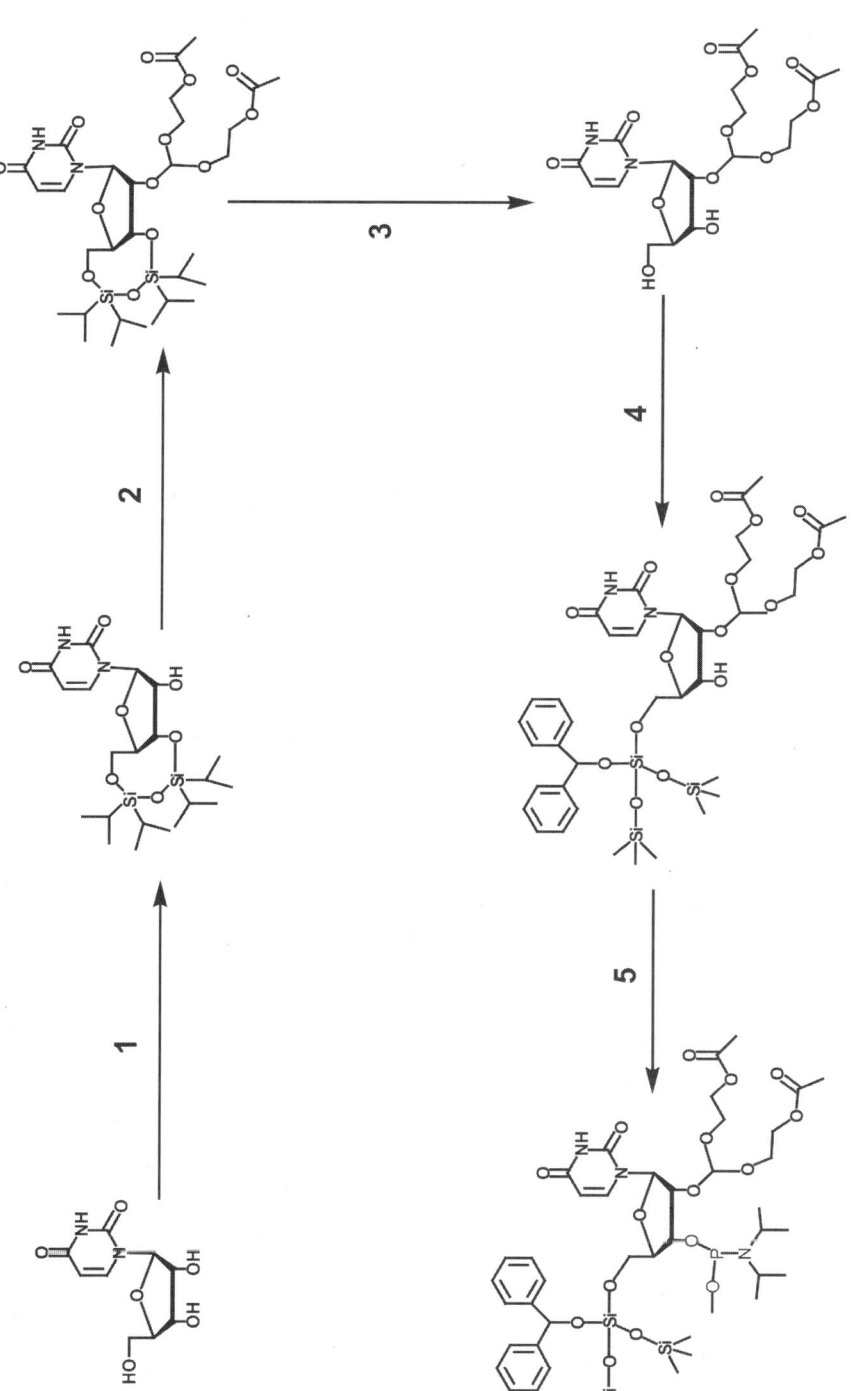

Fig. 3. Synthesis of phosphoramidite monomers from standard base-protected ribonucleosides. *Step 1*: 1,1,3,3-tetraisopropyldisiloxane chloride; *step 2*: Tris(acetoxyethoxy)orthoformate, butyldimethylsilylpentadione, pyridinium toluenesulfonate *step 3*: triethylamine trihydrogenfluoride; *step 4*: benzhydryloxy *bis*(trimethylsilyloxy) chlorosilane; *step 5*: *bis*(*N,N'*,-diisopropylamino) methoxyphosphine.

Table 1
0.2-μmol Synthesis Steps Using 5'-O-Silyl-2'-ACE Chemistry

Synthesis Step	Reagent	Delivery time (s)	Reaction time
Deblock[a]	3% DCA in DCM	35	
Activator	0.5M S-ethyl-tetrazole	6.0	
Coupling	0.1M amidite 0.5M S-ethyl-tetrazole	8.0 30	30
Repeat Coupling		8.0	30
Oxidation	*t*-Butyl hydroperoxide	20	10
Repeat Oxidation Delivery		10	
Capping	1-methylimidazole and acetic anhydride	12	10
Desilylation	TEAHF	35	

[a]The deblock step with 3% DCA is only used for the initial detritylation of the polystyrene bound 3' nucleoside.

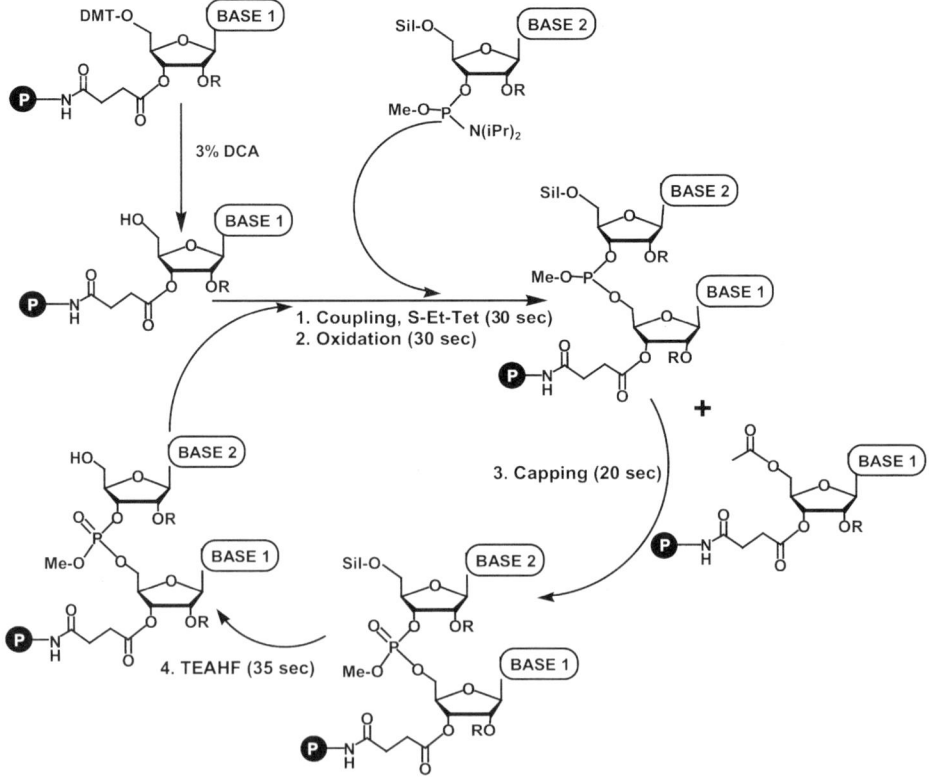

Fig. 4. The 5'-silyl, 2'-ACE RNA synthesis cycle.

The 5-ethylthio-1*H*-tetrazole activates coupling by protonating the diisopropyl amine attached to the trivalent phosphorous. Nucleophilic attack of the 5-ethylthio-1*H*-tetrazole leads to the formation of the tetrazolide intermediate that reacts with the free 5'-OH of the support-bound nucleoside forming the internucleotide phosphite linkage.

2. *Oxidation:* In the next step of chain elongation, the phosphorous(III) linkage is oxidized to the more stable and ultimately desired P(V) linkage using *t*-butylhydroperoxide. Although this reaction is rapid, a reaction time of 10 s is recommended to ensure complete oxidation. Typically, when synthesizing long oligonucleotides, it may be necessary to increase the oxidation reaction time to 20 s.

3. *Capping:* Although delivery of excess activator and phosphoramidite increases coupling efficiency, a small percentage of unreacted nucleoside may remain support-bound. To prevent the introduction of mixed sequences, the unreacted 5'-OH are "capped" or blocked by acetylating the primary hydroxyl. This acetylation is achieved through simultaneous delivery of 1-methylimidazole and acetic anhydride.

4. *5'-Desilylation*: Before the next nucleoside in the sequence can be added to the growing oligonucleotide chain, the 5'-silyl group is removed with fluoride ion. This requires the delivery of triethylamine trihydrogenfluoride for 45 s. The desilylation is rapid and quantitative and no wait step is required.

The cycle is repeated for each subsequent nucleotide until the desired sequence is constructed.

3.3. Oligonucleotide Deprotection

A two-stage rapid deprotection strategy is employed to remove phosphate backbone protection, release the oligonucleotide from the solid support, and remove the exocyclic amine protecting groups on A, G, and C. The treatment also removes the acetyl moiety from the acetoxyethyl orthoester, resulting in the 2'-*bis*-hydroxyethyl protected intermediate that is now 10 times more labile to final acid deprotection.

In the first deprotection step, S_2Na_2 is used to selectively remove the methyl protection from the internucleotide phosphate, leaving the oligoribonucleotide attached to the polystyrene support *(8)*. This configuration allows any residual reagent to be thoroughly washed away before proceeding. Although the following method describes a manual process for this deprotection, a multi-column, manifold approach can also be used.

3.3.1. S_2Na_2 Deprotection

1. Attach a syringe barrel to one of the two luer fittings on the synthesis column. Draw 2 mL of the S_2Na_2 reagent into a second syringe and attach to the opposite side of the synthesis column. Gently push the S_2Na_2 reagent through the column

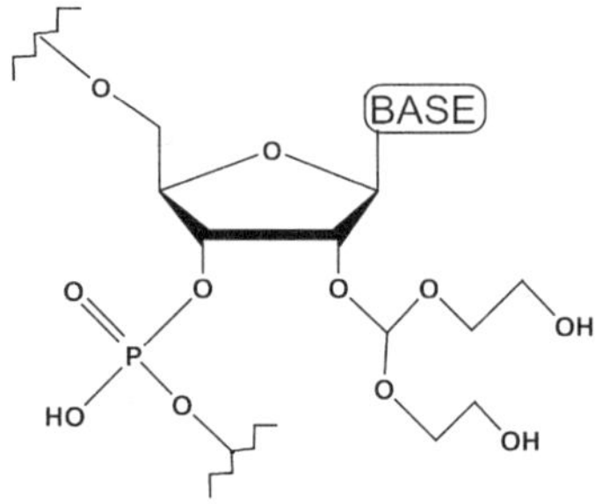

Fig. 5. 2'-ACE protected RNA immediately prior to 2'-deprotection.

and into the empty syringe barrel continuing back and forth several times. Fill the column with reagent, and allow it to sit at room temperature for 10 min.
2. Remove the S_2Na_2 reagent from the column and discard both syringes. Using a clean syringe, wash the column thoroughly with water.

In the second deprotection step, 40% 1-methylamine in water *(9,10)* is used to free the oligoribonucleotide from the solid support, deprotect the exocyclic base amines, and deacylate the 2'-orthoester leaving the deprotected species illustrated in **Fig. 5**.

3.3.2. N-Methylamine Deprotection

1. Transfer the solid support resin from the column into a 4-mL vial (*see* **Note 3**).
2. Add 2 mL 40% methylamine and heat for 12 min at 60°C. *Caution:* Allow deprotection solution to cool to room temperature and open slowly.
3. Remove the methylamine and transfer into a fresh vial. Discard the support.
4. Evaporate the oligonucleotide solution to dryness in a SpeedVac® or similar device.

3.4. Oligonucleotide Analysis

After the phosphate deprotection, the crude oligonucleotide should be analyzed for yield and purity. Typically, oligonucleotide yields are measured using an ultraviolet (UV) spectrophotometer (absorbance at 260 nm). Purity can be determined via anion-exchange HPLC (**Fig. 6A**), mass spectrometry (**Fig. 6B**), or polyacrylamide gel electrophoresis (PAGE) (**Fig. 7**).

The 2'-ACE orthoester protecting groups are still present at this point, and serve two purposes: First, they confer resistance to nuclease attack, and second they prevent the formation of higher order structures. As a result, the oligo can be more readily handled during analysis and/or stored beforehand without risk of degradation. The structure of 2'-ACE protected RNA is shown in **Fig. 5**. The hydrophilic character of the ACE protecting groups ensures that 2'-ACE protected RNA is readily soluble in water, regardless of length or sequence. The ability to conduct yield and purity analysis on protected oligos in aqueous

Table 2
Recommended Gradients (%B) According to Oligo Length (Bases)

Length (bases)	Gradient (%B)
0–5	0–30
6–10	10–40
11–16	20–50
17–32	30–60
33–50	40–70
>50	50–80

solution is one of the major benefits of 2'-ACE chemistry over alternative technologies.

3.4.1. Mass Spectrometry Analysis

1. Using the Perseptive Biosystems Voyager-DE BioSpectrometry Workstation, follow the referenced protocol for mass spectrometry analysis *(7)*. Briefly, a 3-hydroxy picolinic acid matrix is used for sample crystallization. It is prepared by mixing (10:1:1) 3-HPA:picolinic acid:ammonium hydrogen citrate where each component is dissolved in 30% aqueous acetonitrile at a concentration of 50 mg/mL.
2. One optical density unit (ODU) of oligonucleotide is dissolved in 20 µL of matrix and heated at 55°C for 10 min. 1 µL of the sample is spotted on a MALDI plate, allowed to dry, and analyzed accordingly *(7) (see* **Note 4**).

3.4.2. Anion-Exchange HPLC Purification and Analysis

1. Using the Dionex DNA-Pac–PA-100 column, a gradient is employed using HPLC buffer A and HPLC buffer B. Inject 0.5 ODUs of sample that has been dissolved in 100 µL H_2O or Tris buffer, pH 7.5. The gradient employed is based on oligonucleotide length and can be applied according to **Table 2**.
2. Use **Table 3** to program a linear gradient on the HPLC analyzer.

3.4.3. PAGE Purification and Analysis

Gel purification and analysis of 2'-ACE protected RNA follows standard protocols for denaturing PAGE *(11)*.

1. Resuspend the 2'-ACE protected oligo in 200 mL of gel loading buffer.
2. Assemble a standard gel casting apparatus, using 2.0-mm spacers if less than 75 ODUs will be loaded and 4.0-mm spacers if 75 to 150 ODUs will be loaded.
3. Prepare an acrylamide gel solution using 15% polyacrylamide for oligos of 10 to 25 bases and 8% polyacrylamide for 26 or more bases. Add the appropriate amounts of TEMED and 10% APS (200 mL TEMED, 600 mL APS for 2.0-mm gels; 400 mL TEMED, 1.2 mL APS for 4.0-mm gels), and mix thoroughly before pouring into the gel casting apparatus.

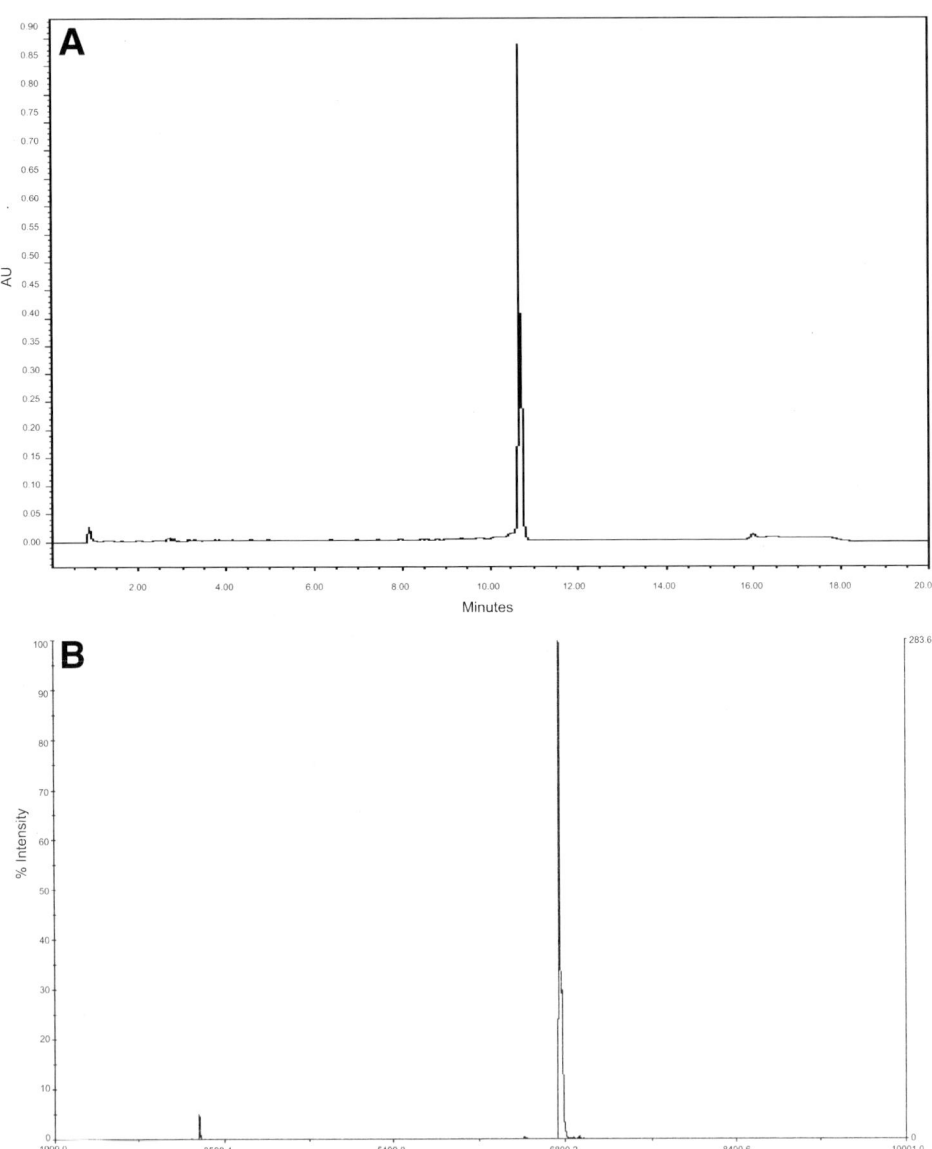

Fig. 6. Analytical evaluation of a crude 21-nucleotide sequence produced using ACE technology. Base content: A: 4.3%, G: 19%, C: 9.5%, U: 28.5%. (**A**) Anion-exchange HPLC; retention time: 10.72 min. (**B**) MALDI-TOF of the same sample measured as % intensity (*y*-axis) vs *m/z* (*x*-axis) where m = mass and z = charge; theoretical mass: 6729 and observed mass: 6748. Note: The peak at 3374 represents the half mass or the $z = {}^+2$ charged species.

Table 3
Recommended Parameters for Programming a Linear Gradient on an HPLC Analyzer

Time (min)	Flow (mL/min)	% Buffer A	% Buffer B
0	1.5	100	0
1	1.5	100	0
3	1.5	70^a	30^a
15	1.5	40^a	60^a
15.5	2.5	0	100
17	2.5	0	100
17.25	2.5	100	0
23	2.5	100	0
23.1	1.5	100	0
24	1.5	100	0
25	0.1	100	0

^{a}Modify the gradient here, based on the values in **Table 2**.

4. Pour the gel solution, insert the comb, and allow the gel to polymerize for 30 min. Once polymerized, remove the comb, add 1X TBE buffer to the electrophoresis reservoirs, and prewarm the apparatus to 40°C for 20 min.
5. Rinse all wells with 1X TBE buffer to remove unpolymerized acrylamide, and load samples. Load tracking dye into separate wells to monitor gel migration.
6. Electrophorese the gel at 50 to 100 W, maintaining the apparatus at 40°C. When complete, shut off power, pry apart plates, and place plastic wrap over the gel.
7. Place the gel on a thin-layer chromatography plate and expose to ultraviolet (UV) light at 254 nm to visualize the RNA using UV shadowing. Excise the desired gel band with a clean razor blade, and place it in a 50-mL conical tube. Crush the gel slice with a glass or plastic rod.
8. Add 30 mL 0.3*M* NaOAc elution buffer to the gel particles, and soak overnight.
9. Decant and filter eluant through sterile siliconized glass wool, and desalt on reverse phase such as SepPak or on a Sephadex column such as Nap-10 or Nap-25. Quantitate and evaporate the RNA solution to dryness.

3.5. 2'-ACE Deprotection

The 2'-protecting groups are the last to be removed. Typically, they are removed immediately prior to use in a biological assay using 100 m*M* acetic acid, pH 3.8. Hydrolysis occurs via the mechanism shown in **Fig. 8** *(12,13)*.

Because the 2'-protecting groups are very hydrophilic and readily solvated by water, acid-catalyzed hydrolysis proceeds to completion regardless of RNA sequence or length. The formic acid and ethylene glycol byproducts are volatile and are removed during evaporation.

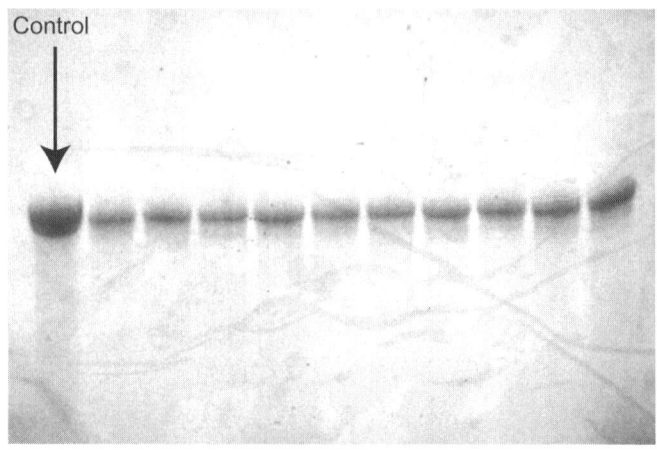

Fig. 7. Polyacrylamide gel electrophoresis of 10 different syntheses of a 75-base sequence. *Lane 1* contains a 75-base control.

2'-ACE deprotection is accomplished with the following method:

1. Reconstitute lyophilized RNA in 400 µL of 100 m*M* acetic acid, adjusted to pH 3.8 with TEMED.
2. Vortex for 10 s to completely dissolve the oligonucleotide pellet.
3. Incubate at 60°C for 30 min.
4. Lyophilize or SpeedVac to dryness before use.

After this final deprotection step, the oligonucleotides are ready for use. At this point, great care must be taken in handling the product because native 2'-OH RNA is prone to form secondary structures and/or be degraded by ribonucleases.

3.6. Conclusion

In the past decade advancements in oligonucleotide synthesis technologies have primarily focused on DNA analogs with altered biophysical properties. The principle driver for these efforts has been in the development of DNA-based genetic medicines. In addition, the major focus of high-throughput synthesis efforts has focused on DNA because of the breadth of molecular biology applications that employ DNA oligonucleotides (polymerase chain reaction [PCR], sequencing, etc.). Subsequently the focus and demand for synthetic RNA has historically been smaller and for highly specialized applications. However, the recent discovery that siRNA induces gene suppression in mammalian cells has contributed to a realization of the enormous biological significance of RNA. Thus, reliable, efficient, and high-quality RNA synthesis

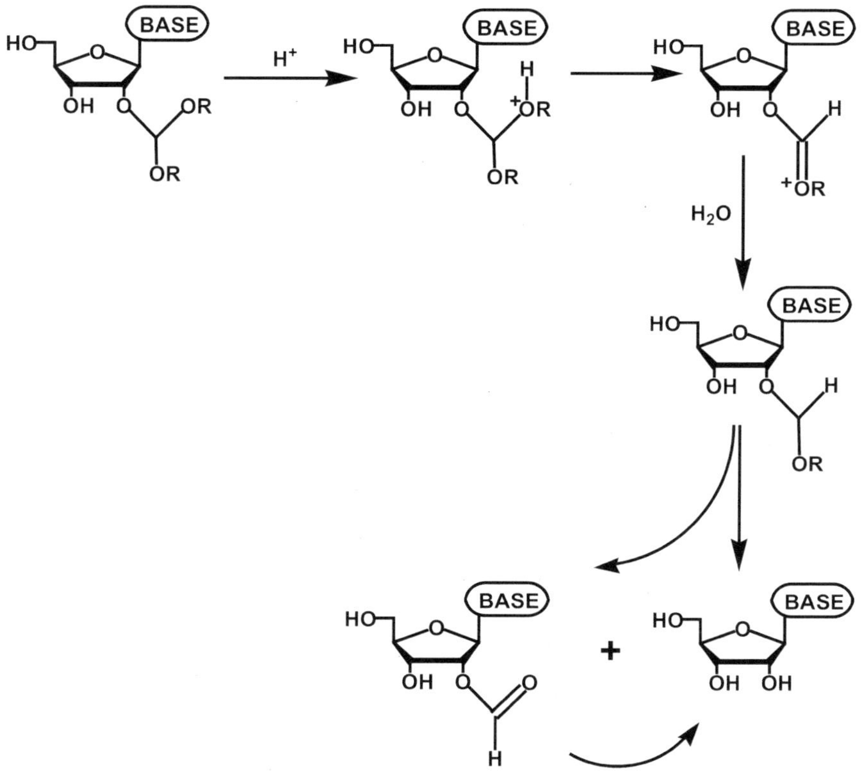

Fig. 8. Removal of the 2'-protecting groups via acid hydrolysis.

methods are now critically important for enabling research in drug discovery and therapeutic intervention.

Chemical synthesis of RNA using 5'-silyl 2'-ACE chemistry provides many favorable properties allowing for high-throughput RNA production. First, 2'-ACE chemistry provides superior oligonucleotide quality; 99% coupling efficiencies are routinely achieved with coupling times of only 30 s. This produces a crude product of very high purity. Using 2'-ACE chemistry, microgram to multigram quantities of material can be produced for high-throughput or in vivo studies. Second, 2'-ACE is the only RNA chemical synthesis method that can be used to synthesize long RNA sequences in excess of 70 bases with high quality and yield. Third, the technology enables the incorporation of virtually any chemical modification. For certain biophysical studies and in vivo applications, such chemical modifications are required to provide a readily monitored label or increase RNA stability in biological environments. Finally, 2'-ACE synthesis technology provides a protecting group strategy designed specifically for RNA synthesis rather than an alteration to previous DNA synthesis methods. The *bis* hydroxyethyl orthoester intermediate has superior handling properties relative to other methods and renders the oligoribonucleotide resistant to nuclease and other forms of degradation during handling and purification. Terminal deprotection can then be carried out under mild, aqueous conditions. Taken together, the favorable properties of 2'-ACE chemistry make it the method of choice for most RNA synthesis applications.

4. Notes

1. Controlled pore glass (CPG) is not compatible with the triethylamine trihydrogenfluoride-assisted desilylation.
2. A 5'-*O*-DMT nucleoside is bound to the support to simplify quantitation of support loading (calculated in µmole/gram).
3. Be sure that the vial can be tightly sealed.
4. The 3-HPA removes the 2'-ACE groups and the mass spectral data retrieved represents the fully deprotected oligonucleotide. If the sample is not heated, partially protected product of higher mass will be observed. Alternatively, heating the sample in matrix for too long will result in an increase in depurinated products.

References

1. Elbashir, S. M., Harborth, J., Lendeckel, W., Yakin, A., Weber, K., and Tuschl, T. S. (2001) Duplexes of 21-nucleotide RNAs mediate RNA interference in cultured mammalian cells. *Nature* **411,** 494–498.
2. Usman, N. O., Jiang, M. Y., and Cedergren, R. J. (1987) The automated chemical synthesis of long oligoribuncleotides using 2'-O-silylated ribonucleoside 3'-O-phosphoramidites on a controlled-pore glass support: synthesis of a 43-nucleotide sequence similar to the 3'-half molecule of an Escherichia coli formylmethionine tRNA. *J. Am. Chem. Soc.* **109,** 7845.
3. Caruthers, M. H. (1985) Gene synthesis machines: DNA chemistry and its uses. *Science* **230,** 281–285.
4. Wu, X. and Pitsch, S. (1998) Synthesis and pairing properties of oligoribonucleotide analogues containing a metal-binding site attached to beta-D-allofuranosyl cytosine. *Nucleic Acids Res.* **26,** 4315–4323.
5. Scaringe, S. A. (2000) Advanced 5'-silyl-2'-orthoester approach to RNA oligonucleotide synthesis. *Methods Enzymol.* **317,** 3–18.
6. Scaringe, S. A. (2001) RNA oligonucleotide synthesis via 5'-silyl-2'-orthoester chemistry. *Methods* **23,** 206–217.
7. Van Ausdall, D. A. and Marshall, W. S. (1998) Automated high-throughput mass spectrometric analysis of synthetic oligonucleotides. *Anal. Biochem.* **256,** 220–228.
8. Dahl, B. J., Bjergarde, K., Henriksen, L., and Dahl, O. (1990) A highly reactive, odorless substitute for thiophenol/triethylamine as a deprotection reagent in the synthesis of oligonucleotides and their analogs. *Acta Chem. Scand.* **44,** 639–641.

9. Reddy, M. P., Hanna, N. B., and Farooqui, F. (1994) Fast cleavage and deprotection of oligonucleotides. *Tetrahedron Lett.* **25,** 4311–4314.
10. Wincott, F., DiRenzo, A., Shaffer, C., et al. (1995) Synthesis, deprotection, analysis, and purification of RNA and ribozymes. *Nucleic Acids Res.* **23,** 2677–2684.
11. Ellington, A. and Pollard, J. D. (1998) Purification of oligonucleotides using denaturing polyacrylamide gel electrophoresis. In *Current Protocols in Molecular Biology,* Chanda, V. B., ed., Wiley, New York, pp. 2.12.1–2.12.7.
12. Griffin, B. E., Jarman, M., Reese, C. B., and Sulston, J. E. (1967) The synthesis of oligoribonucleotides. *Tetrahedron* **23,** 2301–2313.
13. Griffin, B. E., Jarman, M., Reese, C. B., and Sulston, J. E. (1967) Methoxymethylidene derivatives of ribonucleosides and 5'-ribonucleotides. *Tetrahedron* **23,** 2315–2331.

4

Dimethylthiarum Disulfide

New Sulfur Transfer Reagent in Phosphorothioates Oligonucleotide Synthesis

Zhiwei Wang, Quanlai Song, and Yogesh S. Sanghvi

Summary

Dimethylthiarum disulfide (DTD) has been developed as a new and efficient sulfur-transfer reagent for automated synthesis of phosphorothioate oligonucleotides using phosphoramidite chemistry. The traditional four-step automated oligonucleotide synthesis has been compressed to three-step protocol using DTD. This improvement allowed an overall 20% reduction in the solvent consumption and reduced the total synthesis time by 25%. The large-scale application of DTD has been successfully demonstrated by synthesis of therapeutically useful 20-mer phosphorothioate antisense oligonucleotides with excellent yield and purity.

Key Words: Dimethylthiarum disulfide (DTD); phenylacetyl disulfide (PADS); sulfur-transfer reagent; phosphorothioate oligonucleotides; phosphoramidite chemistry; antisense; solid-phase synthesis.

1. Introduction

The use of antisense oligonucleotides represents a powerful new strategy in the development of therapeutic agents that act by the selective inhibition of gene expression. With increasing numbers of antisense products in the clinic and successful outcome of these trials, large-scale production of oligonucleotides at low cost would be mandatory. Therefore developing low-cost approaches for oligonucleotide synthesis is of importance and has gained widespread interest. Most antisense oligonucleotides possess phosphorothioate backbones in which one nonbridging oxygen atom of the natural internucleotide phosphate is replaced by sulfur because oligonucleotide phosphorothioates exhibit increased resistance to cleavage by nucleases. Automated synthesis on solid supports using phosphoramidite chemistry has been the most successful method for large-scale synthesis of oligonucleotide phosphorothioate-based drugs *(1)*. Normally, detritylation, coupling, sulfurization, and capping are the four key steps in automated synthesis of phosphorothioate oligonucleotides in a four-cycle synthesis. During the sulfurization step, oxidation of phosphite P(III) to phosphorothioate triester P(V) is accomplished via a sulfur-transfer reagent. Our effort to develop new sulfur-transfer reagents led us to dimethylthiuram disulfide (DTD) and the possibility of combining sulfurization and capping in a single step (three-step cycle synthesis).

2. Materials

1. Primer 200 or HL 30 solid support (Amersham Biosciences).
2. Dichloroacetic acid (DCA) in toluene (10% v/v).
3. $0.2M$ β-cyanoethyl diisopropyl phosphoramidites (in anhydrous acetonitrile).
4. $1H$-tetrazole in $0.45M$ anhydrous acetonitrile.
5. DTD (Shasun Chemicals and Drugs, India).
6. Cap A solution: N-methylimidazole/pyridine/CH_3CN, 2:3:5, v/v.
7. Cap B solution:
    a. For three-step synthesis cycle: $0.4M$ DTD in acetic anhydide/CH_3CN/THF (1:2:2,v/v). Recipe for 1.0 L of the solution: 85 g of DTD, 200 mL of acetic anhydride, 400 mL of acetonitrile, and 400 mL of THF (*see* **Note 1**).
    b. For four-step synthesis cycle: 20% acetic anhydride in acetonitrile (v/v).
8. Phenylacetyl disulfide (PADS): $0.2M$ in acetonitrile/3-picoline, 1:1 v/v (*see* **Note 2**).
9. Triethylamine.
10. Concentrated ammonium hydroxide.
11. Ethanol.
12. Deionized water (DI water).
13. Methanol.
14. Sodium acetate.
15. Waters HC-C18 HA BondaPak (37–55 µm, 125A) radial compression column.
16. Acetic acid.
17. Sodium hydroxide

3. Methods

In this report, we describe a method to prepare oligonucleotide phosphorothioates on solid support via a three-step synthesis cycle of detritylation, coupling and combined sulfurization-capping using DTD as sulfur-transfer reagent. For comparison reasons, we describe the more classical four-step cycle synthesis using PADS as sulfurizing reagent.

3.1. Three-Step Cycle Automated Synthesis Using Phosphoramidite Chemistry and DTD

The synthesis is performed on an OligoPilot II or AKTA 100 automated synthesizer (Amersham Biosciences) at 0.2 mmol to 1.0 mmol scales. The cycle for solid-phase synthesis using amidite chemistry and DTD is described in detail in **Fig. 1** and **Table 1**. All reactions take place in a fixed-bed reactor vessel on amino-derivatized crosslinked polystyrene beads (Amersham Biosciences) to which an appropriate 5'-O-DMT-protected nucleoside is attached through a succinyl linker. The reactor is connected to pumps that deliver preprogrammed volumes of synthesis reagents and solvents.

3.1.1. Detritylation

Each cycle of the solid-phase synthesis commences with removal of the acid labile 5'-O-4,4'-dimethoxytrityl (DMT) protecting group of the 5'-terminal nucleoside of the support-bound oligonucleotide (or nucleoside for first cycle). During this step, the DMT protecting group is removed from 5'-hydroxy function by treatment with a large excess (50–90 molar equivalents based on the synthesis scale) of DCA in toluene for 1.5–3.0 min. Detritylation plays a crucial role in each synthesis cycle to produce high yield and quality of oligonucleotides. A common side reaction during detritylation is depurination, which is owing to lability of purine nucleotides and nucleosides under acidic environments required for DMT group removal. Increasing contact time with acid will lead to depurination to an extent that could compromise oligonucleotide quality. On the other hand, however, incomplete detritylation can generate n-1 mer and shorter sequence impurities. Fine-tuning the detritylation condition is often necessary to achieve high-quality oligonucleotides (*see* **Note 3**). Following detritylation, the support-bound nucleotide is washed with acetonitrile and the 5'-end hydroxyl group is free for the next coupling reaction with a nucleoside phosphoramidite.

3.1.2. Coupling

The chain elongation step is achieved by using standard phosphoramidite-coupling chemistry. During this step, the 5'-hydroxy groups of support-bound nucleotides (or nucleoside for the first step) reacts with a solution of protected nucleoside phosphoramidite ($0.2M$ in acetonitrile; 1.7–2.0 molar equivalents based on the synthesis scale) in the presence of $1H$-tetrazole (5.8–7.5 molar equivalents based on the scale) for 2–5 min (*see* **Note 4**). This results in the formation of a phosphite triester linkage (P(III)) between the incoming nucleotide synthon and the support-bound oligonucleotide chain. Excess reagents are washed away from the column reactor with acetonitrile in preparation for next reaction.

3.1.3. Combined Sulfurization–Capping

There are two objectives in this combined step. The first is to oxidize phosphite triesters (P(III)) formed in the coupling step to phosphorothioate (PV), using DTD as sulfur transfer reagent. The second is to cap the small proportion

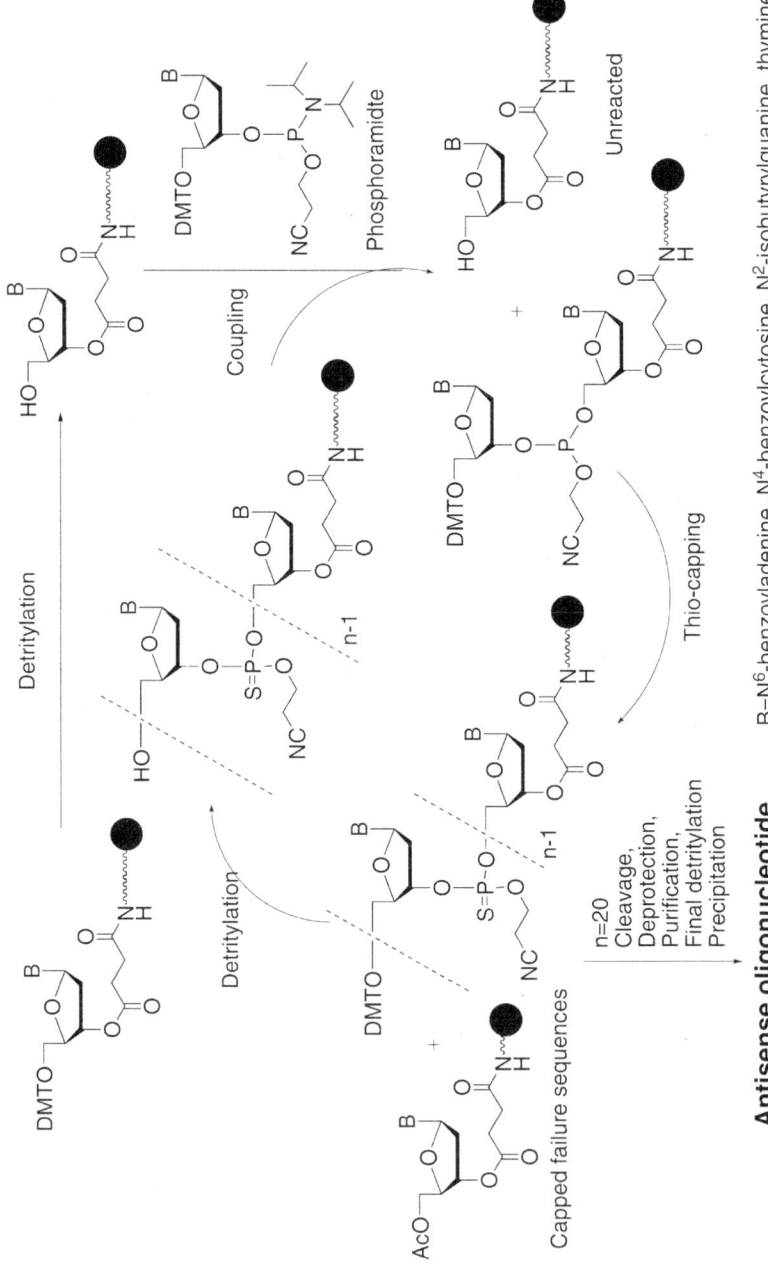

Fig. 1. Three-step cycle phosphorothioate oligonucleotide synthesis. B = N^6-benzoyladenine, N^4-benzoylcytosine, N^2-isobutyrylguanine, thymine

Table 1
Synthesis Parameters Used on AKTA 100 at 1.0 mmol Scale

Step	Reagent	Volume (mL)	Time (min)
Detritylation	10% DCA/toluene	72	1.5
Coupling	0.2M amidite in acetonitrile	10	2.0
	0.45M tetrazole in acetonitrile	15	
Thio-capping	Cap A: N-methylimidazole/pyridine/CH$_3$CN, 2:3:5, v/v/v	36 4.0	
	Cap B: 0.4M DTD in acetic anhydide/CH$_3$CN/THF, 1:2:2, v/v/v	36	

of 5'-hydroxy groups that fail to extend in the coupling step. Although the coupling reaction proceeds in very high yield, it is not quantitative. To prevent the small fraction of 5'-hydroxy groups from reaction during the next coupling cycle these reaction sites are capped with acetic anhydride and N-methylimidazole to produce so called DMT-off failure sequences. To accomplish sulfurizaton and capping in one step, 0.4M DTD is added to 20% acetic anhydride in CH$_3$CN and THF (Composition of a 1-L solution: 85 g DTD, 200 mL acetic anhydride, 400 mL acetonitrile, and 400 mL THF) (*see* **Note 1**). Standard capping solution A is unaltered (N-methylimidazole/pyridine/CH$_3$CN, 2:3:5, v/v). Contact time during synthesis for the combined step was 4 min, and 1.5 column volume each of Cap A and Cap B solutions were used. Following combined sulfurization–capping steps, excess of reagent is removed from the support by washing with acetonitrile in preparation for next reaction.

The three basic cycles (detritylation, coupling, and combined sulfurization–capping) are repeated using the appropriate protected nucleoside phosphoramidite for assembly of the desired oligonucleotide sequence with a DMT group at the 5' terminus. Cleavage of oligonucleotide from solid support and deprotection of the backbone and nucleobases are achieved by incubation with ammonium hydroxide at 55°C overnight. The support is removed by filtration and washed with a solution of ethanol in water (1:1, v/v). The combined filtrate and washings are concentrated to give crude DMT-on oligonucleotide solution for purification by reverse-phase high performance liquid chromatography (RP-HPLC) (*see* **Subheading 3.3.**). A few drops of triethylamine may be added to the solution to stabilize DMT groups.

3.2. Four-Step Cycle Automated Synthesis Using Phosphoramidite Chemistry and PADS

Using PADS as a sulfurizing reagent, sulfurization and capping were done separately. Attempts to combine the two steps failed owing to formation of unacceptable levels of phosphodiesters. Detritylation and coupling steps remain the same as those of the three-step cycle synthesis (*see* **Subheadings 3.1.1.** and **3.1.2.**).

3.2.1. Sulfurization With PADS

The phosphite triester newly formed in the coupling step is converted to the corresponding phosphorothioate triester by treatment with $0.2M$ solution of PADS in acetonitrile and 3-picoline (1:1 v/v). Typically, a 1.5-column volume of PADS solution is delivered, and sulfurization is complete within 3 min, at which time excess of reagent is removed from the reaction vessel by washing with acetonitrile.

3.2.2. Capping

Cap A solution used in this step is the same as that of three-step cycle synthesis (*N*-methylimidazole/pyridine/CH_3CN, 2:3:5 v/v). Cap B solution consists of 20% of acetic anhydride in acetonitrile. An equal volume of the two solutions (1 column volume each) is pumped through a reaction vessel for 2 min. Excess reagent is removed by an acetonitrile wash before the next detritylation cycle.

The four basic cycles (detritylation, coupling, sulfurization, and capping) are repeated until the desired oligonucleotide sequence is assembled. Then, following the procedure mentioned previously (*see* **Subheading 3.1.3.**) the oligonucleotide is cleaved from support to afford crude DMT-on oligonucleotide solution ready for purification.

3.3. Purification by Reverse-Phase High Performance Liquid Chromatography

The hydrophobic DMT group is retained at the 5'-end of the oligonucleotide to facilitate separation of DMT-on full-length oligonucleotide from DMT-off shorter failure sequences (from the incomplete coupling followed by capping) using RP-HPLC. DMT-on fractions, however, contain not only DMT-on full-length (n-mer) product, but also DMT-on shorter or longer sequences owing to depurination, unusual chain growth, incomplete detritylation, and capping, which may be reduced to the minimum levels by refining the chemistry cycle.

The HPLC conditions are as follows. Depending on capacity of the column, a suitable amount of sample is injected and chromatographed using the appropriate wavelength (260–300 nm) for detection. DMT-on fractions are collected. A typical chromatography is shown in **Fig. 2**.

1. System: BioCAD-700E.
2. Column: Waters BondaPak HC18HA (37–55 μM, 125A) with radial compression chamber.

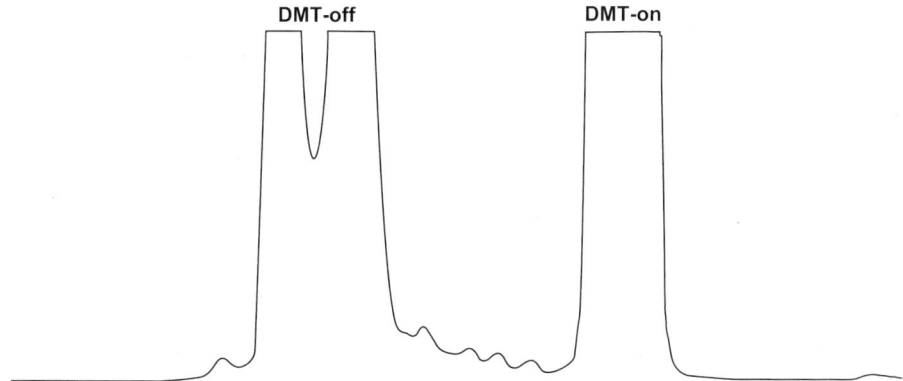

Fig. 2. A typical chromatography during RP-HPLC purification.

3. Buffer A: DI water; Buffer B: MeOH; Buffer C: NaOAc in DI water.
4. Gradient: 20% to 100% B; Buffer C is used to maintain pH at 7.0 (8%). Normally, DMT-on material is eluted out with 50 to 90% B and 0% C.

3.4. Final Detritylation, Precipitation, and Lyophilization

DMT-on fractions obtained from RP-HPLC purification are pooled. The next step is to remove the DMT group. Although final removal of DMT groups from the 5'-end of oligonucleotides is a routine operation at laboratory scale, a publication discussing this subject in detail has recently appeared (2). Controlling the final detritylation step to achieve minimal depurination is crucial to produce high-quality oligonucleotides. The procedure is summarized here.

1. DMT-on oligonucleotide is precipitated out from the HPLC fraction by adding three times the volume of ethanol and collecting by centrifugation or sedimentation.
2. Dissolve the precipitation in DI water to achieve a concentration of 1750 optical density (OD)/mL.
3. Add three times the volume of $0.01M$ NaOAc solution, pH 3.0, to adjust pH to between 3.0–4.0 and record time.
4. Take 50 µL of the above solution at t min and quench by 1.0 mL of $3.0M$ NaOAc, pH 9.0. Analyze the sample by RP-HPLC (column: Luna 5 µ C18, 250 × 4.6 mm, Phenomenex. Buffer A: $0.1M$ triethylammonium acetate. Buffer B: acetonitrile. Gradient: 0% to 40% of B in 25 min, then keep 40% for 5 min.) to determine the percentage of DMT-on oligonucleotide and calculate half-life time for DMT removal according to the following equation:

$$t_{1/2} = [\ln(0.5)/\ln(\text{DMT-on \%})]t$$
$$\text{DMT-on \%} = (\text{peak area of DMT-on})$$
$$(\text{peak area of DMT-on} + \text{peak area of DMT-off})$$

This DMT removal can be consider as pseudo-first-order kinetics, so removal of 99.997% of the DMT group from the oligonucleotide requires 15 half-lives at which time the reaction is defined as completion.

5. Adjust pH value of reaction mixture to >5.5 by adding 3.0M NaOAc, pH 9.0 when reaction is completed.
6. Add three times the volume of ethanol to the above solution to precipitate out oligonucleotide, which is collected by centrifugation or sedimentation.
7. Dissolve the precipitation with minimal DI water; adjust pH t between 7.2 and 7.5 with 1.0M NaOH.
8. Precipitate oligonucleotide with three times the volume of ethanol. Collect by centrifugation or sedimentation.
9. Dissolve the product in minimal DI water and lyophilize to dryness

3.5. Discussion

In this section, sulfur-transfer reagents DTD and PADS and three-step synthesis vs four-step synthesis are compared.

3.5.1. Sulfur-Transfer Reagents

Elemental sulfur is used for early phosphite triester sulfurization, but it is poorly soluble in the organic solvents commonly used for automated synthesis. Also, conversion of P(III) into P(V) product is slow with elemental sulfur. It is crucial that the sulfur transfer step be highly efficient to minimize formation of phosphodiesters. For these reasons, many sulfur-transfer reagents were developed during the past decade *(3–12)*. Among the dozen reagents described in the literature, 3*H*-1,2-benzodithiol-3-one 1,1-dioxide (Beaucage reagent) *(13)* and phenylacetyl disulfide (PADS) *(14)* are widely used reagents at the present time for large-scale transfer. Both are soluble in organic solvents and have efficient and rapid sulfurization kinetics. Although these two reagents are currently used on an industrial scale, they have certain limitations. For example, Beaucage reagent is expensive, its long-term stability in solution is not optimal, and the cyclic sulfoxide byproduct formed during sulfurization is an efficient oxidizing agent, which increases levels of phosphodiesters during packed-bed solid-phase synthesis. Sulfurization with PADS is accomplished in combination with 3-picoline, a mixture with an obnoxious odor, whose efficiency depends on hold time. Our search for an alternative low-cost sulfur-transfer reagent led us to DTD as the reagent of choice.

The preparation (*see* **Note 5**) and structure of DTD are shown in **Fig. 3**. DTD is an analog of tetraethylthiuram disulfide (TETD) a known sulfur-transfer reagent *(15)*. Stability and solubility of DTD were tested in various solvents for extended periods of time. DTD is highly soluble in THF (>1M) and relatively less so in CH$_3$CN (0.3M). Addition of base to a solution of DTD rapidly

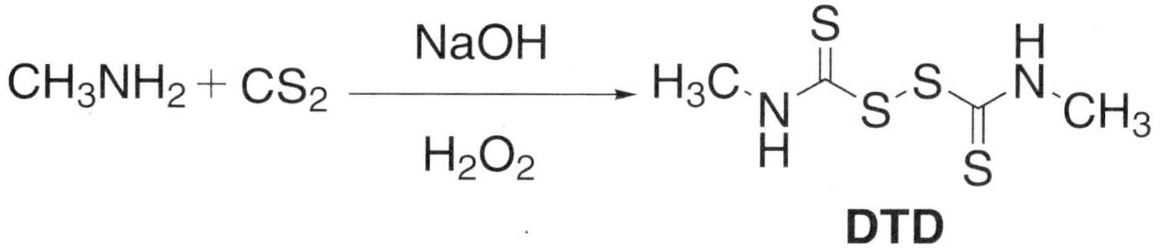

$$CH_3NH_2 + CS_2 \xrightarrow[H_2O_2]{NaOH} \underset{\textbf{DTD}}{H_3C\text{-}NH\text{-}C(=S)\text{-}S\text{-}S\text{-}C(=S)\text{-}NH\text{-}CH_3}$$

Fig. 3. Synthesis of dimethylthiuram disulfide (DTD).

degrades the material, whereas addition of an acid such as 4-chlorophenol stabilized a 0.3M solution of DTD in THF for 3 d. Similarly, addition of acetic acid to a solution of DTD significantly increased stability of the reagent (>30 d) without compromising sulfur-transfer ability. The exceptional stability of DTD in the presence of acetic acid encouraged us to dissolve it in capping solution B, which contains acetic anhydride in CH_3CN, hoping that sulfurization and capping steps could be performed simultaneously. A shorter protocol (three-steps vs four-steps) could result in reducing solvent wash, saving time in the overall synthesis cycle, and making the DTD solution stable.

3.5.2. Result Comparison of Three-Step vs Four-Step Synthesis

Syntheses of ISIS14803 and ISIS 3521/Affinitak™, two 20-mer phosphorothioate oligonucleotides, were undertaken using the new three-step protocol on an OligoPilot II synthesizer at an 169–193-μmol scale. Identical sequences were made using PADS for sulfurization. The data on crude oligonucleotide made using DTD and PADS as sulfur-transfer reagents is summarized in **Table 2**. The data indicate that there was no obvious difference between the syntheses. Subsequent purification of two of the oligonucleotides further confirmed that both were equally pure and furnished expected yields.

The synthesis was repeated with DTD at a larger scale (1 mmol) 20-mer phoshorothioate oligonucleotide ISIS3521 (Affinitak). The same oligonucleotide using PADS was also synthesized as a control. The data in **Table 2** indicate that DTD works as well as PADS at a larger scale. More important, overall solvent (CH_3CN) consumption for synthesis with DTD is reduced by 20% compared to synthesis done with PADS, owing to elimination of a synthesis/wash cycle. The results indicate that quality and yield of the product is not compromised when DTD is employed in the three-step mode as described previously. The products were analyzed by HPLC, CGE, ^{31}P NMR, and liquid chromatography mass spectrometry (**Fig. 4**).

In summary, use of DTD as a sulfur-transfer reagent may provide the following advantages over PADS and other conventional reagents.

Table 2
Synthesis Data on Crude Oligonucleotides

S-Transfer Reagent	Oligo	Scale (μ mol)	OD/ μ mol	Trityl- on %	$p = 0\%$	CH$_3$CN Use (L)	Synthesis Time (min)
PADS	14803	173	148	74.7	0.34^a	2.5	470
DTD	14803	169	149	76.2	0.27^a	2.0	350
PADS	Af Affinitak™	193	151	72.0	8.7	5.4	540
DTD	Affinitak™	193	152	73.0	4.9	4.5	450
PADS	Af Affinitak	1000	119	73.0	6.2	11.5	370
DTD	Affinitak	1000	119	81.0	4.1	10.5	300

^{a}by NMR.
ISIS 14803 sequence: CTGCTCATGGTGCACGGTCT
Affinitak sequence: GTTCTCGCTGGTGAGTTTCA

1. DTD is inexpensive compared to other sulfur-transfer reagents ($10/kg).
2. Overall usage of CH$_3$CN is reduced by 20%, which is a significant cost saving.
3. Owing to reduced solvent consumption, there is less waste solvent to dispose.
4. The total synthesis time was reduced by 25%, which increases the throughput.
5. The atom efficiency *(16)* with DTD is excellent owing to its small size.
6. Use of DTD is safe and convenient for large-scale transfers.

4. Notes

1. Recently, a solution of 85 g of DTD in 200 mL acetic anhydride, 300 mL THF, and 500 mL toluene was found to be more stable than the solution in acetonitrile. The DTD solution in toluene/THF was stable for 3 wk at room temperature (20–25°C) with detectable degradation of 2–3% DTD. To quickly dissolve DTD, first add DTD in acetic anhydride, followed by THF and half the amount of toluene required, and then add the other half. DTD was analyzed by HPLC under the following experimental conditions.
   a. Column: YMC ODS-AQ S3 120A, 4.6 × 100 mm.
   b. Flow rate: 1.0 mL/min.
   c. Detector: UV at 244 nm.
   d. Sample: 1–2 mg in 1 mL of glacial acetic acid; inject 10 μL.
   e. Buffer A: 0.2% acetic acid; Buffer B: acetonitrile.
   f. Linear gradient: 50% to 90% B in 15 min.
   g. Retention time: 5.4 min.
2. There is a long-lasting obnoxious smell. Avoid contact with skin. Good ventilation is needed to prepare the solution.
3. Ideally, detritylation should be performed as quickly as possible. In practice, however, there are some limitations in doing so. During synthesis on a fixed-bed reactor, the solid support is tightly packed in a column-type reaction vessel where high back pressure could be generated during high-flow rates resulting in synthe-

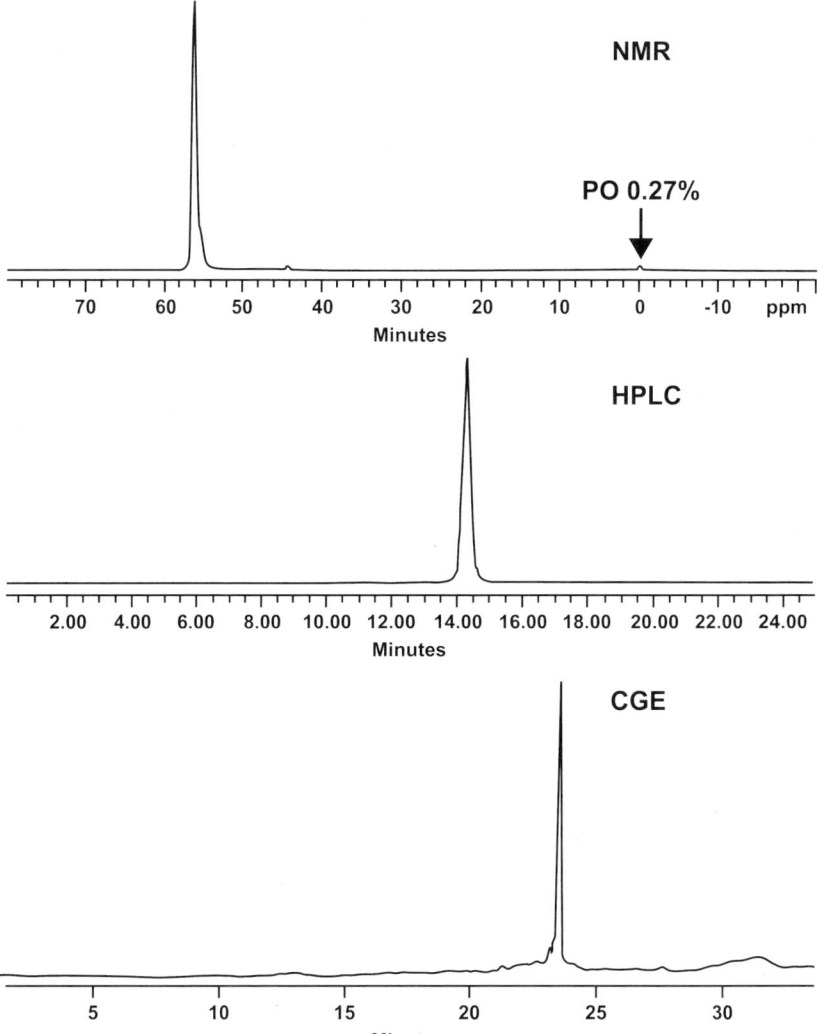

Fig. 4. HPLC, CGE, ^{31}P NMR data of purified ISIS3521 (Affinitak).

sis failure. The back pressure profile may vary from one solid support to another. In addition, pump capacity is another factor to be considered in + large-scale transfers. We recommend 3 column volumes and 2 min contact time when Primer 200 solid support is used.

4. For best results, both amidite and tetrazole solutions were dried over molecular sieves (4A) prior to use. The two solutions were delivered in the ratio of 40/60.
5. Preparation of DTD: NaOH (80g, 2 mol) was dissolved in H_2O (500 mL) and the solution was cooled to 0°C. THF (200 mL), methylamine (40% in H_2O, 170 mL,

2 mol) and carbon disulfide (120 mL, 2 mol) were added over a period of 30 min to aqueous NaOH solution while stirring at 0°C. Crushed ice (1.5 kg) was added to the reaction mixture, followed by acetic acid (300 mL). Hydrogen peroxide (30%, 100 mL, 1 mol) was added gradually while stirring for 15 min and maintaining the reaction temperature below 5°C (*see* **Note 6**). Heptane (800 mL) was added to the reaction mixture and stirred for an additional 30 min. The product was filtered off and washed with aqueous acetic acid (2%, 5 × 200 mL) and heptanes (2 × 200 mL). The product was air-dried until constant weight was obtained (1 d) to furnish an off-white solid (205 g, 97%). M.p. 98–100°C.

6. The addition of acetic acid and hydrogen peroxide is exothermic. Crushed ice or efficient cooling is required to maintain the reaction at low temperature.

Acknowledgments

We are thankful to Dr. Douglas L. Cole for his support and encouragement throughout this project.

References

1. Sanghvi, Y. S., Andrade, M., Deshmukh, R. R., Holmberg, L., Scozzari, A. N., and Cole, D. L. (1999) Chemical synthesis and purification of phosphorothioate antisense oligonucleotides. In *Manual of Antisense Methodology.* Hartmann, G. and Endres, S., eds. Kluwer Academic, New York, pp. 3–23.
2. Krotz, A. H., McElroy, B., Scozzari, A. N., Cole, D. L., and Ravikumar, V. T. (2003) Controlled detritylation of antisense oligonucleotides. *Org. Process R&D* **7,** 47–52.
3. Rao, M. V., Reese, C. B., and Zhao, Z. (1992) Dibenzoyl tetrasulfide: a rapid sulfur-transfer agent in the synthesis of phosphorothiaote analogues of oligonucleotides. *Tetrahedron Lett.* **33,** 4839–4842.
4. Stec, W. J., Uznanski, B., and Wilk, A. (1993) *Bis(O,O*-diisopropoxy phosphinothioyl)disulfide: a highly efficient sulfurizing reagent for cost-effective synthesis of oligo(nucleoside phosphorothioate)s. *Tetrahedron Lett.* **34,** 5317–5320.
5. Rao, M. V. and Macfarlane, K. (1994) Solid phase synthesis of phosphorothioate oligonucleotides using benzyltriethylammonium tetrathiomolybdate as rapid sulfur transfer reagent. *Tetrahedron Lett.* **35,** 6741–6744.
6. Efimov, V. A., Kalinkina, A. L., Chakhmakhcheva, O. G., Hill, T. S., and Jayaraman, K. (1995) New efficient sulfurizing reagents for preparation of oligodeoxyribonucleotide phosphorothioate analogues. *Nucleic Acids Res.* **23,** 4029–4033.
7. Xu, Q., Musier-Forsyth, K., Hammer, R. P., and Barany, G. (1996) Efficient introduction of phosphorothioates into RNA oligonucleotides by 3-ethoxy-1,2,4-dithiazoline-5-one (EDITH). *Nucleic Acids Res.* **24,** 1602–1607.
8. Zhang, Z., Nichols, A., Tang, J. X., Han, Y., and Tang, J. Y. (1999) Solid phase synthesis of oligonucleotide phosphorothioate analogues using 3-methyl-1,2,4-dithiazoli-5-one (MEDITH) as a new sulfur transfer reagent. *Tetrahedron Lett.* **40,** 2095–2098.

9. Zhang, Z., Nichols, A., Alsbeti, M., Tang, J. X., and Tang, J. Y. (1998) Solid phase synthesis of oligonucleotide phosphorothioate analogues using *bis*(ethoxythiocarbonyl)tetrasulfide as a new sulfur-transfer reagent. *Tetrahedron Lett.* **39**, 2467–2470.
10. Tang, J. Y., Han, Y., Tang, J. X., and Zhang, Z. (2000) Large-scale synthesis of oligonucleotide phosphorothioates using 3-amino-1,2,4-dithiazole-5-thione as an efficient sulfur-transfer reagent. *Org. Process R&D* **4**, 194–198.
11. Ju, J. and McKenna, C. E. (2002) Synthesis of oligodeoxyribonucleoside phosphorothioates using Lawesson's reagent for sulfur transfer step. *Bioorg. Med. Chem. Lett.* **12**, 1643–1645.
12. Cheruvallath, Z. S., Krishna Kumar, R., Rentel, C., Cole, D. L., and Ravikumar, V. T. (2003) Solid phase synthesis of phosphorothioate oligonucleotides utilizing diethyldithiocarbonate disulfide (DDD) as an efficient sulfur transfer reagent. *Nucleosides, Nucleotides Nucleic Acids* **22**, 461–468.
13. Iyer, R. P., Phillips, L. R., Egan, W., Regan, J. B., and Beaucage, S. L. (1990) The automated synthesis of sulfur-containing oligodeoxyribonucleotides using 3*H*-1,2-benzodithiol-3-one 1,1-dioxide as a sulfur-transfer reagent. *J. Org. Chem.* **55**, 4693–4699.
14. Cheruvallath, Z. S., Carty, R. L., and Moore, M. N. (2000) Synthesis of antisense oligonucleotides: replacement of 3*H*-1,2-benzodithiol-3-one 1,1-dioxide (Beaucage reagent) with phenylacetyl disulfide (PADS) as efficient sulfurization reagent: from bench to bulk manufacturing of active pharmaceutical ingredient. *Org. Process R&D* **4**, 199–204.
15. Vu, H. and Hirschbein, B. L. (1991) Internucleotide phosphite sulfurization with tetraethylthiuram disulfide: phosphorothioate oligonucleotide synthesis via phosphoramidite chemistry. *Tetrahedron Lett.* **32**, 3005–3008.
16. Trost, B. M. (2002) On inventing reactions for atom economy. *Acc. Chem. Res.* **35**, 695–705.

5

Assay for Evaluating Ribonuclease H-Mediated Degradation of RNA–Antisense Oligonucleotide Duplexes

Annie Galarneau, Kyung-Lyum Min, Maria M. Mangos, and Masad J. Damha

Summary

Ribonucleases H are complex enzymes whose functions are not clearly understood, further compounded by the fact that multiple forms of the enzyme are present in various organisms. They are known to recognize and degrade the ribonucleic acid (RNA) strand of numerous deoxyribonucleic acid (DNA)–RNA duplex substrates, and so may provide a unique mode of therapeutic intervention at the genetic level of virtually any disease. We have therefore set out detailed procedures for conducting routine assays with almost any one of this family of enzymes by a straightforward assay aimed at identifying novel enzyme-activating antisense oligonucleotides (AONs). The procedures described herein should enable easy identification of potent AON molecules, provided that the RNA is appropriately labeled for subsequent visualization following the guidelines set forth in this protocol.

Key Words: Ribonuclease H; antisense oligonucleotide; RNA cleavage; polyacrylamide gel electrophoresis; DNA–RNA hybrid; 2'-fluoroarabinonucleic acid.

1. Introduction

Ribonuclease H (RNase H) enzymes play critical roles in various biological processes in both eukaryotic and prokaryotic organisms, one of the best studied examples being in the life cycles of certain retroviruses *(1)*. All enzymes within this family ensure the accurate replication or transcription of deoxyribonucleic acid (DNA) through, for example, the excision of ribonucleotides that are sometimes incorporated in the DNA template. During pro- and eukaryotic replication, this occurs via the removal of ribonucleic acid (RNA) primers generated during synthesis of the lagging DNA strand *(2)*. RNases H also participate in DNA repair events *(3,4)* and retroviral transcription *(5)* as well, in which the enzyme removes RNA from duplex DNA–RNA intermediates to enable the integration of the viral RNA genome into the host gene pool.

Mammalian RNase H was the first of the family to be discovered *(6)* and was termed a class I enzyme to designate it as the first to display this unique RNA cleaving property. A second human enzyme later emerged on the basis of different physical, biochemical, and serological characteristics and was subsequently named RNase HII to distinguish it from the former *(7,8)*. Later, the amino acid sequence for human RNase HI was determined and shown to display significant homology to the class II RNase H of *Escherichia coli,* which itself corresponds to the least active of the two isoforms present in the bacterial kingdom *(9)*. In fact, all known mammalian RNases HI and HII are now known to be sequence homologs of bacterial RNases HII and HI, respectively *(10)*.

These and other discrepancies in the general nomenclature of the mammalian and prokaryotic enzymes, coupled with recent discoveries of new RNase H genes, have prompted the creation of at least two major categories of RNases H based either on the enzymes' biochemical properties or sequence homologies. Currently, prokaryotic and eukaryotic RNases H are widely designated by class (class I or II) to characterize members with comparable serological, physical, and biochemical characteristics including similarities in peptide molecular weights, isoelectric points, sensitivity to sulfur blocking reagents, and ability to use Mn^{2+} as a cofactor. *E. coli* RNase HI is the best characterized of this group and is also closely related in sequence to the RNase H domain of HIV-1 reverse transcriptase *(11)*.

Alternatively, a newer system of classification specifies the enzymes by type (type 1 or 2), which corresponds to those members that display close sequence homologies rather than similar biochemical attributes. Because of various intricate discrepancies between these two classification modes and the ensuing confusion when drawing comparisons across different organisms, some authors have preferred to designate the enzymes simply by molecular weight, as for example with *Saccharomyces cerevisiae* RNase H(35). Here, the number in parentheses refers to the molecular weight of the protein in kilodaltons; in this example, the 35 kDa yeast enzyme is also widely termed as a class I, type 2 enzyme *(12)*. Similarly, human RNase HI is a class I, type 2 enzyme, whereas *E. coli* RNase HI belongs to the type 1 category. By this account, the type and class designations poorly correlate with one another at best, and a more accurate way of placing the enzymes may need to draw on several additional criteria.

Regardless of origin, all RNases H display almost identical mechanistic profiles in that they specifically cleave the RNA moiety of RNA–DNA hybrids, yielding products with 5'-phosphates and 3'-hydroxyl termini. This is exemplified in **Fig. 1**, which highlights a similar ability for both HIV-1 RT and bacterial RNase H isotypes to act on a particular antisense oligonucleotide (AON)–RNA hybrid duplex. The degradation reaction requires divalent metal cations (magnesium or manganese), and a free 2'-hydroxyl group in the sense

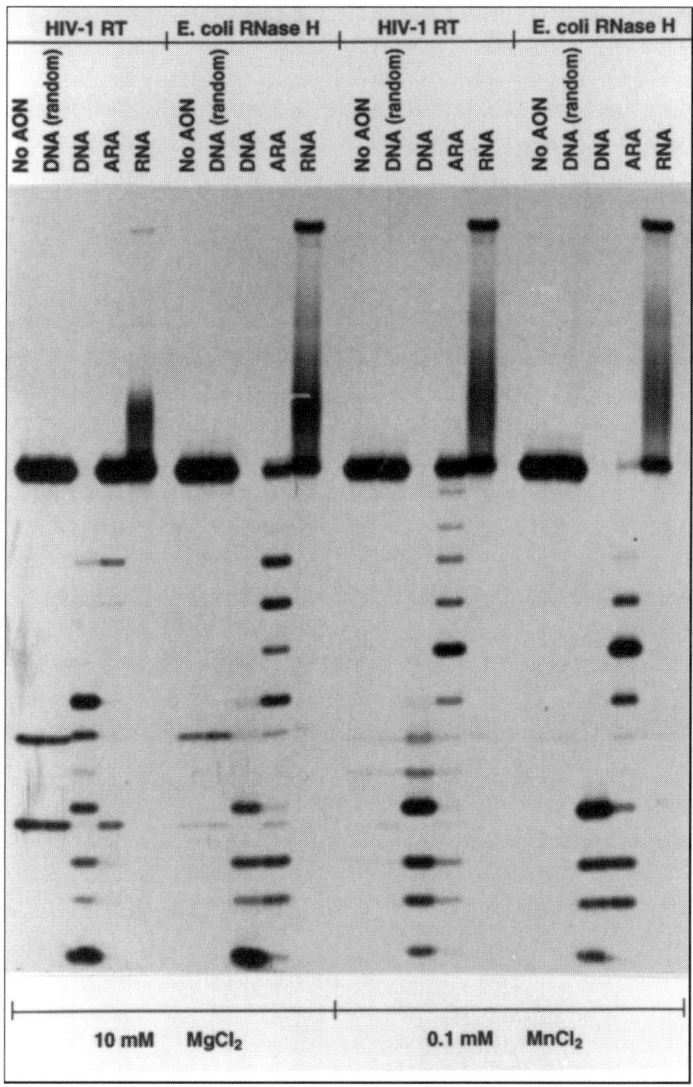

Fig. 1. Ribonuclease H (RNase H) degradation of various 18-bp oligonucleotide hybrid duplexes. An 18-nt 5'–^{32}P-labeled target RNA (5'-GAU CUG AGC CUG GGA GCU-3') was preincubated with complementary 18-nt DNA, ARA (arabinonucleic acid, ANA), and RNA, as well as with a noncomplementary DNA sequence (DNA random), and then added to reaction assays containing either HIV-1 RT or *E. coli* RNase H at the indicated divalent cationic metal concentrations. The slowly migrating band in the RNA lanes is the RNA–RNA duplex, which withstands the denaturing conditions of the gel electrophoresis and the heating of the sample to 100°C prior to being loaded onto the gel. (Reprinted with permission from **ref. *18*.**)

(RNA) strand. The reaction first begins with an endolytic cleavage at a ribonucleotide involved in the duplex and then processes exolytically in a 3' to 5' direction *(13,14)* with respect to the DNA component of the paired strands. Hydrolysis of the target RNA is accomplished by binding of the enzyme to the minor groove of the duplex, away from any steric influence of the nucleic acid heterocycles, which themselves project into the major groove cavity. As such, RNase H-mediated cleavage usually proceeds with minimal sequence specificity, implying that almost any DNA–RNA hybrid can serve as a suitable substrate for the enzyme. This also offers a powerful approach for selectively inhibiting the production of an encoded disease-causing protein, to enable reversal of the disease phenotype and ultimately restore normal cellular function.

Accordingly, the use of AONs as selective inhibitors of gene expression offers a rational approach for the prevention and treatment of many genetic disorders. The RNase H-mediated degradation of a target messenger RNA (mRNA) is considered to be the most effective mode of action of AONs *(15,16)* for it is simply more potent than those that are limited to interfering with the splicing or translational machinery. This is because the antisense-bound RNA is rapidly and permanently disabled by RNase H, thereby freeing the intact antisense for successive rounds of inhibition.

There are currently very few antisense analogs capable of supporting RNase H-induced RNA cleavage. In fact, of hundreds of modified AONs tested, only a handful retain enzyme activity. These include DNA modified with phosphorothioate internucleotide linkages (PS-DNA), and various sugar-modified derivatives, such as the arabinonucleic acids (ANAs), 2'-fluoro-arabinonucleic acids (2'F-ANAs), and cyclohexene nucleic acids (CeNAs) *(17–20)*. In fact, our discovery that ANA and 2'F-ANA are potent RNase H activators when hybridized with RNA has led to several interesting observations regarding the role of duplex structure on the ensuing enzyme activity *(21–23)*.

It appears that duplex flexibility may also be a key element in terms of RNase H recruitment such that incorporation of an acyclic linker along the antisense DNA or 2'F-ANA strands renders the target more amenable to hydrolysis by RNase H *(24,25)*. The general structures of some of these derivatives are shown in **Fig. 2A**. Remarkable differences in the amount of RNA scission occur with these analogs, with butyl-modified 2'F-ANA supporting higher levels of cleavage than all-2'F-ANA analogs. The trend is not as clear for antisense constructs with a DNA backbone, and seems to depend in part on the length and structure of the target RNA (**Figs. 2** and **3**). As such, understanding the mechanisms of catalytic function and substrate recognition for RNase H is critical in

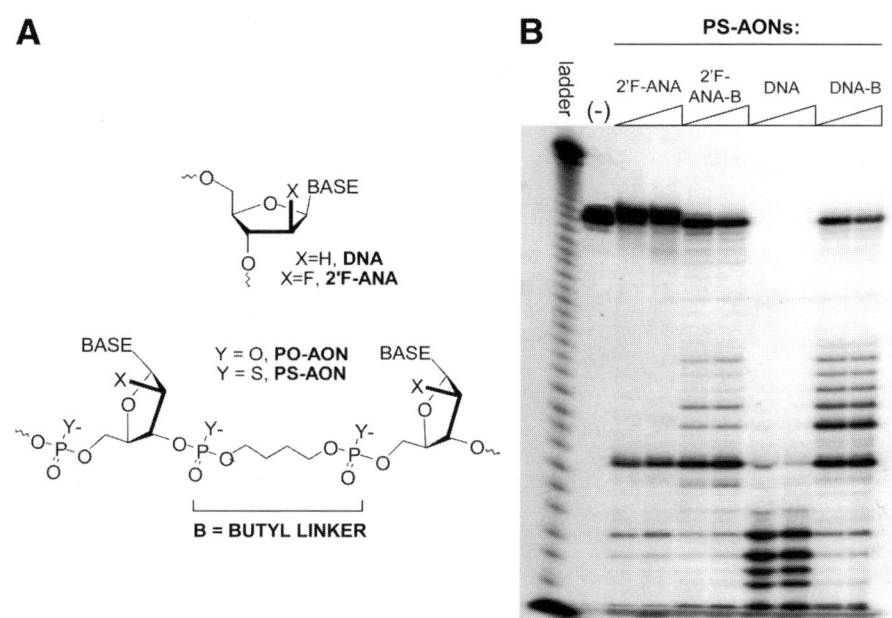

Fig. 2. **(A)** Chemical structures of some enzyme activating antisense oligonucleotides used in the assays. Incorporation of a butyl linker in the center of an all-2' F-ANA AON accelerates the RNase H cleavage reaction in duplexed RNA targets, whereas a slight reduction in hydrolysis occurs with the butyl-containing DNA substrates relative to all-DNA. **(B)** RNase H-mediated cleavage of a 40-mer Ha-*ras* RNA target (1 pmol) duplexed with mixed base all-2'F-ANA and DNA phosphorothioate (PS) oligonucleotides (3 pmol) with or without a butyl linker at the center of the antisense 18 mer. Complementary strands were annealed in buffer (50 µL final volume) containing 60 m*M* Tris-HCl (pH 7.8), 60 m*M* KCl, 2.5 m*M* MgCl$_2$, then heated at 90°C for 5 min and slowly cooled. The substrate solutions were further incubated for 10 and 20 min at room temperature in the presence of 6 ng of purified human RNase HII. Samples were quenched by the addition of an equal volume of denaturing loading buffer (98% deionized formamide, 10 m*M* EDTA, 1 mg/mL bromophenol blue, and 1 mg/mL xylene cyanol). Reaction products were heat denatured at 100°C for 5 min, followed by electrophoretic resolution on 16% polyacrylamide sequencing gels containing 7*M* urea.

the design of future potential AONs, as well as for the design of effective inhibitors.

The protocols described in this chapter are intended for those who wish to conduct routine enzyme assays using radiolabeling of the target RNA for easy visualization of the cleavage reaction. Although only human, HIV, and *E. coli*

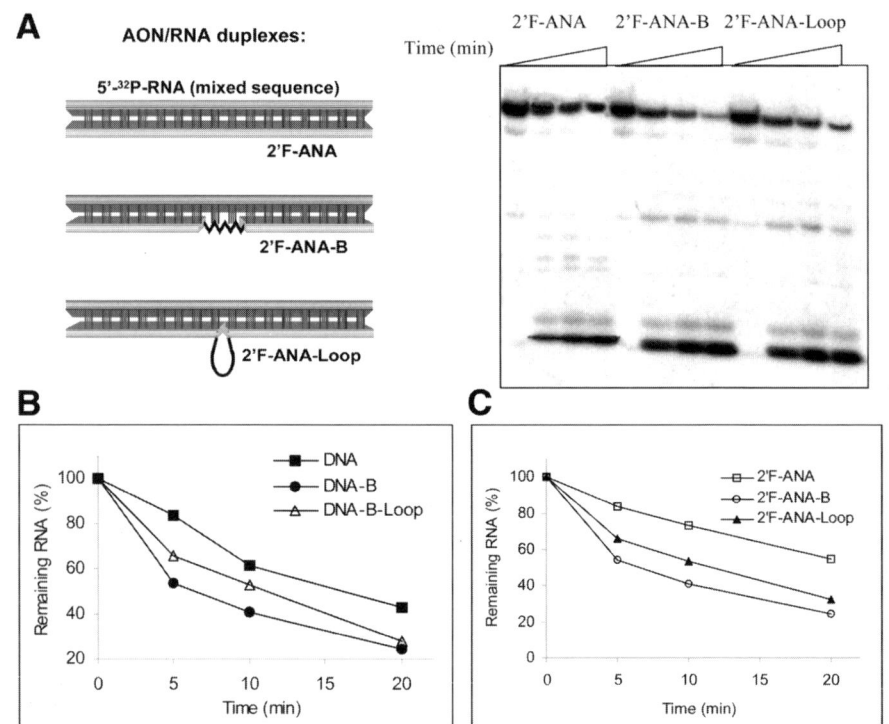

Fig. 3. Human RNase HII assays. (**A**) Polyacrylamide gel electrophoresis results showing RNA cleavage products generated from full-length 5'–^{32}P-labeled-RNA (5'-GGG AAA GAA AAA AUA UAA, 1 pmol) incubated with 3 pmol of each PO–AON. Complementary strands were mixed in a buffer (50 µL final volume) containing 60 mM Tris-HCl (pH 7.8), 60 mM KCl, 2.5 mM MgCl$_2$, then heated at 90°C for 5 min and slowly cooled. The substrate solutions were further incubated for 0, 5, 10, and 20 min at room temperature in the presence of 10 ng of purified human RNase HII. Samples were quenched by the addition of an equal volume of denaturing loading buffer (98% deionized formamide, 10 mM EDTA, 1 mg/mL bromophenol blue, and 1 mg/mL xylene cyanol). Reaction products were denatured by heating at 100°C for 5 min, followed by electrophoretic resolution on 16% polyacrylamide sequencing gels containing 7M urea. A general schematic of AON–RNA duplexes used in the assays is shown on the left; complete chemical structures are depicted in **Fig. 2B**. (**B**) Degradation profiles of 5'–^{32}P-RNA generated in the presence of human RNase HII and complementary DNA-containing AONs as compared to kinetic cleavage products obtained with 2'F-ANA-containing AONs (**A** and **C**). In either case, insertion of a butyl linker in the AON accelerates the RNA hydrolysis reaction, whether stretched along the helix as in DNA-B and 2'F-ANA-B hybrids with RNA, or looped away from the duplex axis (DNA-B-Loop and 2'FANA-Loop).

Assay for Evaluating Ribonuclease H

enzymes are extensively presented, alternate sources of RNases H may be used with minimal modifications to the protocol. Last, the methods discussed herein are not limited to the accompanying examples (i.e, DNA and 2'F-ANA phosphodiester and phosphorothioate AONs) and may be used to measure the ability of any potential AON candidate to elicit enzyme activity.

2. Materials

1. RNase H sources: *E. coli* RNase H, human RNase H, retrovirus reverse transcriptase (*see* **Note 1**).
2. Target RNA sense strand (*see* **Note 2**).
3. AONs (natural or synthetic).
4. Diethyl pyrocarbonate (DEPC), a toxic compound (*see* **Note 3**).
5. DEPC-treated water: add 0.1% DEPC to distilled water, stir for 1 h, and autoclave for 1 h.
6. 10X annealing buffer: 100 mM Tris-HCl, pH 8.0, 500 mM NaCl.
7. 5X RNase H reaction buffer: 300 mM Tris-HCl, pH 7.8, 300 mM KCl, 12.5 mM MgCl$_2$.
8. 10X PNK buffer: 500 mM Tris-HCl, pH 7.6, 100 mM MgCl$_2$, 50 mM dithiothreitol (DTT), 1 mM spermidine, 1 mM ethylenediaminetetraacetic acid (EDTA).
9. T4 polynucleotide kinase (PNK).
10. 5X RNA ligase buffer: 250 mM HEPES, pH 8.3, 50 mM MgCl$_2$, 15 mM DTT, 0.5 mM adenosine triphosphate (ATP).
11. T4 RNA ligase.
12. 10X calf intestine alkaline phosphatase (CIAP) buffer: 100 mM Tris-HCl, pH 7.6, 100 mM MgCl$_2$.
13. CIAP.
14. Alkaline hydrolysis buffer: 50 mM sodium carbonate, pH 9.0, 1 mM EDTA.
15. 10X Tris-borate-EDTA (TBE) buffer: 890 mM Tris-borate, pH 8.3, 20 mM EDTA (54 g Tris base, 27.5 g boric acid, 20 mL of 500 mM EDTA, pH 8.0, to 400 mL of dH$_2$O. Make up to 500 mL and store at room temperature).
16. Sigmacote (other silanizing agents can be used). Toxic compound.
17. Ethanol.
18. 40% w/v acrylamide: 19/1 ratio of acrylamide/*bis*-acrylamide (can be prepared fresh or purchased as such).
19. 10% w/v ammonium persulfate solution in water.
20. Formamide should be deionized with a mixed bed ion-exchange resin (AG 501-X8 resin, Bio-Rad).
21. Gel loading dye: 98% deionized formamide, 10 mM EDTA, 1 mg/mL bromophenol blue, 1 mg/mL xylene cyanol.
22. 0.5M EDTA, pH 8.0: 7.3 g EDTA, 50 mL H$_2$O. Dissolve EDTA, and adjust the pH by using pH paper. Add 0.1% DEPC before autoclaving the solution.

23. [γ–^{32}P]-ATP (5000 Ci/mmol, 10 μCi/μL). This is a radiation hazard, and radiation safety equipment is required.
24. Glass plates for sequencing.
25. Vertical electrophoresis tank.
26. Gel combs and spacers.
27. Power supply.
28. NAP Sepharose columns (Amersham-Pharmacia).
29. SpeedVac equipment.
30. Radioactive safety equipment.
31. Autoradiography equipment.
32. X-ray film.

3. Methods
3.1. ^{32}P-Labeling of RNA Target Sense Strand
3.1.1. 5'-End Labeling

Prepare the following mixture:

1. 2 μL RNA (0.1–1 μg).
2. 25 μL [γ–^{32}P] ATP (approx. 250 μCi).
3. 3.5 μL 10X PNK buffer.
4. 2 μL T4 polynucleotide kinase (20 U).
5. Complete to 35 μL with DEPC-treated H$_2$O.

Incubate for approx 40–60 min at 37°C, followed by another incubation for 20 min (not more) at 70°C to inactivate the kinase enzyme. Concentrate by using SpeedVac system and redissolve in 10 μL of gel loading buffer. Purify the labeled RNA on a denaturing polyacrylamide gel.

3.1.2. 3'-End Labeling

This procedure describes an alternative labeling protocol for visualizing the target RNA (*see* **Note 4**). Prepare the following mixture:

1. 6 μL 5X RNA ligase buffer.
2. 3 μL dimethyl sulfoxide (DMSO).
3. 3.5 μL glycerol.
4. 5 μL ^{32}P-pCp (50 μCi).
5. 1 μL RNA (0.1–1 μg).
6. 9 μL DEPC-treated water.
7. 2.5 μL RNA ligase (10 U).
8. Total volume should be 30 μL.

Incubate for 24 h at 4°C, then stop the reaction at 65°C for 10 min. Dry by SpeedVac and redissolve with 10 μL gel loading buffer. The labeled RNA can then be purified on a denaturing polyacrylamide gel.

3.1.3. Synthesis of Fresh [5'–^{32}P]pCp

The compound is easy to synthesize and is fresher than the commercial product.

1. Prepare the mix:
    a. 20 µL 100 mM Tris-HCl, pH 8.0.
    b. 5 µL 100 mM MgCl$_2$.
    c. 5 µL 50 mM DTT.
    d. 5 µL 15 mM spermidine.
    e. 10 µL 750 µL cytidine-3'-phosphate.
    f. 5 µL [γ–^{32}P]ATP (50 µCi).
    g. 2 µL T4 PNK (20 U).
2. Incubate overnight at 37°C.
3. Heat for 5 min at 95°C.
4. Centrifuge for 5 min and recover the supernatant.
5. Store at –20°C until needed.

3.2. Denaturing Polyacrylamide Gel Electrophoresis of RNA

3.2.1. Preparation of Glass Plates

1. Clean a pair of glass plates first with 70% ethanol, then 100% ethanol, using a soft tissue.
2. Allow the plates to dry before siliconizing the front plate with Sigmacote.
3. Spread a few drops (200-500 µL) of Sigmacote evenly over the surface with a soft tissue, being careful not to spread to the sides (it may cause leaks during the subsequent gel polymerizing stage).
4. Place the plates together with the side spacers in place and clamp the sides.
5. Seal the bottom and both sides of the plate with electric tape, taking care with the bottom corners.

3.2.2. Casting the Gel

1. Add the acrylamide and TBE volumes according to the desired concentration to the urea and complete to 60 mL with distilled water. The list of compounds and quantities required is provided as follows in **Table 1**:
2. Mix until all the urea has dissolved. De-gas using a vacuum tap to remove as much air as possible from the solution (to reduce air bubbles when pouring the gel).
3. Add ammonium persulfate and TEMED.
4. Hold the glass plates at 45° to the vertical, and slowly and continuously inject the acrylamide solution between the plates using a 50-mL pipet. Take great care not to introduce air bubbles while pouring the solution.
5. Lay the gel plates horizontally on support and insert the comb, taking care not to introduce bubbles around the teeth. Clamp the top part of the plates and/or apply constant pressure during the polymerization.
6. Leave to polymerize for 40–60 min at room temperature.

Table 1
Typical Denaturing PAGE Compositions Used for Detecting RNA Hydrolysis

Final acrylamide (%)	40% Acrylamide stock	10% ammonium persulfate	TEMED	Urea	10X TBE
8	12 mL				
12	18 mL	0.5 mL	30 µL	30 g	3 mL
16	24 mL				

Table 2
The Sizes (in Base Pairs) of RNA That Comigrates With the Marker Dye

% Polyacrylamide gel	Bromophenol blue	Xylene cyanol
8	19 bp	76 bp
12	8 bp	28 bp
16	6–7 bp	25 bp

3.2.3. Loading and Running the Gel

1. Remove the comb carefully from the polymerized gel.
2. Rinse out urea and unpolymerized acrylamide from the wells with 0.5X TBE.
3. Remove the sealing tape and then place the plates into a vertical electrophoresis apparatus.
4. Fill the tank with 0.5X TBE buffer and hook to the power supply.
5. Preelectrophorese for 30 min at constant voltage (1500 V).
6. Heat samples in loading buffer for 5 min at 90°C and then cool on ice.
7. Turn off power to gel and wash the wells with 0.5% TBE, using a syringe, until no urea can be seen floating in the buffer.
8. Load the samples carefully in the bottom of the wells, taking care not to spill to neighboring wells.
9. Electrophorese at constant voltage (2000 V) for 1.5–2 h. The appropriate running time should be chosen by the position of comigrating marker dye (shown in **Table 2**).

3.2.4. Detection of RNA in Denaturing Polyacrylamide Gel

1. Take the gel plates out of the electrophoresis chamber and lay them flat.
2. Insert a metal spatula between the two plates and carefully pry apart, being careful to leave the gel on the bottom plate (the plates can be flipped to accommodate the gel).
3. Lay the glass plate on the horizontal surface with the gel facing up.
4. Place a used X-ray film carefully on the entire gel surface, avoiding air pockets, and create a tight seal between the gel and the film.
5. Remove any air pockets with a soft tissue, pushing them to the sides.

6. Slowly lift one corner of the film. The gel should stick to it and lift off the plate completely. If the gel does not stick, a little bit of pressure applied with a soft tissue should help.
7. Cover the gel with plastic wrap.
8. Expose the gel to X-ray film.
9. For longer exposures, use a film cassette and keep in a freezer (–80°C or –20°C) to prevent the diffusion of the samples in the gel.

3.3. Purification and Desalting of the Labeled RNA Strand

The purification of short RNA sequences (18–25 nt) is straightforward and requires only a sterile working environment to prevent undesired degradation of the target sequence (*see* **Notes 3** and **4**). However, for longer sequences (ca. 40 nt or greater), it may become essential to employ an RNase inhibitor during routine sample manipulations (*see* **Note 5**).

1. Run the labeled RNA on a 10–16% denaturing polyacrylamide gel to purify it from undesired oligonucleotides.
2. Expose on X-ray film for about 30 s and develop the film.
3. Overlap the developed film on the wrap-covered gel. This allows identification of the correct position of the desired RNA on the gel.
4. Cut out the part of the gel containing the labeled RNA, put it into an Eppendorf tube, and add 1 mL of DEPC-treated H_2O.
5. After crushing the gel, incubate for 3 to 16 h at 37°C to liberate RNA from the gel.
6. Briefly centrifuge the sample, remove the supernatant (1 mL) and load it on a NAP-10 Sepharose column to desalt the purified/recovered oligonucleotide. (The NAP column should be washed with 15 mL of DEPC-treated H_2O in advance).
7. Elute RNA with 1.5 mL of DEPC-treated H_2O and dry in a SpeedVac.
8. Resuspend the desalted RNA in the desired volume of water and store at –20°C until usage.

3.4. Annealing of the RNA Sense Strand to the Antisense Oligonucleotide

1. Prepare the following mixture:
    a. 5 µL 10X annealing buffer.
    b. 50 pmol cold (unlabeled) RNA.
    c. 5 pmol hot (^{32}P labeled) RNA.
    d. 100–500 pmol antisense oligonucleotide.
    e. Add DEPC-treated water to obtain a total volume of 50 µL.
    The amounts of hot and cold RNA can be varied to suit particular needs, for better-looking pictures, longer or shorter exposition times, and so forth.
2. This mixture is heated to 85°C for 2 min and then slowly cooled down to room temperature.
3. If the radioactivity of the labeled RNA is low, prepare the mixture containing more of the hot RNA. In the case of low duplex stability, use a high ratio of antisense/RNA. When pyrimidine-rich RNA is used to anneal with antisense, use a low ratio of antisense/RNA to avoid triplex formation (*see* **Note 6**).

3.5. RNase H-Mediated Degradation Assay

1. Prepare RNase H assay mixture:
   a. RNase H enzyme (appropriate amount)   1 µL
   b. Annealed duplex solution   7 µL
   c. 5X RNase H assay buffer   2 µL
2. The mixture is incubated at 37°C for the desired time (a few minutes to a few hours) and stopped with an equal volume of gel-loading dye (10 ng).
3. After the samples are heat-denatured at 85°C, they are loaded on an 8-16% denaturing polyacrylamide gel (typically, 4 µL per well is loaded), electrophoresed at 1500-2000V and visualized by autoradiography.

3.6. RNA Ladder Preparation for Gel Electrophoresis

3.6.1. For Low-Molecular-Weight RNA Ladder by Alkaline Hydrolysis

1. Put 5 µL of 1–10 pmol of end-labeled RNA (5'-end or 3'-end labeling) in a tube.
2. Add 10 µL of alkaline hydrolysis-buffer and heat the tube to 80°C.
3. After 5–10 min, add 15 µL of gel-loading buffer.

3.6.2. Prepare a High-Molecular-Weight RNA Ladder

3.6.2.1. Dephosphorylation Step

1. Prepare the following mixture:
   8 µL   RNA ladder from commercial source, 4 µg.
   3 µL   CIAP.
   2 µL   10X CIAP buffer.
   7 µL   DEPC-treated water.
   20 µL Total
2. Incubate at 37°C for 45 min.
3. Extract once with phenol and twice with chloroform.
4. Precipitate RNA with sodium acetate (final 0.4M) and ethanol (final 70%).
5. Dissolve in 10 µL of DEPC-treated water.

3.6.2.2. Labeling Step

1. Prepare the following mixture:
   a. 1 µL dephosphorylated RNA ladder.
   b. 5 µL [γ–^{32}P]-ATP (5000 Ci/mmol, 10 µCi/µL).
   c. 1 µL 10X PNK buffer.
   d. 1 µL T4 polynucleotide kinase (10 U).
   e. 2 µL DEPC treated water.
      Total volume: 10 µL
2. Incubate at 37°C for 40 min.
3. Stop the reaction by adding 10 µL of gel-loading dye.

3.7. RNase H-Mediated Cleavage of RNA

Typical RNA strand cleavage patterns in the RNase H-mediated reaction are as shown in **Fig. 1**. The exact locations of the RNase H-mediated cleavage sites in the sequence can be determined by examining the degradation products adjacent to an alkaline-cleaved ladder. On treatment with base, sugars containing 2'-OH groups are susceptible to internucleotide cleavage, which occurs via an in-line attack on the internucleotide phosphorus center, followed by hydrolysis of the 2',3'-cyclic phosphate intermediate to form a mixture of 2'/3'-monophosphate and 5'-hydroxyl products *(26)*. It is important to note that there exists a slight shift in mobility between enzymatically hydrolyzed 5'–^{32}P-labeled fragments and chemically hydrolyzed 5'–^{32}P-labeled fragments (RNA ladder). This shift originates from the fact that RNase H hydrolyzes an RNA strand leaving 5'-phosphate and 3'-hydroxyl oligonucleotide products *(13,14)*, whereas chemical hydrolysis results in 5'-hydroxyl and 2'/3'-phosphate products *(27)*.

4. Notes

1. Although the enzymes have been purified as RNase-free, it is possible to reintroduce various contaminating RNases during general experimentation via aqueous solution or from the environment. Autoclaving will kill contaminating bacteria, but RNases will be liberated in the solution. Ungloved fingers can introduce bacteria into solutions, which results in RNase contamination.
2. RNA is more susceptible to degradation than DNA, owing to the 2' hydroxyl groups adjacent to the phosphodiester linkages in RNA that act as intramolecular nucleophiles. Furthermore, deoxyribonucleases (DNases) require metal ions for activity and can therefore easily be inactivated with chelating agents such as EDTA, whereas many ribonucleases (RNases) have no cofactor requirements.
3. RNase A enzymes need histidine residues within the active site for catalytic activity. Thus, the alkylating agent, DEPC, which modifies these residues can block their activity permanently *(28)*. RNase contamination can effectively be avoided by using RNase-free solutions and following laboratory procedures:
   - Always wear gloves when working with RNA.
   - Maintain a separate area for RNA work. This is especially important if your work requires the use of RNases.
   - Sterile disposable plasticware is RNase-free and should be used.
   - Baking the glassware at 180°C for several hours can inactivate contaminating RNases.

   As DEPC will attack polycarbonate (e.g., centrifuge tubes) or polystyrene (e.g., standard microtiter plates), decontamination of these materials can be achieved by soaking in 3% hydrogen peroxide for 10 min. Residuals of peroxide can be removed by extensively rinsing with RNase-free water *(28,29)*.

4. Decontamination of all buffers and solutions from general RNase activity is accomplished by adding 0.1 mL DEPC per 100 mL of solution, stirred for 1 h, and autoclaving for 1 h to allow any remaining DEPC to hydrolyse into CO_2 and ethanol. Because compounds with amine groups (e.g., Tris) will react with DEPC, Tris buffers (e.g., 5X RNA ligase buffer) should be prepared by dissolving Tris base in DEPC-treated and autoclaved water. After adjusting the pH, Tris buffer can be reautoclaved to sterilize. Solutions of thermolabile materials (e.g., DTT, nucleotides, manganese salts) should be prepared by dissolving the solid in DEPC-treated water and passing the solution through a 0.2 μM filter to sterilize *(29,30)*.
5. As many RNases are resistant to autoclaving, a very convenient and effective way to protect RNA from degrading by contaminating ribonucleases involves the use of an RNase inhibitor. An RNase inhibitor (e.g., RNasin) is a protein that binds noncovalently to RNases and thereby protects the RNA from degradation.
6. As mentioned previously, any modified oligonucleotide may be used to examine its ability to induce degradation of a complementary target RNA molecule through recruitment of RNase H. However, it is important to run the same assay on an already established enzyme-activating AON–RNA hybrid such as the DNA–RNA or 2'F-ANA–RNA hybrids mentioned here when testing a potential AON for activity, to ensure that the enzyme preparation is viable and that assay conditions are favorable. These and other guidelines set forth herein should enable even those with minimal experience to conduct various RNase H assays both routinely and with high efficiency.

References

1. Walder, R. Y. and Walder, J. A. (1978) Role of RNase H in hybrid-arrested translation by antisense oligonucleotides. *Proc. Natl. Acad. Sci. USA* **75,** 5011–5015.
2. Turchi, J. J., Huang, L., Murante, R. S., Kim, Y., and Bambara, R. A. (1994) Enzymatic completion of mammalian lagging-strand DNA replication. *Proc. Natl. Acad. Sci. USA* **91,** 9803–9807.
3. Arudchandran, A., Cerritelli, S. M., Narimatsu, S. K., et al. (2000) The absence of ribonuclease H1 or H2 alters the sensitivity of *Saccharomyces cerevisiae* to hydroxyurea, caffeine and ethyl methanesulphonate: implications for roles of RNases H in DNA replication and repair. *Genes Cells* **5,** 789–802.
4. Rydberg, B. and Game, J. (2002) Excision of misincorporated ribonucleotides in DNA by RNase H (type 2) and FEN-1 in cell-free extracts. *Proc. Natl. Acad. Sci. USA* **99,** 16,654–16,659.
5. Moelling, K., Bolognesi, D. P., Bauer, H., Büsen, W., Plassmann, H. W., and Hausen, P. (1971) Association of viral reverse transcriptase with an enzyme degrading the RNA moiety of RNA–DNA hybrids. *Nat. New Biol.* **234,** 240–243.
6. Stein, H. and Hausen, P. (1969) Enzyme from calf thymus degrading the RNA moiety of DNA–RNA hybrids: effect on DNA-dependent RNA polymerase. *Science* **166,** 393–395.

7. Eder, P. S. and Walder, J. A. (1991) Ribonuclease H from K562 human erythroleukemia cells: purification, characterization and substrate specificity. *J. Biol. Chem.* **266,** 6472–6479.
8. Frank, P., Albert, S., Cazenave, C., and Toulmé, J.-J. (1994) Purification and characterization of human ribonuclease HII. *Nucleic Acids Res.* **22,** 5247–5254.
9. Frank, P., Braunshofer-Reiter, C., Wintersberger, U., Grimm, R., and Büsen, W. (1998) Cloning of the cDNA encoding the large subunit of human RNase HI, a homologue of the prokaryotic RNase HII. *Proc. Natl. Acad. Sci. USA* **95,** 12,872–12,877.
10. Frank, P. Braunshofer-Reiter, C., Pöltl, A., and Holzmann, K. (1998) Cloning, subcellular localization and functional expression of human RNase HII. *Biol. Chem.* **379,** 1404–1412.
11. Johnson, M. S., McClure, M. A., Feng, D. F., Gray, J., and Doolittle, R. F. (1986) Computer analysis of retroviral pol genes: assignment of enzymatic functions to specific sequences and homologies with nonviral enzymes. *Proc. Natl. Acad. Sci. USA* **83,** 7648–7652.
12. Frank, P., Braunshofer-Reiter, C., and Wintersberger, U. (1998) Yeast RNase H(35) is the counterpart of the mammalian RNase HI, and is evolutionarily related to prokaryotic RNase HII. *FEBS Lett.* **421,** 23–26.
13. Berkower, L., Leis, J., and Hurwitz, J. (1973) Isolation and characterization of an endonuclease from *Escherichia coli* specific for ribonucleic acid in ribonucleic acid–deoxyribonucleic acid hybrid structures. *J. Biol. Chem.* **248,** 5914–5921.
14. Crouch, R. J. and Dirksen, M. L. (1982) *Ribonuclease H.* In Linn, S. M. and Roberts, R. J. (Eds.), *Nucleases.* Cold Spring Harbor Laboratory Press, Cold Spring Harbor, NY, pp. 211–241.
15. Minshull, J. and Hunt, T. (1986) The use of single-stranded DNA and RNase H to promote quantitative "hybrid arrest of translation" of mRNA–DNA hybrids in reticulocyte lysate cell-free translations. *Nucleic Acids Res.* **14,** 6433–6451.
16. Crooke, S. T. (1999) Molecular mechanism of action of antisense drugs. *Biochim. Biophys. Acta* **1489,** 31–44.
17. Damha, M. J., Wilds, C. J., Noronha, A., et al. (1998) Hybrids of RNA and arabinonucleic acids (ANA and 2'F-ANA) are substrates of ribonuclease H. *J. Am. Chem. Soc.* **120,** 12,976–12,977.
18. Noronha, A. M., Wilds, C. J., Lok, C.-N., Viazovkina, K., Arion, D., Parniak, M. A., and Damha, M. J. (2000) Synthesis and biophysical properties of arabinonucleic acids (ANA): circular dichroic spectra, melting temperatures and ribonuclease H susceptibility of ANA:RNA hybrid duplexes. *Biochemistry* **39,** 7050–7062.
19. Wilds, C. J. and Damha M. J. (2000) 2'-deoxy-2'-fluoro-β-D-arabinonucleosides and oligonucleotides (2'F-ANA): synthesis and physicochemical studies. *Nucleic Acids Res.* **28,** 3625–3635.
20. Wang, J., Verbeure, B., Luyten, I., et al. (2000) Cyclohexene nucleic acids (CeNA): serum stable oligonucleotides that activate RNase H and increase duplex stability with complementary RNA. *J. Am. Chem. Soc.* **122,** 8595–8602.

21. Trempe, J. F., Wilds, C. J., Denisov, A. Y., Pon, R. T., Damha, M. J., and Gehring, K. (2001) NMR solution structure of an oligonucleotide hairpin with a 2'F-ANA–RNA stem: implications for RNase H specificity toward DNA–RNA hybrid duplexes. *J. Am. Chem. Soc.* **123,** 4896–4903.
22. Damha, M. J., Noronha, A. M., Wilds, C. J., Trempe, J.-F., Denisov, A., and Gehring, K. (2001) Properties of arabinonucleic acids (ANA & 2'F-ANA): implications for the design of antisense therapeutics that invoke RNase H cleavage of RNA. *Nucleosides Nucleotides Nucleic Acids* **20,** 429–440.
23. Minasov, G., Teplova, M., Nielsen, P., Wengel, J., and Egli, M. (2000) Structural basis of cleavage by RNase H of hybrids of arabinonucleic acids and RNA. *Biochemistry* **39,** 3525–3532.
24. Mangos, M. M. and Damha, M. J. (2002) Flexible and frozen sugar-modified nucleic acids—modulation of biological activity through furanose ring dynamics in the antisense strand. *Curr. Top. Med. Chem.* **2,** 1147–1171.
25. Mangos, M. M., Min, K.-L., Viazovkina, E., et al. (2003) Efficient RNase H-directed cleavage of RNA promoted by antisense DNA or 2'F-ANA constructs containing acyclic nucleotide inserts. *J. Am. Chem. Soc.* **125,** 651–659.
26. Roberts, G. C., Dennis, E. A., Meadows, D. H., Cohen, J. S., and Jardetzky, O. (1969) The mechanism of action of ribonuclease. *Proc. Natl. Acad. Sci. USA* **62,** 1151–1153.
27. Yazbeck, D. R., Min, K.-L., and Damha, M. J. (2002) Molecular requirements for degradation of a modified sense RNA strand by *Escherichia coli* ribonuclease H. *Nucleic Acids Res.* **30,** 3015–3025.
28. Blackburn, P. and Moore, S. (1982) *The Enzymes.* Academic Press, New York.
29. Blumberg, D. D. (1987) Creating a ribonuclease-free environment. *Methods Enzymol.* **152,** 20–24.
30. Sambrook, J., Fritsch, E. F., and Maniatis, T. (1989) *Molecular Cloning: A Laboratory Manual.* 2nd ed. Cold Spring Harbor Press, Cold Spring Harbor, NY.

6

Di- and Oligonucleotide Synthesis Using *H*-Phosphonate Chemistry

Jacek Stawinski and Roger Strömberg

Summary

In this chapter, a concise account of the synthesis of oligonucleotides using the *H*-phosphonate methodology is given. It includes various methods for the preparation of the starting material, nucleoside 3'-*H*-phosphonate monoesters, their conversion into dinucleoside *H*-phosphonate diesters, oxidative transformations of dinucleoside *H*-phosphonates into the corresponding phosphate and phosphorothioate derivatives, and protocols for the synthesis of oligonucleotides and their phosphorothio analogs.

Key Words: *H*-phosphonate approach; *H*-phosphonate monoesters; PCl3/imidazole, *H*-pyrophosphonate, dinucleoside *H*-phosphonate diesters, *H*-Phosphonate diesters, oxidation, sulfurization, oligonucleotide synthesis, phosphorothioate oligonucleotide, diphenyl *H*-phosphonate.

1. Introduction

Although the most common method today for the synthesis of oligonucleotides and their analogs on solid support is the phosphoramidite approach *(1)*, the *H*-phosphonate methodology *(2–7)* can often be a viable alternative. The latter is not only well suited to solid phase synthesis but it appears to be a preferred means for the preparation of dinucleotides *(8,9)* and oligonucleotides *(10,11)* in solution.

Formation of an internucleotide bond via the *H*-phosphonate approach consists of condensation of protected nucleoside *H*-phosphonate monoesters with a nucleoside in the presence of a coupling agent to produce the corresponding dinucleoside *H*-phosphonate diesters, which can be subsequently oxidized to the target phosphodiester linkage *(2,3,12)* (**Scheme 1**). The *H*-phosphonate and the phosphoramidite methods are making use of a reactive

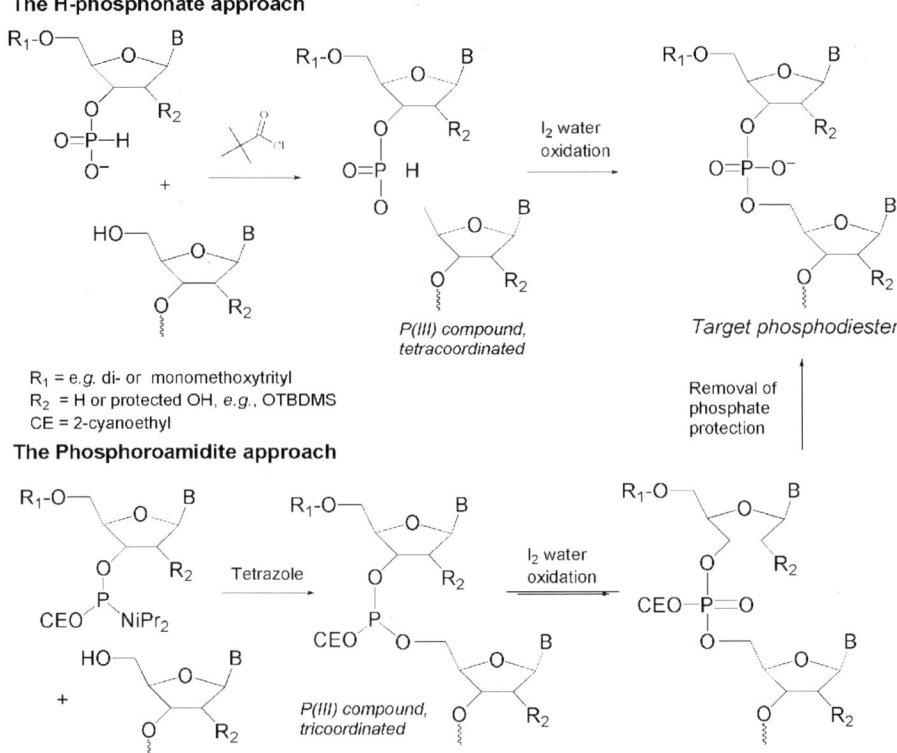

Scheme. 1. Synthesis of oligonucleotides via *H*-phopshonate and phosphoramidite methods.

P(III) intermediate to get fast and efficient condensation with the 5'-OH group of a nucleoside moiety. The main difference between these approaches is that in the *H*-phosphonate method reactive P(III) intermediates are tetracoordinated species and are generated from stable *H*-phosphonate monoesters using a condensing agent, whereas in the phosphoramidite method P(III) intermediates are tricoordinated and are generated via acid-promoted activation of often unstable, trivalent nucleoside phosphoramidites. In the latter approach, products of the condensation are phosphite triesters bearing a phosphate-protecting group that has to be removed after oxidation

The most important advantages of the *H*-phosphonate method *(13–15)* are that the starting materials (nucleoside *H*-phosphonate monoesters) are hydrolytically stable, easy to prepare and handle, and because excess of a condensing agent is usually used, the synthesis is less sensitive to adventitious water. The latter aspect is particularly important for solution-phase synthesis and in preparations in which only a small excess of building blocks is used.

Di- and Oligonucleotide Synthesis

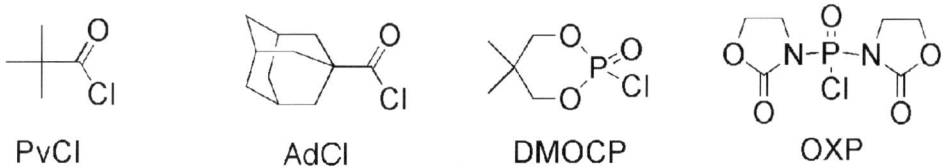

Fig. 1. Examples of the most common condensing agents in *H*-phosphonate couplings.

The condensation step of the elongation cycle, that is, the formation of an internucleotide *H*-phosphonate linkage between a nucleoside 3'-*H*-phosphonate monoester and the support-bound 5'-hydroxylic component, is usually carried out in pyridine–acetonitrile (MeCN) mixtures. Out of various condensing agents initially tested *(13–15)* pivaloyl chloride (PvCl) gave the best results in automated solid support synthesis of oligonucleotides, and it is still the most frequently used reagent. The reaction in pyridine or MeCN–pyridine mixtures using two to five equivalents of PvCl is usually fast and goes to completion in less than 1 min.

A large number of other condensing agents are available and can, in some instances, be superior to PvCl, especially for solution synthesis. Most notable of these are 5,5-dimethyl-2-oxo-2-chloro-1,3,2-dioxaphosphinane (DMOCP) *(16)*, bis(2-oxo-3-oxazolidinyl)phosphinic chloride (OXP) *(16)* and adamantane carbonyl chloride (AdCl) *(17)* (**Fig. 1**).

Because both *H*-phosphonate and phosphoramidite methods are based on P(III) phosphorus compounds, they also provide an easy access to various phosphate analogs, or P(V) compounds, by introducing the appropriate changes to the oxidation step *(15,18)*.

In this chapter, preparation of nucleoside *H*-phosphonate monoesters (*see* **Subheadings 3.1.1.– 3.1.3.**), synthesis of dinucleoside *H*-phosphonate diesters (*see* **Subheading 3.2.1.**) and their oxidative transformations to the corresponding phosphate (*see* **Subheading 3.2.2.**) and thiophosphate (*see* **Subheading 3.2.3.**) derivates, and synthesis of oligonucleotides (*see* **Subheading 3.3.1.**) via the *H*-phosphonate approach are presented.

2. Materials

1. 10% Aqueous sodium thiosulfate ($Na_2S_2O_3$).
2. 2% I_2 in pyridine–water (98:2, v/v).
3. 5% Aqueous sodium bicarbonate ($NaHCO_3$).
4. DMOCP or OXP.
5. MeCN, anhydrous, kept over 3Å sieves.
6. Dichloroacetic acid (DCA) or trichloroacetic acid.
7. Diethyl ether, anhydrous.
8. Diphenyl *H*-phosphonate (DPHP).

9. Hexane.
10. Imidazole.
11. Methanol.
12. Methylene chloride (CH_2Cl_2), freshly distilled from CaH_2.
13. Phosphonic acid (H_3PO_3).
14. Phosphorus trichloride (PCl_3), freshly distilled.
15. PvCl, freshly distilled.
16. Pyridine, anhydrous, kept over 4Å sieves.
17. Suitably protected nucleosides.
18. Sulfur.
19. Silica gel (Merck silica gel 60).
20. Triethylamine, distilled and stored over CaH_2.
21. 0.5–2M triethylammonium bicarbonate (TEAB, pH = 7.5); prepared by passing CO_2 through a suspension of water containing the appropriate amount of triethylamine, until the pH value has reached 7.5.
22. Polystyrene support (or controlled pore glass support) from Applied Biosystems loaded (20– 30 µmol/g) with a 5'-DMT or MMT nucleoside succinate (or equivalent nucleoside loaded solid support); typical scale 1–2 µmol (ca. 50 mg support).
23. Dichloroethane (CH_2ClCH_2Cl), anhydrous, kept over 4Å sieves.
24. Protected nucleoside 3'-*H*-phosphonate building blocks (*see* **Subheadings 3.1.1.– 3.1.3.**) 15 µmol/coupling and 15 µmol extra for margins and priming of solutions.
25. Concentrated (28–32%) aqueous ammonia– ethanol (3:1) or concentrated ammonia (aqueous) or methanol saturated with ammonia.

3. Methods

3.1. Synthesis of Nucleoside 3'-H-Phosphonate Monoesters (see Scheme 2)

Nucleoside 3'-*H*-phosphonate monoesters (triethylammonium [TEAH$^+$] or 1,8-diazabicyclo[5.4.0]undec-7-ene [DBUH$^+$] salts) are stable solids, resistant to air oxidation and to hydrolysis by air moisture. In this respect, they offer distinct advantages over nucleoside phosphoramidites as precursors in the synthesis of nucleotides and their analogs *(13,15,19,20)*. Nucleoside *H*-phosphonate monoesters can be stored for several months at ambient temperature or in a refrigerator without noticeable decomposition.

Phosphonylation with the PCl_3/imidazole reagent system (*see* **Subheading 3.1.1.**) is a straightforward procedure, applicable for the preparation of both deoxyribo- and ribonucleoside 3'-*H*-phosphonates. There are several variants of this protocol that differ in the kinds of azolides, external bases, or solvents used for the reaction *(3,6,8,21)*. Formation of symmetrical *H*-phosphonate diesters under the reaction conditions is negligible. If present, these uncharged species are easily removed during chromatography. A possible inconvenience of using this reagent system is that it must be generated *in situ* prior to synthesis.

Pyridinium *H*-pyrophosphonate (HPP) (*see* **Subheading 3.1.2.**) *(22)* is a convenient phosphonylating reagent. It can be generated *in situ* from phosphonic acid and a suitable condensing agent or prepared and stored as a stock solution in pyridine. The moderate reactivity of HPP makes it possible to leave the reaction mixtures containing a nucleoside and five equivalents of pyrophosphonate overnight without the danger of side reactions occurring with the heterocyclic base residues. Because the condensing agent is completely consumed in the activation process of phosphonic acid and does not participate in the reaction of a nucleoside with *H*-pyrophosphonate, its chemical nature is of minor importance for the final yield of the *H*-phosphonate monoesters. Indeed, practically the same results are obtained using DHOCP *(23)* as condensing agents. Mild reaction conditions, reasonably high yields, and simple experimental procedure make this method convenient both for small- and large-scale preparations. Another advantage of this method is that starting materials are stable, cheap, and commercially available. Because HPP reacts very slowly with 2'-*O*-protected ribonucleosides, this reagent is less useful for the preparation of ribonucleoside 3'-*H*-phosphonates. Other methods for the preparation of nucleoside *H*-phosphonate monoesters, which make use of phosphonic acid and a condensing agent, have also been reported *(24,25)*.

Subheading 3.1.3., which involves transesterification of diphenyl *H*-phosphonate (DPHP) with suitable protected nucleosides *(26)*, probably represents the most convenient approach for the preparation of both deoxyribo- and ribonucleoside *H*-phosphonate monoesters. DPHP is an inexpensive, commercially available, stable, and easy to handle reagent that affords *H*-phosphonate monoesters of purity usually better than 95% even without column chromatography *(26)*. No side reactions involving the heterocyclic bases were detected after treatment of fully protected nucleosides with 20 molar excess of diphenyl *H*-phosphonate during 8 h in pyridine. Although the reagent in pyridine undergoes disproportionation *(27)*, this process is much slower than the transesterification reaction, and the side products formed (phenyl *H*-phosphonate and triphenyl phosphite) are easily removed during a regular work up. For a large-scale synthesis it is possible to significantly reduce the excess of diphenyl *H*-phosphonate used (to 2–3 equivalents) by adding the reagent in a dropwise manner to the solution of a nucleoside.

Other reagents useful in the preparation of nucleoside *H*-phosphonates include, 2-cyanoethyl (*N*-isopropylamino)chlorophosphine *(2)*, salicylchlorophosphite *(28)*, *bis*(*N*,*N*-di-isopropylamino)chlorophosphine *(28)*, 2-cyanoethyl *H*-phosphonate *(29)*, and aryl *H*-phosphonate monoesters *(30)*.

3.1.1. Synthesis of Nucleoside 3′-H-Phosphonate Monoesters 2 Using PCl₃/Imidazole/Triethylamine Reagent System (PCl₃/Im)

Imidazole (5.8 g, 86 mmol) (*see* **Note 1**) is dissolved in 200 mL CH_2Cl_2 and the solution kept on an ice-water bath (*see* **Note 2**). Under vigorous stirring, (2.45 mL, 28 mmol PCl_3) is added, followed by 12.5 mL, 90 mmol triethylamine dissolved in 10 mL CH_2Cl_2. The mixture is stirred for 15 min and then a solution of a suitably protected nucleoside (*see* **Note 3**) (8 mmol) in 200 mL CH_2Cl_2 is added dropwise for 30 min. After the addition is completed, the cooling bath is removed, and the mixture is kept at room temperature with continuous stirring for about 1 h (thin-layer chromatography [TLC] analysis) (*see* **Note 4**). The reaction mixture is quenched by the addition of 20 mL 0.1M TEAB, pH 7.5, and extracted with 2M TEAB, pH 7.5, (3× 200 mL). The organic layer is separated, dried over Na_2SO_4, and after evaporation of the solvent the crude product is purified by a short column chromatography using a stepwise gradient of MeOH in CH_2Cl_2 (containing 0.1% Et_3N) (0–10%). The fractions containing the desired product are combined, concentrated, and after dissolving in a small amount of CH_2Cl_2, precipitated from hexane–diethyl ether mixture (1:1, v/v). The protected nucleoside 3′-*H*-phosphonate (triethylammonium salts) (*see* **Note 5**) are obtained as a white, microcrystalline powder. Yields 70–95%. ^{31}P nuclear magnetic resonance (NMR) (in pyridine): δ_P = 1–4 ppm ($^1J_{PH}$ = 600–630 Hz).

3.1.2. Synthesis of Nucleoside 3′-H-Phosphonate Monoesters Using H-Pyrophosphonate (HPP)

A suitably protected nucleoside **1** (2 mmol) is rendered anhydrous by evaporation of added pyridine and the residue is dissolved in 10 mL of 2M stock solution of phosphonic acid (20 mmol) in pyridine (*see* **Note 6**). To the stirred reaction mixture, 11 mmol DMOCP, or 11 mmol PvCl is added (*see* **Notes 7 and 8**), and progress of the reaction is monitored by TLC (*see* **Note 9**). After approx 3 h, the reaction mixture is quenched with 2 mL 1M TEAB and then partitioned between 75 mL CH_2Cl_2 and 50 mL 2M TEAB buffer, pH 7.5. The organic layer is evaporated, the residue dissolved in CH_2Cl_2 and purified by flash column chromatography on silica gel using a stepwise gradient of MeOH in CH_2Cl_2 (containing 1% of pyridine) (0–10%, v/v). After chromatography, the product is again partitioned between dichloromethane and 0.05M TEAB buffer to remove inorganic salts, if present (*see* **Note 10**). After precipitation from hexane-diethyl ether (1:1, v/v), white solids of nucleoside *H*-phosphonates and triethylammonium salts (*see* **Note 11**) are obtained. Yield 85–95%. ^{31}P NMR (in pyridine): δ_P = 1–4 ppm ($^1J_{PH}$ = 600–630 Hz).

3.1.3. Synthesis of Nucleoside 3'-H-Phosphonate Monoesters Using DPHP

To the solution of a suitably protected nucleoside **1** (1 mmol) (*see* **Note 12**) in 5 mL pyridine, DPHP (7 mmol, 1.34 mL) is added (*see* **Notes 13–15**). After 15 min (TLC analysis) (*see* **Note 16**) the reaction mixture is quenched by the addition of 2 mL H_2O (*see* **Note 17**) and is left standing for 15 min. The solvent is evaporated and the residue is partitioned between 20 mL CH_2Cl_2 and 20 mL 5% aqueous $NaHCO_3$. The organic layer is extracted two times with 20 mL 5% aqueous $NaHCO_3$ (*see* **Note 18**), dried over Na_2SO_4, and evaporated to an oil. The residue is dissolved in 3 mL methylene chloride and purified (*see* **Note 19**) by chromatography on silica gel using a stepwise gradient of methanol (0–10%) in methylene chloride (containing 0.5% of triethylamine). The fractions containing product are combined, concentrated, and after dissolving in a small amount of CH_2Cl_2, precipitated from hexane–diethyl ether mixture (1:1, v/v). The protected nucleoside 3'-*H*-phosphonate (triethylammonium salts) (*see* **Note 20**) is obtained as a white, microcrystalline powder. Yields *(26)* 80–90%. ^{31}P NMR (in pyridine): δ_P = 1–4 ppm ($^1J_{PH}$ = 600–630 Hz).

3.2. Synthesis of Dinucleoside H-Phosphonate Diesters and Their Conversion Into Phosphate and Thiophosphate Derivatives (see Scheme 3)

The formation of dinucleoside *H*-phosphonates of type **4** via a condensing agent-promoted coupling of suitably protected nucleoside *H*-phosphonates **2** with nucleosidic components **3** (**Scheme 2**) is a reliable and usually a high yielding reaction. The produced *H*-phosphonate diesters **4** are in general stable enough to be purified by a silica gel chromatography, but for most synthetic applications, they can be transformed into various phosphate analogs without isolation. Owing to the chirality of the phosphorus center, dinucleoside *H*-phosphonates exist as pairs of R_P and S_P diastereomers, which can be separated by silica gel column chromatography and subjected to various stereospecific transformations *(13,15)*.

To ensure fast and efficient coupling to *H*-phosphonate diesters, condensations are usually performed in pyridine or in a mixture of pyridine and another solvent, which most commonly is MeCN. Because preactivation of *H*-phosphonate monoesters **2** with a condensing agent results in formation of tervalent phosphorus species *(31)*, which may give rise to side products formation, it is important to add the condensing agent as the last component to the reaction mixture *(31)* (*see* **Subheading 3.2.1.**).

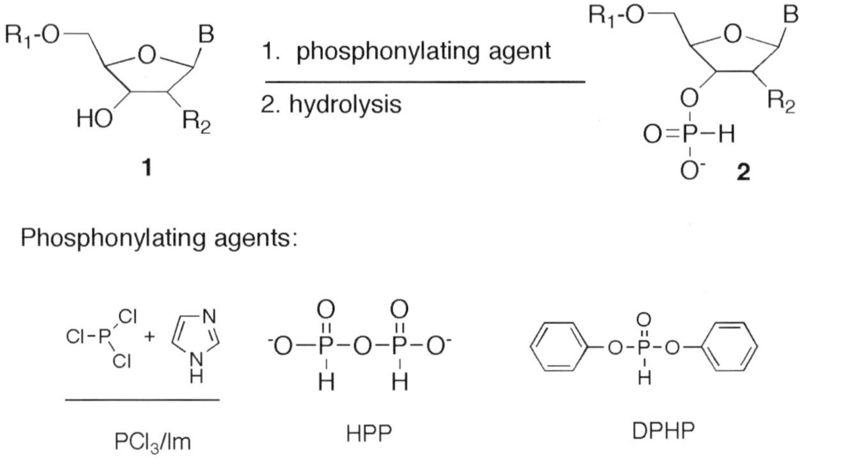

Scheme. 2. Synthesis of nucleoside H-phosphonate monoesters.

B = uracil, thymin or N-protected adenine, cytidine, and guanine moieties.
R_1 = 5'-O-protecting group; R_2 = H or a sutiably protected 2'-OH

Scheme. 3. Formation of an internucleotide bond via the H-phosphonate approach.

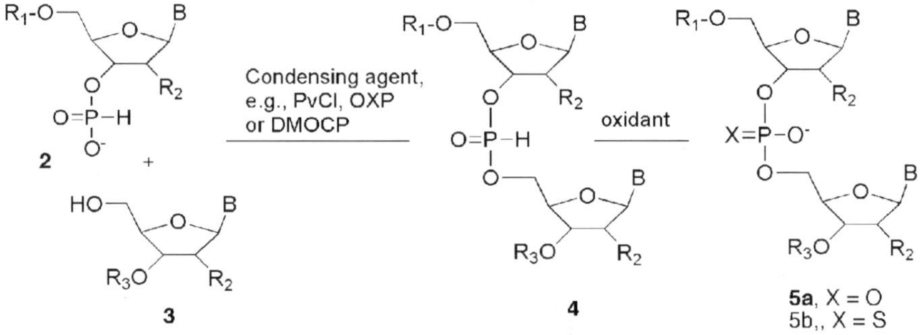

B = uracil, thymin or N-protected adenine, cytidine, and guanine moieties.
R_1, R_3 = e.g. monomethoxy- or dimethoxytrityl; R_2 = H or e.g. O-*t*-butyldimethylsilyl

Side reactions are usually negligible during H-phosphonate diester formation *(32)*. Condensations of H-phosphonate monoesters **2** with a nucleosidic component **3** to produce the corresponding H-phosphonate diesters **4** are rapid, and thus possible competing reactions of condensing agents with the hydroxylic component are not able to compete significantly. However, in the presence of excess of a coupling agent, some subsequent reactions involving the P-H bond (e.g., P-acylation) or reactive heteroaromatic lactam systems (e.g., in the guanine residue), can occur *(32)*. For this reason unnecessary excess of condensing agent should be avoided, and the reaction mixture is best worked up soon after the condensation is complete.

H-phosphonate diesters (e.g., **4**) are more susceptible to oxidation than *H*-phosphonate monoesters (e.g., **2**), and oxidative transformations of these compounds (e.g., oxidation, sulfurization) that lead to the formation of the corresponding phosphate diesters or their analogs are usually quantitative and easy to perform (*see* **Subheadings 3.2.1.2.** and **3.2.2.**). The conversion of *H*-phosphonate diesters into the corresponding phosphate derivatives is most conveniently performed using iodine in aqueous pyridine *(33,34)*. A tertiary base, or a basic solvent (e.g., pyridine), is an indispensable component of the reaction mixture, or else the oxidation is slow. Because oxidation of *H*-phosphonate diesters with iodine involves the corresponding iodophosphates as reactive intermediates, to secure clean formation of phosphate derivatives of type **5a** water has to be present in the reaction mixture to rapidly hydrolyze the produced iodophosphates. With limited amounts of water present, iodophosphates may react with other nucleophilic species present in the reaction mixture—for example, phosphate diesters—and produce relatively stable pyrophosphates as side products. If oxidation according to **Subheading 3.2.2.** would be inefficient for any reason, one can add trimethylsilyl chloride *(33,35)* to the reaction mixture before treatment with iodine. This transforms *H*-phosphonate diesters of type **4** into tricoordinated silyl derivatives that are rapidly oxidized with iodine in aqueous pyridine *(33)*.

Sulfurization of *H*-phosphonate diesters **4** to produce thiophosphate derivatives **5b** is a straightforward transformation *(7,9,36)*. However, because sulfurization in pyridine is a slow reaction (a few hours), one should secure anhydrous reaction conditions to avoid competing hydrolysis of *H*-phosphonate diesters **4** by adventitious water. If hydrolysis of the starting material **4** would become an issue, one can add to the reaction mixture one to two equivalents of trimethylsilyl chloride. This would secure an anhydrous reaction condition and also facilitate sulfurization owing to formation of tricoordinated silyl derivatives from **4**. Sulfurization of dinucleoside *H*-phosphonates is a stereospecific reaction *(37–39)*. To produce separate diastereomers of thiophosphate **5a**, usually the starting *H*-phosphonate diesters **4** are separated into diastereomers, and these are subjected separately to sulfurization *(9,37)*. For other types of oxidative transformations of *H*-phosphonate diesters, for review *see* **refs. 13,15,19**, and **20**.

3.2.1. Synthesis of Dinucleoside-(3'– 5')-H-Phosphonates 4

Nucleoside *H*-phosphonate monoester **2** (1.1 mmol) (*see* **Note 21**) and a suitably protected nucleoside **3** (1 mmol) are rendered anhydrous by repeated evaporation of added pyridine (3 × 5 mL). The gummy residue is dissolved in

10 mL pyridine and while stirring 3 mmol PvCl (*see* **Note 22**) is added in one portion. When TLC analysis indicates that the reaction is over (*see* **Note 23**), 1 mL 2*M* TEAB is added. The reaction mixture is concentrated and the residue partitioned between 30 mL methylene chloride and 20 mL 0.5*M* TEAB, pH 7.5. The organic layer is collected, dried over anhydrous Na_2SO_4, and evaporated. The crude reaction mixture containing *H*-phosphonates **4** (*see* **Note 24**) is then purified on a silica gel column using a stepwise gradient of ethyl acetate in toluene (30–50%) as eluent (*see* **Note 25**). The fractions containing product are combined, concentrated, and after dissolving in a small amount of CH_2Cl_2, precipitated from hexane–diethyl ether mixture (1:1, v/v). The protected dinucleoside *H*-phosphonate diesters **4** are obtained as a white, microcrystalline powder. Yields 85– 95%. ^{31}P (in pyridine): $\delta_P = 7–10$ ppm ($^1J_{PH} = 700–720$ Hz).

3.2.2. Synthesis of Dinucleoside-(3'– 5')-Phosphates 5a

Dinucleoside *H*-phosphonate **4** (1 mmol) (*see* **Note 26**) is dissolved in 10 mL pyridine, and a 2% solution of iodine in pyridine/water (96:4, v/v; 10 mL) is added (*see* **Note 27**). After 5 min (*see* **Note 28**) the reaction mixture is diluted with 30 mL methylene chloride, washed with 10 mL 10% aqueous $Na_2S_2O_3$, and then with 10 mL 1.0*M* TEAB. The organic layer is separated, solvent removed by evaporation under vacuum, and the residue is chromatographed on a silica gel column using a stepwise gradient of methanol (1–10%) in CH_2Cl_2 containing 0.1% triethylamine. After evaporation of the desired fractions, compound **5a** (TEAH$^+$ salt) is obtained as a white solid. Yields: >90%. ^{31}P NMR (in pyridine): $\delta_P = -1–3$ ppm.

3.2.3. Synthesis of Dinucleoside-(3'–5')-Thiophosphate 5b

Dinucleoside *H*-phosphonate diester **4** (0.5 mmol) (*see* **Note 29**) is dissolved in 20 mL pyridine, and 4 equivalents of elemental sulfur are added. The reaction mixture is stirred for 2–3 h (TLC analysis, *see* **Note 28**), pyridine is evaporated, and the residue passed through a short silica gel column using a stepwise gradient of methanol (0–10%) in CH_2Cl_2 containing 0.1% of triethylamine. Yields >90%. ^{31}P NMR (in pyridine): $\delta_P = 59–63$ ppm.

3.3. Synthesis of Oligonucleotides

Oligonucleotide synthesis employing *H*-phosphonates is considerably simpler than when using the phosphotriester or phosphoramidite procedures. The elongation cycle includes only two chemical steps: deprotection of the terminal 5'-OH function of the support-bound oligonucleotide and its coupling with a nucleoside 3'-*H*-phosphonate in the presence of a condensing agent. After completion of the desired number of elongation cycles (i.e., assembly of the oligomeric chain), a single oxidation cycle is performed (whereas this is performed in each elongation cycle for the phosphoramidite method) to convert the internucleoside *H*-phosphonate functions to phosphodiesters (or some analog such as phosphorothioates). Thus, because *H*-phosphonate functions in the oligomeric chain remain intact during all steps in the elongation cycle, the oxidation can be carried out in a single step after assembly of the oligomeric chain has been completed. This can be especially advantageous for the preparation of oligonucleotide analogs—for example, synthesis of phosphorothioates. In this instance the less reactive but inexpensive elemental sulfur can be used and the sulfurization reaction can be carried out outside the machine in a separate reaction vessel.

The efficiency of each elongation step in oligonucleotide synthesis on solid support is high (at least 98–99%), but the procedure may need some adjustment for each particular machine. The most important aspects are (1) time of preactivation of the *H*-phosphonate before it reaches the solid support, (2) concentration of the condensing agent used, and (3) the proportion of pyridine in the solvent mixture.

Because preactivation of *H*-phosphonate monoesters with PvCl (see above) gives *bis*-acyl phosphites *(31)*, this must occur to some extent also when a condensing agent and an *H*-phosphonate are mixed before entering the column in a synthesizer. No side products that could arise from this species have so far been detected during solid-phase synthesis, but it has been shown that too long preactivation substantially slows down the coupling reaction. This can also give some capping of 5'-OH functions *(21)*. Even if condensation with nucleoside *bis*-acyl phosphites gives the correct product on a solid support, the reaction with a nucleoside is substantially slower *(40)* than that of an *H*-phosphonate-carboxylic acid anhydride. Although some formation of *bis*-acyl phosphite is tolerable (because excess of *H*-phosphonate to alcohol is normally used), a clear conclusion is that the preactivation time, before the reaction mixture reaches the column, is best kept at a minimum. This also puts restrictions on the machine design. Apart from minimizing preactivation time one can also adjust the performance and control the extent of preactivation by the amount of condensing agent (which is usually used in excess) and the solvent composition used for the reaction.

To obtain high-purity crude oligonucleotides, the oxidation must be virtually quantitative. Any remaining *H*-phosphonate functions are instantaneously cleaved in the subsequent step of ammonia treatment *(41)*. Incomplete oxidation will produce shorter oligomers (with dangling *H*-phosphonate functions) rather than modified oligomers, and this will reduce the yield and purity of the product. For oligomer synthesis it is therefore advisable to use a longer oxidation time than that used for oxidation of dinucleoside *H*-phosphonates (less than 2 min) *(33)*. For example, an oxidation time of 10 min with 2% iodine in pyridine–water (98:2) seems to be sufficient for synthesis of 20 mers *(4,5)*, but because oxidation needs to be carried out only once during the synthesis, to be on safe side one can extend the oxidation time even further (20–30 min) with only a marginal increase in the total synthesis time.

An interesting feature of *H*-phosphonate-based ribonucleic acid (RNA) synthesis is that the condensation is stereoselective *(42,43)*. This has been exploited (together with the stereospecific sulfurization *(9)* and nuclease P1 treatment) for synthesis of phosphorothioate RNA fragments that contain only Rp-thiophosphate linkages *(42,43)*.

3.3.1. Machine-Assisted Oligonucleotide Synthesis

Charge the synthesizer with solvents and detritylation solution (these can usually be kept on the synthesizer until consumed). Dissolve the protected nucleoside 3'-*H*-phosphonates (TEAH$^+$ or DBUH$^+$ salts) in pyridine, and evaporate the solvent twice under reduced pressure. Dissolve the residue in CH$_3$CN-pyridine (3:1, v/v) to a concentration of 50 mM, and transfer to appropriate vessels for attachment to the synthesizer (*see* **Note 31**). Prepare a PvCl solution (225 mM in CH$_3$CN-pyridine [3:1, v/v], 0.2 mL/coupling and a little extra for margins—for example, approx 0.2 mL) directly in the vessel that is to be attached to the synthesizer (*see* **Note 32**). Prime the solutions and connect the column/cartridge filled with nucleoside loaded support (approx 50 mg, 1–2 μmol). The iodine solution is prepared by dissolving 0.4 g of iodine in 20 mL pyridine– water 98:2 just before the oxidation step (*see* **Note 33**). Machine-assisted synthesis is then carried out according to the elongation cycle (*see* **Table 1**). If ordinary phosphodiester linkages are required, then the oxidation cycle (*see* **Table 2**) is performed immediately after completion of the elongation cycles. If a phosphorothioate oligonucleotide is desired, then the sulfurization protocol in **Subheading 3.3.1.1.** is followed instead of the iodine–water oxidation cycle.

3.3.2. The Sulfurization Step (in Synthesis of Oligonucleoside Phosphorothioates)

For synthesis of oligonucleotides with all linkages converted into phosphorothioates, the following sulfurization step is performed instead of the oxidation in **Table 2**.

After completion of the machine-assisted elongation cycles, the cartridge/column is removed from the synthesizer and the support is washed with diethyl ether and air-dried using a syringe with a luer fitting. The support is then removed from the cartridge, transferred into a 2-mL cryovial (with screw cap) and treated with a solution of elemental sulfur in pyridine (1 mL 0.055M S$_8$) at room temperature for 18–20 h. The support is then washed with pyridine (3 × 2

Table 1
The Elongation Cycles

Step	Reagent/solvent	Time	Flow rate
1. Wash	Dichloroethane (DCE)	1.5 min	2 mL/min
2. Detritylation (see **Note 34**)	3.5% DCA in DCE	2.5 min	
3. Wash	DCE, 2 min Acetonitrile (MeCN) 1.5 min MeCN/pyridine 3:1, 1.5 min	5 min	
4. Coupling (see **Note 35**)	50 mM H-phosphonate, 0.1 min 225 mM PvCl, 0.1 min 50 mM H-phosphonate, 0.1 min 225 mM PvCl, 0.1 min 50 mM H-phosphonate, 0.1 min pushing forward the segments with pyridine/MeCN, 0.1 min. Recycling in a closed loop, 1.5 min.	2.1 min	1 mL/min 2 mL/min
5. Wash	MeCN/Pyridine 3:1, 2 min MeCN, 1 min	3 min	2 mL/min
	Total time for one elongation cycle:	14.1 min	

Abbr: Dichloroethane, DCE; dichloroacetic acid, DCA.

Table 2
The Final Oxidation Cycle[a]

Step	Reagent/solvent	Time	Flow rate
1. Wash	DCE	1.5 min	2 mL/min
2. Detritylation	3% DCA in DCE	2.5 min	
3. Wash	DCE, 2 min MeCN, 1 min	3 min	
4. Oxidation	2% I_2 in Pyridine-H_2O (98:2)	30 min	1 mL/min
5. Wash	Pyr/MeCN, 4 min MeCN, 4 min	8 min	2 mL/min

[a]Performed once after all elongation cycles are completed.

mL), MeCN (2 × 2 mL), and dichloromethane (3 × 2 mL). After drying under reduced pressure the support is then subjected to ammonolysis as described in the next section.

3.3.3. Cleavage From Support After Completion of Synthesis and Base Deprotection

After completion of the machine-assisted synthesis, the cartridge/column is removed from the synthesizer, and the support is washed with diethyl ether and air-dried using a syringe with a luer fitting. The support is then removed from the cartridge and transferred into a 1.5-mL cryovial (with screw cap) to which the ammonolysis solution is added (about 1 mL for up to a 1-mmol scale). The ammonolysis conditions for deoxyribonucleic acid (DNA) or 2'-O-alkyl RNA synthesis are usually concentrated aqueous ammonia at 60°C for 16–20 h (for deprotection and cleavage from support, the reaction time can vary depending on the kind of base protecting groups used). In RNA synthesis with 2'-O-TBDMS protection and N^2-phenoxyacetyl protection for guanosine, N^6-butyryl for adenosine, and N^4-propionyl for cytidine, ammonolysis is done with concentrated ammonia (aqueous)-ethanol (3:1) at 20–25°C for 8–16 h (longer time for longer sequences, e.g., 50–60 mers).

The support is then filtered off, washed with one portion of concentrated aqueous ammonia or concentrated aqueous ammonia–ethanol (3:1), concentrated and purified by ion exchange, and reversed-phase high-pressure liquid chromatography (for oligoribonucleotide synthesis with 2'-O-TBDMS protection an additional deprotection step is done first) (*see* **Note 36**).

4. Notes

1. Imidazole was made anhydrous by repeated evaporation of added anhydrous MeCN.
2. For guanosine derivatives the reaction is preferably carried out at lower temperature (to avoid side reactions on the base, which lower the yield)—for example, –40 or –78°C. The solution is not brought to room temperature, but instead is kept for 1 h at the low temperature after addition of all reagents. Then the cold reaction mixture is poured directly into the TEAB solution in a separating funnel.
3. The nucleoside was rendered anhydrous by consecutive evaporation of added pyridine and MeCN. For the preparation of suitably protected deoxy- and ribonucleoside, *see* Chapters **refs. 8–11**.
4. Formation of a base-line material (TLC on silica gel plates, solvent: chloroform–methanol 9:1, v/v).
5. On some occasions, DBUH$^+$ salts **(6)** of *H*-phosphonate monoesters can be more useful than TEAH$^+$ salts. The transformation can be effected by washing a solution of nucleoside *H*-phosphonate monoester (triethylammonium salt) in CH_2Cl_2 with 0.2*M* DBU bicarbonate buffer (pH 8.5, prepared analogously to TEAB buffer).
6. Stock solution of 2*M* phosphonic acid in pyridine was prepared by evaporation of added pyridine to the appropriate amount of phosphonic acid and dissolving the residue in anhydrous pyridine.

7. One should not exceed this amount of a condensing agent, otherwise more reactive species can be generated from phosphonic acid *(44)*.
8. Instead of generating *in situ* H-pyrophosphonate from phosphonic acid and a condensing agent in pyridine, one can prepare a 1*M* stock solution of this phosphonylating reagent by adding 0.5 equivalents of PvCl or DMOCP to 2*M* stock solution of phosphonic acid in pyridine. The stock solution of HPP in pyridine is stable for several months at room temperature.
9. Formation of a trityl positive baseline material (TLC on silica gel plates, solvent: chloroform–methanol 9:1, v/v).
10. This, in principle, may happen when DMOCP is used as a condensing agent.
11. *See* **Note 5** in **Subheading 3.1.1.**
12. The nucleoside was rendered anhydrous by evaporation of added pyridine.
13. Diphenyl *H*-phosphonate was a commercial grade from Aldrich and contained ca. 10% of phenol.
14. An excess of phosphonylating agents is necessary to avoid the formation of symmetrical dinucleoside *H*-phosphonate diesters.
15. In the instance of 2'-*O*-protected ribonucleosides, the amount of diphenyl *H*-phosphonate can be reduced to 3-mol equivalents.
16. Disappearance of the starting nucleosidic material (TLC on silica gel plates, solvent: chloroform–methanol 9:1, v/v).
17. In the case of ribonucleosides, a mixture of water–triethylamine (1:1 v/v, 2 mL) was used to speed up hydrolysis of the intermediate nucleoside phenyl *H*-phosphonate.
18. This extraction removes most phenol and phenyl *H*-phosphonate formed during the reaction and from hydrolysis of diphenyl *H*-phosphonate.
19. Precipitation of crude reaction products (after extraction with aqueous $NaHCO_3$) from CH_2Cl_2 into hexane–diethyl ether (1:1, v/v) usually afforded nucleoside *H*-phosphonates of purity better than 95%.
20. *See* **Note 5** in **Subheading 3.1.1.**.
21. *H*-phosphonate monoesters can be used as a $TEAH^+$ or $DBUH^+$ salt.
22. Instead of PvCl, other condensing agents—for example, DMOCP or OXP—can be used.
23. Reaction is usually over within 1 min, but to make sure we suggest waiting for 3–5 min before the first TLC analysis. A baseline material (*H*-phosphonate **2**) is converted into a compound with a higher R_f value (TLC on silica gel plates, solvent: chloroform–methanol 9:1, v/v). In the instance of using DMOCP or OXP as coupling agents, the reaction is usually completed within 40 and 20 min, respectively, but can in some cases be up to 2 h.
24. Owing to the chirality at the phosphorus center, *H*-phosphonate diesters of type **4** are formed as a mixture of R_P and S_P diastereomers.
25. This solvent system usually permits separation of *H*-phosphonates **4** diastereomers, if so desired.
26. Synthesis of a dinucleoside *H*-phosphonate **4** and its oxidation by iodine to produce dinucleoside phosphate **5a**, can be performed as a one-pot reaction without isolation of **4**. The aqueous iodine solution is then simply poured directly into the

coupling reaction mixture in **Subheading 3.2.2.** as soon as TLC analysis reveals complete condensation.

27. During oxidation, water has to be present in the reaction mixture at the moment of addition of iodine, otherwise significant amounts of stable pyrophosphate derivatives can be formed.
28. Formation of a baseline material (TLC on silica gel plates, solvent: chloroform–methanol 9:1, v/v).
29. If a single *H*-phosphonate diastereomer (*see* **Notes 4** and **5** in **Subheading 3.2.1.**) is used, then only one isomer of the phosphorothioate is obtained because the sulfurization is stereospecific. Synthesis of a dinucleoside *H*-phosphonate **4** and its oxidation with elemental sulfur to produce dinucleoside thiophosphate **5b** can be performed as a one-pot reaction without isolation of **4**. The latter procedure gives a mixture of the two phosphorothioate stereoisomers.
30. Formation of a baseline material (TLC on silica gel plates, solvent: chloroform–methanol 9:1, v/v).
31. When preparing and transferring the *H*-phosphonate solutions it is important for high yields in coupling that moisture be avoided as much as possible. It is advisable to transfer under a flow of dry gas (nitrogen or argon) or by syringes through septa for the best results. To some extent this can be compensated by use of a higher concentration of PvCl. For some modified *H*-phosphonate building blocks there may be solubility problems, but for these several alternative solvent mixtures, MeCN/pyridine 1:1 (v/v), neat pyridine, and THF/pyridine 1:1 (v/v), can be used without compromising the coupling yield too much.
32. The quality of the PvCl is important for the outcome of the synthesis. Because the reagent decomposes with time, it should be distilled regularly (e.g., once every 1–2 mo). The distilled material, divided into a number of smaller vials (to avoid too frequent opening and closing of the same bottle), can be safely stored in a freezer. When preparing and transferring the PvCl solutions it is crucial that moisture be avoided as much as possible. It is advisable to transfer under a flow of dry gas (nitrogen or argon) or by syringes through septa for the best results. It is also advisable to prepare a fresh PvCl solution for each synthesis, but if one wishes to keep the solution for a couple of days on the synthesizer it is best to omit the pyridine in the PvCl solution and compensate it by increasing the pyridine content in the *H*-phosphonate solutions. If there are problems with moisture one can to some extent compensate by using a higher concentration of PvCl (e.g., 300 mM, 4 equivalents, or 375 mM, 5 equivalents) but usually 225 mM and very dry conditions gives slightly better results.
33. The oxidation solution is usually prepared while the elongation cycles are running to have this as fresh as possible because basic aqueous iodine solutions are disproportionate, which can give somewhat slower oxidation. Although there is probably enough of a time margin in this protocol, it may be better to play it safe.
34. 1% trifluoroacetic acid can be used instead of dichloroacetic acid in synthesis of RNA fragments or for sequences not containing deoxypurine nucleosides (e.g., dG and dA). Trichloroacetic acid 2–3% is also commonly used.

35. In the coupling step, *H*-phosphonate and PvCl are taken up in alternating segments of 100 µL. The volume of the tubing between valve and column (including the pump hose of the peristaltic pump through which the reagents flow) is 0.4 mL, which means that the front of the condensation mixture reaches the column when the last segment is taken up from the reagent bottles. The segments are then pushed into a loop that includes the column, the loop is closed, and the condensation mixture is recycled, giving a total condensation cycle time of 2.1 min (effective condensation time ca. 1.6 min). The effective time of preactivation for the nucleoside *H*-phosphonate when it first reaches the column is about 0.4 min, and at the end of condensation it totals to 2.0 min. An alternative setup is possible if the machine allows for reversed flow. The segments are then passed through the column at 1 mL/min for 0.4 min, the flow is lowered to 0.5 mL/min, and the direction reversed for 1 min, giving a total condensation cycle time of 1.9 min (effective condensation time 1.5 min). The effective time of preactivation of the nucleoside *H*-phosphonate when it first reaches the column is about 0.3 min for this setup, and at the end of condensation it totals to 1.8 min.
36. After filtration the solution is lyophilized and, to remove the 2'-O-TBDMS, the residue is dissolved in triethylamine trihydrofluoride (0.3 mL, neat), and the mixture is left for 8 h (or up to 16 h for convenience) at room temperature; 30 µL H_2O and 1 mL *n*-butanol is added, the mixture is left at –20°C for 1 h and the precipitate is collected by centrifugation.

Acknowledgments

Financial support from the Swedish Research Council is gratefully acknowledged.

References

1. Beaucage, S. L. and Iyer, R. P. (1992) Advances in the synthesis of oligonucleotides by the phosphoramidite approach. *Tetrahedron* **48,** 2223–2311.
2. Garegg, P. J., Regberg, T., Stawinski, J., and Strömberg, R. (1985) Formation of internucleotidic bond via phosphonate intermediates. *Chem. Scr.* **25,** 280–282.
3. Garegg, P. J., Regberg, T., Stawinski, J., and Strömberg, R. (1986) Nucleoside hydrogenphosphonates in Oligonucleotide Synthesis. *Chem. Scr.* **26,** 59–62.
4. Garegg, P. J., Lindh, I., Regberg, T., Stawinski, J., Strömberg, R., and Henrichson, C. (1986) Nucleoside H-phosphonates: III, chemical synthesis of oligodeoxyribonucleotides by the hydrogenphosphonate approach. *Tetrahedron Lett.* **27,** 4051–4054.
5. Garegg, P. J., Lindh, I., Regberg, T., Stawinski, J., Strömberg, R., and Henrichson, C. (1986) Nucleoside H-phosphonates: IV, automated solid phase synthesis of oligoribonucleotides by the hydrogenphosphonate approach. *Tetrahedron Lett.* **27,** 4055–4058.
6. Froehler, B. C., Ng, P. G., and Matteucci, M. D. (1986) Synthesis of DNA via deoxynucleoside H-phosphonate intermediates. *Nucleic Acids Res.* **14,** 5399–5407.

7. Froehler, B. C. (1986) Deoxynucleoside H-phosphonate diester intermediates in the synthesis of internucleotide phosphate analogues. *Tetrahedron Lett.* **27,** 5575–5578.
8. Stawinski, J., Strömberg, R., Thelin, M., and Westman, E. (1988) Studies on the t-butyldimethylsilyl group as 2'-O-protection in oligoribonucleotide synthesis via the H-phosphonate approach. *Nucleic Acids Res.* **16,** 9285–9298.
9. Almer, H., Stawinski, J., Strömberg, R., and Thelin, M. (1992) Synthesis of diribonucleoside phosphorothioates via stereospecific sulfurization of H-phosphonate diesters. *J. Org. Chem.* **57,** 6163–6169.
10. Reese, C. B. and Song, Q. (1999) The H-phosphonate approach to the synthesis of oligonucleotides and their phosphorothioate analogues in solution. *J. Chem. Soc. Perkin Trans.* **1,** 1447–1486.
11. Reese, C. B. and Yan, H. B. (2002) Solution phase synthesis of ISIS 2922 (Vitravene) by the modified H-phosphonate approach. *J. Chem. Soc. Perkin Trans.* **1,** 2619–2633.
12. Hall, R. H., Todd, A., and Webb, R. F. (1957) Nucleotides: part XLI, mixed anhydrides as intermediates in the synthesis of dinucleoside phosphates. *J. Chem. Soc.* 3291–3296.
13. Stawinski, J. (1992) Some aspects of H-phosphonate chemistry. In *Handbook of Organophosphorus Chemistry* (Engel, R., ed.). Dekker, New York, pp. 377–434.
14. Stawinski, J. and Thelin, M. (1992) 3H-1,2-Benzothiaselenol-3-one: a new selenizing reagent for nucleoside H-phosphonate and H-phosphonothioate diesters. *Tetrahedron Lett.* **33,** 7255–7258.
15. Stawinski, J. and Strömberg, R. (1993) H-phosphonates in oligonucleotide synthesis. *Trends Org. Chem.* **4,** 31–67.
16. Strömberg, R. and Stawinski, J. (1987) Evaluation of some new condensing reagents for hydrogenphosphonate diester formation. *Nucleic Acids Sym. Ser.* **18,** 185–188.
17. Andrus, A., Efcavitch, J. W., McBride, L. J., and Giusti, B. (1988) Novel activating and capping reagent for improved hydrogen-phosphonate DNA synthesis. *Tetrahedron Lett.* **29,** 861–864.
18. Beaucage, S. L. and Iyer, R. P. (1993) The synthesis of modified oligonucleotides by the phosphoramidite approach and their applications. *Tetrahedron* **49,** 6123–6194.
19. Kers, A., Kers, I., Kraszewski, A., et al. (1996) Nucleoside phosphonates: development of synthetic methods and reagents. *Nucleosides Nucleotides* **15,** 361–378.
20. Bollmark, M. and Stawinski, J. (1998) Nucleotide analogues containing the P-F bond: an overview of the synthetic methods. *Nucleosides Nucleotides* **17,** 663–680.
21. Gaffney, B. L. and Jones, R. A. (1988) Large-scale oligonucleotide synthesis by the H-phosphonate method. *Tetrahedron Lett.* **29,** 2619–2622.
22. Stawinski, J. and Thelin, M. (1990) Nucleoside H-phosphonates: XI, a convenient method for the preparation of nucleoside H-phosphonates. *Nucleosides Nucleotides* **9,** 129–135.

23. McConnell, R. L. and Coover, H. W. (1959) Phosphorus-containing derivatives of 2,2-dimethyl-1,3-propanediol. *J. Org. Chem.* **24,** 630–635.
24. Sekine, M. and Hata, T. (1975) Phenylthio group as a protecting group of phosphates in oligonucleotide synthesis via phosphotriester approach. *Tetrahedron Lett.* 1711–1714.
25. Bhongle, N. N. and Tang, J. Y. (1995) A convenient synthesis of nucleoside 3'-H-phosphonate monoesters using triphosgene. *Tetrahedron Lett.* **36,** 6803–6806.
26. Jankowska, J., Sobkowski, M., Stawinski, J., and Kraszewski, A. (1994) Studies on aryl H-phosphonates: I, efficient method for the preparation of deoxyribo- and ribonucleoside 3'-H-phosphonate monoesters by transesterification of diphenyl H-phosphonate. *Tetrahedron Lett.* **35,** 3355–3358.
27. Kers, A., Kers, I., Stawinski, J., Sobkowski, M., and Kraszewski, A. (1996) Studies on aryl H-phosphonates: 3, mechanistic investigations related to the disproportionation of diphenyl H-phosphonate under anhydrous basic conditions. *Tetrahedron* **52,** 9931–9944.
28. Marugg, J. E., Tromp, M., Kuyl-Yeheskiely, E., van der Marel, G. A., and van Boom, J. H. (1986) A convenient and general approach to the synthesis of properly protected d-nucleoside-3' hydrogenphosphonates via phosphite intermediates. *Tetrahedron Lett.* **27,** 2661–2664.
29. Szabó, T., Almer, H., Strömberg, R., and Stawinski, J. (1995) 2-cyanoethyl H-phosphonate: a reagent for the mild preparation of nucleoside H-phosphonate monoesters. *Nucleosides Nucleotides* **14,** 715, 716.
30. Ozola, V., Reese, C. B., and Song, Q. L. (1996) Use of ammonium aryl H-phosphonates in the preparation of nucleoside H-phosphonate building blocks. *Tetrahedron Lett.* **37,** 8621–8624.
31. Garegg, P. J., Regberg, T., Stawinski, J., and Strömberg, R. (1987) Nucleosides H-phosphonates: V, the mechanism of hydrogenphosphonate diester formation using acyl chlorides as coupling agents in oligonucleotide synthesis by the hydrogenphosphonate approach. *Nucleosides Nucleotides* **6,** 655–662.
32. Regberg, T., Stawinski, J., and Strömberg, R. (1988) Nucleoside H-phosphonates: IX, possible side-reactions during hydrogenphosphonate diester formation. *Nucleosides Nucleotides* **7,** 23–35.
33. Garegg, P. J., Regberg, T., Stawinski, J., and Strömberg, R. (1987) Nucleoside H-phosphonates: VII, studies on the oxidation of nucleoside hydrogenphosphonate esters. *J. Chem. Soc. Perkin Trans.* **1,** 1269–1273.
34. Garegg, P. J., Regberg, T., Stawinski, J., and Strömberg, R. (1987) Studies on the oxidation of nucleoside hydrogenphosphonates. *Nucleosides Nucleotides* **6,** 429–432.
35. Hata, T. and Sekine, M. (1974) Silyl phosphites: I, the reaction of silyl phosphites with diphenyl disulphides: synthesis of S-phenyl nucleoside phosphorothioates. *J. Am. Chem. Soc.* **96,** 7363–7364.
36. Fujii, M., Ozaki, K., Kume, A., Sekine, M., and Hata, T. (1986) Acylphosphonates: 5, a new method for stereospecific generation of phosphorothioate via aroylphosphonate intermediates. *Tetrahedron Lett.* **26,** 935–938.

37. Seela, F. and Kretschmer, U. (1991) Diastereomerically pure Rp and Sp dinucleoside H-phosphonates. The stereochemical course of their conversion into P-methylphosphonates, phosphorothioates and [18O] chiral phosphates. *J. Org. Chem.* **56,** 3861–3869.
38. Stawinski, J., Strömberg, R., and Thelin, M. (1991) Some chemical and stereochemical aspects of ribonucleoside H-phosphonate and H-phosphonothioates diester synthesis. *Nucleosides & Nucleotides* **10,** 511–514.
39. Battistini, C., Brasca, M. G., Fustinoni, S., and Lazzari, E. (1992) An efficient and stereoselective synthesis of 2',5'-oligo-(sp)-thioadenylates. *Tetrahedron* **48,** 3209–3226.
40. Efimov, V. A. and Dubey, I. Y. (1990) Modification of the H-phosphonate oligonucleotide synthesis on polymer support. *Bioorg. Khim.* **16,** 211–218.
41. Hammond, P. R. (1962) A simple preparation of alkyl ammonium phosphonates and some comments on the reaction. *J. Chem. Soc.* 2521–2522.
42. Almer, H., Stawinski, J., and Strömberg, R. (1994) Synthesis of stereochemically homogeneous oligoribonucleoside all-Rp-phosphorothioates by combining H-phosphonate chemistry and enzymatic digestion. *J. Chem. Soc. Chem. Commun.* 1459–1500.
43. Almer, H., Stawinski, J., and Strömberg R. (1996) Solid support synthesis of all Rp-oligoribonucleoside phosphorothioates. *Nucleic Acids Res.* **24,** 3811–3820.
44. Stawinski, J. and Thelin, M. (1990) Studies on the activation pathway of phosphonic acid using acyl chlorides as activators. *J. Chem. Soc. Perkin Trans.* **2,** 849–853.

7

Solid-Phase Synthesis of Circular Oligonucleotides

Enrique Pedroso, Nuria Escaja, Miriam Frieden, and Anna Grandas

Summary

A protocol for the straightforward preparation of small circular oligodeoxyribonucleotides (2–28 nt) is reported. The assembly of the oligonucleotide chain (standard phosphoramidite chemistry) and cyclization by the phosphotriester method take place on a tailor-made nucleotide-derivatized solid support. Although cyclization yields are moderate, the procedure exploits a synthesis design that allows selective cleavage of the circular oligonucleotide from the support, which facilitates isolation of the target molecule by simple filtration.

Key Words: Circular oligonucleotides; macrocycles; cyclization; antisense; solid-phase synthesis.

1. Introduction

Over the last decade hundreds of oligonucleotide analogs have been designed with the aim of improving their properties for a variety of applications. Circular oligonucleotides attract increased attention *(1,2)* because they are more resistant to nucleases than their linear counterparts *(3)*, show unique deoxyribonucleic acid (DNA) recognition properties *(4)*, and are efficient templates for DNA and ribonucleic acid (RNA) polymerases *(5–9)*. They also have interesting properties in several diagnostic *(10,11)* and therapeutic applications *(12–16)*, and they have been used in the study of noncanonical DNA structural motifs *(17–20)*.

The synthesis of single-stranded DNA circles, where each nucleotide furnishes six atoms to the circular backbone, is a challenging problem owing to the unfavorable entropy of the macrocycle formation. Nevertheless, large circles (>28 nt) are relatively easy to synthesize by chemical or enzymatic intramolecular ligation from unprotected oligonucleotides *(21–25)*. These methods take advantage of the remarkable capacity of recognition between nucleobases and preorganize the linear oligonucleotide as a duplex *(21,22)*, triplex *(23,24)*, or quadruplex *(25)* on a template. The template is an oligonucleotide sequence (DNA splint) either contained in the linear precursor (internal pairing) or, more frequently, a complementary oligonucleotide that joins the two reactive ends of the linear precursor. These template-directed approaches are limited because minimal base-pairing between the template and the linear precursor is required for structural stability. As a result, with one exception *(26)*, small circles cannot be obtained by these methods. In addition, sequence-induced secondary structures of the linear precursor may prevent the target circle from being obtained.

Several solution *(27–29)* and solid-phase *(30–33)* methods have been developed for the preparation of small circular oligonucleotides (<28 nt) from partially protected linear precursors. Solid-phase procedures are almost universally considered to be superior in terms of speed and effectiveness, and they also provide the ability to simulate pseudodilution conditions for the crucial cyclization step *(34)*. Different solutions have been proposed for the cyclization of immobilized linear precursors. De Napoli et al. have used the exocyclic amine groups of cytosine as the anchor point, and the phosphotriester method for the cyclization *(30,31)*, and Kool and coworkers have recently reported the use of automated synthesis on commercial controlled pore glass (CPG) supports and an S_N2 reaction for the preparation of circles bearing a phosphorothioate linkage *(32)*. However, a common feature of these methods is that, in the absence of the template-directed structural preorganization, the yield of cyclization will decrease as the size of the circle increases. Then, if the cyclization yield is low, the identification and isolation of the target circle from a mixture containing other circles, unreacted linear precursors, and other impurities may be a very difficult task.

The method *(8,33)* described here addresses these challenges because (1) it allows the synthesis of small-to-medium-sized circular oligodeoxyribonucleotides (2–28 nt), reaching the size under which the template-directed methods are no longer effective; and (2) the efficiency of the method is based on a design that exploits the main advantage of solid-phase procedures, that is, the ease of separation of a molecule (in this case the circle) from reagents and side products by simple filtration (*see* **Subheading 3.5.**).

This method can also be used for the preparation of circular oligoribonucleotides, but the presence of the 2'-OH protecting groups makes cyclization more difficult *(35,36)*. The corresponding experimental details are not included in this chapter.

2. Materials
2.1. Anchoring the First Nucleotide to the Support

1. Polyethyleneglycol–polystyrene copolymer, TentaGel N-NH$_2$ (0.24 mmol/g), from Rapp Polymere.
2. Polypropylene syringes fitted with a porous polyethylene disk and equipped with Teflon two-way stopcocks (RTV SF2, Shimadzu Scientific Research), and Teflon stir rods are used for the manual solid-phase syntheses.
3. Dry acetonitrile (ACN): distilled over CaH$_2$ powder and stored over CaH$_2$ lumps.
4. Dichloromethane (DCM): neutralized by elution through a basic Al$_2$O$_3$ column.
5. Dry DCM: neutralized, distilled over CaH$_2$ powder and stored over CaH$_2$ lumps.
6. N,N-dimethylformamide (DMF): freed of volatile amines by N$_2$ bubbling and stored over 4 Å molecular sieves.
7. Dry pyridine: stored over CaH$_2$ lumps. For the preparation of the ninhydrin reagent, pyridine is distilled over ninhydrin and stored over CaH$_2$ lumps.
8. Deionized water: Water is deionized and filtered through a Milli-Q system (Millipore).
9. Dichloromethane-d$_2$ (CD$_2$Cl$_2$), deuterochloroform (CDCl$_3$), ethyl acetate (AcOEt), hexane, methanol (MeOH), perchloric acid (HClO$_4$), and absolute ethanol (EtOH) are used without further purification.
10. 1H-tetrazole is purified and dried by sublimation (110°C at 0.05 mm Hg).
11. N-Fmoc-6-aminohexanoic acid, dicyclohexylcarbodiimide (DCC), diisopropylcarbodiimide (DIPC), acetic anhydride (Ac$_2$O), 1-hydroxybenzotriazole (HOBt), 2,4,5-trichlorophenol, 3-chloro-4-hydroxyphenylacetic acid, 5'-O-DMT-thymidine 3'-N,N'-diisopropyl(2-cyanoethyl)-phosphoramidite, 6M ^{t}BuOOH in decane (stored over 4Å molecular sieves), piperidine, trichloroacetic acid (TCA), triethylamine (TEA), diisopropylethylamine (DIEA), NaHCO$_3$, MgSO$_4$, Na$_2$SO$_4$, P$_2$O$_5$, KOH, and NaCl are of commercial origin and used without further purification.
12. Ninhydrin reagent: The ninhydrin test (37) detects free primary amines on a solid support. A small aliquot of dry resin (0.5–2 mg) is treated with three drops of reagent A and one drop of reagent B and heated to 110°C for 3 min. A blue color indicates the presence of free primary amines, whereas a yellow color indicates a negative test.
    a. Reagent A: A solution of 40 g of phenol in 10 mL of absolute EtOH and a solution of 2 mL of 10 mM KCN in 100 mL of distilled pyridine are, separately, vigorously shaken with 4 g of Amberlite resin MB-3 (Merck) for 45 min, filtered, and mixed.
    b. Reagent B: 2.5 g of ninhydrin is dissolved in 50 mL of absolute EtOH, and the solution is stored in an amber glass container.
13. Thin-layer chromatography (TLC) is performed on silica gel sheets 60F$_{254}$, 0.2 mm (Merck).
14. Flash column chromatography is performed on Chromatogel 60Å CC, 230–240 mesh (SDS).

15. ^{1}H and ^{13}C-nuclear magnetic resonance (NMR) spectra are recorded with tetramethylsilane as the internal standard. ^{31}P-NMR spectra are recorded with 85% H_3PO_4 in D_2O as the internal standard. Special tubes, containing a plunger that concentrates the sample in the irradiation zone, are used to record gel-phase ^{31}P-NMR spectra (Shigemi, Japan); 85% H_3PO_4 in D_2O is used as the external standard.

2.2. Chain Elongation

1. 5'-O-DMT-nucleoside-3'-O-N,N-diisopropyl(2-cyanoethyl)-phosphoramidites, the activator solution (0.5M tetrazole in ACN), capping solutions A (Ac_2O/lutidine/tetrahydrofuran (THF) 1:1:8), and B (10% 1-methylimidazole in THF) and the detritylation solution (3% TCA in DCM) are purchased from Glen Research. The oxidizing solution (1M ^{t}BuOOH) is prepared by diluting a 6M ^{t}BuOOH solution in decane over 4Å molecular sieves (Fluka, anhydrous ^{t}BuOOH) with dry DCM.
2. The solvents and noncommercially available solutions used in the DNA synthesizer (Applied Biosystems 308B) must be previously filtered (nylon filters, 25 mm, 0.45 µm).

2.3. Cyclization, Cleavage, and Deprotection

1. 1-mesitylenesulfonyl-3-nitro-1,2,4-triazole (MSNT), triethylamine, 2-pyridincarbaldehyde oxime, 1,1,3,3-tetramethylguanidine (TMG), and concentrated aqueous ammonia 33% are from commercial suppliers and used without further purification.
2. Dioxane is freed of peroxides by elution through basic alumina.

2.4. Purification and Characterization

2.4.1. Chromatography

1. Oligonucleotide desalting by gel filtration is carried out on a Sephadex G-10 (Pharmacia) column (l = 800 mm, Ø = 20 mm) using systems that combine one peristaltic pump (Pharmacia), a fixed wavelength ultraviolet (UV)-Vis detector (Uvicord 2158 SD, LKB), an automatic fraction collector (Ultrorac, LKB), and a recorder (Pharmacia-LKB REC 101).
2. High-performance liquid chromatography (HPLC): Shimadzu system with two LC-10AS pumps (high pressure mixing), SIL-9A autoinjector, SPD-10A variable wavelength UV-Vis detector and Chromatopac C-R5A recorder. Eluents: A = 0.01M triethylammonium acetate (TEAA) (or 0.01M $AcONH_4$), pH 7.0 buffer, and B = ACN/H_2O 1:1.
3. 0.01M acetate buffers are freshly prepared from 2M stock solutions, pH 7.0, stored at 4°C. The TEAA solution is obtained by mixing the required amounts of TEA, AcOH, and water and adjusting the pH. 2M ammonium acetate buffer, pH 7.0, is prepared by dissolving, with vigorous stirring, solid $AcONH_4$ in the required volume of water. The pH is adjusted by addition of acetic acid.

4. 2M triethylammonium bicarbonate (TEAB) buffer, pH 8.0. A CO_2 current is passed through a suspension of the required amount of TEA in water until a pH of 7.5–8.0 is reached (the initial suspension becomes a solution as the pH approaches the final value). The buffer is stored at 4°C.
5. Sep-Pak C_{18} cartridges are from Waters, Micro Bio-Spin columns are from Bio-Rad, and the strong cation-exchange resin Dowex 50W × 4 (200–400 mesh, 4.8 meq Na^+/g dry resin) is from Fluka.

2.4.2. Polyacrylamide Gel Electrophoresis

1. Polyacrylamide gel electrophoresis (PAGE) is performed in a Hoeffer SE410 apparatus (Pharmacia), with 18 × 24.5 cm vertical glass plates connected to a voltage source LKB 2197 BROMMA.
2. PAGE reagents: acrylamide, *bis*-acrylamide, urea, N,N,N',N'-tetramethylethylenediamine (TEMED) and ammonium persulfate (PA) are from Serva. The TBE buffer 10X, 1.3M tris(hydroxymethyl)aminomethane (Tris), 0.45M boric acid, 25 mM ethylenediaminetetraacetic acid, is from Bio-Rad. Formamide and the Stains All reagent are from Sigma.
3. 20% acrylamide/*bis*-acrylamide 38:2, 8M urea: a mixture of 190 g acrylamide, 10 g *bis*-acrylamide, 450 g urea, and 100 mL of TBE 10X buffer is filled up to 1 L with deionized water and vigorously stirred until the chemicals are totally dissolved. The final solution is filtered through a 0.45 μm pore filter and stored at 4°C in an amber glass container. *Caution:* acrylamide and *bis*-acrylamide are hazardous chemicals. Work in a well-ventilated fume hood taking extreme care in minimizing contact.
4. 10% PA solution: 1 g of PA is dissolved in 10 mL H_2O. This solution has to be prepared just prior to use.
5. Formamide loading buffer with marker dyes: Prepare a 0.05% (w/v) bromophenol blue and 0.05% (w/v) xylene cyanol solution in formamide. Store at –20°C.

2.4.3. Enzymatic Assays

1. Alkaline phosphatase (AP), (EC 3.1.3.1, Sigma, 0.23 U/mL).
2. Snake venom phosphodiesterase (SVPD), (EC 3.1.4.1 from *Crotalus durissus*, Sigma or Boehringer-Mannheim, 1.5 U/500 μL). SVPD incubation buffer: 50 mM Tris-HCl, 10 mM $MgCl_2$, pH 8.0.
3. Calf spleen phosphodiesterase (SpPD), (EC 3.1.16.1, Sigma, 0.0337 U/mL). SpPD incubation buffer: 200 mM aqueous $AcONH_4$, pH 5.4, prepared from a 2M stock solution (*see* **Subheading 2.4.1.**).
4. Nuclease S1 (EC 3.1.30.1 from *Aspergillus oryzae*, Pharmacia Biotech, 332 U/μL). Nuclease S1 incubation buffer (5X): 50 mM NaCl, 50 mM NaAcO, 5 mM $ZnCl_2$, pH 4.6.

2.4.4. Mass Spectrometry

The following, 2,4,6-trihydroxyacetophenone (THAP), ammonium citrate (AC) and 3-hydroxypicolinic acid (HPA), are from commercial suppliers and used without further purification.

3. Methods

The methods described below outline (1) the anchoring of the first nucleotide to the support, (2) the elongation of the oligonucleotide chain, (3) the cyclization on the solid support of the linear oligonucleotide, the cleavage and deprotection of the circular product, and (4) the purification and characterization of the circular oligonucleotide.

3.1. Anchoring the First (3'-End) Nucleotide to the Support

The steps described in **Subheadings 3.1.1.– 3.1.3.** outline the procedure for the anchoring, through the 3-chloro-4-hydroxyphenylacetic acid linker (**Fig. 1, 1**), of the first nucleotide (exemplified with B = T) onto the solid support (**Fig. 1**). These steps include the derivatization of the support with the 6-aminohexanoic acid spacer to afford (**Fig. 1, 4**), the synthesis of the nucleotide-linker (**Fig. 1, 3**), its anchoring to the derivatized support (**Fig. 1, 5**) to obtain the nucleotide resin (**Fig. 1, 6**), and then the phosphate deprotected nucleotide-resin **7**.

The four natural nucleotides can be anchored to the support, but if the target circle contains one thymidine it is advisable to start the synthesis of the linear precursor with this nucleoside.

Solid-phase synthesis of circular oligonucleotides is typically carried out on the polyethyleneglycol–polystyrene copolymer TentaGel N-NH$_2$ (240 µmol/g), which gives the best results, although other supports can also be used (*see* **Note 1**).

3.1.1. Support Derivatization

The 6-aminohexanoic acid spacer is incorporated into the solid support to separate the 3'-end of the growing oligonucleotide chain from the solid support and thus facilitate the cyclization reaction. This reaction is also used to reduce the initial loading of the solid support to a more appropriate value of approx 100 µmol DMT/g for **6** and **7**.

1. Before derivatization, place the TentaGel support into a polypropylene syringe fitted with a polyethylene disk, and wash it with the following solvents or reagents 2× each for 2 min.
   a. DCM
   b. 20% TCA in DCM
   c. DCM
   d. Piperidine/DMF 1:1
   e. DMF
   f. DCM
2. Add to the support solutions of *N*-Fmoc-6-aminohexanoic acid (1 equivalent with respect to the initial loading) in DCM, DCC (1 equivalent) in DCM, and HOBt

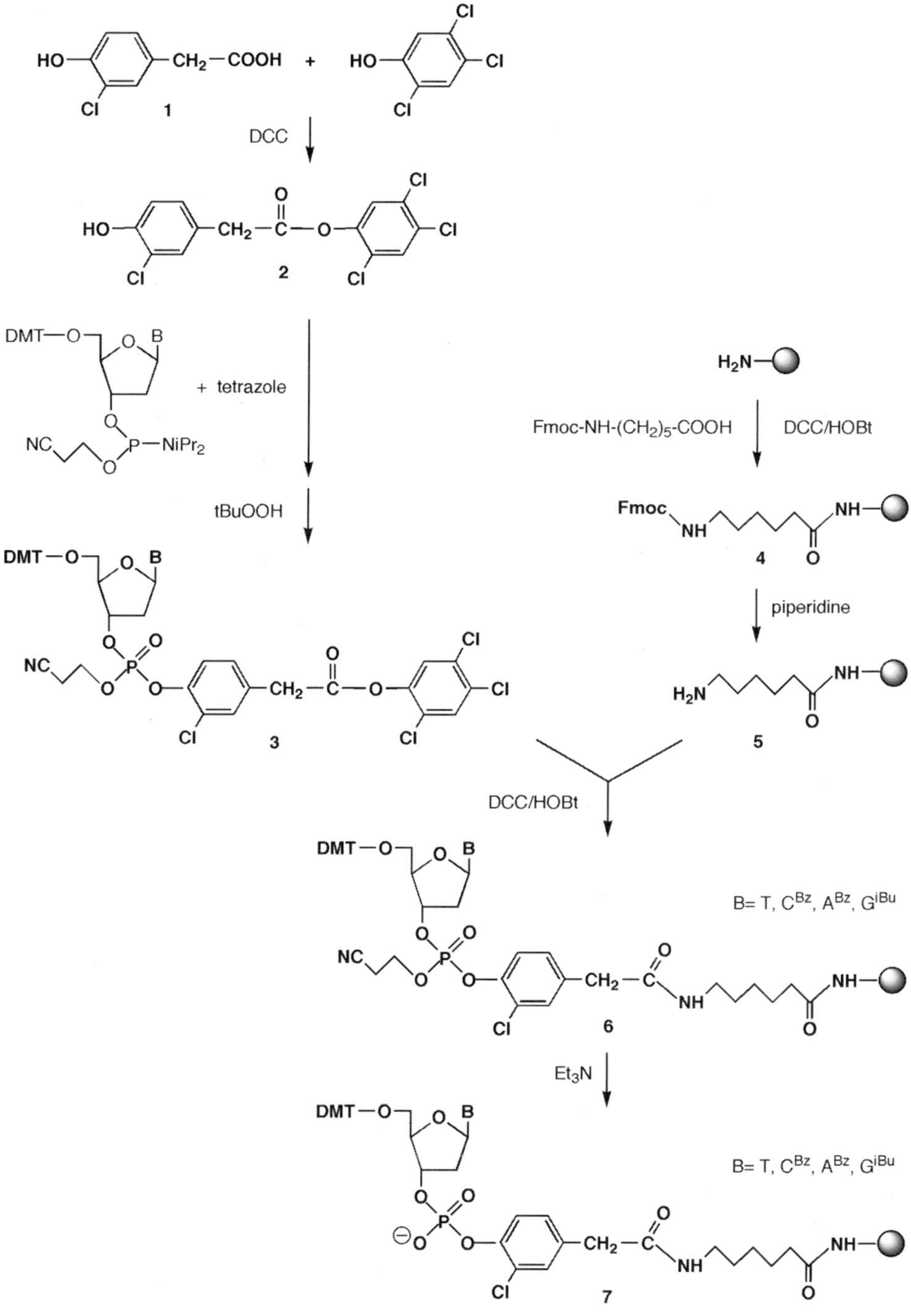

Fig. 1. Anchoring the 3'-end nucleotide to the solid support.

(1 equivalent) in DMF, and allow to react for 5–8 h at room temperature (DCC can be replaced by DIPC). All the reagents have to be dissolved in the minimum amount of solvent so that a thick suspension of the solid support is obtained. Wash the resin with DCM (2× for 2 min), DMF (2× for 2 min), and MeOH (1× for 1 min), and dry in a desiccator.

The anchoring of the spacer can be qualitatively controlled by comparing the result of the ninhydrin test (*see* **Subheading 2.1.**) before (clearly positive) and after (slightly positive) this step.

3. Determine the loading of the derivatized solid support (**Fig. 1, 4**) by measuring the amount of *N*-(9-fluorenylmethyl)piperidine (Fmp) released on deprotection of the amine groups.

   Accurately weigh an aliquot of dried resin (approx 10–12 mg) and treat it as follows:
   a. DCM 4× for 1 min.
   b. DMF 4× for 1 min.
   c. 200 µL 20% piperidine in DMF, 3× for 3 min.
   d. 100 µL DMF 1× for 1 min.

   Collect filtrates from **steps C** and **D** and additional washings with DCM, pour into a volumetric flask, and fill it with DCM. Measure the UV absorbance of the resulting solution at $\lambda = 300$ nm (ε_{300}(Fmp) = 7800). The loading (*f*) of the support is determined by applying the following equation:

   $$f(\mu mol/g) = \frac{A \cdot V \cdot 10^6}{\varepsilon \cdot l \cdot m}$$

   where A = absorbance, V = volume in mL, ε = molar extinction coefficient (L·mol^{-1}·cm^{-1}), l = cell path (cm), and m = mg of resin. Loadings within the range 120–180 µmol/g are generally obtained. If a considerably lower loading is obtained, **step 2** can be repeated with the whole batch of resin.

4. Cap the unreacted amino groups by acetylation. Treat resin (**Fig. 1, 4**) with 50 equivalents of Ac$_2$O in DMF for 30 min, gently wash it with DCM, DMF, and DCM, and dry it in a desiccator. Complete capping of free amines is assessed by a negative ninhydrin test (*see* **Subheading 2.1.**). Store the derivatized and capped solid support in its Fmoc-protected form at 4°C until needed for the anchoring of the nucleotide linker (*see* **Subheading 3.1.3.**).

3.1.2. Synthesis of Nucleotide Linker

The first step in the synthesis of the nucleotide linker is the preparation of the bifunctional linker 2,4,5-trichlorophenyl 3-chloro-4-hydroxyphenylacetate (**Fig. 1, 2**). This is followed by the synthesis of 3'-(5'-*O*-DMT-thymidine) 2-chloro-4-(2,4,5-trichlorophenoxycarbonylmethyl)phenyl 2-cyanoethyl phosphate (**Fig. 1, 3**).

Circular Oligonucleotides Synthesis

3.1.2.1. SYNTHESIS OF 2,4,5-TRICHLOROPHENYL 3-CHLORO-4-HYDROXY-PHENYLACETATE (SEE FIG. 1, 2)

Slowly add a solution of 3-chloro-4-hydroxyphenylacetic acid (**Fig. 1, 1**) (4.0 g, 21.6 mmol) in 130 mL of AcOEt to a solution of 2,4,5-trichlorophenol (4.7 g, 23.6 mmol) and DCC (4.9 g, 23.6 mmol) in 22 mL of DCM. Stir the resulting mixture at room temperature for 12 h. Filter the N,N'-dicyclohexylurea precipitate formed during the reaction. Cool the filtrate to 0°C, and filter again the new precipitate of N,N'-dicyclohexylurea. Wash the resulting solution with 5% aqueous $NaHCO_3$, dry over anhydrous $MgSO_4$, and evaporate to dryness. Purify the resulting orange oil by silica gel flash column chromatography, eluting with hexanes and increasing amounts of DCM (20–80%). Pure 2,4,5-trichlorophenyl 3-chloro-4-hydroxyphenylacetate (2.8 g) is obtained after elimination of the solvent from the corresponding fractions and precipitation in hexanes. Dry the product in a desiccator. It is important to check by TLC that the purified product does not contain any impurity of unreacted 3-chloro-4-hydroxyphenylacetic acid.

Figure 1, step 2: 34% yield, white solid, mp 77–78°C, R_f 0.33 (hexanes/AcOEt 70:30). ^{1}H-RMN ($CDCl_3$) δ (ppm): 7.53 (s, 1H, Ar-trichlorophenyl H5), 7.34 (d, J_m = 2.19 Hz, 1H, Ar-chlorophenyl H2), 7.25 (s, 1H, Ar-trichlorophenyl H6), 7.18 (dd, J_o = 8.39 Hz, J_m = 2.19 Hz, Ar-chlorophenyl H6), 7.01 (d, J_o = 8.39 Hz, Ar-chlorophenyl H5), 5.60 (s, 1H, OH), 3.82 (s, 2H, CH_2). ^{13}C-RMN ($CDCl_3$) δ (ppm): 167.3 (CO); 149.9, 144.6, 129.7, 124.5 and 118.9 (C arom); 130.0, 128.9, 128.5, 124.2, and 115.5 (CH arom); 38.6 (CH_2). EI-MS: m/z = 366, 198, 168, 141, 97, 77.

3.1.2.2. 3'-(5'-O-DMT-THYMIDINE) 2-CHLORO-4-(2,4,5-TRICHLOROPHENOXY-CARBONYLMETHYL)PHENYL 2-CYANOETHYL PHOSPHATE 3 (B = T)

All glass material must be completely dry to carry out the reaction under anhydrous conditions (maintain on a stove at 120°C for 4–6 h and allow to cool down to room temperature in a desiccator).

Dissolve the 5'-O-DMT-thymidine 3'-O-(2-cyanoethyl)-N,N-diisopropyl-phosphoramidite (141.1 mg, 0.2 mmol) (or any other nucleoside phosphor-amidite) under a dry argon atmosphere in 0.5 mL of dry DCM. Add, via cannula and also under argon atmosphere, 2,4,5-trichlorophenyl 3-chloro-4-hydroxiphenylacetate **2** (65.9 mg, 0.18 mmol) dissolved in 0.5 mL of dry DCM and 1H-tetrazole (13.8 mg, 0.2 mmol) dissolved in 0.5 mL of dry ACN. Stir the mixture at room temperature, under argon atmosphere, for 1 h. Then add 0.25 mL of 6M ^{t}BuOOH in decane (ca. 10 equivalents) to the reaction mixture to oxidize the phosphite to phosphate. This reaction takes place in about 10 min.

Dilute the reaction mixture with 20 mL of DCM, wash with H_2O (2 times in 50 mL) and brine (2 times in 5 mL), dry over anhydrous Na_2SO_4, and evaporate to dryness. The final product is obtained by precipitation over hexanes and used without further purification. The main accompanying impurity is usually the H-phosphonate generated by hydrolysis of the unreacted phosphoramidite and does not interfere with the coupling reaction of the nucleotide linker to the solid support. The product, thoroughly dried in a desiccator, can be stored at –18°C for up to 1 mo.

Figure 1, step 3: (B = T): 95% yield, white solid, R_f 0.51 (DCM/MeOH 90:10). ^{31}P-RMN (CD_2Cl_2) δ (ppm): –7.89, –7.92 (two diastereomers), 8.37 (H-phosphonate impurity).

3.1.3. Preparation of the Nucleotide-Resin

The anchoring of the first nucleotide is carried out by reaction of the aminohexanoic-derivatized solid support **5** and the nucleotide-linker **3** (**Fig. 1**).

3.1.3.1. COUPLING OF THE NUCLEOTIDE-LINKER TO THE SUPPORT

1. Remove the Fmoc protecting group of **4** by treating the resin with (1) DCM four times for 1 min, (2) DMF four times for 1 min, (3) 20% piperidine in DMF (twice for 3 min and once for 5 min), and (4) DMF once for 1 min. After deprotection, repeatedly wash the solid support **5** with DMF, DCM, and dry DCM. If desired, the loading of the whole batch of resin can again be determined as in **Subheading 3.1.1.**, but limited amounts of DMF must then be used so as not to interfere with the UV absorbance of Fmp.
2. Nucleotide-linker **3** (1.2–3 equivalents with respect to NH_2 groups in resin **5**, *see* **Note 2**) is coevaporated twice with dry ACN, dissolved in the minimum amount of dry DCM, and added to the derivatized support **5** in a polypropylene syringe. Add DCC or DIPC (1.2–3 equivalents) and HOBt (1.2–3 equivalents), also dissolved in the minimum volume of dry DCM and DMF, respectively. Cap the syringe with a septum and Parafilm, and keep at room temperature with mechanical shaking for 8–10 h. Wash the resin with DCM (twice for 2 min), DMF (twice for 2 min), and DCM (twice for 2 min), and dry in a desiccator over P_2O_5.
3. Determine the loading of the nucleotide-resin **6** by measuring the amount of 4,4'-dimethoxytrityl cations (DMT^+) released on acid deprotection of the 5'-hydroxyl groups. Submit an aliquot of known weight (approx 10–12 mg) of the dried support to the following treatments: (1) DCM four times for 1 min, (2) 3% TCA in DCM (300 µL) three times for 3 min, and (3) DCM (200 µL) four times for 1 min. Collect the filtrates from **2–3** and evaporate to dryness. Dissolve the orange residue in $HClO_4$/EtOH 3:2, pour into a volumetric flask, and fill it with the same solution. Measure the absorbance of the resulting solution at λ = 498nm ($\varepsilon_{498}(DMT^+) = 79,700$). Determine the loading of the support using the same equation as in **Subheading 3.1.1.** Resins **6** with a loading ranging from 50 to 120

μmol DMT/g are generally obtained. If coupling yield is lower, **step 2** of this section can be repeated with the whole batch of resin.
4. Cap the unreacted amine groups by acetylation. Treat nucleotide resin **6** with a 1:1 Ac$_2$O/DIEA mixture (20 equivalents excess Ac$_2$O with respect to free amines) in DCM for 20 min (twice for 10 min), and then wash with DCM, DMF, and DCM. Complete capping is assessed by a negative ninhydrin test (**Subheading 2.1.**).

3.1.3.2. REMOVAL OF THE CYANOETHYL GROUP

Before chain elongation, remove the 2-cyanoethyl (CNE) protecting group by treating resin **6** with TEA/pyridine 1:1 (three times for 1 h). Wash the resin with DCM and MeOH, and dry it in a desiccator. Complete deprotection of the phosphate group, as well as the homogeneity of the nucleotide-resin, is assessed by gel-phase ^{31}P-NMR (CD$_2$Cl$_2$). Nucleotide-resin **7** (B = T) shows a single signal within the range from –5 to –7 ppm, corresponding to the deprotected 3'-end phosphate (*see* **Note 3**).

3.2. Chain Elongation

Assembly of the oligonucleotide chain **8** is carried out automatically in a DNA synthesizer by the phosphoramidite approach *(38)* and using commercially available deoxyribonucleoside (T, A^{Bz}, C^{Bz}, G^{iBu}) phosphoramidite derivatives (**Fig. 2**). This standard methodology is not described in detail owing to space limitations. Here we report, as examples, the protocols for 2 and 10 μmol synthesis scales on TentaGel supports using an Applied Biosystems 380B synthesizer (**Table 1**). Reaction times, excess reagents, and the size of the reactors have to be adjusted to the synthesis scale. It is not recommended to carry out the syntheses of circular oligonucleotides at a scale lower than 1 μmol, owing to the moderate yields obtained.

The synthesis of the linear precursor **8** ends with a detritylation reaction (**steps 1–4** of the synthesis cycle), followed by careful drying, to get the oligonucleotide resin with both unprotected 5'-OH and 3'-phosphate ends.

3.3. Cyclization, Cleavage, and Deprotection

The steps described in **Subheadings 3.3.1.–3.3.4.** outline the procedures for the transformation of the immobilized linear precursor **8** into the circular oligonucleotide **10** (**Fig. 2**).

3.3.1. Cyclization

The cyclization of the linear precursor, by reaction of the 5'-OH with the activated 3'-phosphate, must be carried out in strictly anhydrous conditions, and it is best performed in the DNA synthesizer. It is important to check the purity of the coupling reagent, MSNT, prior to its use by measuring the melting

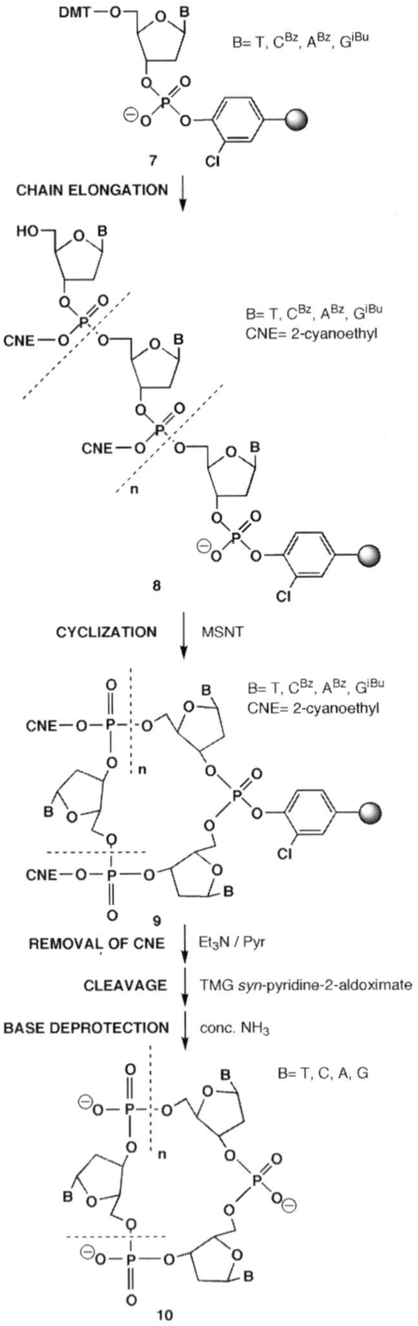

Fig. 2. Chain elongation, cyclization, cleavage, and deprotection.

Circular Oligonucleotides Synthesis

Table 1
Protocol for 2 and 10 µmol Oligonucleotide Synthesis, Using Reactors of 400 µL and 1200 µL, Respectively

Step	Solvents/reagents	Time (s) 2 µmol scale	Time (s) 10 µmol scale
1. Washing	ACN	20	65
2. Detritylation	3% TCA in DCM	120	2 × 120 + 1 × 90
3. Washing	ACN	200	120
	DMF	30	210
	ACN	20	270
	Dry ACN	45	300
4. Drying	Argon	45	90
5. Coupling	0.1M phosphoramidite in dry DCM (10 eq)+	25	60
	0.5M tetrazole in ACN	wait: 900	wait: 1200
6. Washing	ACN	30	60
7. Acetylation	Ac$_2$O/lutidine/THF 1:1:8	30	100
	10% 1-methylimidazole in THF	wait: 120	wait: 180
8. Oxidation	1M ^{t}BuOOH in anhydrous DCM/decane	30	150
		wait: 60	wait: 120
9. Washing	DCM	60	90
	DMF	60	120
	ACN	2 × 30	180

point (138°C). MSNT is first coevaporated with dry pyridine three times and then dissolved in the same solvent under argon atmosphere. The 0.1–0.15M MSNT solutions in dry pyridine (10–20-fold excess with respect to the oligonucleotide resin), must be freshly prepared before each treatment. The procedure for the cyclization is shown in **Table 2**.

For the largest circular oligonucleotides (>15 mer) it is recommended to increase the number of treatments with MSNT (four times for 2 h and once for 12 h).

After the cyclization step, the resin is removed from the reactor and placed again in a syringe fitted with a disk where the deprotection and cleavage reactions will be further performed (*see* **Note 4**).

3.3.2. Deprotection of the Phosphate Groups

Treat the resin **9** with TEA/pyridine 1:1 (three times for 1 h) to remove the 2-cyanoethyl protecting groups. Wash the resin with pyridine (twice for 1 min), DCM (four times for 1 min), and MeOH (twice for 1 min), and dry it. It is commonly observed that during this reaction the resin acquires an intense reddish brown color.

Table 2
Protocol for the Cyclization Reaction

Step	Solvent/Reagent	Time (s)
1. Drying	Argon	90
2. Washing	Anhydrous pyridine	180
	Anhydrous ACN	240
3. Drying	Argon	90
4. Coupling	0.15M MSNT in dry pyridine	2 × 4 h and 1 × 12 h[a]
5. Washing	Anhydrous pyridine	180

[a]Repeat **steps 1–3** before each MSNT treatment.

3.3.3. Cleavage of the Circular Oligonucleotide

Treat the resin (twice for 4 h and overnight) with a 0.2M solution of 1,1,3,3-tetramethylguanidinium *syn*-pyridine-2-aldoximate in dioxane/H_2O 1:1 (approx 50-fold excess with respect to the oligonucleotide resin loading). Oximate selectively cleaves phosphotriesters *(39)* (*see* **Subheading 3.5.**). The filtrates from the different treatments are combined and evaporated to dryness.

3.3.4. Nucleobase Deprotection

Treat the residue obtained in the previous step with a concentrated aqueous ammonia solution (NH_3 33%) at 55°C for 12 h in a capped tube sealed with Parafilm, to deprotect the nucleobases (Bz for dA and dC, and ^{i}Bu for dG). After deprotection, the solution containing the fully deprotected circular oligonucleotide is evaporated to dryness.

3.4. Purification and Characterization

The steps described in **Subheadings 3.4.1.–3.4.3.** outline the procedures for the analysis and purification of the circular oligonucleotide. Only the most relevant procedures are described in detail owing to space limitations.

3.4.1. Desalting the Crude Product

The residue obtained after the deprotection and cleavage steps contains, in addition to the desired circle, a large amount of salts and small molecules that have to be removed by gel filtration before the final purification.

1. Dissolve the residue that contains the circular oligonucleotide in 2–3 mL H_2O and elute through a Sephadex G-10 column (*see* **Note 5**) with 0.05M TEAB at a flow rate of 0.8 mL/min, λ = 260 nm, collecting fractions of 3–5 mL. The first peak, at the dead-end volume, contains the oligonucleotide.

2. Analyze the collected fractions by UV. The peak containing the circle has the typical UV spectrum of an oligonucleotide, with a maximum close to 260 nm.
3. Analyze the different fractions by HPLC in a 250 × 4.6 mm Nucleosil C_{18} reversed-phase column (10 µm), gradient from 10% to 40% B in 20 min (A: $0.01M$ TEAA or $AcONH_4$, B: ACN/H_2O 1:1), flow rate = 1 mL/min, λ = 260 nm. See **Subheading 3.4.2.2.** for PAGE purification and analysis.
4. Combine and lyophilize the fractions that contain the circular oligonucleotide. Discard the fractions that contain low amounts of product and do not give satisfactory HPLC profiles.
5. Dissolve the product in 1–2 mL of H_2O and, on the basis of the calculated ε *(40)* of the circle and the synthesis scale, prepare a diluted solution with an expected UV_{260} absorbance lower than 1. Record the UV spectrum and quantify the amount of crude circle. This is usually given in optical density units, $(OD)_{260}$ (*see* **Notes 6 and 7**):

$$OD_{260} = A_{260} \cdot V$$

where A_{260} = absorbance of the oligonucleotide solution at 260 nm, and V = total volume in mL.

3.4.2. Purification

The method of choice for the purification of circular oligonucleotides depends basically on their size. For small circles, up to a 12 mer, the best alternative is HPLC. For larger oligonucleotides, HPLC profiles are more complicated and do not always show a clear major peak, especially if the circle tends to fold into secondary structures. In these cases, analysis and purification of the circle are best achieved by PAGE under denaturing conditions.

3.4.2.1. HPLC PURIFICATION

HPLC purification of circles can be carried out using an analytical C_{18} column for small amounts of product, as described in the previous section, or a semipreparative column. We report the procedure we have used for the purification of milligram amounts of small circles.

1. Dissolve the sample in water or in $0.01M$ TEAA buffer.
2. Inject increasing volumes of the sample solution in a 305 × 7 mm semipreparative polystyrene-*co*-divinylbenzene column (Hamilton PRP-1, 10 µm), and elute with a gradient from 10% to 40% B in 20 min (A: $0.01M$ TEAA, B: ACN/H_2O 1:1), flow rate = 2 mL/min, λ = 260 nm. The gradient can be adapted to the HPLC profile of each crude product.
3. Collect the desired fractions (often the major peak) either manually or with an automated fraction collector. Combine and lyophilize the fractions that contain the circular oligonucleotide.
4. Quantify the amount of purified circular oligonucleotide as described at the end of **Subheading 3.4.1.** (*see* **Note 8**).

3.4.2.2. PAGE Purification

1. Prepare a vertical 1.5-mm thick, denaturing 20% polyacrylamide gel. Assemble glass plates (18 × 24.5 cm), according to manufacturer's instructions, using 1.5-mm uniform thickness spacers. Mix 60 mL of 20% acrylamide:*bis*-acrylamide 38:2 solution (**Subheading 2.4.2.**), 400 µL 10% PA solution, and 13 µL TEMED immediately before pouring the gel. Put gel solution into a 60-mL syringe, avoiding bubbles, and slowly expel the gel between plates. Insert the flat side of a 1.5-mm gel comb into the solution being careful to avoid bubbles. Allow to polymerize for 30 min. Remove the gel comb and rinse wells thoroughly with 1X TBE buffer to remove debris. *See* **Note 9** for analytical PAGE conditions.
2. Place the gel in the gel electrophoresis apparatus, using 1X TBE buffer in both top and bottom buffer reservoirs, and prerun 1 h at 750 V, room temperature.
3. Rinse wells with 1X TBE buffer just prior to loading samples to remove urea that has leached into them. Reserve two wells for marker dyes (20 µL of formamide loading buffer) (*see* **Subheading 2.4.2.**).
4. Prepare oligonucleotide samples by mixing 10 µL of crude circle in water (approx 2 OD in 10 µL H_2O) and 10 µL of the formamide loading buffer or (*see* **Subheading 2.4.2.**). H_2O/formamide 1:1. Heat samples 2 min at 110°C and place them on ice. Load samples in separate adjacent wells, 20 µL of sample per well.
5. Run the gel at 700 V, room temperature, and observe migration of marker dyes (on a 20% gel bromophenol blue migrates as an 8-nt linear DNA and xylene cyanol as a 28-nt linear DNA) to determine length of electrophoresis.
6. Disassemble the gel apparatus. Sandwich the gel between two pieces of UV-transparent plastic wrap and view under UV light with fluorescent white background (a silica TLC sheet for example) to visualize the bands. Electrophoretic mobility of most circles (up to approx 25 nt) is higher than that of a linear DNA of the same length, whereas the relative mobility is inverted for the largest circles.
7. Cut out the desired bands with a razor blade, and transfer the pieces to a 15-mL centrifuge tube. Crush the gel pieces thoroughly with a glass stir rod, add 5–10 mL of a 0.2*M* NaCl solution (or 2*M* TEAA), and shake for 12 h. Filter by centrifugation (1–5 min) and keep the filtrate. Resuspend the gel in 5 mL H_2O, and centrifuge and filter again. Combine the filtrates and desalt the purified product by dialysis (*see* **Note 10**) or, for low amounts of product, with a Sep-Pak C_{18} cartridge or a Micro Bio-Spin 6 column (follow manufacturer's instructions).
8. Quantify the amount of purified circular oligonucleotide as described at the end of **Subheading 3.4.1.** (*see* **Note 8**).

3.4.2.3. Conversion to the Sodium or Ammonium Salt

This step is only required for certain applications such as NMR studies.

1. Place an appropriate quantity of strong cation-exchange resin Dowex 50W × 4 in a polypropylene syringe fitted with a porous polyethylene disk, and regenerate as follows:

a. 200 mL H$_2$O
   b. 200 mL of 10% HCl
   c. H$_2$O until neutral pH is obtained
   d. 500 mL of 1M NaOH (or 1M NH$_4$OH)
   e. H$_2$O until neutral pH is obtained.
2. Dissolve the oligonucleotide in 1.5 mL of H$_2$O, load onto the regenerated resin, and slowly elute with water. Collect fractions of approx 0.5 mL until no absorbance at 260 nm is observed (approx 15–20 mL). Combine the fractions containing the circle and lyophilize.

3.4.3. Characterization

Three enzymatic assays *(24,41)* can be performed to check the nucleotide composition and assess the circularity of the product.

3.4.3.1. TO VERIFY THE NUCLEOTIDE COMPOSITION: DIGESTION WITH SVPD AND AP *(41)*

1. Prepare the enzyme solution by dissolving 5 μL of the commercial AP solution and a spatula tip of SVPD (Sigma) in 100 μL of the buffer solution described in **Subheading 2.4.3.**
2. Incubate a lyophilized aliquot (0.5–1 OD$_{260}$) of the cyclic oligonucleotide, contained in an Eppendorf tube, with the enzyme solution (50 μL/OD$_{260}$) at 37°C overnight. Analyze the resulting mixture by reversed-phase HPLC, with a Spherisorb ODS2 column (10 μm), 250 × 4.6 mm, flow rate: 1 mL/min, using the following gradient: isocratic 10% B 5 min, 10–30% B in 15 min, 30–100% B in 5 min.
3. Identify the nucleosides by comparison with the retention time of the nucleosides of a reference sample mixture (dC: 4.0, dG: 6.3, T: 6.7, and dA: 13.1 min).
4. Calculate the relative proportion of each nucleoside by integration of the corresponding HPLC peaks, taking into account the molar extinction coefficient at 260 nm, as follows:
   (i) Average area/ε (X) determination   (ii) Nucleosides proportion determination

$$\frac{\Sigma(Area/\varepsilon)_i}{N} = X \qquad\qquad \frac{(Area/\varepsilon)_i}{X} = n_i$$

where i = nucleoside, *Area* = peak area for nucleoside i, ε = molar extinction coefficient value of nucleoside i, N = theoretical total number of bases, and n_i is the experimental value calculated for nucleoside i.

3.4.3.2. TO CONFIRM THE CIRCULARITY OF THE OLIGONUCLEOTIDE: SPPD DIGESTION *(41)*

1. Prepare the enzyme solution by dissolving 30 μL of the commercial SpPD solution in 100 μL of 0.2M AcONH$_4$, pH 5.4.
2. Incubate a lyophilized aliquot (0.5–1 OD$_{260}$) of cyclic oligonucleotide, in an Eppendorf tube, with the enzyme solution (50 μL/OD$_{260}$) at 37°C overnight.

Analyze the resulting mixture by reversed-phase HPLC in the conditions described in **Subheading 3.4.1.** or by PAGE (*see* **Subheading 3.4.2.2.**) (*see* **Note 9**).

3.4.3.3. TO CONFIRM THE CIRCULARITY OF THE OLIGONUCLEOTIDE: NUCLEASE S1 DIGESTION *(24)*

This assay is only useful for the largest circles (approx >25-mer) because smaller circles degrade so fast that it is impossible to observe the full-length linear precursor and hence confirm the circularity of the product.

1. To 0.2 OD_{260} of the lyophilized oligonucleotide samples (two samples of the circular oligonucleotide and two samples of the corresponding linear precursor) in 0.5-mL Eppendorf tubes add (a) control samples: 0.6 µL Nuclease S1 buffer 5X and 2.5 µL H_2O; (b) circles: 0.6 µL Nuclease S1 buffer 5X, 2.0 µL H_2O and 0.5 µL Nuclease S1 diluted solution (0.5 µL of the commercial solution diluted with 373 µL H_2O).
2. Incubate the samples at 37°C overnight, and stop the reaction by addition of 3.10 µL of loading buffer. Analyze the reaction mixture by PAGE (*see* **Note 9**). Nuclease S1 digestion of the circle initially produces linear oligonucleotides of the same length as the control oligonucleotides. This analysis is complicated by the fact that the electrophoretic mobility of the largest circles (approx >25 nt) is lower than that of the corresponding linear DNA, whereas the relative mobility is inverted in small circles.

3.4.3.4. MASS SPECTROMETRY

Two different mass spectrometry techniques, matrix-assisted laser desorption/ionization time-of-flight (MALDI-TOF) and electrospray ionization mass spectrometry (ESI–MS), are particularly useful to assess the circularity of the oligonucleotides. MALDI-TOF can be used for the whole range of circles (2–30 nt). In our hands, ESI–MS only gives good results with the smallest circles (up to 16 nt), although the mass accuracies obtained are higher than with MALDI-TOF.

1. Desalt the HPLC- or PAGE-purified circular oligonucleotide as indicated at the end of **Subheading 3.4.2.2.** (Sep-Pak or Micro Bio-Spin) to avoid the formation of cation adducts.
2. Determine the molecular weight (mass) of the purified circle by recording the MALDI-TOF or ESI spectrum, both in the negative-ion mode (*see* **Note 11**). Prepare the sample as follows:
   a. MALDI-TOF: Add 1 µL of ammonium citrate solution (50 mg/mL in water) to a 1-µL sample of oligonucleotide (25–50 µ*M* in water), and allow the mixture to react for a few seconds to facilitate cation-exchange, then add 1 µL of freshly prepared matrix solution. Deposit 1 µL of this mixture on the plate and allow to dry for some minutes.

b. Recommended matrices: 10 mg/mL 2,4,6-trihydroxyacetophenone (2,4,6-THAP) *(42,43)* in ACN/H$_2$O 1:1, for circles up to 15 nt, and 50 mg/mL HPA *(44)* in ACN/H$_2$O 1:1 for larger circles.
c. ESI–MS: Prepare a 1 to 10 µ*M* solution of purified circle in ACN/H$_2$O 1:1.

3.5. Discussion

This method is based on a cyclization reaction carried out by the phosphotriester method. This is a well-established approach for the stepwise solid-phase assembly of oligonucleotide sequences in which an excess (10- to 20-fold) monomer synthon is reacted with the 5'-end of a nascent oligonucleotide chain. In contrast with this methodology, the cyclization is an intramolecular reaction with a 1:1 stoichiometry between the phosphorylating agent and the 5'-OH. For this reason, the cyclization reaction with MSNT is repeated several times, and long reaction times are needed to push the reaction to completion. Nevertheless, after this cyclization step there are several types of oligonucleotide chains anchored to the support (**Fig. 3**)—the target circular molecule, the unreacted linear precursor, and polymeric sequences arising from intermolecular condensations between a 3'-phosphodiester group with the 5'-OH of a different chain—as well as the standard impurities (not shown in **Fig. 3**) originating in stepwise elongation procedures (capped or deleted linear sequences, etc.). When the cleavage reaction of the oligonucleotide resin is performed with TMG pyridinaldoximate, the phosphate triesters are cleaved, but not the phosphate diesters (*see* arrows). Then, in the subsequent filtration, the main impurities accompanying the circle remain attached to the support, whereas a fairly pure circle is obtained in the filtrate. In our hands this is the result in most cases. HPLC profiles of crude small circles are similar to those obtained for standard oligonucleotide syntheses (80–90% purity). However, when the size of the circle increases, the yields decrease, and the ratio of the target circle to impurities in the cleavage products is also lower.

In some cases the main impurity accompanying the circle is the linear oligonucleotide attached to the linker, HO-[nucleoside-*O*-PO(O$^-$)-*O*]$_n$-C$_6$H$_3$Cl-CH$_2$-COOH (*see* **Note 11**). It is worth mentioning that if the cyclization step is suppressed, 15% is the maximum amount of side product that can be cleaved from the resin in the oximate treatment of **8**. Therefore, this product, which is formed by oximate-mediated cleavage of the amide bond that binds the linker to the support *(45)*, is only observed in the synthesis of the large, most difficult circles or when the cyclization is not properly carried out. It is also important to realize that if the amide bond linking a circular molecule to the support is cleaved, the circle-linker product will not be observed among the products detached from the resin because the oximate treatment will also cleave—and much faster—the phosphate triester linking the two moieties.

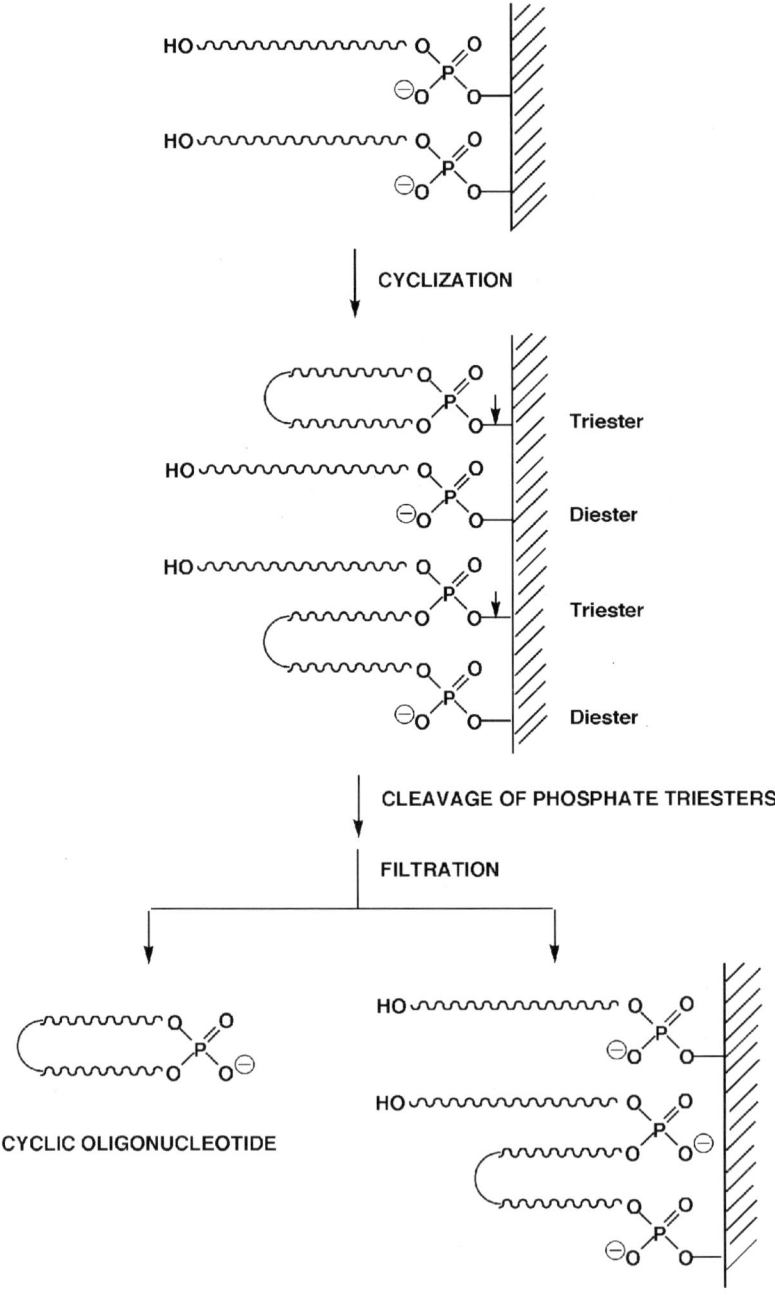

Fig. 3. Schematic description of the basis for the easy separation of circles and side products.

In summary, this protocol allows synthesis of small circular oligonucleotides (2–28 nt), without any sequence restriction, by a simple and efficient solid-phase method. The isolation of the circles, which are obtained in moderate, but reasonable, yields, would have been almost impossible to envisage without the high selectivity attained in the oximate reaction that cleaves the circles from the support but leaves most of the side products still anchored. This is accomplished by the use of a bifunctional linker molecule, 3-chloro-4-hydroxyphenylacetic acid, that (1) provides the anchoring point for the oligonucleotide to the support, (2) protects the 3'-end as a chlorophenyl phosphate diester throughout the oligonucleotide assembly, and (3) allows to selectively release the circle from the support as a result of the cyclization-mediated conversion of the 3' phosphate diester of the linear precursor into a chlorophenyl phosphate triester-linked circle.

4. Notes

1. The 4-methylbenzhydrilamine polystyrene-*co*-1%-divinylbenzene resin (MBHA-PS), can also be used as solid support, but not CPG. The main changes that have to be made in the reported protocols when using MBHA-PS are (1) **Subheading 3.1.1.** Use only 0.5 equivalents of both *N*-Fmoc-aminohexanoic acid and DCC to derivatize the resin (0.56 mmol/g); (2) **Subheading 3.2.** Use DCM instead of ACN for the washing steps and replace ACN by THF in **steps 5** and **6** (**Table 1**); and (3) **Subheading 3.3.3.** Cleave the circle from the support with a $0.2M$ oximate solution in dioxane/water 2:1.
2. The amount of nucleotide linker used in the preparation of the nucleotide resin depends on the final loading to be attained and on the substitution degree of the aminohexanoic resin. Using 1.2 equivalents of nucleotide-linker, loadings of approx 70 µmol/g are typically obtained, whereas 3 equivalents are needed to obtain higher loadings (approx 120 µmol/g). In the conditions described, about 35% of the nucleotide linker is coupled to the support.
3. The homogeneity of the nucleotide resin is usually assessed before and after deprotection of the 3'-end phosphate. Before deprotection, it is possible to observe the signals of two diastereomers with a chemical shift 1 ppm lower than that of the 3'-end phosphodiester.
4. The three steps of cleavage and deprotection should be performed in the prescribed order, despite the fact that ammonia removes 2-cyanoethyl as well as *o*-chlorophenyl protecting groups. The reason is that the oximate-mediated deprotection of *o*-chlorophenyl-protected phosphates is more selective than the ammonia-mediated reaction. In addition, direct oximate treatment of the fully protected circle-resin **9** may result in some internucleosidic cleavage of CNE-protected phosphate triesters *(39)*.
5. Store the Sephadex G-10 column with 10% NaN_3 solution, and prior to use wash with water and equilibrate with $0.05M$ TEAB.

6. One OD_{260} corresponds to the absorbance at 260 nm of a 1-mL oligonucleotide solution measured in a 1-cm path length cell.
7. Crude yields range from 40–50% for circles up to 8 nt to less than 10% for the largest circles (>20 nt).
8. Isolation yields depend on the homogeneity of the crude product, which is usually lower as the size of the circle increases (*see* **Subheading 3.5.**), and on the recovery yields associated with the method of purification. This yield does not exceed 60% for HPLC and is much lower for PAGE purification.
9. For analytical PAGE, spacers and combs of 0.75 mm are used, and the gel is run at 1000 V. 0.1–0.2 OD of sample are typically loaded. The bands can either be visualized under a UV lamp or dyed with Stains All reagent. When using Stains All, the gel is previously soaked in water to remove the rest of the urea. The water is removed and then the dye is added (20 mg of Stains All in 30 mL of formamide and diluted with water to 200 mL). After 30 min the gel is gently washed with water, placed under an infrared lamp to discolor the gel background, and finally dried in a gel drier (placed over porous paper and covered with a wet cellophane membrane).
10. Transfer the oligonucleotide solution to a dialysis tube with one clamped end (Spectra Por, MWCO 1000). Tightly secure the other end of the bag and dialyze against 2 L ultrapure water for 6 h. Repeat three times.
11. A product with a mass 186–187 units higher than the expected circle is sometimes observed corresponding to the full-length linear oligonucleotide esterified at its 3'-end with the linker, $HO\text{-}[\text{nucleoside-O-PO(O}^-)\text{-O}]_n\text{-}C_6H_3Cl\text{-}CH_2\text{-}COOH$ (*see* discussion). Such side products have, in most cases (<25 nt), a PAGE mobility lower than that of the circles, are digested with SVPD giving the correct nucleoside composition, but in contrast to the circles are not resistant to SpPD digestion.

Acknowledgments

This work was supported by funds from the Spanish Ministerio de Ciencia y Tecnología (grant BQU2001-3693-C02-01) and the Generalitat de Catalunya (2001SGR49 and Centre de Referència de Biotecnologia).

References

1. Kool, E. T. (1998) Recognition of DNA, RNA and proteins by circular oligonucleotides. *Acc. Chem. Res.* **31,** 502–510.
2. Kool, E. T. (1996) Circular oligonucleotides: new concepts in oligonucleotide design. *Annu. Rev. Biophys. Biomol. Struct.* **25,** 1–28.
3. Rumney, S., IV and Kool, E. T. (1992) DNA recognition by hybrid oligoether-oligodeoxynucleotide macrocycles. *Angew. Chem. Int. Ed. Engl.* **31,** 1617–1619.
4. Prakash, G. and Kool. E. T. (1992) Structural effects in the recognition of DNA by circular oligonucleotides. *J. Am. Chem. Soc.* **114,** 3523–3527.
5. Fire, A. and Xu, S. (1995) Rolling replication of short DNA circles. *Proc. Natl. Acad. Sci. USA* **92,** 4641–4645.

6. Liu, D., Daubendiek, S. L., Zillman, M. A., Ryan, K., and Kool, E. T. (1996) Rolling circle DNA synthesis: small circular oligonucleotides as efficient templates for DNA polymerases. *J. Am. Chem. Soc.* **118,** 1587–1594.
7. Daubendiek, S. L. and Kool, E. T. (1997) Generation of catalytic RNAs by rolling transcription of synthetic DNA nanocircles. *Nat. Biotechnol.* **15,** 273–277.
8. Frieden, M., Pedroso, E., and Kool, E. T. (1999) Tightening the belt on polymerases: evaluating the physical constraints on enzyme substrate size. *Angew. Chem. Int. Ed. Engl.* **38,** 3654–3657.
9. Kuhn, H., Demidov, V. V., and Frank-Kamenetskii, M. D. (2002) Rolling-circle amplification under topological constraints. *Nucleic Acids Res.* **30,** 574–580.
10. Nilsson, M., Krejci, K., Koch, J., Kwiatkowski, M., Gustavsson. P., and Landegren. U., (1997) Padlock probes reveal single-nucleotide differences, parent of origin and *in situ* distribution of centromeric sequences in human chromosomes 13 and 21. *Nat. Genet.* **16,** 252–254.
11. Lizardi, P. M., Huang, X., Zhu, Z., Bray-Ward, P., Thomas, D. C., and Ward, D. C. (1998) Mutation detection and single-molecule counting using isothermal rolling-circle amplification. *Nat. Genet.* **19,** 225–232.
12. Clusel, C., Ugarte, E., Enjolras, N., Vasseur, M., and Blumenfeld, M. (1993) *Ex vivo* regulation of specific gene expression by nanomolar concentration of double-stranded dumbbell oligonucleotides. *Nucleic Acids Res.* **21,** 3405–3411.
13. Mazumder, A., Uchida, H., Neamati, N., et al. (1997) Probing interactions between viral DNA and human immunodeficiency virus type I integrase using dinucleotides. *Mol. Pharmacol.* **51,** 567–575.
14. Abe, T., Takai, K., Nakada, S., Yokota, T., and Takaku, H. (1998) Specific inhibition of influenza virus RNA polymerase and nucleoprotein gene expression by circular dumbbell RNA/DNA chimeric oligonucleotides containing antisense phosphodiester oligonucleotides. *FEBS Lett.* **425,** 91–96.
15. Rowley, P. T., Kosciolek, B. A., and Kool, E. T. (1999) Circular antisense oligonucleotides inhibit growth of chronic myeloid leukemia cells. *Mol. Med.* **5,** 693–700.
16. Roulon, T., Hélène, C., and Escudé, C. (2001) A ligand-modulated padlock oligonucleotide for supercoiled plasmids. *Angew. Chem. Int. Ed. Engl.* **40,** 1523–1526.
17. Ippel. J. H., Lanzotti, V., Galeone, A., et al. (1995) Slow conformational exchange in DNA minihairpin loops: a conformational study of the circular dumbbell d<pCGCTTGCGTT>. *Biopolymers* **36,** 681–694.
18. Salisbury, S. A., Wilson, S. E., Powell, H. R., et al. (1997) The bi-loop, a new general four-stranded DNA motif. *Proc. Natl. Acad. Sci. USA* **94,** 5515–5518.
19. Escaja, N., Pedroso, E., Rico, M., and González, C. (2000) Dimeric solution structure of two cyclic octamers: four-stranded DNA structures stabilized by A:T:A:T and G:C:G:C tetrads. *J. Am. Chem. Soc.* **122,** 12,732–12,742.
20. Escaja, N., Gelpí, J. L., Orozco, M., Rico, M., Pedroso, E., and González, C. (2003) Four-stranded DNA structure stabilized by a novel G:C:A:T tetrad. *J. Am. Chem. Soc.* **125,** 5654–5662.
21. Ashley, G. W. and Kushlan, D. M. (1991) Chemical synthesis of oligodeoxynucleotide dumbbells. *Biochemistry* **30,** 2927–2933.

22. Dolinnaya, N. G., Blumenfeld, M., Merenkova, I. N., et al. (1993) Oligonucleotide circularization by template-directed chemical ligation. *Nucleic Acids Res.* **21,** 5403–5407.
23. Maksimenko, A. V., Volkov, E. M., Bertrand, J.-R., et al. (2000) Targeting of single-stranded DNA and RNA containing adjacent pyrimidine and purine tracts by triple helix formation with circular and clamp oligonucleotides. *Eur. J. Biochem.* **267,** 3592–3603.
24. Diegelman, A. M. and Kool, E. T. (2000) Chemical and enzymatic methods for preparing circular single-stranded DNAs. In *Current Protocols in Nucleic Acid Chemistry,* vol. 2 (Beaucage, S. et al., eds.). John Wiley, New York, Chapter 5.2.
25. Li, T., Liu, D., Chen, J., Lee, A. H. F., Qi, J., and Chan, A. S. C. (2001) Construction of circular oligodeoxyribonucleotides on the new structural basis of i-motif. *J. Am. Chem. Soc.* **123,** 12,901–12,902.
26. Liu, D., Chen, J., Lee, A. H. F., Chow, L. M. C., Chan, A. S. C., and Li, T. (2003) Small circular oligodeoxynucleotides achieved from self-assembling entities. *Angew. Chem. Int. Ed. Engl.* **42,** 797–799.
27. Capobianco, M., Carcuro, A., Tondelli, L., Garbesi, A., and Bonora, G. M. (1990) One pot solution synthesis of cyclic oligodeoxyribonucleotides. *Nucleic Acids Res.* **18,** 2661–2669.
28. Vroom, E., Broxterman, H. J. G., Sliedregt, L. A. J. M., van der Marel, G. A., and van Boom, J. H. (1988) Synthesis of cyclic oligonucleotides by a modified phosphotriester approach. *Nucleic Acids Res.* **16,** 4607–4620.
29. Reese, C. B. and Song, Q. (1999) The *H*-phosphonate approach to the solution phase synthesis of linear and cyclic oligoribonucleotides. *Nucleic Acids Res.* **27,** 963–971.
30. De Napoli, L., Montesarchio, D., Piccialli, G., et al. (1991) Solid-phase synthesis of cyclic oligodeoxyribonucleotides: intermolecular *versus* intramolecular coupling. *Gazz. Chim. Ital.* **121,** 505–508.
31. De Napoli, L., Galeone, A., Mayol, L., Messere, A., Montesarchio, D., and Picciali, G. (1995) Automated solid phase synthesis of cyclic oligonucleotides: a further improvement. *Bioorg. Med. Chem.* **3,** 1325–1329.
32. Smietana, M. and Kool, E. T. (2002) Efficient and simple solid-phase synthesis of short cyclic oligodeoxynucleotides bearing a phosphorothioate linkage. *Angew. Chem. Int. Ed. Engl.* **41,** 3704–3707.
33. Alazzouzi, E., Escaja, N., Grandas, A., and Pedroso, E. (1997) A straightforward solid-phase synthesis of cyclic oligonucleotides. *Angew. Chem. Int. Ed. Engl.* **36,** 1506–1508.
34. Mazur, S. and Jayalekshmy, P. (1979) Chemistry of polymer-bound *o*-benzyne: frequency of encounter between substituents on cross-linked polystyrenes. *J. Am. Chem. Soc.* **101,** 677–683.
35. Frieden, M., Grandas, A., and Pedroso, E. (1999) Making cyclic RNAs easily available. *Chem. Commun.* 1593, 1594.
36. Micura, R. (1999) Cyclic oligoribonucleotides (RNA) by solid-phase synthesis. *Chem. Eur. J.* **5,** 2077–2082.

37. Kaiser, E., Colescott, R. L., Bossinger, C. D., and Cook, P. I. (1970) Color test for detection of free amino groups in the solid-phase synthesis of peptides. *Anal. Biochem.* **34,** 595–598.
38. Caruthers, M. H., Barone, A. D., Beaucage, S. L., et al. (1987) Chemical synthesis of deoxyoligonucleotides by the phosphoramidite method. *Methods Enzymol.* **154,** 287–313.
39. Reese, C. B., Titmas, R. C., and Yau, L. (1978) Oximate ion promoted unblocking of oligonucleotide phosphotriester linkages *Tetrahedron Lett.* 2727–2730.
40. Brown T. and Brown D. J. S. (1990) Modern machine-aided methods of oligodeoxyribonucleotide synthesis. In *Oligonucleotides and Analogues: A Practical Approach* (Eckstein, F., ed.). IRL Press, Oxford, U.K., p. 20.
41. Adams, R. L. P., Knowler, J. T., and Leader, D. P. P. (1992) *The Biochemistry of the Nucleic Acids,* 11th ed. Chapman & Hall, London, pp. 87–108.
42. Pieles, U., Zürcher, W., and Moser, H. E., (1993) Matrix-assisted laser desorption ionization time-of-flight mass spectrometry: a powerful tool for the mass and sequence analysis of natural and modified oligonucleotides. *Nucleic Acids Res.* **21,** 3191–3196.
43. Zhu, Y. F., Chung, C. N., Taranenko, N. I., et al. (1996) The study of 2,3,4-trihydroxyacetophenone and 2,4,6-trihydroxyacetophenone as matrices for DNA detection in matrix-assisted laser desorption/ionization time-of-flight spectrometry. *Rapid Commun. Mass Spectrom.* **10,** 383–388.
44. Tang, K., Fu, D., Köter, S., Cotter, R. J., Cantor, C. R., and Köster, H. (1995) Matrix-assisted laser desorption/ionisation mass spectrometry of immobilized duplex DNA probes. *Nucleic Acids Res.* **23,** 3126–3131.
45. Patel, T. P., Chauncey, M. A., Millican, T. A., Bose, C. C., and Eaton, M. A. (1984) A rapid deprotection procedure for phosphotriester DNA synthesis. *Nucleic Acids Res.* **12,** 6953–6859.

8

Locked Nucleic Acid Synthesis

Henrik M. Pfundheller, Anders M. Sørensen, Christian Lomholt, Anders M. Johansen, Troels Koch, and Jesper Wengel

Summary

Methods and protocols for automated synthesis and purification of locked nucleic acid (LNA), a class of oligonucleotides obeying the Watson–Crick base-pairing rules but displaying unprecedented binding affinities toward complementary deoxyribonucleic acid (cDNA) and ribonucleic acid (RNA), is described. LNA and LNA–DNA chimeras containing phosphordiester or phosphorothioate linkages, or a mixture thereof, can be assembled by standard DNA synthesizers using 2-cyanoethyl DNA phosphoramidites and 2-cyanoethyl LNA phosphoramidites. Compared to the standard protocols used for DNA synthesis, slightly longer coupling time and oxidation time are needed for efficient oligomerization of LNA phosphoramidites. When the LNA has been assembled, it is removed from the solid support as the 5'-end-*O*-dimethoxytrityl (DMT) protected LNA oligomer by treatment with concentrated aqueous ammonia that also removes the phosphate and nucleobase protecting groups. The crude DMT protected LNA product can be purified using, for example, reversed-phase chromatography.

Key Words: LNA; locked nucleic acid; LNA oligonucleotide; oligonucleotide; phosphordiester LNA; phosphorothioate LNA; LNA–DNA chimera.

1. Introduction

This chapter describes solid-phase synthesis and purification of locked nucleic acid (LNA), a class of oligonucleotides obeying the Watson–Crick base-pairing rules and displaying unprecedented binding affinities toward complementary deoxyribonucleic acid (DNA) and ribonucleic acid (RNA) *(1–5)* (*see* **Note 1**). LNA containing phosphordiester or phosphorothioate linkages, or a mixture thereof, can be assembled using standard DNA synthesizers, 2-cyanoethyl DNA phosphoramidites, and 2-cyanoethyl LNA phosphoramidites (*see* **Fig. 1** and **Note 2**). Procedures are described for synthesis of fully modi-

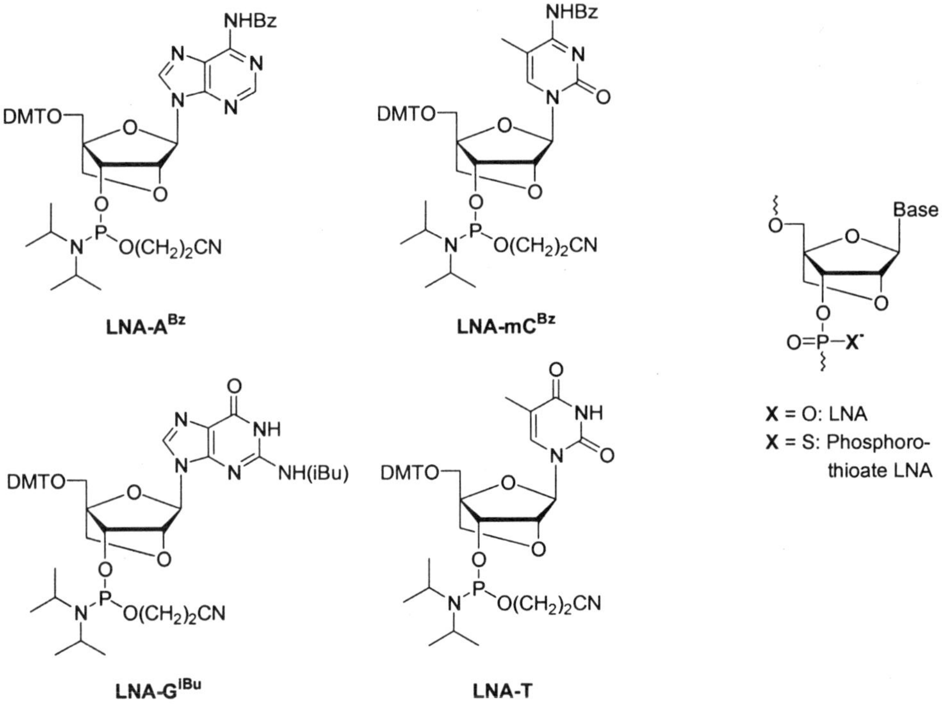

Fig. 1. Structure of LNA phosphoramidites, LNA monomers, and phosphorothioate LNA monomers (*see* **Note 6**).

fied LNA and for synthesis of LNA–DNA chimeric oligonucleotides containing phosphordiester (described in **Subheadings 3.1.–3.7.**) and phosphorothioate linkages (described in **Subheadings 3.8.–3.14.**; *see* **ref. 6** and **Notes 3 and 4**). Compared to the protocols used for DNA synthesis, slightly longer coupling time and oxidation time are needed for efficient oligomerization of LNA phosphoramidites. LNA oligonucleotides are sequentially assembled in the 3'- to 5'-direction on solid supports (synthesis columns) containing the first 5'-*O*-dimethoxytrityl (DMT) protected LNA (or DNA) nucleoside. Each coupling cycle consists of five steps: removal of the acid labile DMT group (detritylation), coupling of the phosphoramidite to the 5'-OH group of the growing chain (coupling), capping of unreacted 5'-hydroxy groups by acylation (capping), oxidation of the phosphite triester (oxidation), and additional capping. When the LNA has been assembled, it is removed from the solid support as the 5'-end DMT protected LNA oligomer by treatment with concentrated aqueous ammonia that also removes the phosphate and nucleobase protecting groups. The crude DMT protected LNA product can be purified using reversed-phase

(RP) cartridges or reversed-phase-high-performance liquid chromatography (RP-HPLC), and analyzed using strong anion-exchange (SAX) HPLC or mass spectrometry (*see* **Note 5**).

2. Materials

2.1. Synthesis of 5'-DMT Protected LNA

1. Expedite 8909 DNA synthesizer equipped with a MOSS (Perseptive Biosystems) (*see* **Note 6**).
2. Expedite 8909 reagent kit (acetonitrile [ACN] wash, deblocking solution, activator, oxidizer, cap A and B, amidite diluent) (Perseptive Biosystems). Tetrazole was used as activator but 4,5-dicyanoimidazole (DCI) (Proligo) can be used instead.
3. Protected LNA phosphoramidites: $A^{Bz}/{}^{Me}C^{Bz}/G^{iBu}/T$ (Exiqon or Proligo) (*see* **Note 7**).
4. Protected DNA phosphoramidites: $dA^{Bz}/dC^{Bz}/dG^{iBu}/T$ (ABI).
5. LNA synthesis columns: 5'-DMT-LNA-CPG (1000 Å, 30–40 µmol/g) (Proligo).
6. DNA synthesis columns: 5'-DMT-DNA-CPG (500 Å, approx 40 µmol/g) (ABI).
7. Anhydrous tetrahydrofuran (THF) (Aldrich).
8. Anhydrous ACN (Aldrich).
9. 3Å or 4Å molecular sieves.
10. Anhydrous Ar, He, or N_2 gas.
11. Synthesizer vials with caps.

2.2. Deprotection and Cleavage From the Solid Support of 5'-DMT Protected LNA

1. Concentrated aqueous ammonia (29–32%, w/w) (*see* **Note 8**).
2. 1.5-mL microcentrifuge tube.

2.3. Cartridge Purification of LNA

1. Extraction cartridges, 0.2 µmol (OASIS, Waters).
2. 0.10*M* Aqueous triethylammonium acetate (TEAA), pH 8.0.
3. 0.05*M* TEAA, pH 7.4.
4. ACN (analytical grade).
5. 10% ACN in 0.10*M* aqueous TEAA (v/v), pH 8.0.
6. 20% ACN in 0.05*M* aqueous TEAA (v/v), pH 7.4.
7. 3% Aqueous trifluoroacetic acid (TFA).
8. Milli-Q water.
9. 1.5-mL microcentrifuge tubes with caps.

2.4. RP-HPLC Purification of 5'-DMT Protected LNA

1. RP-HPLC system (e.g., Agilent 1100 series, Agilent Technologies).
2. Xterra™ RP_{18} column, 5 µm 4.6 × 150 mm (Waters).

3. 0.1 μM Membrane filter (Gelman Sciences).
4. 0.05 M Aqueous TEAA, pH 7.4.
5. ACN (analytical grade).
6. 1.5-mL Microcentrifuge tubes with caps.

2.5. Removal of the 5'-DMT Group From RP-HPLC-Purified LNA

1. 80% Aqueous acetic acid (w/w) (analytical grade).
2. Milli-Q water.
3. Materials described in **Subheading 2.4.** (except the membrane filter).

2.6. Quantification of LNA

1. Ultraviolet spectrophotometer (e.g., Lambda 40, Perkin Elmer) and quartz cuvets (e.g., Hellma).
2. Milli-Q water.

2.7. Analysis of LNA

1. PA-100 column, 13 μm, 4.0 × 250 mm (Dionex).
2. 20 mM Aqueous NaOH (analytical grade).
3. 2.0M Aqueous NaCl (analytical grade).
4. Milli-Q water.

2.8. Synthesis of 5'-DMT Protected Phosphorothioate LNA

1. Materials described in **Subheading 2.1.**
2. DCI (used as activator).
3. 3H-1,2-benzodithiol-3-one 1,1-dioxide (Beaucage reagent; thiolation reagent).

2.9. Deprotection and Cleavage From the Solid Support of 5'-DMT Protected Phosphorothioate LNA

1. Concentrated aqueous ammonia (29–32%, w/w) (*see* **Note 8**).
2. 3-mL Plastic tubes with screw caps.

2.10. RP-HPLC Purification of 5'-DMT Protected Phosphorothioate LNA

1. RP-HPLC system (e.g., Agilent 1100 series, Agilent Technologies).
2. Xterra RP$_{18}$ 5 μm column, 7.8 × 50 mm (Waters).
3. 0.1M Aqueous ammonium acetate, pH 10.0 (analytical grade).
4. ACN (HPLC grade).
5. Diluted aqueous ammonia: 0.1% (w/w) concentrated aqueous ammonia (32%, w/w) in Milli-Q water.
6. Sample vials.
7. 3-mL Microcentrifuge tubes.

2.11. Removal of the 5'-DMT Group From RP-HPLC-Purified Phosphorothioate LNA

1. 80% (w/w) aqueous acetic acid (analytical grade).
2. Milli-Q water.
3. Diethylether (technical grade).
4. TFA (analytical grade).
5. Transfer pipets.
6. Tubes for ether extraction.

2.12. Quantification of Phosphorothioate LNA

1. Materials described in **Subheading 2.6.**

2.13. Ion-Exchange HPLC Analysis of Phosphorothioate LNA

1. Anion-exchange HPLC system (e.g., Agilent 1100 series, Agilent Technologies).
2. Resource Q column, 1 mL, 13 µm, 4 × 250 mm. (Pharmacia Biotech).
3. Buffer A: 10 mM aqueous NaOH + 2.0M aqueous NaCl (analytical grade).
4. Buffer B: 10 mM aqueous NaOH (analytical grade).
5. Sample vials.

2.14. Analysis of Phosphorothioate LNA by Mass Spectrometry

1. HPLC system connected to an electrospray ionization mass spectrometer (ESI-MS) (e.g., Agilent 1100 series, Agilent Technologies) (*see* **Note 9**).
2. Resource S column, 1 mL, 13 µm, 4 × 250 mm. (Pharmacia Biotech).
3. Methanol (HPLC grade).
4. Milli-Q water.
5. Sample vials.

3. Methods

This section describes the step-by-step synthesis and purification of LNA, that is, fully modified LNA and LNA–DNA chimeric oligonucleotides (phosphordiester or phosphorothioate oligonucleotides). Also described is cleavage from the solid support, removal of protecting groups, purification, and analysis.

3.1. Synthesis of 5'-DMT Protected LNA

The synthesis scale is 0.2 µmol, but the protocol can easily be adapted to 0.05- and 1.0-µmol scales. Modifications of the default protocol are necessary (*see* **Tables 1** and **2**). For LNA phosphoramidites the coupling time has been increased from 120 s to 250 s and the oxidation time to 45 s. The synthesized LNA is cleaved from the solid support by treatment with concentrated aqueous ammonia, which also removes the phosphate and nucleobase protecting groups.

Table 1
Synthesis Cycle for LNA Phosphoramidites

LNA-cycle (0.2μmole scale)

```
/* ---------------------------------------------------------------------- */
/*      Function              Mode      Amount   Time(sec)  Description   */
/*                                      /Arg1    /Arg2                    */
/* ---------------------------------------------------------------------- */
$Deblocking
   12 /*Wsh A               */ PULSE     10       0        "Pre cycle wash"
  144 /*Index Fract. Coll.  */ NA         1       0        "Event out ON"
    0 /*Default             */ WAIT       0       1.5      "Wait"
  141 /*Trityl Mon. On/Off  */ NA         1       1        "START data collection"
   16 /*Dblk                */ PULSE     10       0        "Dblk to column"
   16 /*Dblk                */ PULSE     50      49        "Deblock"
   38 /*Diverted Wsh A      */ PULSE     40       0        "Flush system with Wsh A"
  141 /*Trityl Mon. On/Off  */ NA         0       1        "STOP data collection"
  144 /*Index Fract. Coll.  */ NA         2       0        "Event out OFF"
$Coupling
    1 /*Wsh                 */ PULSE     10       0        "Flush system with Wsh"
    2 /*Act                 */ PULSE      5       0        "Flush system with Act"
   18 /*A + Act             */ PULSE      5       0        "Monomer + Act to column"
    2 /*Act                 */ PULSE      1       0        "Couple monomer"
    2 /*Act                 */ PULSE      3      75        "Couple monomer"
    1 /*Wsh                 */s/ PULSE    7     175        "Couple monomer"
    1 /*Wsh                 */ PULSE      8       0        "Flush system with Wsh"
$Capping
   12 /*Wsh A               */ PULSE     20       0        "Flush system with Wsh A"
   13 /*Caps                */ PULSE      8       0        "Caps to column"
   12 /*Wsh A               */ PULSE      6      45        "Cap"
   12 /*Wsh A               */ PULSE     14       0        "Flush system with Wsh A"
$Oxidizing
   15 /*Ox                  */ PULSE     15       0        "Ox to column"
   12 /*Wsh A               */ PULSE     10      45        "Slow pulse to Ox"
   12 /*Wsh A               */ PULSE      4       0        "Flush system with Wsh A"
$Capping
   13 /*Caps                */ PULSE      7       0        "Caps to column"
   12 /*Wsh A               */ PULSE     30       0        "End of cycle wash"
```

Bottle/position codes:

```
   18 /*A + Act             */
   19 /*C + Act             */
   20 /*G + Act             */
   21 /*T + Act             */
```

The crude LNA is purified using either RP cartridges or RP-HPLC. The purified LNA is analyzed by anion-exchange chromatography.

1. The DNA amidites are dissolved in ACN according to the manufacturer's protocol. The bottles (dA, dC, dG, and T phosphoramidites) are installed on the synthesizer in positions 5, 6, 7, and 8, respectively, following the instructions in the manufacturer's protocol.
2. The LNA phosphoramidites are dissolved in 3 mL to give 0.07M solutions according to the following:
   a. 0.07M Amidite.
   b. 186 mg LNA-A^{Bz}.

Table 2
Synthesis Cycle for DNA Phosphoramidites

```
DNA-cycle (0.2µmole scale)
/* ------------------------------------------------------------------------ */
/*        Function              Mode       Amount     Time(sec)   Description                  */
/*                                         /Arg1      /Arg2                                    */
/* ------------------------------------------------------------------------ */
$Deblocking
   12 /*Wsh A                */ PULSE       10         0          "Pre cycle wash"
  144 /*Index Fract. Coll.   */ NA           1         0          "Event out ON"
    0 /*Default              */ WAIT         0         1.5        "Wait"
  141 /*Trityl Mon. On/Off   */ NA           1         1          "START data collection"
   16 /*Dblk                 */ PULSE       10         0          "Dblk to column"
   16 /*Dblk                 */ PULSE       50        49          "Deblock"
   38 /*Diverted Wsh A       */ PULSE       40         0          "Flush system with Wsh A"
  141 /*Trityl Mon. On/Off   */ NA           0         1          "STOP data collection"
  144 /*Index Fract. Coll.   */ NA           2         0          "Event out OFF"
$Coupling
    1 /*Wsh                  */ PULSE       10         0          "Flush system with Wsh"
    2 /*Act                  */ PULSE        5         0          "Flush system with Act"
   23 /*6 + Act              */ PULSE        5         0          "Monomer + Act to column"
    2 /*Act                  */ PULSE        1         0          "Couple monomer"
    2 /*Act                  */ PULSE        3        36          "Couple monomer"
    1 /*Wsh                  */ PULSE        7        84          "Couple monomer"
    1 /*Wsh                  */ PULSE        8         0          "Flush system with Wsh"
$Capping
   12 /*Wsh A                */ PULSE       20         0          "Flush system with Wsh A"
   13 /*Caps                 */ PULSE        8         0          "Caps to column"
   12 /*Wsh A                */ PULSE        6        45          "Cap"
   12 /*Wsh A                */ PULSE       14         0          "Flush system with Wsh A"
$Oxidizing
   15 /*Ox                   */ PULSE       15         0          "Ox to column"
   12 /*Wsh A                */ PULSE       15         0          "Flush system with Wsh A"
$Capping
   13 /*Caps                 */ PULSE        7         0          "Caps to column"
   12 /*Wsh A                */ PULSE       30         0          "End of cycle wash"

Bottle/position codes:

   23 /*6 + Act              */
   24 /*7 + Act              */
   25 /*8 + Act              */
   26 /*9 + Act              */
```

c. 162 mg LNA-T.
d. 184 mg LNA-mC^{Bz}.
e. 179 mg LNA-G^{DMF}.

Except for the LNA-mC amidite, the LNA amidites are dissolved in anhydrous ACN. For the LNA-mC amidite it is necessary to use a 15–25% (v/v) solution of THF in ACN; For example, the LNA-mC amidite is dissolved in anhydrous THF (0.75 mL), and ACN (2.25 mL) is added (*see* **Note 10**). Molecular sieves are added and the solutions are allowed to stand under inert gas for >16 h (*see* **Notes 11** and **12**). The bottles (LNA-A, LNA-mC, LNA-G, and LNA-T amidites) are installed on the synthesizer in the positions A, C, G, and T, respectively, following the instructions in the manufacturer's protocol.

3. All reagents and amidites should be primed twice on column position 1, and once on column position 2 if used, following the instructions in the manufacturer's

protocol. Finally both column positions are primed with Wsh A.
4. The appropriate column (solid support) is installed on the synthesizer and is primed with Wsh A and subsequently Gas B.
5. The synthesis protocols are modified for both DNA and LNA phosphoramidites. The synthesis cycle for LNA phosphoramidites (0.2-µmol scale; cycle A, C, G, and T) is shown in **Table 1** (*see* **Note 13**). The synthesis cycle for DNA phosphoramidites (cycle 6, 7, 8, and 9) is shown in **Table 2** (*see* **Note 14**).
   Enter the desired sequence to be synthesized using the following scheme:

Monomer	Nomenclature
LNA-A	A
LNA-mC	C
LNA-G	G
LNA-T	T
DNA-a	6
DNA-c	7
DNA-g	8
DNA-t	9

6. Select Final DMT "ON" and start the synthesis according to the manufacturer's protocol.
7. The efficiency of the different coupling steps can be followed on the Trityl monitor.

3.2. Deprotection and Cleavage From the Solid Support of 5'-DMT Protected LNA

1. The solid support is removed from the column and transferred to a 1.5-mL microcentrifuge tube.
2. 1 mL of concentrated aqueous ammonia is added and the tube is sealed tightly and heated for 4 h at 60°C (*see* **Notes 15** and **16**) *(7)*.
3. When the deprotection is complete the tube is cooled to room temperature and carefully opened.
4. To obtain the crude 5'-DMT-protected LNA, the solution is evaporated under vacuum using low vacuum first to remove ammonia (*see* **Note 17**).

3.3. Cartridge Purification of LNA

1. The crude 5'-DMT-protected LNA (*see* **Subheading 3.2.**, **step 4**) is dissolved in 1 mL 0.1*M* aqueous TEAA, pH 8.0.
2. The cartridge is flushed twice with 1 mL of ACN and twice with 1 mL of 0.1*M* aqueous TEAA, pH 8.0 (*see* **Note 18**).
3. The dissolved crude LNA is transferred to the top of the cartridge. *Important:* When a solution is applied to the cartridge, it should be allowed to completely enter the packing material before the next step is initiated. Example LNA sequence (*see* **Fig. 2**): 5'-aGccAatCcaCggCtcCaaAgtCaaAagAg (capital letters, LNA monomers; lower case letters, DNA monomers).

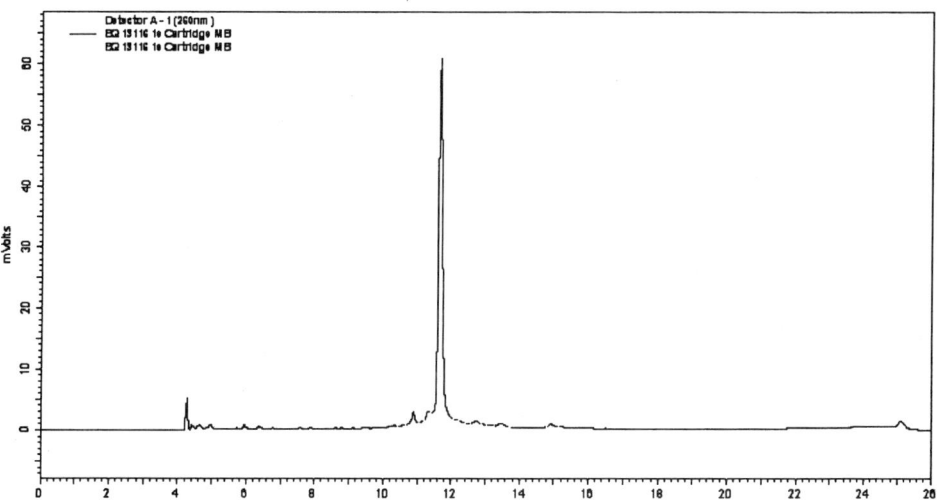

Fig. 2. Analysis of cartridge-purified (*see* **Subheading 3.3.**) LNA (80% purity) [5'-aGccAatCcaCggCtcCaaAgtCaaAagAg].

4. 1 mL 10% ACN in 0.1M aqueous TEAA, pH 8.0, is added twice to elute shorter non-DMT containing sequences.
5. 1 mL 3% aqueous TFA is added and the cartridge is left for 10 min (*see* **Note 19**). Additional 1 mL 3% aqueous TFA is then added.
6. 1 mL Milli-Q water is added.
7. The LNA product is eluted by addition of 1 mL 20% ACN in 0.05M aqueous TEAA, pH 7.4, and the product is collected in a 1.5-mL microcentrifuge tube (*see* **Note 20**).
8. The solution is evaporated under vacuum (for purity analysis, *see* **Fig. 2**).

3.4. RP-HPLC Purification of 5'-DMT-Protected LNA

1. Prepare the HPLC system by heating the column oven to 60°C, setting the detector to 300 nm (*see* **Note 21**), and equilibrating the column with a 5% solution of ACN in 0.05M aqueous TEAA, pH 7.4, for at least 1.5. h at a rate of 2 mL/min.
2. The crude 5'-DMT-protected LNA (*see* **Subheading 3.2.**, **step 4**) is dissolved in 1 mL 0.05M aqueous TEAA, pH 7.4. Example LNA sequence (*see* **Figs. 2**, **4**, and **5**): 5'-aCaaGccGgaAagCcaAtcCacGgcTccAg (capital letters, LNA monomers; lower case letters, DNA monomers).
3. The solution is filtered through a 0.2 µM membrane filter into a microcentrifuge tube, and the filter is washed with approx 0.5 mL of 0.05M aqueous TEAA, pH 7.4.
4. The mixture is loaded to the column, and the following gradient is applied at a flow rate of 2 mL/min: 5–50% ACN in 0.05M aqueous TEAA, pH 7.4 for 0–15

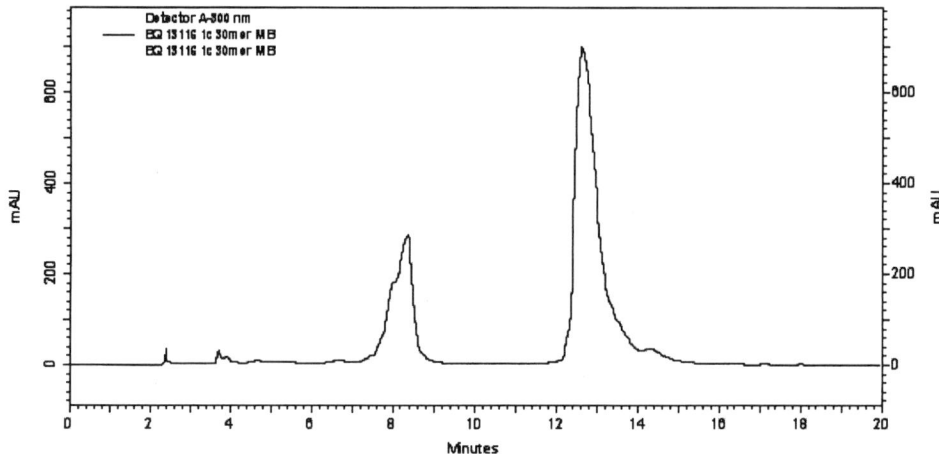

Fig. 3. RP-HPLC chromatogram of unpurified 5'-DMT LNA.

min; 50–80% ACN in 0.05 M aqueous TEAA, pH 7.4 for 15–16 min; 80–5% ACN in 0.05M aqueous TEAA, pH 7.4 for 16–20 min; 5% ACN in 0.05M aqueous TEAA, pH 7.4 for 21–30 min. The product is collected in a 1.5-mL microcentrifuge tube. The 5'-DMT protected LNA product elutes after approx 12.5 min depending on its length (see **Fig. 3**).

5. To obtain the purified 5'-DMT-protected LNA product, the eluted fraction is evaporated under vacuum.

3.5. Removal of the 5'-DMT Group From RP-HPLC-Purified LNA

1. The 5'-DMT-protected LNA (see **Subheading 3.4., step 5**) is dissolved in approx 0.5 mL 80% aqueous acetic acid and is allowed to stand for 30 min at room temperature.
2. The solution is evaporated under vacuum.
3. The HPLC system is prepared as described under **Subheading 3.4., step 1** and the residue obtained is dissolved in 1 mL 0.05M aqueous TEAA, pH 7.4.
4. The mixture is loaded to the column and the gradient described under **Subheading 3.4., step 4** is applied. The product is collected in a 1.5-mL microcentrifuge tube. The purified LNA product elutes after approx 7.5 min depending on its length (see **Fig. 4**).
5. To obtain the purified LNA product, the eluted fraction is evaporated under vacuum (for purity analysis, see **Fig 5**).

LNA Synthesis

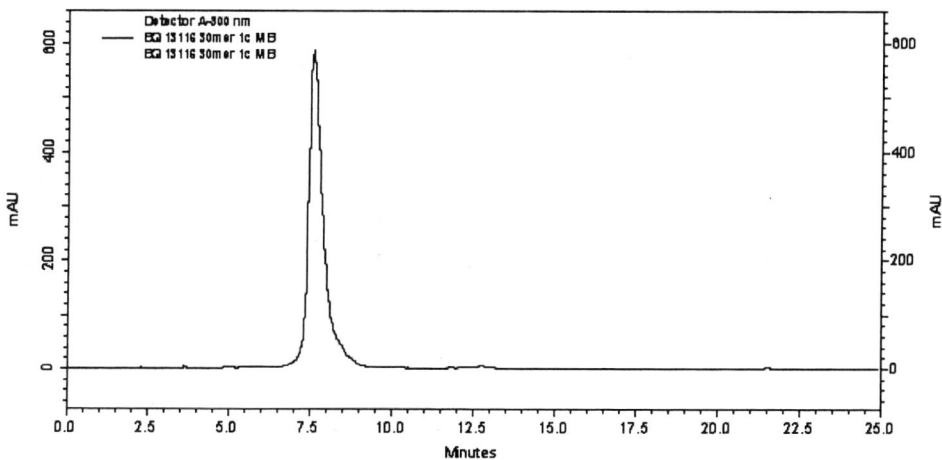

Fig. 4. RP-HPLC chromatogram of LNA after purification.

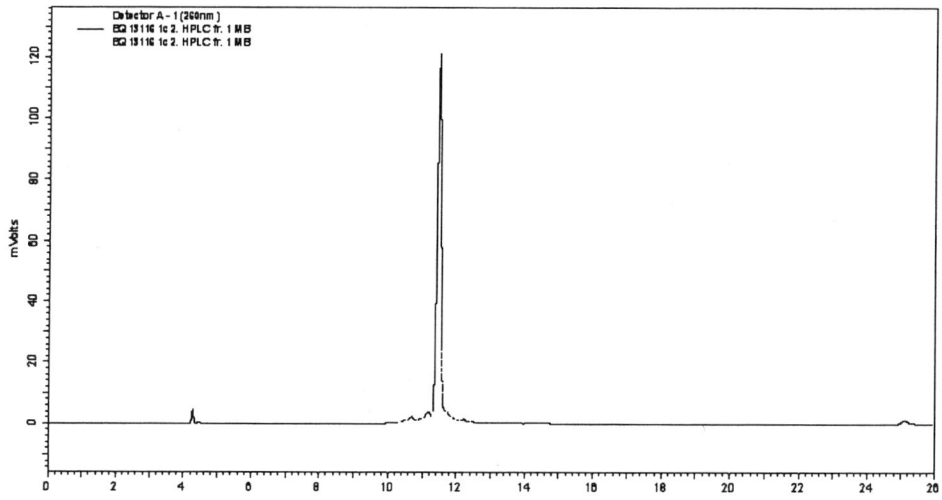

Fig. 5. Analysis of RP-HPLC-purified (*see* **Subheadings 3.4.–3.7.**) LNA (91% purity) [5'-aCaaGccGgaAagCcaAtcCacGgcTccAg].

3.6. Quantification of LNA

1. The purified LNA product obtained under either **Subheading 3.3.**, **step 8** or **Subheading 3.5.**, **step 5** is dissolved in 1 mL Milli-Q water; 10 µL is removed and diluted to 1 mL with Milli-Q water.
2. The absorbance is measured at 260 nm and the amount of LNA is calculated in A_{260} U (*see* **Note 22**).

3.7. Analysis of LNA

1. Prepare the HPLC system by setting the detector to 260 nm and equilibrate the column with a 25% solution of 2.0M aqueous NaCl in 2.0 mM aqueous NaOH for at least 1 h at a rate of 1 mL/min at 60°C.
2. 0.05–0.1 optical density (OD) of LNA (*see* **Subheading 3.6.**, **step 2**) is loaded to the column, and the following gradient is applied at a flow rate of 1 mL/min (*see* **Note 23**): 25% 2.0M aqueous NaCl in 2.0 mM aqueous NaOH for 0–3 min; 25–40% 2.0M aqueous NaCl in 2.0 mM aqueous NaOH for 3–18 min; 40–80% 2.0M aqueous NaCl in 2.0 mM aqueous NaOH for 18–21 min; 80% 2.0M aqueous NaCl in 2.0 mM aqueous NaOH for 21–22.9 min; 80–15% 2.0M aqueous NaCl in 2.0 mM aqueous NaOH for 22.9–23 min; 15–0% 2.0M aqueous NaCl in 2.0 mM aqueous NaOH for 23–26 min.
3. The purity of the LNA can be calculated after integration (*see* **Figs. 4** and **5**).
4. Mass spectrometry can be applied for verification of the molecular weight of the synthesized LNA (*see* **Subheading 3.14.**).

3.8. Synthesis of Phosphorothioate 5'-DMT-Protected LNA (see Note 24)

This section describes step-by-step the synthesis and purification of LNA (and LNA–DNA chimeric oligonucleotides) containing phosphorothioate linkages or a mixture of phosphorothioate and phosphordiester linkages. The synthesis scale is 1.0 µmol, but the protocol can easily be adapted to a 0.05- and a 1-µmol scale. Modifications of the default protocol are necessary (*see* **Tables 3** and **4**). For coupling of LNA phosphoramidites, the coupling time is increased from 100 s to 400 s. For both the LNA and DNA cycles, the oxidation time has been increased to 70 s and 50 s, respectively, and a 60-s wait period is added in the thiolation cycles. The synthesized phosphorothioate LNA oligonucleotide is deprotected and cleaved from the solid support, purified by RP-HPLC, and analyzed by anion-exchange chromatography or mass spectrometry.

1. Follow **Subheading 3.1., step 1**.
2. The LNA phosphoramidites are dissolved in anhydrous ACN in 0.1M solutions, except for the LNA-^{Me}C phosphoramidite for which it is necessary to use a 15–25% (v/v) solution of THF in ACN. Otherwise follow **Subheading 3.1., step 2**.
3. A 0.2M solution of Beaucage reagent in anhydrous ACN is prepared (*see* **Notes 25** and **26**).
4. Follow **Subheading 3.1., steps 3–4**.
5. The default synthesis protocols are modified for both LNA and DNA cycles. Both the thiolation (for incorporation of phosphorothioate LNA monomers) and standard oxidation (for incorporation of phosphordiester LNA monomers) synthesis cycles for LNA phosphoramidites are shown in **Table 3** (*see* **Note 27**). Both the thiolation (for incorporation of phosphorothioate DNA monomers) and

Table 3
Synthesis Cycles for LNA Phosphoramidites (Thiolation and Standard Oxidation)

LNA thiolation cycle (1.0μmole scale)

```
/* ------------------------------------------------------------------------ */
/*      Function              Mode    Amount   Time(sec)  Description       */
/*                                    /Arg1    /Arg2                        */
/* ------------------------------------------------------------------------ */
$Deblocking
 144 /*Index Fract. Coll.   */ NA       1        0       "Event out ON"
   0 /*Default              */ WAIT     0        1.5     "Wait"
  16 /*Dblk                 */ PULSE   20        0       "Dblk to column"
 141 /*Trityl Mon. On/Off   */ NA       1        1       "START data collection"
  16 /*Dblk                 */ PULSE   20        0       "Dblk to column"
  16 /*Dblk                 */ PULSE   40       40       "Deblock"
  38 /*Diverted Wsh A       */ PULSE   20       20       "Deblock"
  38 /*Diverted Wsh A       */ PULSE   60        0       "Flush system with Wsh A"
 141 /*Trityl Mon. On/Off   */ NA       0        1       "STOP data collection"
 144 /*Index Fract. Coll.   */ NA       2        0       "Event out OFF"
$Coupling
   1 /*Wsh                  */ PULSE    5        0       "Flush system with Wsh"
   2 /*Act                  */ PULSE    5        0       "Flush system with Act"
  41 /*Gas B                */ PULSE    1        5       "Gas B"
  23 /*A + Act              */ PULSE   10        0       "Monomer + Act to column"
   2 /*Act                  */ PULSE    8        0       "Chase with Act"
   0 /*Default              */ WAIT     0       30       "Wait"
   1 /*Wsh                  */ PULSE   20      400       "Slow pulse to couple"
   1 /*Wsh                  */ PULSE   20        0       "Flush system with Wsh"
$Oxidizing
  17 /*Aux                  */ PULSE   30        0       "SOx to column"
   0 /*Default              */ WAIT     0       60       "Wait"
  12 /*Wsh A                */ PULSE   20      120       "Slow pulse to thioate"
$Capping
  13 /*Caps                 */ PULSE   30        0       "Caps to column"
  12 /*Wsh A                */ PULSE   10       30       "Slow pulse to cap"
  12 /*Wsh A                */ PULSE   20       15       "Slow pulse to cap"
  12 /*Wsh A                */ PULSE   60        0       "End of cycle wash"
```

LNA oxidation cycle (1.0μmole scale)

```
/* ------------------------------------------------------------------------ */
/*      Function              Mode    Amount   Time(sec)  Description       */
/*                                    /Arg1    /Arg2                        */
/* ------------------------------------------------------------------------ */
$Deblocking
  As for LNA thiolation cycle
$Coupling
  As for LNA thiolation cycle
$Capping
  13 /*Caps                 */ PULSE   30        0       "Caps to column"
  12 /*Wsh A                */ PULSE   10       30       "Slow pulse to cap"
  12 /*Wsh A                */ PULSE   20       15       "Slow pulse to cap"
$Oxidizing
  15 /*Ox                   */ PULSE   30        0       "Ox to column"
   0 /*Default              */ WAIT     0       20       "Wait"
  12 /*Wsh A                */ PULSE   20       50       "Chase with Wsh A"
  12 /*Wsh A                */ PULSE  120        0       "End of cycle wash"
```

Bottle/position codes:

```
  23 /*A + Act              */
  24 /*C + Act              */
  25 /*G + Act              */
  26 /*T + Act              */
```

Table 4
Synthesis Cycles for DNA Phosphoramidites (Thiolation and Standard Oxidation)

DNA thiolation cycle (1.0μmole scale)

```
/* ---------------------------------------------------------------------- */
/*       Function              Mode    Amount   Time(sec)   Description   */
/*                                     /Arg1    /Arg2                     */
/* ---------------------------------------------------------------------- */
$Deblocking
  144 /*Index Fract. Coll.    */ NA       1       0       "Event out ON"
    0 /*Default               */ WAIT     0       1.5     "Wait"
   16 /*Dblk                  */ PULSE   20       0       "Dblk to column"
  141 /*Trityl Mon. On/Off    */ NA       1       1       "START data collection"
   16 /*Dblk                  */ PULSE   20       0       "Dblk to column"
   16 /*Dblk                  */ PULSE   40      40       "Deblock"
   38 /*Diverted Wsh A        */ PULSE   20      20       "Deblock"
   38 /*Diverted Wsh A        */ PULSE   60       0       "Flush system with Wsh A"
  141 /*Trityl Mon. On/Off    */ NA       0       1       "STOP data collection"
  144 /*Index Fract. Coll.    */ NA       2       0       "Event out OFF"
$Coupling
    1 /*Wsh                   */ PULSE    5       0       "Flush system with Wsh"
    2 /*Act                   */ PULSE    5       0       "Flush system with Act"
   41 /*Gas B                 */ PULSE    1       5       "Gas B"
   18 /*6 + Act               */ PULSE    9       0       "Monomer + Act to column"
    2 /*Act                   */ PULSE    4       0       "Chase with Act"
    0 /*Default               */ WAIT     0      30       "Wait"
    1 /*Wsh                   */ PULSE    4       0       "Chase with Wsh"
    1 /*Wsh                   */ PULSE   20     100       "Slow pulse to couple"
    1 /*Wsh                   */ PULSE   20       0       "Flush system with Wsh"
$Oxidizing
   17 /*Aux                   */ PULSE   30       0       "SOx to column"
    0 /*Default               */ WAIT     0      60       "Wait"
   12 /*Wsh A                 */ PULSE   20     120       "Slow pulse to thioate"
$Capping
   13 /*Caps                  */ PULSE   30       0       "Caps to column"
   12 /*Wsh A                 */ PULSE   10      30       "Slow pulse to cap"
   12 /*Wsh A                 */ PULSE   20      15       "Slow pulse to cap"
   12 /*Wsh A                 */ PULSE   60       0       "End of cycle wash"
```

DNA oxidation cycle (1.0μmole scale)

```
/* ---------------------------------------------------------------------- */
/*       Function              Mode    Amount   Time(sec)   Description   */
/*                                     /Arg1    /Arg2                     */
/* ---------------------------------------------------------------------- */
$Deblocking
  As for DNA thiolation cycle
$Coupling
  As for DNA thiolation cycle
$Capping
   13 /*Caps                  */ PULSE   30       0       "Caps to column"
   12 /*Wsh A                 */ PULSE   10      30       "Slow pulse to cap"
   12 /*Wsh A                 */ PULSE   20      15       "Slow pulse to cap"
$Oxidizing
   15 /*Ox                    */ PULSE   20       0       "Ox to column"
   12 /*Wsh A                 */ PULSE   20      50       "Chase with Wsh A"
   12 /*Wsh A                 */ PULSE  120       0       "End of cycle wash"
```

Bottle/position codes:

```
   18 /*6 + Act               */
   19 /*7 + Act               */
   20 /*8 + Act               */
   21 /*9 + Act               */
```

LNA Synthesis

standard oxidation (for incorporation of phosphordiester DNA monomers) synthesis cycles for DNA phosphoramidites are shown in **Table 4** (*see* **Note 28**).
6. Follow **Subheading 3.1.6.–3.1.7.**

3.9. Deprotection and Cleavage From the Solid Support of 5'-DMT-Protected Phosphorothioate LNA

1. The solid support is removed from the column and transferred to a 3-mL plastic tube with screw cap.
2. Follow **Subheading 3.2.**, steps 2–4.

3.10. RP-HPLC Purification of 5'-DMT-Protected Phosphorothioate LNA

1. Prepare the RP-HPLC system by setting the detector to 260 nm, and equilibrate the column with a 5% solution (v/v) of ACN in $0.1M$ aqueous ammonium acetate, pH 10, at a rate of 5 mL/min at ambient temperature.
2. The crude phosphorothioate LNA (*see* **Subheading 3.9.**, **step 2**) is dissolved in 300 µL H_2O made basic by addition of diluted aqueous ammonia, pH 8.0. For example, phosphorothioate LNA sequence (*see* **Fig. 6**): 5'-TCCAaagcaccaAACA-3' (fully thiolated) (capital letters: LNA monomers; lower case letters: DNA monomers).
3. The solution is heated to 65°C for 3 min in a tightly closed tube and then cooled to room temperature. The solution is centrifuged and transferred to an appropriate sample vial.
4. The solution is loaded to the column and the following gradient is applied: 5% ACN in $0.1M$ aqueous ammonium acetate, pH 10.0 for 0–5 min; 5–35% ACN in $0.1M$ aqueous ammonium acetate, pH 10.0 for 5–12 min; 35–100% ACN in $0.1M$ aqueous ammonium acetate, pH 10.0 for 12–16 min; 100% ACN for 16–19 min; 100–5% ACN in $0.1M$ aqueous ammonium acetate, pH 10.0 for 19–21 min. The product is collected in suitable tubes. The product elutes after approx 10 min depending on its length and design (*see* **Fig. 6**).
5. To obtain the purified phosphorotioate 5'-DMT–LNA, the eluted fractions are evaporated under vacuum or lyophilized.

3.11. Removal of the 5'-DMT-Group From RP-HPLC-Purified Phosphorothioate LNA

1. The purified 5'-DMT-protected phosphorothioate LNA (*see* **Subheading 3.10.5.**) is dissolved in approx 0.5 mL 80% aqueous acetic acid and is allowed to stand for 30 min at room temperature.
2. The acidic solution is evaporated under vacuum or lyophilized.
3. The residue is dissolved in 1 mL Milli-Q water and extracted three times with 3 mL ether (*see* **Note 29**).
4. To obtain the phosphorothiate LNA, the aqueous phase is evaporated to dryness under vacuum or lyophilized.

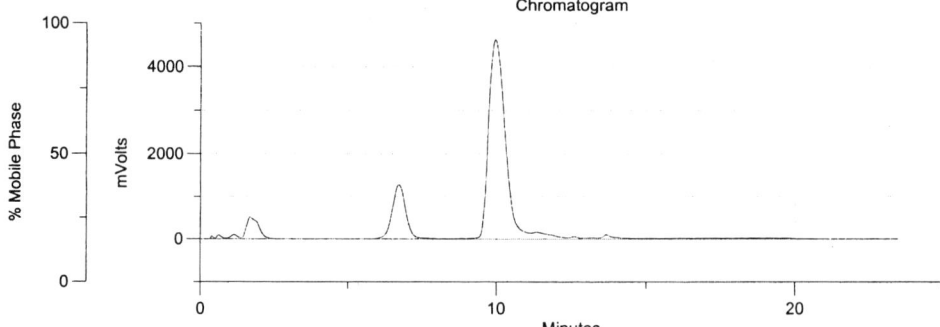

Fig. 6. RP-HPLC chromatogram of unpurified 5'-DMT phosphorothioate LNA [5'-TCCAaagcaccaAACA- 3'; fully thiolated].

3.12. Quantification of Phosphorothioate LNA

1. The completely deprotected phosphorothioate LNA oligonucleotide (*see* **Subheading 3.11.**, **step 4**) is dissolved in 200 µL Milli-Q water and heated to 65°C for 5 min. From the stock solution, 2.0 µL is removed and diluted to 1 mL with Milli-Q water.
2. Follow **Subheading 3.6.**, **step 2**.

3.13. Ion-Exchange HPLC Analysis of Phosphorothioate LNA

1. Prepare the HPLC system by setting the detector to 260 nm and equilibrate the column with 20% of buffer A and 80% of buffer B at a flow rate of 0.5 mL/min at ambient temperature.
2. 5.0 µL of the stock solution of phosphorothioate LNA (*see* **Subheading 3.12.**, **step 1**) is diluted to 35 µL with Milli-Q water. Example phosphorothioate LNA sequence (*see* **Fig. 7**): 5'-TCCAaagcaccaAACA-3' (fully thiolated) (capital letters: LNA monomers; lower case letters: DNA monomers).
3. Approximately 10 µL of the solution is loaded to the column and the following linear gradient is applied: 20–100% buffer A for 0–22 min.
4. The purity of the phosphorothioate LNA can be quantified by integration (*see* **Fig. 7**).

3.14. Analysis of Phosphorothioate LNA by Mass Spectrometry (see Note 9)

1. The liquid chromatography mass spectrometry instrumentation must be equipped with a Resource S 1-mL cation-exchange column. The cation-exchange column has to be charged with ammonium ions ($1M$ ammonium acetate) prior to use. The column has to be "recharged" regularly.
2. Equilibrate the column using 90% MeOH in Milli-Q water.
3. Approximately 2.0 µL of the diluted stock solution of phosphorothioate LNA (*see* **Subheading 3.12.**, **step 1**) is loaded to the column.

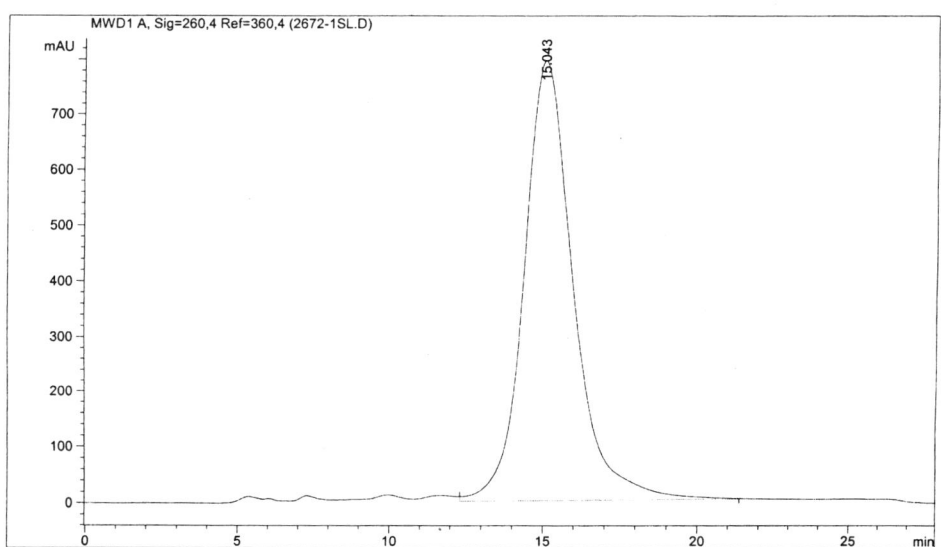

Fig. 7. Ion-exchange HPLC chromatogram of purified phosphorothioate LNA [5'-TCCAaagcaccaAACA-3'; fully thiolated].

4. Elute with 90% MeOH in Milli-Q water at 0.27 mL/min.
5. The mass of the phosphorothioate LNA can be found by deconvolution of the recorded mass data.

4. Notes

1. LNA is defined as an oligonucleotide containing one or more LNA monomers (i.e., 2'-O,4'-C-methylene-β-D-ribofuranosyl nucleotide monomers).
2. LNA phosphoramidites and solid supports for LNA synthesis are commercially available (Exiqon or Proligo).
3. The latter has shown superior properties for antisense applications.
4. LNA–RNA chimeric oligonucleotides have also been synthesized (4).
5. For relatively short LNAs (up to 20 nucleotides long) requiring a purity of no more than 75–90%, the 5'-end DMT group can be removed at the end of the synthesis and the LNA product subsequently isolated by standard ethanol precipitation or gel filtration.
6. With slight modifications according to manufacturers' protocols other DNA synthesizers can be used, for example, an ABI3900 (www.exiqon.com).
7. LNA-G^{DMF} can be purchased from Exiqon.
8. It should be stored in the refrigerator.
9. Analysis by matrix-assisted laser desorption/ionization mass spectroscopy is another useful approach for verification of the molecular weight of LNA, including phosphorothioate LNA.

10. Instead of anhydrous THF, anhydrous dichloromethane can be used; we recommend to use THF as it evaporates less easily than dichloromethane.
11. Addition of molecular sieves to the LNA phosphoramidite solutions is recommended but is not essential.
12. The amidite should be dissolved in a dry synthesizer vial (dried in an oven at 100°C for at least 3 d and then cooled under inert atmosphere).
13. The synthesis cycle has been written for position A. If positions C, G, or T are used, the third line in the coupling step has to be changed.
14. The synthesis cycle has been written for position 6. If positions 7, 8, or 9 are used, the third line in the coupling step has to be changed.
15. If the DNA-g phosphoramidite has dimethyl formamide (DMF) as protecting group, the deprotection time can be reduced to 1 h.
16. It is recommended not to use methylamine for deprotection of LNA containing mC as this has been reported to result in introduction of N^4-methyl modifications for other oligonucleotides (*see* **ref. 7**).
17. A fast introduction of high vacuum may cause bumping of the liquid resulting in loss of material.
18. A needle can be added to the bottom of the cartridge to increase the elution rate.
19. This treatment removes the DMT group at the 5'-end of the LNA.
20. It is recommended to keep all eluted solutions until elution of the LNA product has been verified.
21. The detector will be saturated at lower wavelengths.
22. The extinction coefficient, ε_{260}, is calculated using $\varepsilon_{DNA} = \varepsilon_{LNA}$.
23. This gradient is used for LNA up to 30 nucleotides long.
24. Phosphorothioate LNA is defined herein as an oligonucleotide (e.g., a fully modified LNA or a chimeric LNA–DNA oligonucleotide) containing at least one phosphorothiate linkage, either connected to a DNA or an LNA monomer.
25. Beaucage reagent is stabile in solution for a maxium of 2 d. Discard the solution if precipitation is observed.
26. Use silanized bottles according to the manufacturer's protocol (e.g., Glen Research), and avoid use of ACN dried over sieves as basic conditions promote decomposition of Beaucage reagent.
27. The synthesis cycles have been written for position A. If positions C, G, or T are used, the fourth line in the coupling step has to be changed.
28. The synthesis cycles have been written for position 6. If positions 7, 8, or 9 are used, the fourth line in the coupling step has to be changed.
29. Add 5–10 drops of TFA to 1.5 mL of each ether phase. Red coloration shows presence of the DMT cation. Continue extractions until no coloration is seen.

Acknowledgments

Mette Bjørn, Marianne B. Mogensen, and Emine Bilgin are thanked for technical assistance.

References

1. Koshkin, A. A., Singh, S. K., Nielsen, P., et al. (1998) LNA (locked nucleic acids): synthesis of the adenine, cytosine, guanine, 5-methylcytosine, thymine and uracil bicyclonucleoside monomers, oligomerisation, and unprecedented nucleic acid recognition. *Tetrahedron* **54**, 3607–3630.
2. Obika, S., Nanbu, D., Hari, Y., et al. (1998). Stability and structural features of the duplexes containing nucleoside analogues with a fixed *N*-type conformation, 2'-*O*,4'-*C*-methyleneribonucleosides. *Tetrahedron Lett.* **39**, 5401–5404.
3. Kumar, R., Singh, S. K., Koshkin, A. A., Rajwanshi, V. K., Meldgaard, M., and Wengel, J. (1998) The first analogues of LNA (locked nucleic acids): phosphorothioate-LNA and 2'-thio-LNA. *Bioorg. Med. Chem. Lett.* **8**, 2219–2222.
4. Singh, S. K. and Wengel, J. (1998) Universality of LNA-mediated high-affinity nucleic acid recognition. *Chem. Commun.* 1247, 1248.
5. Petersen, M. and Wengel, J. (2003) LNA: a versatile tool for genomics and therapuetics. *Trends Biotechnol.* **21**, 74–81.
6. Fluiter, K., ten Asbroek, A. L. M. A., de Wissel, M. B., et al. (2003) In vivo tumor growth inhibition and biodistribution studies of locked nucleic acid (LNA) antisense oligonucleotides. *Nucleic Acids Res.* **31**, 953–962.
7. Reddy, M. P., Hanna, N. B., and Farooqui, F. (1994) Fast cleavage and deprotection of oligonucleotides. *Tetrahedron Lett.* **35**, 4311–4314.

9

Synthesis of DNA Mimics Representing HypNA–pPNA Hetero-Oligomers

Vladimir A. Efimov and Oksana G. Chakhmakhcheva

Summary

The methods for the synthesis and purification of negatively charged peptide nucleic acid (PNA)-relative deoxyribonucleic acid (DNA) mimics containing alternating residues of phosphono peptide nucleic acid (pPNA) monomers and PNA-like monomers on the base of *trans*-4-hydroxy-L-proline are described. Examples of the chimeric oligomers hybridization with complementary DNA and ribonucleic acid fragments are demonstrated.

Key Words: DNA mimics; peptide nucleic acids; phosphono-PNAs; *trans*-4-hydroxy-L-proline; HypNA–pPNA hetero-oligomers; hybridization properties; translation inhibitors.

1. Introduction

In recent years, a large number of nucleic acid analogs and mimics have been synthesized, including such successful examples as peptide nucleic acids (PNAs; for recent reviews *see* **ref. *1***) and morpholino phosphorodiamidate oligonucleotide analogs *(2,3)*. Among these compounds, deoxyribonucleic acid (DNA) mimics representing negatively charged analogs of PNAs, in which monomer units are connected with phosphono–ester bonds, were designed *(4–6)*. Thus, the synthesis of phosphono–PNA oligomers (pPNAs) with *N*-(2-hydroxyethyl)-phosphono glycine, or *N*-(2-aminoethyl)-phosphono glycine backbone were developed using a solid-phase technique, which is similar to the phosphotriester oligonucleotide synthesis with the *O*-nucleophilic intramolecular catalysis *(4)*. Later, a set of chimeras composed of pPNA and PNA monomers (PNA–pPNAs) was synthesized *(7)*, as well as *trans*-4-hydroxy-L-proline (HypNA)–pPNA hetero-oligomers consisting of pPNA monomers and PNA-like monomers (**Fig. 1**) *(8)*. A HypNA monomer represents a chiral PNA

From: *Methods in Molecular Biology, vol. 288: Oligonucleotide Synthesis: Methods and Applications*
Edited by: P. Herdewijn © Humana Press Inc., Totowa, NJ

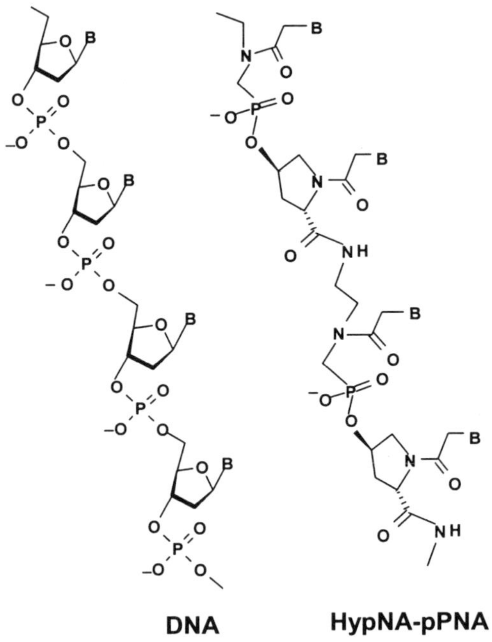

Fig. 1. General structures of DNA and HypNA–pPNA mimic.

analog having the constrained conformation with β-C atom of a hydroxyethyl unit and α-C atom of a glycyl unit of the backbone bridged by the methylene group. The evaluation of physicochemical and biological properties of these types of DNA mimics revealed that they have excellent solubility in water and are fully stable to the action of nucleases and proteases. The thermal stability of complexes formed by pPNA and PNA–pPNA mimics with the complementary DNA (or ribonucleic acid [RNA]) targets was lower than the stability of corresponding complexes formed by classic PNAs *(7)*. At the same time, HypNA–pPNA mimics composed from alternating pPNA and HypNA residues demonstrated the strong binding to complementary DNA and RNA strands with the stability of complexes very close to that of PNA–DNA/RNA complexes *(8)* (**Fig. 2**). Similar to other PNA-related mimics, homopyrimidine sequences of these chimeras formed triple helices with complementary templates, whereas oligomers with mixed sequences gave duplexes. The assays based on the hybridization technique revealed a high potential of these mimics as biomolecular probes for the solution and solid-phase nucleic acids analysis *(9)*. It was shown that 16–18-mer HypNA–pPNA probes can effectively discriminate between single base mismatches in the target sequence and that they have similar detection sensitivity as natural oligonucleotides. Thus, the intro-

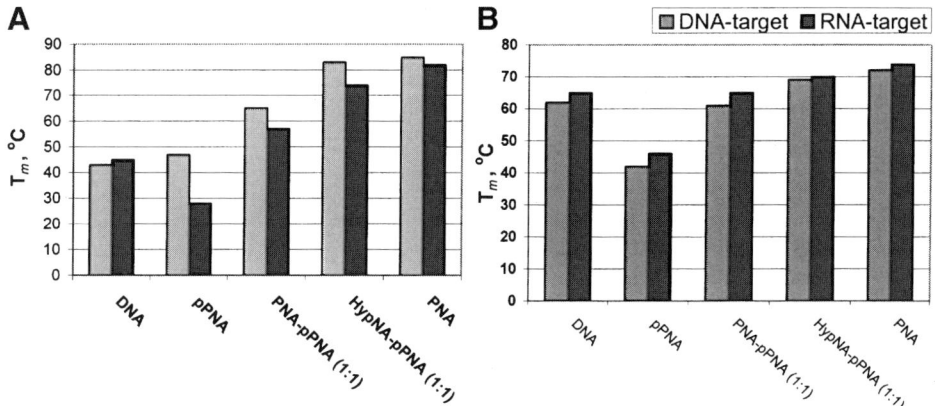

Fig. 2. Comparison of melting temperatures of complexes formed by various PNA-related 16-mer oligomers with complementary DNA, or RNA, targets: (**A**) Homo-Thy sequence; (**B**) a mixed nucleobase sequence (ctgcaaaggacaccat).

duction of one mismatch in the center of a sequence gives a drop in the melting temperature of 17–23°C, whereas oligomers with two separated mismatches are not able to form stable complexes with the targets. No difference in background was observed between mimic probes and DNA probes. At the same time, the shelf-life of the mimics is considerably longer than natural oligonucleotides. Excellent hybridization properties of HypNA–pPNA hetero-oligomers stimulated the investigation on their effectiveness as molecular biology tools. One of the examples is their use for messenger RNA isolation from cells *(10)*. Also, it was shown that HypNA–pPNA chimeric oligomers are able to penetrate into the living cells and distribute in the cytoplasm *(11)*. First experiments on the application of these mimics as in vivo translation inhibitors have shown that 18-mer HypNA–pPNA oligonucleotide analogs can bind to the target genes without side effects, which means that they can be used as specific and effective antisense agents *(12)*. This chapter describes the methods used for the synthesis and purification of HypNA–pPNA oligomers as well as some illustrations of their hybridization properties.

2. Materials

1. Organic solvents (acetonitrile [ACN], pyridine, triethylamine, dichloromethane [DCM], methanol, dioxane) high-pressure (performance) liquid chromatography (HPLC) grade.
2. 1,8-diazabicyclo[5.4.0]undec-7-ene (DBU).
3. Ethylenediamine.
4. Diphenylphosphite.

5. 1,3-dicyclohexylcarbodiimide (DCC).
6. 30% formaldehyde.
7. 1-(2,4,6-triisopropylbenzenesulfonyl)-3-nitro-1,2,4-triazole (TPSNT).
8. 2,4,6-triisopropylbenzenesulfonyl chloride (TPSCl).
9. N,N-diisopropylethylamine (DIEA).
10. N,N-dimethylformamide (DMF).
11. Triphenyl phosphine.
12. N,N-dimethylacetamide.
13. Benzoyl chloride.
14. Isobutyric anhydride.
15. L-4-hydroxyproline methyl ester hydrochloride.
16. 4-monomethoxytrityl chloride (MMTrCl).
17. 4,4'-dimethoxytrityl chloride (DMTrCl).
18. Thymine-1-acetic acid.
19. Cytosine.
20. Adenosine.
21. Guanosine.
22. Picric acid.
23. Thiophenol.
24. Merck silica gel 60.
25. Aqueous concentrated ammonia.
26. Trichloroacetic acid (TCA).
27. 1M Triethylammonium acetate (TEAA).
28. 1M Triethylammonium bicarbonate (TEAB).
29. Universal support 1000 (Glen Research) or standard T, C, A, or G-modified controlled pore glass (CPG) support.
30. Dowex-50 (PyH$^+$).
31. 1-ethyl-2-[3-(1-ethylnaphtho[1,2-d]thiazolin-2-ylidene)-2-methylpropenyl]-naphtho[1,2-d]thiazolium bromide (Stains All).
32. Polyacrylamide slab gel electrophoresis equipment.
33. 15% denaturing and native polyacrylamide slab gels.
34. Thin-layer chromatography (TLC) Merck Silica Gel 60 F_{254} plates.
35. TLC solvents: $CHCl_3/CH_3OH$ (9:1, v/v) (A) or $CHCl_3/CH_3OH$/triethylamine (8.4: 1.5: 0.1, v/v/v) (B).
36. Silica gel column chromatography equipment.
37. Automated DNA synthesizer.
38. Fast-protein liquid chromatography (FPLC) System (Pharmacia).
39. Mono-Q and Pep-RPC FPLC columns (Pharmacia).
40. Apparatus for vertical gel slab electrophoresis.
41. Ultraviolet (UV)/VIS spectrophotometer equipped with a thermo-cell holder and a Peltier temperature control accessory.

3. Methods

The following methods outline steps in the synthesis of four pPNA monomer units, four HypNA monomers, and the general scheme for the synthesis of

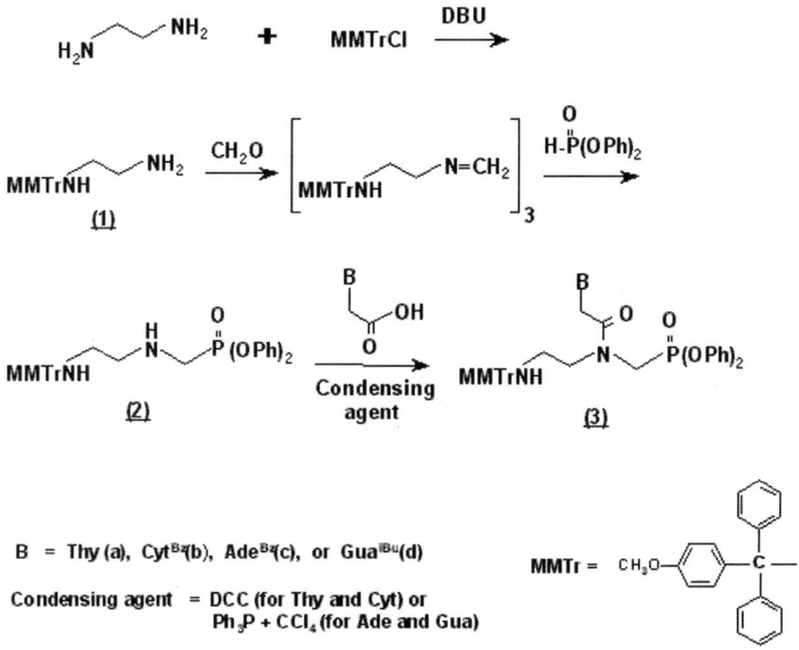

Synthesis of DNA Mimics

Scheme 1. Synthesis of a pPNA monomer unit for the construction of HypNA–pPNA dimers.

HypNA–pPNA dimers as building units for the construction of corresponding oligomers. Also, the conditions for the solid-phase synthesis of HypNA–pPNA oligomers and their purification procedures as well as characterization in hybridization assays are presented.

3.1. General Methods for Preparation of pPNA Monomers

The synthesis of pPNA monomers (*see* **Scheme 1**) was accomplished via a common intermediate representing N-[2-(4-monomethoxytrityl)-aminoethyl]-aminomethyl diphenyl-phosphonate (**2**), which can be modified with any properly protected nucleobase-acetic acid. Thymine-N^1-acetic acid can be purchased from Aldrich; N^4-benzoylcytosine-N^1-acetic acid (*see* **Note 1**), N^4-benzoyladenine-N^9-acetic acid (*see* **Note 2**), and N^2-isobutyrylguanine-N^9-acetic acid (*see* **Note 3**) can be obtained according to the procedures described in **Subheading 4**.

3.1.1. N-[2-(4-Monomethoxytrityl)-Aminoethyl]-Aminomethyl Diphenylphosphonate

1. Ethylenediamine (3.4 mL, 50 mmol) was mixed with 3.1 mL, 20 mmol DBU in 10 mL DCM. The solution was cooled to –50°C, and a solution of 3.1 g, 10 mmol

MMTrCl in 40 mL DCM was added. After 30–40 min stirring at room temperature, 100 mL H_2O was added to the reaction mixture, and the organic layer was separated and washed with 100 mL 5% $NaHCO_3$. The organic fraction was evaporated under reduced pressure to dryness.

2. The residue was dissolved in 100 mL DCM and applied to a column filled with silica gel (150 mL). The column was eluted with a gradient of 0–5% methanol in DCM containing 3 L 0.1% triethylamine. Control by TLC in system A. Fractions containing 9.5 mmol N-(monomethoxytrityl)-ethylenediamine (compound **1**) were combined and concentrated to a gum, which was dissolved in 30 mL ACN.

3. The solution was treated with 1.2 mL 30% formaldehyde for 30 min. The reaction mixture was evaporated to dryness, the residue was coevaporated with 100 mL ACN. The residue was dissolved in 30 mL benzene, and 2.5 mL 11 mmol diphenylphosphite was added. The mixture was heated at 70°C for 15 min and evaporated to an oil-like residue. Yield of the crude intermediate (**2**) was about 95% (R_f = 0.5 in system B; ^{31}P-nuclear magnetic resonance (^{31}P-NMR) δ = 20.5).

3.1.2. Synthesis of Pyrimidine Containing pPNA Diphenylphosphonates (3a and 3b)

1. Thymine-, or N^4-benzoylcytosine-N^1-acetic acid, (13 mmol) was mixed with a solution of approx 10 mmol crude intermediate (**2**) in 40 mL ACN–pyridine (1:1 v/v). Then 13 mmol DCC was added, and the reaction mixture was shaken for 1–2 h. The reaction was terminated by the addition of 5 mL H_2O. After 1 h, the precipitate formed was removed by filtration, and the filtrate was evaporated to a gum.

2. The gum was dissolved in 200 mL DCL, and the desired diphenylester of pPNA (**3**) was isolated by chromatography on a silica gel column (300 mL) using a gradient of methanol (0–5%) in CH_2Cl_2 containing 5 L 1% triethylamine. Fractions containing the desired compound were combined, and the solvent was evaporated in a vacuum. Yield 85–90%. For Thy-derivative (**3a**): R_f = 0.6 (solvent A); ^{31}P-NMR δ = 15.4; m/z (ES) = 806.4 (M + H)$^+$ ($C_{43}H_{45}N_5O_9P_1$). For Cyt-derivative (**3b**): R_f (solvent A) = 0.69, ^{31}P-NMR δ = 15.3; m/z (ES) = 895.7 (M + H)$^+$ ($C_{49}H_{48}N_6O_9P_1$).

3.1.3. Synthesis of Purine Containing pPNA Diphenyl Phosphonates (3c and 3d)

It was accomplished according to the following general procedure.

1. N^2-isobutyrylguanine-, or N^4-benzoyladenine-N^9-acetic acid (15 mmol) was mixed with a 10-mmol solution of the intermediate (**2**) in 100 mL CH_3CN–CCl_4 (9:1, v/v) containing 50 mmol DIEA. Triphenyl phosphine (30 mmol) was added, and the mixture was shaken for 30 min at room temperature. The reaction was terminated by the addition of 5 mL methanol. Then 100 mL 0.5M TEAB was added, and the mixture was extracted with CH_2Cl_2 (2 × 100 mL).

2. The combined organic fractions were evaporated to a gum containing the desired diphenylphosphonate (**3**), which was isolated by chromatography on a silica gel column (300 mL) using a gradient of methanol (0–5%) in CH_2Cl_2 containing 1% triethylamine (5 l) (yield 65–70%). For Ade derivative (**Scheme 1, 3c**): R_f (solvent A) = 0.65, ^{31}P-NMR δ = 15.3; *m/z* (ES) = 919.8 (M + H)$^+$ ($C_{50}H_{48}N_8O_8P_1$). For Gua derivative (**3d**): R_f (solvent A) = 0.55, ^{31}P-NMR δ = 15.3; *m/z* (ES) = 901.8 (M + H)$^+$ ($C_{47}H_{46}N_8O_9P_1$).

3.2. Synthesis of HypNA Monomers

The monomers of type (**7**): 4-*O*-(4,4'-dimethoxytrityl)-*N*-(thymine-1-ylacetyl)-L-hydroxyproline (**7a**), 4-*O*-(4,4'-Dimethoxytrityl)-*N*-(N^4-benzoylcytosine-9-ylacetyl)-L-hydroxyproline (**7b**), 4-*O*-(4,4'-dimethoxytrityl)-*N*-(N^6-benzoyladenine-9-ylacetyl)-L-hydroxyproline (**7c**), and 4-*O*-(4,4'-dimethoxytrityl)-*N*-(N^2-isobutyrylguanine-9-ylacetyl)-L-hydroxyproline (**7d**) were synthesized in three steps according to the general procedure (*see* **Scheme 2**) starting from 4-hydroxy-L-proline methyl ester (**4**). The first step was the addition of the corresponding protected nucleobase-acetic acid to obtain compound (**5**). Then the DMTr-protective group was introduced on the 4-OH function (compound **6**), and the last synthetic step included the removal of the methyl protective group from the carboxyl function.

1. 4-hydroxyproline methyl ester hydrochloride (**4**) (10 mmol) in 40 mL of pyridine–ACN mixture (1:1) containing 1.5 mL of triethylamine was mixed with the corresponding thymine-1-acetic acid or other *N*-protected nucleobase-acetic acid (13 mmol), and 14 mmol DCC was added to the mixture. After stirring for 3 h at room temperature, the reaction was terminated by the addition of 3 mL H_2O and allowed to stand overnight. The precipitate of 1,3-dicyclohexylurea was removed by filtration, and the filtrate was evaporated in a vacuum. The residue containing compound (**5**) was dried by coevaporation with pyridine (2 × 30 mL) and dissolved in 40 mL pyridine.
2. Then 15 mmol DMTrCl was added to the pyridine solution, and the reaction mixture was incubated at 50°C for 1 h, cooled to room temperature, and treated with 100 mL 5% $NaHCO_3$. The product (**6**) was extracted from the mixture with DCM (2 × 100 mL), and the combined organic fractions were dried over Na_2SO_4. The solvent was evaporated in a vacuum.
3. The residue containing the intermediate (**6**) was dissolved in 50 mL methanol, and 2*M* KOH in 15 mL methanol–H_2O (1:1 v/v) was added (for alternative procedure, *see* **Note 4**). After 30 min, the alkali was neutralized by the addition of some amount of Dowex-50 (PyH$^+$). The resin was filtered off and washed with 50% aqueous pyridine (2 × 50 mL). Triethylamine (2.3 mL) was added to the combine filtrate. The solvent was evaporated in a vacuum, and the resulting oil was dried and coevaporated with toluene (2 × 50 mL). The crude product (**7**) was isolated by chromatography on a silica gel column (200 mL) in a gradient of

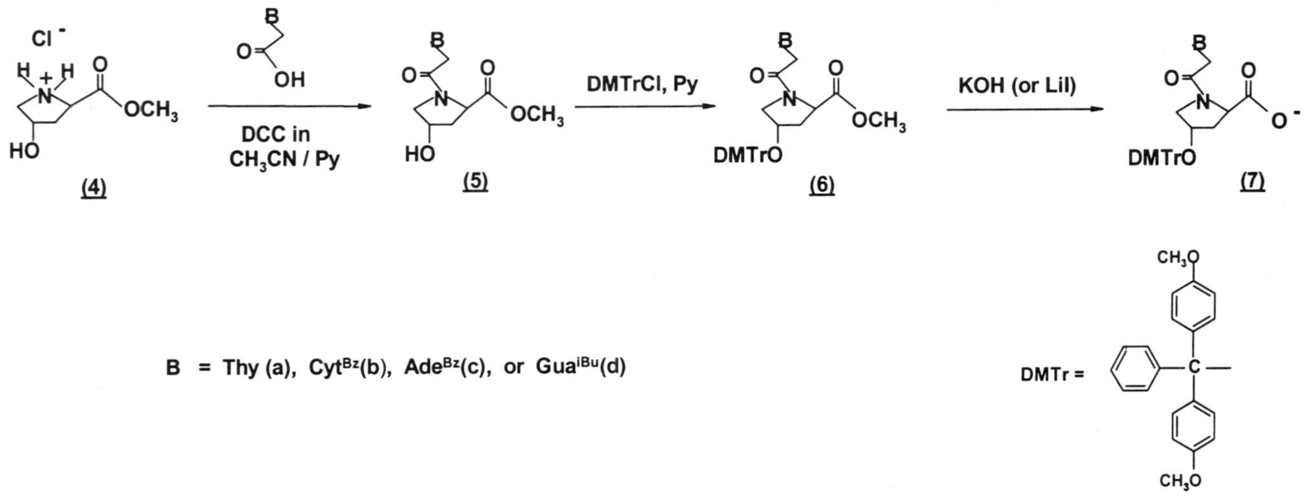

Scheme 2. Synthesis of a HypNA monomer unit for the construction of HypNA–pPNA dimers.

methanol 0–8% in 4 L DCM containing 1% triethylamine. The fractions containing the title compound (**7**) were combined, and the solvent was evaporated in a vacuum to obtain a foam. Yield 45–50%. For Thy-derivative (**7a**): $R_f = 0.28$ (system B); m/z 601.2 $(M + H)^+$, $C_{33}H_{34}N_3O_8$. For Cyt-derivative (**7a**): $R_f = 0.32$ (system B); m/z 689.5 $(M + H)^+$ $C_{39}H_{37}N_4O_8$. For Ade-derivative (**7c**): $R_f = 0.34$ (system B). m/z 713.4 $(M + H)^+$ $C_{40}H_{37}N_6O_7$. For Gua-derivative (**7d**): $R_f = 0.25$ (system B). m/z 695.2 $(M + H)^+$ $C_{37}H_{39}N_6O_8$.

3.3. General Method for the Synthesis of HypNA–pPNA Dimers

The HypNA–pPNA dimers, which are the building units for the construction of oligo-mimics, were synthesized via the formation of the amide bond between the free amino function of a pPNA monomer of type (**4**) and the carboxyl function of a HypNA monomer (**7**). Then, the catalytic protective group was introduced to the dimer (**8**) in two steps as it is shown in **Scheme 3**.

1. To remove the *N*-protective monomethoxytrityl group, the 2 mmol diphenylphosphonate derivative (**3**) was treated with a 2.1-mmol solution of picric acid in 10 mL ACN–H$_2$O (95:5) for 15 min. The reaction mixture was evaporated to dryness in vacuum, coevaporated with ACN (2 × 15 mL), and the residue was dissolved in 8 mL dry pyridine.
2. A 1.9-mmol solution of a monomer (**7**) in 8 mL pyridine and then 2.5 mmol DCC were added. After 1 h at room temperature, 2 mL H$_2$O and then 1.25 mL 8 mmol DBU were added to the resulting crude dimer (**8**) to remove one of the phenyl protective groups. After 1 h, the precipitate formed was removed by filtration, 40 mL 0.5M TEAB was added to the filtrate, and the monophenyl ester of a dimer (**9**) was extracted with DCM (2 × 50 mL). The combined organic fractions were evaporated to dryness, and the residue was dried by coevaporation with toluene (2 × 25 mL).
3. The monophenyl ester (**9**) was converted to the compound (**10**) by the introduction of a catalytic protective group. Compound (**9**) (1 mmol) was allowed to react with 1.2. mmol 1-oxido-4-methoxy-2-pyridinemethanol in 10 mL of ACN–pyridine mixture (4:1, v/v) in the presence of 1.5 mmol TPSNT for 0.5 h. The reaction was terminated by the addition of 10 mL 5% NaHCO$_3$. The solution obtained was extracted with CH$_2$Cl$_2$ (2 × 20 mL), and the combined organic fractions were evaporated to dryness.
4. To remove the phenyl protective group from the phosphonate function of compound (**10**), the residue was treated with 10 mL of 0.5M DBU solution in ACN–H$_2$O (9:1, v/v) for 1 h. After evaporation, the desired product (**11**) was isolated by the chromatography on a silica gel column (50 mL) using 0–8% gradient of methanol in CH$_2$Cl$_2$/1% triethylamine. The fractions containing a dimer of type (**11**) were combined and concentrated by evaporation in a vacuum to a colorless foam. Yield 50–60%. $R_f = 0.10$–0.12 (system B); ^{31}P-NMR $\delta = 15.2$ and 15.8.

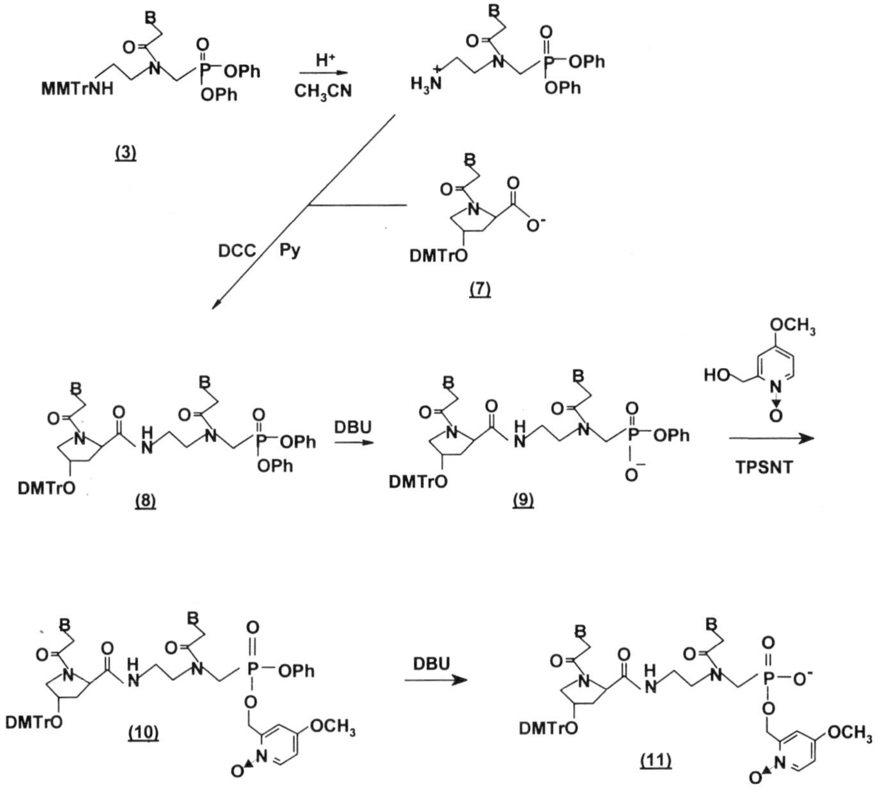

Scheme 3. Construction of a HypNA–pPNA dimer building unit.

3.4. Solid-Phase Automated Synthesis and Purification of HypNA–pPNA Oligo Mimics

1. Chain elongation in the solid-phase synthesis of HypNA–pPNA oligomers was carried out by the sequential addition of dimers (**11**) using the standard CPG support modified with properly protected deoxyribonucleosides or a universal CPG support (Glen Research) in the reaction column at the 0.2–10-µmol scale according to the DNA synthesizer instructions (*see* **Scheme 4**). The steps in the synthetic cycle are shown in **Table 1**.
2. After the completion of chain elongation, the removal of protecting groups from an oligomer was carried out in the same reaction column in a flowthrough mode. At first, the terminal dimethoxytrityl protecting group was cleaved by the action of 2% TCA for 1 min with the following wash of the support with 3 mL ACN. Then, the removal of catalytic phosphonate protective groups was carried out by the action of thiophenol– triethylamine–dioxane (1:2:2 v/v/v; 2 mL/30 mg of the support) for 3–4 h at room temperature. Then the column was washed with 5 mL

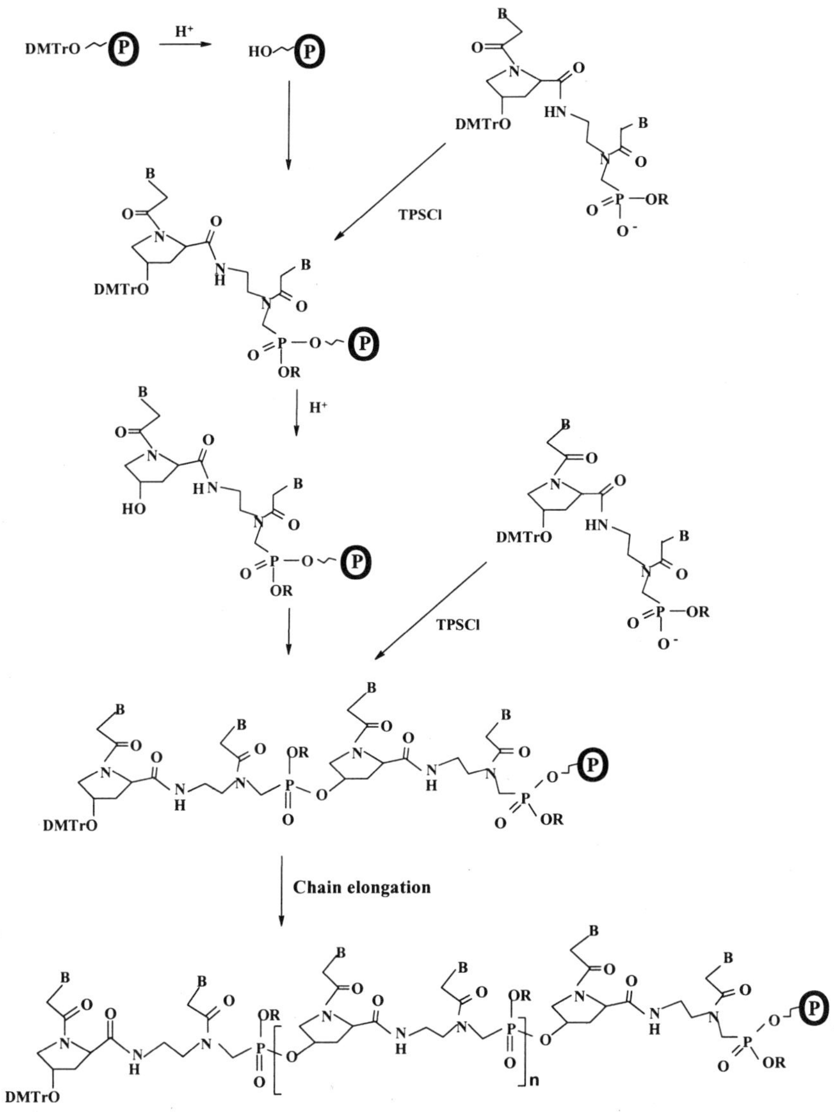

R = 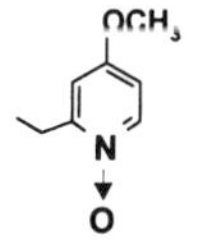

B = Thy, Cyt^{Bz}, Ade^{Bz}, or Gua^{iBu}

TPSCl = 2,4,6-triisopropylbenzenesulfonyl chloride

DMTr = 4,4'-dimethoxytrityl

Scheme 4. Elongation of a chain in the synthesis of a HypNA–pPNA oligomer using dimer building units.

Table 1
Elongation Cycle for the Solid-Phase Synthesis of HypNA–pPNA Oligomers

Step	Solvents and reagents	Time (min)
1. Detritylation	2% TCA in DCM	1.5–3.0
2. Wash	ACN	0.5
3. Wash	ACN–pyridine (4:1, v/v)	1.0
4. Coupling	0.05M P-component; 0.15 M TPSCl in ACN–pyridine (3:1, v/v)	5.0
5. Wash	ACN–pyridine	1.0
6. Capping	Ac_2O–1-methylimidazole–pyridine–ACN (1:1:2:6, v/v/v/v)	1.0
7. Wash	ACN	1.5

ACN, and the support was removed from the column and treated with 1 mL of concentrated aqueous ammonia for 3–5 h at 55°C in the hermetically closed 1.5-mL tube with a screw cap to remove N-protecting groups from heterocycles and to achieve the oligomer cleavage from the support. Then the tube was cooled to room temperature, and the 1 mL ammonia solution of oligomer was taken off and applied to a Pharmacia NAP-10 column. The column was washed with water. The crude oligomer was eluted in the first 1.5 mL of H_2O. The amount of oligomer was estimated by measuring UV-absorption at 260 nm, and the solution obtained was stored at –20°C.

3. HypNA–pPNA mimic oligomers were purified, in a method similar to natural oligonucleotides, by anion-exchange chromatography and/or denaturing gel electrophoresis. Ion-exchange FPLC was performed in a linear gradient of NaCl (0–1.2M) in 0.02M NaOH, pH 12.0, on a Mono-Q column (Pharmacia) for 20 min at the flow rate of 1 mL/min. The purity of the oligomers was confirmed by reversed-phase FPLC, which was performed on a Pharmacia Pep-RPC column in a linear gradient of ACN (1–30%) in 0.1M TEAA, pH 8.0, for 20 min at the flow rate of 1 mL/min. The identity of the oligomers obtained can be confirmed by matrix-assisted laser desorption/ionization time-of-flight.

4. The purification of deprotected oligomers by gel electrophoresis was performed using the standard 10–15% denaturing acrylamide slab gels (20 × 20 × 0.3 cm) containing 0.05M Tris-HCl-borate buffer, pH 8.3. A sample of oligomer (usually 30–50 A_{260} U) in 500 µL of H_2O containing 7M urea and marker dyes is applied in the gel slot (7–10-cm wide). The electrophoresis was performed in the same buffer at 400–500 V for 3–5 h. The experimental procedure is, in general, identical to the procedure of the natural oligonucleotide isolation. After the removal of glasses from the gel, the visualization of oligomer bands on the gel was by UV-shadowing under a 254-nm light source using a Saran wrap-covered TLC plate (*see* **Fig. 3**). The desired band (usually the least mobile) was cut out, the gel was crushed, the oligomer was eluted from the gel by 0.05M TEAB, pH 8.0, for 10–12 h at room temperature. After the removal of the gel by filtration (or centrifugation), the desired oligomer was desalted by gel filtration. For this purpose, the solution

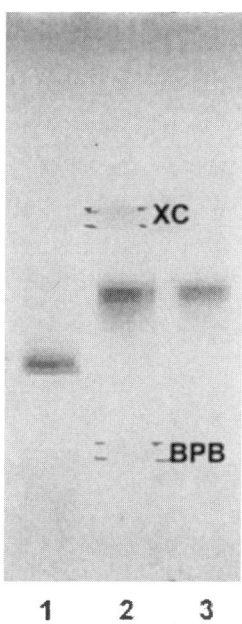

Fig. 3. *Lane 1*: Electrophoresis in 15% denaturing polyacrylamide gel of a 16-mer natural oligonucleotide d(TACTGAGTTACTTACT). *Lane 2*: crude deprotected HypNA–pPNA oligomer with the same sequence. *Lane 3*: mimic oligomer after the purification by anion-exchange FPLC. BPB and XC are marker dyes. Visualization is by UV shadowing.

of gel-eluted material was evaporated in a vacuum to reduce the volume to 2.5 mL and passed through a NAP-25 column (Pharmacia). The desired desalted oligomer was eluted in the next 3.5 mL of H_2O and stored at –20°C. The recovery of oligomer from the gel was 60–75%.

3.5. Hybridization Properties of HypNA–pPNA Oligomers

Described below are the procedures that can be used for the demonstration of the ability of oligo-mimics to form complexes with complementary DNA (or RNA) fragments and for the determination of the stability of these complexes.

3.5.1. Melting Temperature Determinations

Melting temperature (Tm) studies were performed by measuring the change in absorbance at 260 nm using a UV/VIS spectrophotometer equipped with a heated sample holder and a Peltier temperature control accessory. Determinations were performed in a 1-mL stoppered quartz cuvet. To obtain a complex (duplex or triplex) of a mimic oligomer with the complementary target, 3-mM solutions of a mimic and its complementary DNA (or RNA) target were prepared in the buffer containing 150 mM NaCl, 10 mM Tris-HCl, pH 7.5, (or 10 mM sodium phosphate, pH 7.0), and 10 mM MgCl$_2$, then 400 µL of these solutions were mixed. The mixture was kept at 95°C for 2 min and allowed to cool slowly to 20°C. Data were collected with the help of the corresponding to the spectrophotometer UV Thermal software from 20 to 95°C using the heating rate 0.5°C/min with an equilibration time of 0.2 min at each temperature.

3.5.2. Gel Mobility Assays

The corresponding DNA, or RNA, target (50 µM) was mixed with the complementary mimic oligomer in 1:1 molar ratio in a buffer containing 200 mM NaCl, 0.01M Tris-HCl, pH 7.0, 1 mM ethylenediaminetetraacetic acid, and 10 mM MgCl$_2$. The mixture was heated at 90°C for 2 min and slowly cooled to 10°C. Aliquots of the mixtures (20–50 µL) were resolved at 10°C in a 15% native polyacrylamide slab gel containing 95 mM Tris-borate buffer, pH 8.3, and 5 mM MgCl$_2$. The bands were detected by UV-shadowing and visualized by staining with a symmetrical cyanine dye, 1-ethyl-2-[3-(1-ethylnaphtho[1,2-d]thiazolin-2-ylidene)-2-methylpropenyl]naphtho[1,2-d]thiazolium bromide (Stains All) (*see* **Fig. 4**).

4. Notes

1. N^4-benzoylcytosine-N^1-acetic acid was synthesized according the following procedure. Cytosine (40 mmol) was suspended in 120 mL of DMF, NaH (60% dispersion, 42 mmol) was added while stirring, and the mixture was stirred for 1 h. Then 42 mmol methyl bromoacetate was added, and the stirring was continued to obtain clear (usually red) solution. After 2–3 h, 100 mmol N-methylimidazole and 50 mmol benzoyl chloride were added. The mixture was stirred for 30–40 min at room temperature and then cooled to −20°C in a freezer before the addition of 200 mL 1M NaOH solution in methanol–water (3:2 v/v). Preliminary cooled to 0°C in an ice bath. The temperature of the reaction mixture should be about 5°C. The removal of the methyl group from the carboxyl function was complete in 15 min. Then, 350 mL ice water was added, and the pH of the reaction mixture was adjusted to 2.0–3.0 with concentrated HCl. The product precipitation took place on this step. The precipitate of the title compound was isolated as a pink powder by filtration and sequentially washed with 200 mL H$_2$O, 200 mL dioxane, and ether (2 × 200 mL) to remove the excess of benzoic acid. Yield 65–75%.

2. N^6-benzoyladenin-N^9-acetic acid was synthesized according to the following procedure. Adenine (100 mmol) was suspended in 250 mL of DMF, 105 mmol NaH (60% dispersion) was added while stirring. After 0.5 h, 120 mmol methyl bromoacetate was added dropwise while stirring for about 1 h, and the stirring

Synthesis of DNA Mimics

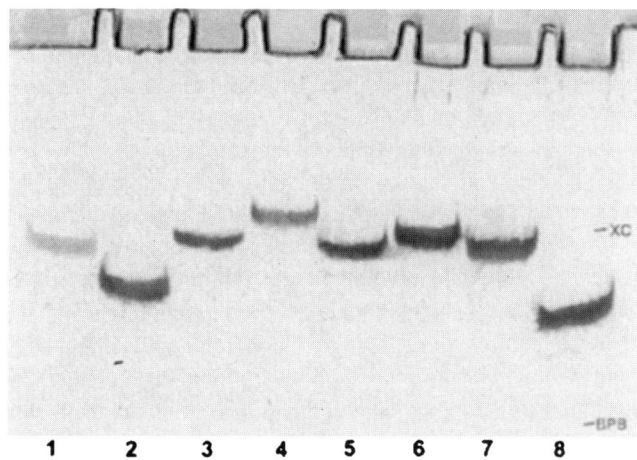

Fig. 4. Electrophoresis of complexes formed by HypNA–pPNA 16-mer mimic with the sequence (tactgagttacttact) and the corresponding natural oligonucleotide d(TACTGAGTTACTTACT) with the complementary DNA and RNA targets in a 15% native polyacrylamide gel (visualization by staining in Stains All). *Lane 1*: HypNA–pPNA oligomer; *Lane 2*: complementary 16-mer deoxyribooligonucleotide; *Lane 3*: HypNA–pPNA/DNA duplex; *Lane 4*: HypNA–pPNA/RNA duplex; *Lane 5*: DNA–DNA duplex; *Lane 6*: DNA–RNA duplex; *Lane 7*: complementary 16-mer ribooligonucleotide; *Lane 8*: corresponding natural 16-mer deoxyribooligonucleotide.

was continued for 3 h. Sometimes it needs more methyl bromoacetate. The solvent was evaporated in a vacuum, the gum obtained was washed with pentane (2 × 100 mL) and mixed with 150 mL brine. The precipitate of N^6-adenin-9-ylacetic acid methyl ester was filtered off, washed with 100 mL brine, ice water (2 × 100 mL) and dried. Yield 80%. Trimethylchlorosilane (220 mmol) was added to the 80 mmol suspension of N^6-adenin-9-ylacetic acid methyl ester in 250 mL of pyridine. After stirring at room temperature for 30 min, 220 mmol benzoyl chloride was added, and the reaction mixture was heated at 60°C for 30 min. Pyridine was evaporated in a vacuum, and 400 mL H_2O was added to the residue. The product was extracted with DCM (2 × 300 mL), combined organic fractions were dried over Na_2SO_4 for 2 h, then the solvent was removed by evaporation. The residue was dried by coevaporation with toluene and suspended in 400 mL of ether–pentane mixture (2:1 v/v). The precipitate of dibenzoyladenin-9-ylacetic acid methyl ester was isolated by filtration, washed with the 200 mL of the same mixture, and dried. Yield 75–80%. Dibenzoyladenin-9-ylacetic acid methyl ester (50 mmol) was dissolved in the 300 mL mixture of dioxane:methanol (2:1 v/v). The solution was cooled in an ice bath and 80 mL of 2*M* NaOH was added in MeOH:water mixture (1:1 v/v). After 0.5 h, 18–20 mL concentrated HCl was added to pH 2.0–3.0, the mixture was evaporated to dryness, and the residue was coevaporated with ACN (2 × 100 mL). Then, 30 mL methanol was added to the solid residue, and 400–500 mL ether was poured into suspension. The precipitate was filtered off and washed with 100 mL cold H_2O and 200 mL ether. The yield of 45 mmol N^6-benzoyladenin-9-ylacetic acid was 90%.

3. The following procedure was used for the preparation of N^2-isobutyrylguanin-N^9-acetic acid. For the introduction of the isobutyryl blocking group, 200 mmol guanine was suspended in 300 mL of dimethylacetamide. Isobutyric anhydride. (100 mL) was added, and the mixture was stirred for 4–5 h at 150–160°C to obtain a clear solution. The solvent was evaporated in a vacuum, then 800 mL of ethanol–water mixture (1:1 v/v) was added to the solid residue. The mixture was refluxed for 30–40 min, the precipitate formed during the reaction was filtered off, and sequentially washed with 200 mL ethanol–water (1:1 v/v) and 200 mL ethanol and ether (2 × 150 mL). The yield of N^2-isobutyrylguanine was 90–95%. Separately, benzhydryl ester of bromoacetic acid was synthesized. Bromoacetic acid (130 mmol), 120 mmol benzhydrol, and 2 mmol 4-dimethylaminopyridine were mixed in 350 mL of toluene, then 120 mmol DCC was added. After the incubation for 1 h at room temperature, the mixture was filtered to remove precipitated dicyclohexyl urea, and the filtrate was washed with 400 mL of H_2O. The organic layer was dried over Na_2SO_4, and the solvent was removed by evaporation in a vacuum. The oily product was used for further reaction without purification after oil pump drying. The yield of bromoacetic acid benzhydryl ester was about 90%. N^2-isobutyrylguanine (100 mmol) was dissolved with stirring in 400 mL DMF containing 105 mmol DBU. Bromoacetic acid benzhydryl ester (105 mmol) in 30 mL DMF was added dropwise during 30–40 min, and the mixture was stirred for another 30–40 min. The solvent was removed by evaporation in a vacuum, and 700 mL of H_2O was added to the residue. The precipitate containing the mixture of isomers (7 and 9 positions of guanine) was filtered off, washed with H_2O (2 × 300 mL) and dried. The isomers were separated by crystallization from 2–3 L of toluene–ethyl acetate mixture (8:2 v/v). The mixture was boiled to achieve the dissolving 80–90% material, then it was slowly cooled to room temperature and allowed to stand overnight. The precipitate formed containing pure (**9**) isomer was filtered off, washed with 200 mL toluene, and dried. The yield of isobutyrylguanine-9-ylacetic acid benzhydryl ester was about 50%. To remove the benzhydryl group, 36 mmol isobutyrylguanin-9-ylacetic acid benzhydryl ester was dissolved in a mixture of 60 mL DCM, 40 mL trifluoroacetic acid, and 3 mL mercaptoethanol. After 30–40 min, solvents were removed by evaporation, and the residue was dried by coevaporation with 50 mL ACN and 50 mL toluene. Methanol (10 mL) was added to the residue followed by the addition of the ether–pentane mixture (2:1, v/v, 600 mL). The precipitate of the title compound was isolated with 90% yield after the filtration and ether–pentane washing and drying.

4. The problem in the preparation of HypNA monomers of type (**7**) is a partial removal of N-blocking groups from the heterocycles during the alkali treatment of Ade, Gua, and Cyt derivatives of type (**6**) compound. It is especially dangerous

for cytosine-containing monomers. So we developed an alternative procedure for the removal of the methyl blocking group from the carboxyl function of the fully blocked 4-hydroxy-L-prolin derivative (**6b**) with the use of lithium iodide. The crude compound (**6b**) obtained as described in **Subheading 3.3.** was dissolved in 40 mL pyridine, 40 mmol LiI was added, and the mixture was heated at 90°C (or reflux) for 8–10 h. Then the solution was evaporated to a gum, 100 mL $0.5M$ TEAB solution was added, and the desired product was isolated by the extraction with DCM (3 × 50 mL). Combined organic fractions were washed with 50 mL $0.5M$ TEAB, dried over Na_2SO_4, and the solvent was removed by evaporation. The crude compound (**7b**) was purified as described in **Subheading 3.3.**

References

1. Nielsen, P. and Egholm, M. (1999) *Peptide Nucleic Acids, Protocols and Applications.* Horizon Scientific Press, Norfolk, UK.
2. Summerton, J. and Weller, D. (1997) Morpholino antisense oligomers: design, preparation and properties. *Antisense Nucleic Acid Drug Dev.* **7,** 187–195.
3. Nasevicius, A. and Ekker, S. (2000) Effective targeted gene "knockdown" in zebrafish. *Nat. Genet.* **26,** 216–220.
4. Van der Laan, A., Stromberg, R., van Boom, J., Kuyl-Yeheskiely, E., Efimov, V., and Chakhmakhcheva, O. (1996) An approach toward the synthesis of oligomers containing a N-2-hydroxyethyl-aminomethyl phosphonate backbone: a novel PNA analogue. *Tetrahedron Lett.* **37,** 7857–7860.
5. Peyman, A., Uhlmann, E., Wagner, K., et al. (1996) Phosphonic ester nucleic acids (PHONAs): oligonucleotide analogues with an achiral phosphonic acid ester backbone. *Angew. Chem. Int. Ed. Engl.* **35,** 2636–2638.
6. Kehler, J., Henriksen, U., Vejbjerg, H., and Dahl, O. (1998) Synthesis and hybridization properties of an acyclic achiral phosphonate DNA analogue. *Bioorg. Med. Chem.* **6,** 315–322.
7. Efimov, V., Choob, M., Buryakova, A., Kalinkina, A., and Chakhmakhcheva, O. (1998) Synthesis and evaluation of some properties of chimeric oligomers containing PNA and phosphono-PNA residues. *Nucleic Acids Res.* **26,** 566–575.
8. Efimov, A., Buryakova, A., Choob, M., and Chakhmakhcheva, O. (1999) Phosphonate analogues of peptide nucleic acids and related compounds: synthesis and hybridization properties. *Nucleosides Nucleotides* **18,** 1393–1396.
9. Efimov, V., Buryakova, A., and Chakhmakhcheva, O. (1999) Synthesis of polyacrylamides N-substituted with PNA-like oligonucleotide mimics for molecular diagnostic applications. *Nucleic Acids Res.* **27,** 4416–4426.
10. Phelan, D., Hondorp, K., Choob, M., Efimov, V., and Fernandez, J. (2001) Messenger RNA isolation using novel PNA analogues. *Nucleosides Nucleotides Nucleic Acids,* **20,** 1107–1110.
11. Efimov, V. and Chakhmakhcheva, O. (2002) Phosphono-PNAs: synthesis, properties and applications. *Collection Symp. Series* **5,** 136–144.
12. Urtishak, K., Choob, M., Tian, X., et al. (2003) Targeted gene knockdown in zebrafish using negatively charged peptide nucleic acid mimics. *Dev. Dyn.* **228,** 405–413.

10

Base-Modified Oligonucleotides With Increased Duplex Stability

Pyrazolo[3,4-d] Pyrimidines Replacing Purines

Frank Seela, Yang He, Junlin He, Georg Becher, Rita Kröschel, Matthias Zulauf, and Peter Leonard

Summary

Oligonucleotides incorporating 8-aza-7-dazapurines (pyrazolo[3,4-d]pyrimidines) were synthesized. The corresponding nucleosides were prepared and were converted into phosphoramidites. The oligonucleotide duplex stability was studied and was compared to that of the parent compounds containing the canonical purine nucleosides. The presence of 7-halogeno or 7-alkynyl substituents increases the duplex stability significantly.

Key Words: Nucleosides; phosphoramidites; oligonucleotides; glycosylation; solid-phase synthesis; base modification; pyrazolo[3,4-d]pyrimidines; 8-aza-7-deazapurines; base pair stability; Tm values; hybridization; harmonization of base pair stability.

1. Introduction

Base-modified nucleosides have been isolated from natural sources as monomers *(1)* as well as constituents of nucleic acids *(2,3)*. Several monomeric nucleosides show antiviral and/or anticancer activity *(4,5)*. They are used for deoxyribonucleic acid (DNA) labeling *(6)* or sequencing *(7,8)* in the form of their triphosphates. The base-modified oligonucleotides *(9,10)* are employed for multiple purposes, such as antisense pharmaceuticals *(11,12)*, for hybridization studies in solution *(13,14)*, or on arrays *(15,16)* as well as to construct supramolecular assemblies *(17,18)* and nanostructures *(19)*. Within this framework efforts have been made to increase the DNA or ribonucleic acid (RNA) duplex stability *(20)*. Particular attention has been paid to hybridization studies

From: *Methods in Molecular Biology, vol. 288: Oligonucleotide Synthesis: Methods and Applications*
Edited by: P. Herdewijn © Humana Press Inc., Totowa, NJ

with primers and probes. In general, the duplex stabilization can be accomplished by the strengthening of hydrogen bonds within a base pair and/or by increasing the stacking interactions. Modified bases such as the 5-propynyl derivatives of dU and dC show such favorable properties *(21–28)*. Backbone modification has also been successfully employed leading to novel nucleic acid analogs, such as peptide nucleic acid (PNA) *(29–32)* or locked nucleic acid (LNA) *(33)*.

Our laboratory is involved in studies on oligonucleotide duplexes incorporating 7-deazapurine-pyrrolo[2,3-d]pyrimidine and pyrazolo[3,4-d]pyrimidine–nucleosides carrying substituents of modest size at the 7-position of the modified purine heterocycle (purine numbering is used throughout the discussion). These substituents are well accommodated in the major groove of duplex DNA and have shown to increase the duplex stability significantly *(34–38)*. This report focuses on the duplex stabilization induced by pyrazolo[3,4-d]pyrimidine (8-aza-7-deazapurine) 2'-deoxyribonucleosides, which can be considered as ideal shape mimics of the parent purine nucleosides. Their 7-halogenated and alkynylated derivatives induce a positive effect on the base pair stability *(38)*. Moreover, it was shown that the bidentate dA–dT base pair can be stabilized to the level of a dG–dC pair by incorporation of 7-substituted 8-aza-7-deazapurines in place of purines *(39)*.

Within the group of 8-aza-7-deazapurine nucleoside derivatives a number of phosphoramidites have been synthesized in our laboratory. **Scheme 1** shows typical members of those building blocks, such as **1a** *(40)*, **2a–e** *(41–45)*, **3a, c–e** *(38,41,45–47)*, **4a,c** *(39)*, **5c–e** *(48)*, **6a,c,d**, **7a** *(41,42)*, **8a** *(18)*, **9c** *(49)*, **9d** *(50)*, **10a** *(51)*, **11a** *(52)*. They have been employed in solid-phase oligonucleotide synthesis. This article reports on the synthesis of nucleosides, such as **12c** *(53)* or **13c** *(54)* as well as their phosphoramidites (**2c, 3c**). Oligonucleotides incorporating various 8-aza-7-deazapurine 2'-deoxyribonucleosides are described, and their duplex stability is discussed. The stability is compared to duplexes containing the canonical DNA constituents (*see* **Scheme 1**).

2. Materials

1. Thin-layer chromatography (TLC) was performed on silica gel 60 F_{254} aluminum sheets (0.2 mm, Merck, Germany).
2. Flash chromatography (FC) was carried out in silica gel 60H columns at 0.4 bar.
3. Ultraviolet (UV) spectra were measured on a U-3200 spectrometer (Hitachi, Japan); λ_{max} in nm, ε in L· mol^{-1} cm^{-1}.
4. Nuclear magnetic resonance (NMR) spectra were obtained on AC-250 and AMX-500 spectrometers (Bruker, Germany), δ values are given in ppm downfield from internal SiMe$_4$ (^{1}H, ^{13}C) or external 85% H$_3$PO$_4$ solution (^{31}P).
5. The solid-phase synthesis of oligonucleotides was carried out on an automated DNA synthesizer (Applied Biosystems, ABI 392-08) employing the protocol of phosphoramidite chemistry.

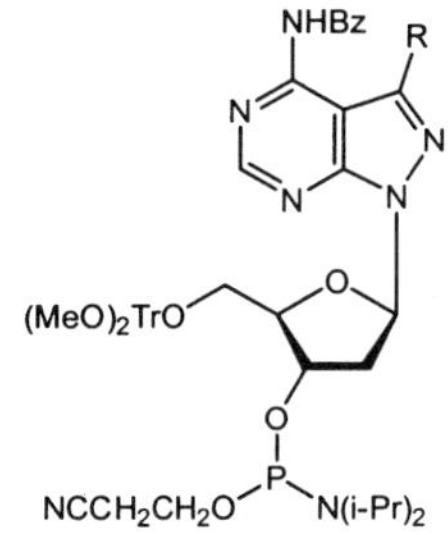

Base-Modified Oligonucleotides

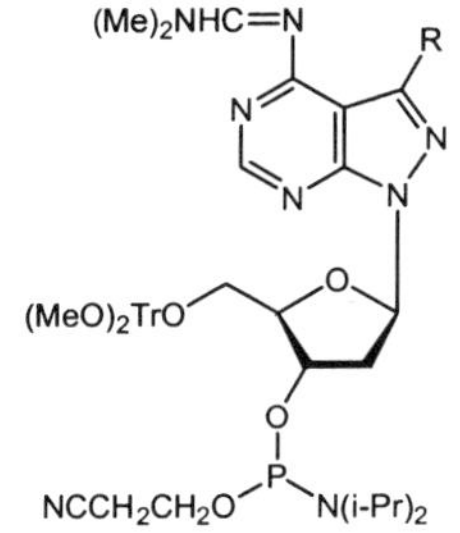

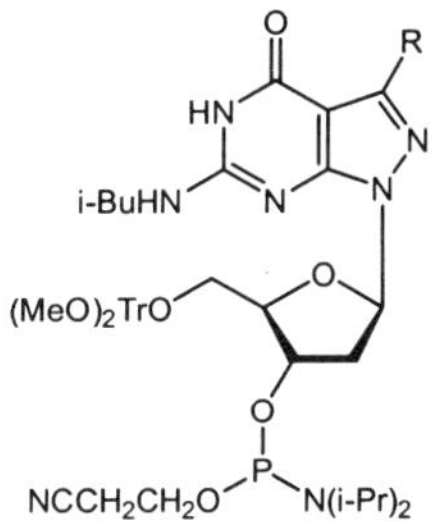

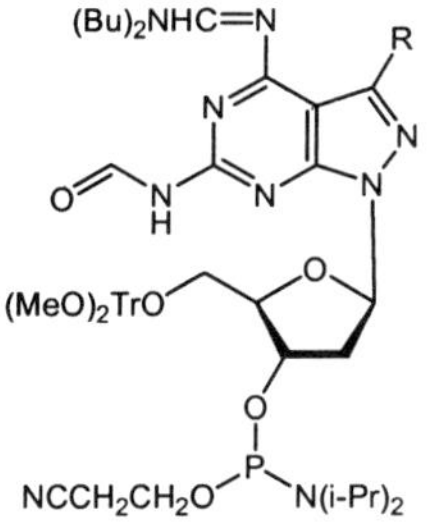

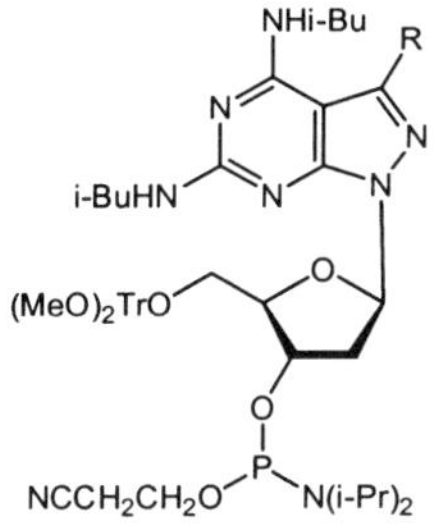

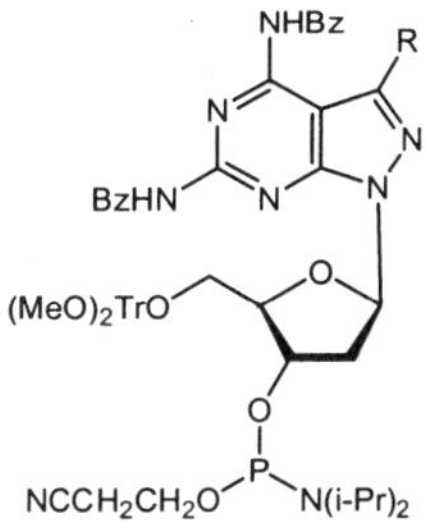

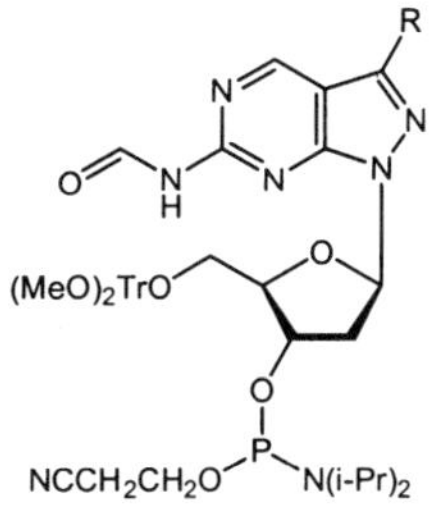

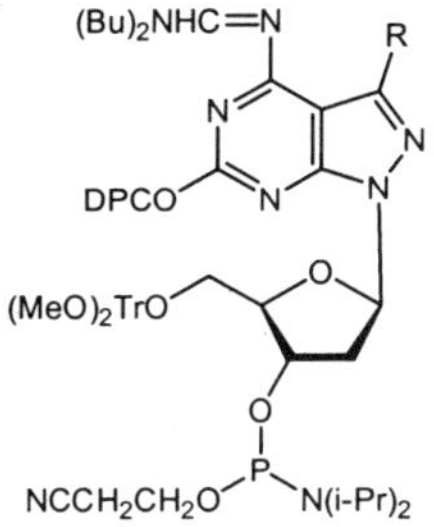

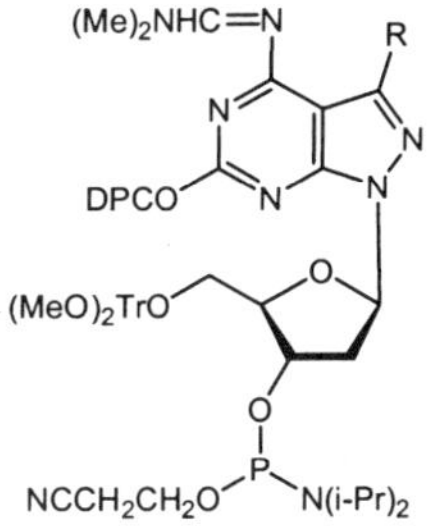

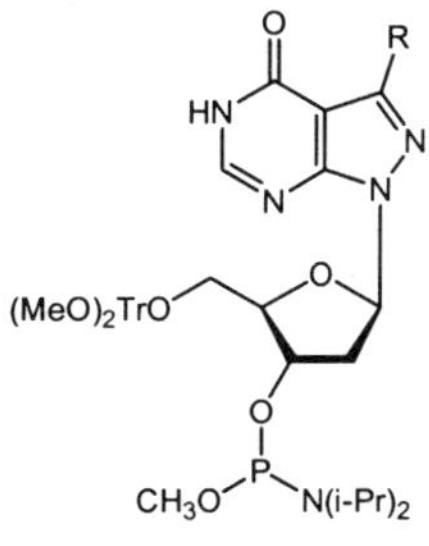

a: R = H, b: R = Cl, c: R = Br, d: R = I, e: R = Propynyl

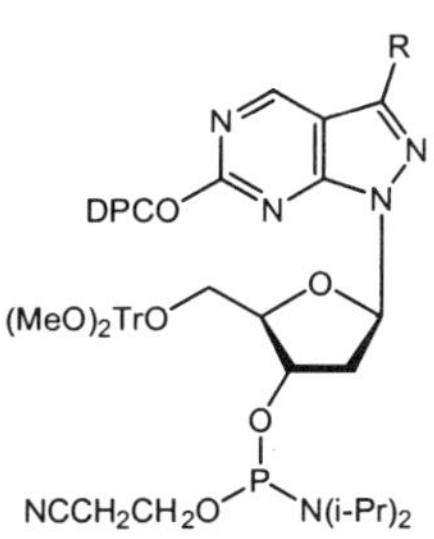

Scheme. 1. Selected phosphoramidites containing 8-aza-7-deazapurines as nucleobases.

6. Purification of oligonucleotides was performed by reverse-phase high-pressure (performance) liquid chromatography (HPLC) on a 250 × 4 mm RP-18 (5 µ) LiChrosorb column (Merck) connected with a Merck–Hitachi HPLC pump (model 655 A-12), a variable-wavelength monitor (model 655-A), a controller (model L-5000), and an integrator (model D-2000).
7. Melting curves were determined using a Cary 1E UV/VIS spectrophotometer (Varian, Australia) equipped with a thermoelectrical controller. The calculation of thermodynamic data was performed with the program MeltWin (version 3.1) using curve fitting of the melting profiles according to a two-state model. The standard errors of thermodynamic data for $\Delta H°$ and $\Delta S°$ obtained from the curve fitting are within an error limit of ±15%.

3. Methods

3.1. Nucleoside Synthesis and Characterization

The 8-aza-7-deazapurine nucleosides discussed in this article have been synthesized by the stereoselective nucleobase anion glycosylation, a method which was developed in our laboratory *(55)*. The syntheses of brominated derivatives of 8-aza-7-deaza-2'-deoxyadenosine (**12c**) and 8-aza-7-deaza-2'-deoxyguanosine (**13c**) are outlined in **Schemes 2** and **3**. Particular nucleobase precursors are used to ensure glycosylation within the pyrazole moiety. The glycosylation reaction of the nucleobases **14** or **15** was performed in MeCN with 2-deoxy-3,5-di-*O*-(*p*-toluoyl)-α-D-erythro-pentofuranosyl chloride (**16**) *(56)* in the presence of a phase-transfer catalyst (TDA-1). While the reaction proceeds in a stereoselective way under the selective formation of β-D anomers, regioisomeric glycosylation products (N-9 and N-8 isomers) are formed, and also debrominated derivatives can be detected under certain reaction conditions *(54)*. They are separated by preparative column chromatography. The N-9 isomers are usually formed in preponderance with a ratio of 2:1. Nevertheless, this depends on the nucleobase structure and the reaction conditions. The nucleosides as well as their intermediates are characterized by elemental analyses, UV, and ^{1}H-NMR (*see* **Subheadings 3.1.1.–3.2.6.**) as well as by ^{13}C-NMR spectra (**Table 1**).

The following Schemes outline the synthetic routes for the preparation of the pyrazolo[3,4-d]pyrimidine-derivative of 2'-deoxyadenosine **12c** (**Scheme 2**) and the 2'-deoxyguanosine analog **13c** (**Scheme 3**). The experimental procedures are described under **Subheadings 3.3.1.–3.2.6.**

3.1.1. 3-Bromo-1-[2-deoxy-3,5-di-O-(p-toluoyl)-β-D-erythro-pentofuranosyl]-4-methoxy-1H-pyrazolo[3,4-d]pyrimidine (17)

1. To a suspension of **14** *(54)* (1.0 g, 4.4 mmol) in MeCN (60 mL), powdered KOH (85%, 470 mg, 7.1 mmol) and TDA-1 {= tris[2-(2-methoxyethoxy)ethyl]amine; 75 µL} were added and stirred for 10 min at room temperature.

Table 1
^{13}C-NMR Chemical Shifts of Pyrazolo[3,4-d] Pyrimidine 2'-Deoxyribofuranosides[a] (38,53,57,58)

[b] [c]	C(7) C(3)	C(5) C(3a)	C(6) C(4)	C(2) C(6)	C(4) C(7a)	OMe	Oi-Pr	
20	118.8	102.5	163.2	156.4	155.3	54.6		
12c	118.9	99.8	157.3	156.9	154.5			
24	120.9	106.1	161.8	156.1	155.0			
25	121.0	106.2	161.8	156.2	155.0	54.9		
23	119.3	96.2	162.5[d]	162.5[d]	158.6[d]		21.6, 69.2	
13c	122.0	98.4	158.1[d]	156.3[d]	156.5[d]			
26	122.3	101.6	154.8[d]	150.7[d]	153.7[d]			
27	122.2	101.8	154.8[d]	150.6[d]	153.5[d]			

	C(1')	C(2')	C(3')	C(4')	C(5')	C = O	Me$_2$	HC = N
20	84.3	37.9	70.7	87.8	62.1			
12c	84.0	37.8	70.8	87.7	62.3			
24	84.0	37.8	70.8	87.6	62.3		34.9, 40.8	157.6
25	83.8	38.0	70.6	85.4	64.2		34.8, 40.7	157.5
23	83.2	37.5	70.8	87.4	62.4			
13c	83.1	37.6	70.8	87.4	62.4			
26	83.7	37.7	70.7	87.7	62.2	180.6		
27	83.9	38.0	70.5	85.7	64.1	180.6		

[a]Measured in DMSO-d_6 at 298 K.
[b]Purine numbering.
[c]Systematic numbering.
[d]Tentative.

2. Then 2-deoxy-3,5-di-O-(p-toluoyl)-α-D-erythro-pentofuranosyl chloride *(56)* (16, 2.1 g, 5.4 mmol) was added, and the stirring was continued for another 20 min.
3. The insoluble material was filtered off, the filtrate was evaporated, and the residue was subjected to flash chromatography (FC). From the faster migrating main zone (petroleum ether/ethyl acetate 2:1) compound **17** was isolated (*see* **Note 1**).
4. Crystallization from petroleum ether/ethyl acetate yielded colorless needles (1.13 g, 44%): melting point (mp) 177–178°C. TLC (petroleum ether/ethyl acetate 1:1): R_f 0.6.
5. UV (MeOH): λ_{max} 240, 270 nm (ε = 34,300, 7500). ^{1}H-NMR ([D$_6$])dimethyl sulfoxide [DMSO]): 2.41, 2.51 (2s, 6 H, 2 CH$_3$); 2.81 (m, 1 H, H$_\alpha$-C[2']); 3.27 (m, 1 H, H$_\beta$-C[2']); 4.14 (s, 3 H, OCH$_3$); 4.48 (m, 2 H, H-C[5']); 4.58 (m, 1H, H-C[4']), 5.81 (m,1 H, H-C[3']); 6.83 (t, J = 6.2 Hz, 1 H, H-C[1']); 7.35, 7.92 (4d, J = 7.8, 7.9 Hz, 8 H, 2 C$_6$H$_4$); 8.67 (s, 1 H, H-C[6]). Analysis calculated for C$_{27}$H$_{25}$BrN$_4$O$_6$: C 55.78, H 4.33, N 9.64; Found: C 55.98, H 4.43, N 9.53.

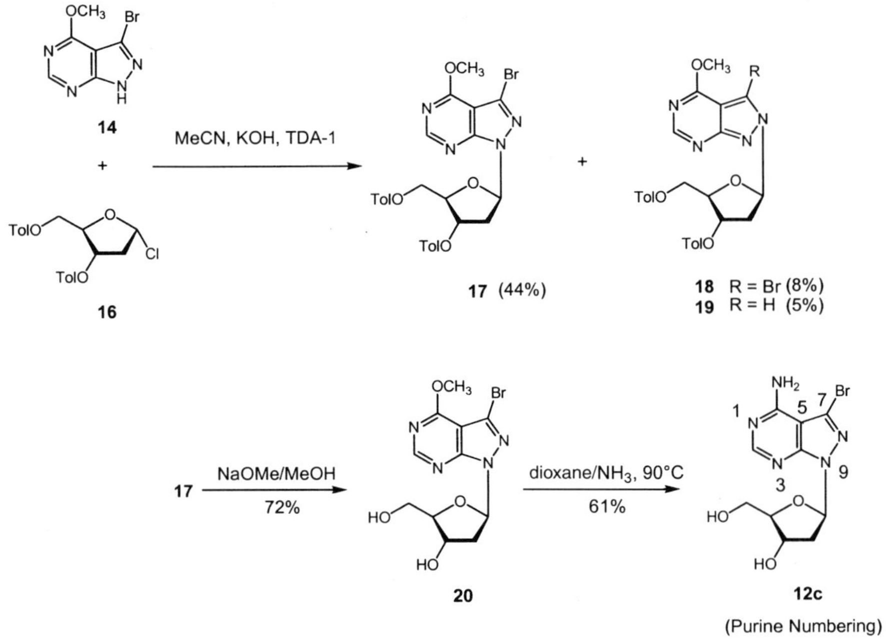

Scheme. 2. Synthesis of the nucleoside **12c** by nucleobase anion glycosylation.

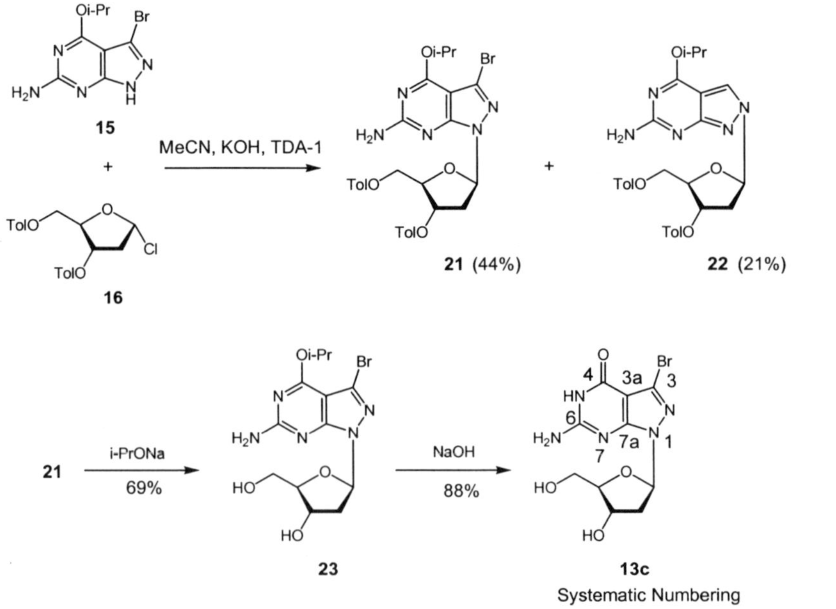

Scheme. 3. Synthesis of the nucleoside **13c** by nucleobase anion glycosylation.

3.1.2. 3-Bromo-1-(2-deoxy-β-D-erythro-pentofuranosyl)-4-methoxy-1H-pyrazolo[3,4-d]pyrimidine (20)

1. A solution of compound **17** (1.0 g, 1.7 mmol) in 0.2M NaOMe in MeOH (100 mL) was stirred for 4 h at room temperature. The reaction was monitored by TLC.
2. When the reaction was finished the solution was evaporated to dryness and the residue subjected to FC (column 10 × 4 cm, CH_2Cl_2/MeOH 95:5). The content of the main zone was collected, and the solution was evaporated to dryness.
3. Crystallization from MeOH afforded colorless crystals of **20** (420 mg, 72%): mp, 148–149°C. TLC (CH_2Cl_2/MeOH 9:1): R_f 0.6.
4. UV (MeOH): λ_{max} 247, 272 nm (ε = 7200, 6600). ^{1}H-NMR ([D_6]DMSO): 2.32 (m, 1 H, H_α-C[2']); 2.80 (m, 1 H, H_β-C[2']); 3.52 (m, 2 H, H-C[5']); 3.84 (m, 1 H, H-C[4']); 4.14 (s, 3 H, OCH_3); 4.45 (m,1 H, H-C[3']); 4.70 (t, J = 5.0 Hz, 1 H, OH-C[5']); 5.30 (d, J = 4.1 Hz, 1 H, OH-C[3']); 6.63 (t, J = 5.8 Hz, 1 H, H-C[1']); 8.66 (s, 1 H, H-C[6]). Analysis calculated for $C_{11}H_{13}BrN_4O_4$: C 38.28, H 3.80, N 16.23; Found: C 38.8, H 4.0, N 16.4.

3.1.3. 4-Amino-3-bromo-1-(2-deoxy-β-D-erythro-pentofuranosyl)-1H-pyrazolo[3,4-d]pyrimidine (12c)

1. A solution of compound **20** (300 mg, 0.87 mmol) in conc. aq NH_3-dioxane (200 mL, 1:1) was stirred for 4 h at 90°C in an autoclave.
2. The solution was evaporated to dryness and the residue was subjected to FC (column 10 × 3 cm, CH_2Cl_2/MeOH 9:1). The combined fractions of the main zone were evaporated.
3. Crystallization from MeCN afforded compound **12c** as colorless crystals (175 mg, 61%): mp, 214°C. TLC (CH_2Cl_2/MeOH 9:1): R_f 0.4.
4. UV (MeOH): λ_{max} 231, 264, 281nm (ε = 4700, 4600, 6100). ^{1}H-NMR ([D_6])DMSO): 2.24 (m, 1 H, H_α-C[2']); 2.74 (m, 1H, H_β-C[2']); 3.45 (m, 2 H, H-C[5']); 3.79 (m, 1H, H-C[4']); 4.39 (m,1 H, H-C[3']); 4.74 (t, J = 5.6 Hz, 1 H, OH-C[5']); 5.26 (d, J = 4.5 Hz, 1 H, OH-C[3']); 6.51 (t, J = 6.1 Hz, 1 H, H-C[1']); 7.0 and 7.95 (2 br s, 2 H, NH_2); 8.23 (s, 1 H, H-C[6]). Analysis calculated for $C_{10}H_{12}BrN_5O_3$: C 36.38, H 3.66, N 21.21; Found: C 36.2, H 3.7, N 21.1.

3.1.4. 6-Amino-3-bromo-1-[2-deoxy-3,5-di-O-(p-toluoyl)-β-D-erythro-pentofuranosyl]-4-isopropoxy-1H-pyrazolo[3,4-d]pyrimidine (21)

1. A suspension of compound **15** *(54)* (1.5 g, 5.5 mmol) in MeCN (100 mL) and powdered KOH (85%, 1.46 g, 22 mmol) was stirred for 15 min at room temperature.
2. Then, TDA-1{= tris[2-(2-methoxyethoxy)ethyl]amine, 200 μL} was introduced, the stirring was continued for another 10 min, and the 2-deoxy-3,5-di-O-(p-toluoyl)-α-D-erythro-pentofuranosyl chloride (**16**, 2.6 g, 6.7 mmol) was added in portions.
3. After 20 min, insoluble material was filtered off and the solvent was evaporated.

4. The residue was subjected to FC (column 10 × 4 cm, CH$_2$Cl$_2$/acetone 100:0 → 95:5). The fast-migrating main zone furnished compound **21** as a colorless foam (1.5 g, 44%) (*see* **Note 2**). TLC (CH$_2$Cl$_2$/acetone 98:2): R_f 0.5.
5. UV (MeOH): λ_{max} 233, 272 nm (ε = 39,300, 11,100). ^{1}H-NMR ([D$_6$])DMSO): 1.33 (d, J = 6.2 Hz, 6 H, CH[CH$_3$]$_2$); 2.36, 2.38 (2s, 2 Ar-CH$_3$); 2.67 (m, 1 H, H$_\alpha$-C[2']); 3.15 (m, 1 H, H$_\beta$-C[2']); 4.50 (m, 3 H, H-C[5', 4']); 5.43 (m, 1 H, OCH), 5.73 (m,1 H, H-C[3']); 6.55 (t, J = 5.9 Hz, 1 H, H-C[1']); 7.05 (s, 2 H, NH$_2$); 7.28–7.92 (4d, J = 8.1 Hz, 8 H, 2 C$_6$H$_4$). Analysis calculated for C$_{29}$H$_{30}$BrN$_5$O$_6$: C 55.78, H 4.84, N 11.21; Found: C 56.11, H 4.84, N 10.93.

3.1.5. 6-Amino-3-bromo-1-(2-deoxy-β-D-erythro-pentofuranosyl)-4-isopropoxy-1H-pyrazolo[3,4-d]pyrimidine (23)

1. A solution of **21** (5 g, 8 mmol) in 0.1*M* i-PrONa in i-PrOH (200 mL) was stirred for 3 h at 40°C.
2. The solution was adsorbed with silica gel, loaded on the top of a silica gel column (10 × 4 cm) and subjected to FC. Elution with CH$_2$Cl$_2$/MeOH 98:2 → 95:5 afforded a main zone. From the combined fractions a colorless solid was obtained after evaporation.
3. Crystallization (water) gave compound **23** as colorless needles (2.15 g, 69%): mp 181°C. TLC (CH$_2$Cl$_2$/MeOH 98:2): R_f 0.2.
4. UV (MeOH): λ_{max} 277 nm (ε = 8700). ^{1}H-NMR ([D$_6$]DMSO): 1.33 (d, J = 6.8 Hz, 6 H, CH[CH$_3$]$_2$); 2.18 (m, 1 H, H$_\alpha$-C[2']); 2.67 (m, 1 H, H$_\beta$-C[2']); 3.35, 3.47 (m, 2 H, H-C[5']); 3.74 (m, 1 H, H-C[4']); 4.34 (m, 1 H, H-C[3']); 4.69 (t, J = 5.7 Hz, OH-C[5']); 5.22 (d, J = 4.3 Hz, OH-C[3']); 5.39 (m, 1 H, OCH), 6.33 (t, J = 6.5 Hz, 1 H, H-C[1']); 6.99 (s, 2 H, NH$_2$). Analysis calculated for C$_{13}$H$_{18}$BrN$_5$O$_4$: C 40.22, H 4.67, N 18.04; Found: C 40.17, H 4.63, N 18.01.

3.1.6. 6-Amino-3-bromo-1-(2-deoxy-β-D-erythro-pentofuranosyl)-1,5-dihydro-4H-pyrazolo[3,4-d]pyrimidin-4-one (13c)

1. A solution of **23** (1.8 g, 4.6 mmol) in 1*M* NaOH (80 mL) was stirred for 2 h at 60°C, cooled and neutralized with 96% AcOH.
2. The resulting precipitate was filtered off and recrystallized from water yielding colorless needles (1.4 g, 88%): mp 221°C. TLC (CH$_2$Cl$_2$/MeOH 9:1): R_f 0.18.
3. UV (MeOH): λ_{max} 257 nm (ε = 10,600). ^{1}H-NMR ([D$_6$]DMSO): 2.12 (m, 1 H, H$_\alpha$-C[2']); 2.62 (m, 1 H, H$_\beta$-C[2']); 3.35,3.44 (m, 2 H, H-C[5']); 3.73 (m, 1 H, H-C[4']); 4.31 (m, 1 H, H-C[3']); 4.71 (t, J = 5.7 Hz, OH-C[5']); 5.19 (d, J = 4.3 Hz, OH-C[3']); 6.23(t, J = 6.5 Hz, 1 H, H-C[1']); 7.01 (s, 2 H, NH$_2$); 10.79 (s, 1 H, NH). Analysis calculated for C$_{10}$H$_{12}$BrN$_5$O$_4$: C 34.70, H 3.49, N 20.23; Found: C 34.50, H 3.56, N 20.12.

3.2. Preparation of Phosphoramidites

The phosphoramidites of the 8-aza-7-deazapurine 2'-deoxynucleosides **12c** and **13c** are prepared according to **Schemes 4** and **5**. The chemical stability of

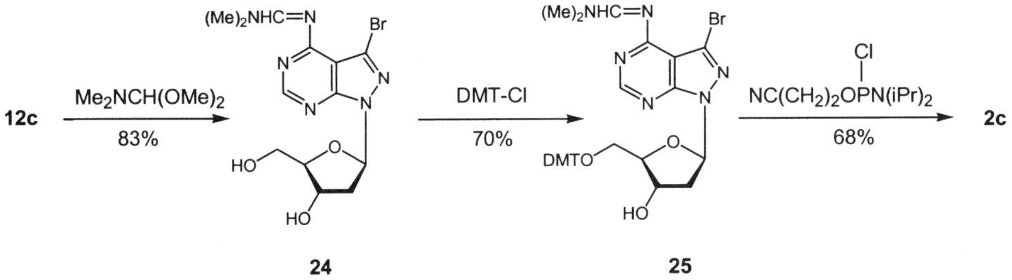

Scheme. 4. Synthesis of phosphoramidite **2c**.

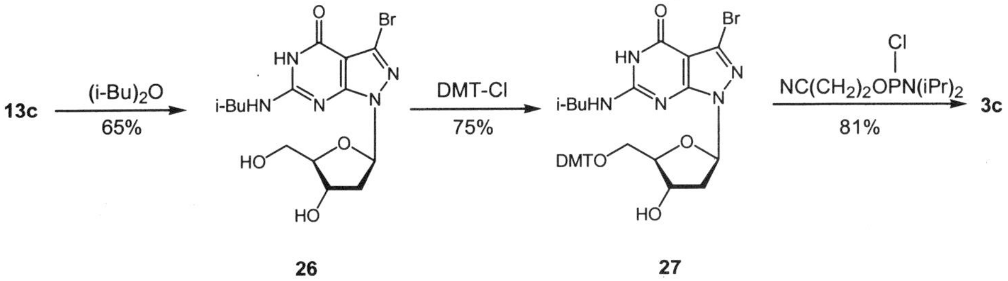

Scheme. 5. Synthesis of phosphoramidite **3c**.

the 8-aza-7-deazapurine compounds is similar to that of the purine counterparts. However, π-electron changes between the purine and the 8-aza-7-deazapurine system result in changes of the protecting group stability. For the amino protecting group of compound **12c** the dimethylaminomethylidene residue was selected, whereas the isobutyryl residue was used for **13c**.

3.2.1. 3-Bromo-1-(2-deoxy-β-D-erythro-pentofuranosyl)-4-([(dimethyl-amino)-methylidene]amino)-1H-pyrazolo[3,4-d]pyrimidine (24)

1. A solution of compound **12c** (330 mg, 1 mmol) in MeOH (20 mL) was stirred with N,N-dimethylformamide dimethyl acetal (2.0 g 16.8 mmol) for 2 h at 40°C.
2. After evaporation, the residue was applied to FC (column 12 × 3 cm, CH_2Cl_2/MeOH 9:1). The title compound was isolated as a colorless foam (320 mg, 83%). TLC (CH_2Cl_2/MeOH 9:1): R_f 0.52.
3. UV (MeOH): λ_{max} 236, 260, 301, and 305 nm (ε = 6200, 4900, 7100, 7000). ^{1}H-NMR ([D_6])DMSO): 2.26 (m, 1 H, H_α-C[2']); 2.78 (m, 1 H, H_β-C[2']); 3.21 and 3.23 (2s, 6 H, Me_2N); 3.45 (m, 2 H, H-C[5']); 3.80 (m, 1 H, H-C[4']); 4.40 (m, 1 H, H-C[3']); 4.75 (br, 1 H, OH-C[5']); 5.27 (br, 1 H, OH-C[3']); 6.55 (t, J = 6.4 Hz, 1 H, H-C[1']); 8.44 (s, 1 H, H-C[6]); 8.94 (s, 1 H, N = CH). Analyzed and calculated for $C_{13}H_{17}BrN_6O_3$: C 40.53, H 4.45, N 21.82; Found: C 40.2, H 4.6, N 20.9.

3.2.2. 3-Bromo-1-[2-deoxy-5-O-(4,4'-dimethoxytrityl)-β-D-erythro-pentofuranosyl]-4-([(dimethylamino)methylidene]amino)-1H-pyrazolo[3,4-d]pyrimidine (25)

1. Compound **24** (300 mg, 0.78 mmol) was dried by coevaporation with anhydrous pyridine.
2. The residue was dissolved in anhydrous pyridine (2 mL) and stirred with 4,4'-dimethoxytrityl chloride (290 mg, 0.86 mmol) (DMTrCl).
3. After being stirred at room temperature for 8 h the mixture was poured into 10 mL ice-cold 3% aqueous $NaHCO_3$ (10 mL) and extracted with CH_2Cl_2 (2 × 100 mL).
4. The combined organic layers were dried over Na_2SO_4, filtered, and coevaporated with toluene three times.
5. The residue was applied to FC (column 15 × 3 cm, CH_2Cl_2/MeOH 9:1). The title compound was isolated as a colorless foam (375 mg, 70%). TLC (CH_2Cl_2/MeOH 9:1): R_f 0.62.
6. UV (MeOH): λ_{max} 230, 262, 282, and 321 nm (ε = 20,700, 7100, 9200, 5200). ^{1}H-NMR ([D_6]DMSO): 2.31 (m, 1 H, H_α-C[2']); 2.82 (m, 1 H, H_β-C[2']); 3.00 (m, 2 H, H-C[5']); 3.21 and 3.23 (2s, 6 H, Me_2N); 3.68 and 3.69 (2s, 6 H, 2 × MeO); 3.93 (m, 1 H, H-C[4']); 4.51 (m,1 H, H-C[3']); 5.30 (d, J = 4.8 Hz, 1 H, OH-C[3']); 6.58 (t, J = 6.5 Hz, 1 H, H-C[1']); 6.73 (m, 4 H, ArH); 7.14–7.31 (m, 9 H, ArH); 8.47 (s, 1 H, H-C[6]); 8.94 (s, 1 H, N = CH). Analysis calculated for $C_{34}H_{35}BrN_6O_5$: C 59.39, H 5.13, N 12.22; Found: C 59.5, H 5.3, N 12.1.

3.2.3. 3-Bromo-1-[2-deoxy-5-O-(4,4'-dimethoxytrityl)-β-D-erythro-pentofuranosyl]-4-([(dimethylamino)methylidene]amino)-1H-pyrazolo[3,4-d]pyrimidine 3'-(2-cyanoethyl diisopropylphosphoramidite) (2c)

1. To a solution of compound **25** (300 mg, 0.44 mmol) in dry THF (2 mL), anhydrous *N,N*-diisopropylethylamine (170 mg 1.32 mmol) and chloro(2-cyanoethoxy)(*N,N*-diisopropylamino)phosphine (137 mg 0.54 mmol) were added under Ar.
2. The reaction mixture was stirred for 30 min and filtered.
3. The filtrate was diluted with 80 mL ethyl acetate and extracted twice with ice-cold aqueous 3% $NaHCO_3$ (2 × 10 mL) and H_2O (2 × 10 mL).
4. The organic phase was dried over Na_2SO_4, filtered, and evaporated to dryness.
5. The residue was applied to FC (column, 12 × 2 cm, petroleum ether/acetone 1:1), and the title compound was isolated (266 mg, 68%). TLC (petroleum ether/acetone 1:1): R_f 0.4 and 0.5. ^{31}P-NMR ($CDCl_3$): 148.9, 149.0.

3.2.4. 3-Bromo-1-(2-deoxy-β-D-erythro-pentofuranosyl)-1,5-dihydro-6-[(2-methylpropanoyl)amino]-4H-pyrazolo[3,4-d]pyrimidin-4-one (26)

1. Compound **13c** (1.0 g, 2.9 mmol) was dried by coevaporation with pyridine.
2. The residue was dissolved in dimethyl formamide (10 mL) and 1,1,1,3,3,3-hexamethyldisilazane (5 mL) was added.
3. Stirring was continued for 3 h, and pyridine (10 mL) and 2-methylpropanoic anhydride (10 mL) were added.

4. The solution was stirred for an additional 12 h, then MeOH (20 mL) was added, and stirring was continued for 3 h.
5. The volume of the solution was reduced, and the solution was adsorbed on silica gel and loaded on a silica gel column (10 × 4 cm). Compound **26** was eluted with CH_2Cl_2/MeOH (95:5).
6. Recrystallization from H_2O/MeOH (2:1) gave colorless crystals of **26** (785 mg, 65%): mp 215°C. TLC (CH_2Cl_2/MeOH 9:1): R_f 0.4.
7. UV (MeOH): λ_{max} 225, 274 nm (ε = 13,000, 11,500). ^{1}H-NMR ([D_6]DMSO): 1.11 (d, J = 6.7, 6 H, Me_2CH); 2.22 (m, 1 H, H_α-C[2']); 2.70 (m, 1 H, H_β-C[2']); 3.36,3.46 (m, 2 H, H-C[5']); 3.78 (m, 1 H, H-C[4']); 4.38 (m, 1 H, H-C[3']); 4.70 (t, J = 5.7 Hz, OH-C[5']); 5.23 (d, J = 4.3 Hz, OH-C[3']); 6.34 (t, J = 6.5 Hz, 1 H, H-C[1']); 11.77 (s, 1 H, NH); 11.93 (s, 1 H, NH). Analysis calculated for $C_{14}H_{18}BrN_5O_5$: C 40.40, H 4.36, N 16.83; Found: C 40.60, H 4.52, N 16.77.

3.2.5. 3-Bromo-1-[2-deoxy-5-O-(4,4'-dimethoxytrityl)-β-D-erythro-pentofuranosyl]-1,5-dihydro-6-[(2-methylpropanoyl)amino]-4H-pyrazolo[3,4-d]pyrimidin-4-one (27)

1. Compound **26** (500 mg, 1.2 mmol) was dried by coevaporation with anhydrous pyridine.
2. The residue was dissolved in anhydrous pyridine (10 mL) and stirred with DMTrCl (540 mg 1.6 mmol) in the presence of N,N-diisopropylethylamine (315 µL, 1.8 mmol) for 1.5 h at room temperature.
3. The solution was poured into 5% aqueous $NaHCO_3$ (100 mL) and then extracted with CH_2Cl_2 (2 × 100 mL). The combined organic extracts were dried (Na_2SO_4) and coevaporated with toluene three times.
4. FC (CH_2Cl_2/acetone 9:1 followed by CH_2Cl_2/acetone 8:2) furnished **27** as colorless foam (645 mg, 75%). TLC (CH_2Cl_2/MeOH 95:5): R_f 0.25.
5. UV (MeOH): λ_{max} 275 nm (ε = 15,500). ^{1}H-NMR ([D_6]DMSO): 1.03 (d, J = 6.7, 6 H, Me_2CH); 2.25 (m, 1 H, H_α-C[2']); 2.75 (m, 1H, H_β-C[2']); 3.05 (m, 2 H, H-C[5']); 3.70 (s, 6H, 2 MeO); 3.95 (m, 1 H, H-C[4']); 4.44 (m, 1 H, H-C[3']); 5.30 (d, J = 4.5 Hz, OH-C[3']); 6.37 (t, J = 6.5 Hz, 1 H, H-C[1']); 6.77–7.16 (2m, 13 H, Arom-H); 11.86 (s, 1 H, NH); 11.97 (s, 1 H, NH). Analysis calculated for $C_{35}H_{36}BrN_5O_7$: C 58.50, H 5.05, N 9.75; Found: C 58.65, H 5.10, N 9.73.

3.2.6. 3-Bromo-1-[2-deoxy-5-O-(4,4'-dimethoxytrityl)-β-D-erythro-pentofuranosyl]-1,5-dihydro-6-[(2-methylpropanoyl)amino]-4H-pyrazolo[3,4-d]pyrimidin-4-one 3'-(2-cyanoethyl diisopropylphosphoramidite) (3c)

1. A solution of **27** (300 mg, 0.42 mmol) in dry CH_2Cl_2 (5 mL) was stirred with anhydrous N,N-diisopropylethylamine (140 µL 0.84 mmol) at room temperature.
2. Then chloro(2-cyanoethoxy)(N,N-diisopropylamino)phosphine (125 µL 0.54 mmol) was added.
3. After 30 min the solution was washed with saturated aqueous $NaHCO_3$, dried over Na_2SO_4, and evaporated.

4. FC (column 10 × 2 cm, CH$_2$Cl$_2$/acetone 9:1) afforded **3c** as colorless foam (312 mg, 81%). TLC (CH$_2$Cl$_2$/acetone 9:1): R_f = 0.6. ^{31}P-NMR (CDCl$_3$): 148.4, 148.5.

3.3. Oligonucleotides

3.3.1. Oligonucleotide Synthesis

1. The oligonucleotides are prepared by solid-phase synthesis under standard conditions *(59)* in a 1-µmol scale. The coupling yields of modified phosphoramidites were always higher than 98%.
2. After the synthesis, the oligonucleotides were deprotected in 25% aqueous NH$_3$ at 60°C for 16 h.
3. The oligonucleotides were purified by reversed-phase HPLC using the following solvent systems: MeCN (A), and 0.1*M* (Et$_3$NH)OAc, (pH 7.0)/MeCN 95:5 (B). They were used in the following order: gradient I, 3 min 15% A in B, 12 min 15–40% A in B, 5 min 40–15% A in B with a flow rate of 1.0 mL/min; gradient II, 20 min 0–20% A in B with a flow rate of 1.0 mL/min; solvent system III, 30 min B with a flow rate of 0.6 mL/min. The oligomers carrying 5'-O-dimethoxytrityl (DMT) residues were purified by HPLC (250 × 4 mm RP-18 column) using gradient I.
4. The DMT-residues were removed by treating the oligomers with 2.5% Cl$_2$CHCOOH /CH$_2$Cl$_2$ for 5 min at room temperature followed by neutralization with Et$_3$N and coevaporation with absolute MeOH.
5. The detritylated oligomers were purified by HPLC (250 × 4 mm RP-18 column) using gradient II.
6. The oligomers separated by RP18-HPLC were desalted on a 125 × 4 mm RP-18 silica gel column with water to elute the salt, whereas the oligomers were eluted by MeOH/H$_2$O (3:2 v/v).
7. The purified oligonucleotides were lyophilized on a SpeedVac evaporator to yield colorless solids that were dissolved in 100 µL of H$_2$O and stored frozen at –18°C.
8. The homogeneity of the oligonucleotides was established by ion-exchange chromatography on a NucleoPac-PA-100 column (4 × 50 mm, Dionex, P/N 043018, USA).
9. The composition of the oligonucleotide was determined by tandem hydrolysis with snake-venom phosphodiesterase (SVPD) and alkaline phosphatase, Typically, the oligonucleotides (0.2 A$_{260}$ U) were dissolved in 200 µL 0.1*M* Tris-HCl buffer, pH 8.3, and treated with SVPD (EC 3.1.4.1, *Crotalus adamanteus,* 3 µL) at 37°C for 45 min and then with alkaline phosphatase (EC. 3.1.3.1, *Escherichia coli*, 3 µL) at 37°C for 30 min. The mixture was analyzed on reversed-phase HPLC (RP-18, solvent system III, at 260 nm). Quantification of the constituents was made on the basis of the peak areas, which were divided by the extinction coefficients of the nucleosides.
10. The molecular mass of oligonucleotides was obtained by matrix-assisted laser desorption/ionization time-of-flight mass spectrometry (MALDI-TOF) (positive mode, matrix: 3-hydroxypicolinic acid).

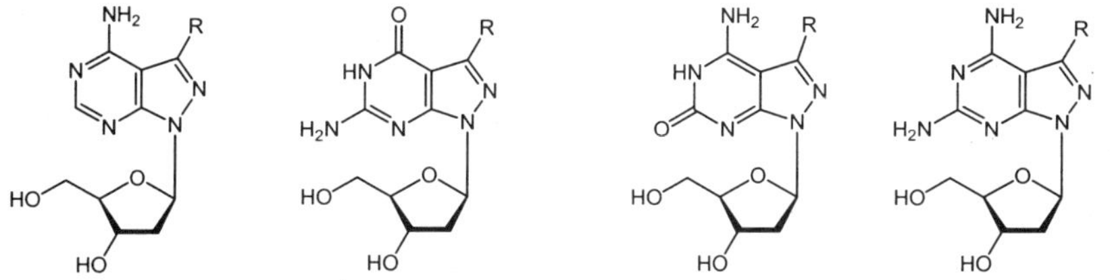

a:R = H, b: R = Cl, c: R = Br, d: R = I, e: R = Propynyl

Scheme. 6. Base-modified nucleosides incorporated in oligonucleotides shown in **Tables 2–6**.

3.3.2. Oligonucleotide Duplex Stability

The stability of oligonucleotide duplexes depicted in **Tables 2–6** was measured UV-spectrophotometrically within a temperature range of 5–85°C. The Tm-values of the duplexes containing the base-modified nucleosides **12, 13, 28**, and **29** were determined through absorbance–temperature profiles measured in 1M NaCl, 100 mM MgCl$_2$, 60 mM Na-cacodylate, pH 7.0, or 0.1M NaCl, 10 mM MgCl$_2$, 10 mM Na-cacodylate buffer, pH 7.0. The thermodynamic data were calculated by the shape analysis of the melting profiles using the program MeltWin *(60)*. The monomeric nucleosides shown in **Scheme 6** were synthesized as reported: **12a–e** *(43,53,61)*, **13a** *(62)*, **13c–e** *(45,54,57,63)*, **28a** *(18,64)*, **28c,d** *(49,50)*, **29a** *(65)*, **29c–e** *(39,48,66)*.

From **Table 2** it is apparent that the replacement of dA residues by the 8-aza-7-deazapurine analog **12a** has no influence on the duplex stability. However, when the 7-bromo, 7-iodo or 7-propynyl derivatives are incorporated, a significant increase of the Tm value is observed. For the standard duplex, **30 · 31**, which is used in those experiments, the Tm-increase corresponds to about 2°C per modification. The effect of the halogen modifications (**12c**:Br vs **12d**:I) is similar; an additional increase is observed in the case of the propynyl derivative **12e** (*see* **Table 2**).

The situation changes in the case of the dG analogs **13a, c–e** (**Table 3**). Already the modification of the purine moiety by the 8-aza-7-deazapurine system increases the base pair stability by about 1°C per replacement duplex **40 · 41** in comparison to **30 · 31**). A further increase is observed when halogeno or propynyl substituents are introduced at the 7-position. The Tm increase amounts to about 2°C per modification. The motifs of the base-modified base pair related to the dA–dT as well as the dG–dC pair are depicted in **Scheme 7**.

A related but different situation is observed when the dG–dC base pair is replaced by an iG$_d$–iC$_d$ pair and the same substituent changes are made on the

Table 2
Tm-Values and Thermodynamic Data for Duplexes Containing dA Derivatives 12a, c–e[a] *(45,58)*

Duplex	Tm [°C]	ΔTm[b] [°C]	ΔH° [kcal/mol]	ΔS° [cal/mol K]	ΔG°$_{310}$ [kcal/mol]
5'-d(TAGGTCAATACT) (**30**) 3'-d(ATCCAGTTATGA) (**31**)	47	—	–84	–236	–10.6
5'-d(T**12a**GG T C**12a12a**T**12a** C T) (**32**) 3'-d(A T CC**12a** G T T **12a**T G A) (**33**)	47	0	–97	–260	–10.9
5'-d(T **12c**G G T C **12c 12c** T **12c**CT) (**34**) 3'-d(A T C C **12c**G T T **12c** T G A) (**35**)	57	1.7	–111	–308	–15.2
5'-d(T **12d**G G T C **12d 12d** T **12d**CT) (**36**) 3'-d(A T C C**12d**G T T **12d** T G A) (**37**)	58	1.8	–95	–261	–13.9
5'-d(T **12e**G G T C **12e12e** T**12e**CT) (**38**) 3'-d(A T C C **12e**G T T **12e** T G A) (**39**)	63	2.7	–99	–269	–15.7

[a]Measured UV-spectrophotometrically at 260 nm in 0.1M NaCl, 10 mM MgCl$_2$, 10 mM Na-cacodylate, pH 7.0. The oligonucleotide concentration is 5 μM + 5 μM. [b]The Tm-increase per modification.

Table 3
Tm-Values and Thermodynamic Data for Duplexes Containing dG Derivatives 13a, c–e[a] *(38,45)*

Duplex	Tm [°C]	ΔTm[b] [°C]	ΔH° [kcal/mol]	ΔS° [cal/mol K]	ΔG°$_{310}$ [kcal/mol]
5'-d(TAG GTC AAT ACT) (**30**) 3'-d(ATC CAG TTA TGA) (**31**)	47	—	–84	–236	–10.6
5'-d(TA**13a13a**T C A A T A C T) (**40**) 3'-d(AT C C A **13a**T TA T**13a**A) (**41**)	50	0.8	–93	–263	–11.8
5'-d(TA**13c13c**T C A A T A C T) (**42**) 3'-d(AT C C A **13c**T TA T**13c**A) (**43**)	55	2	–78	–219	–14.3
5'-d(TA**13d13d**T C A A T A C T) (**44**) 3'-d(AT C C A **13d**T TA T**13d**A) (**45**)	54	1.8	–79	–217	–14.2
5'-d(TA**13e13e**T C A A T A C T) (**46**) 3'-d(AT C C A **13e**T T A T**13e**A) (**47**)	58	2.8	–93	–256	–13.7

[a,b]See **Table 2**.

modified isoguanosine nucleoside **28a** (**Table 4**). In this case the unmodified purine-pyrimidine iG$_d$–iC$_d$ pair is already significantly more stable than that of dG–dC, which has already been reported *(67)*. Therefore, the duplex **48 · 49** has to be used as the reference and not the duplex **30 · 31**, which was used in all other cases. The duplex **50 · 51** incorporating four **28**–iC$_d$ base pairs shows almost the same Tm value as those containing the unmodified purine bases (**48 · 49**). The introduction of the brominated nucleoside residue **28c** increases the stability by 9°C, which corresponds to 2.3°C per modification. The influence of the iodo nucleoside **28d** was significantly higher than that of the bromo compound, which showed only a small change in the other cases (*see* **Table 4**). The propynyl derivative was not studied.

The most significant contribution to the base pair stability came from the halogenated or propynyl derivative of the 2,6-diaminopurine nucleoside analogs **29a–e**. In this case, the total replacement of four dA by four **29c** residues resulted in a Tm change from 47 to 67°C, which equals to 5°C per modification (*see* **Table 5**). This is the result of the formation of a tridentate base pair as shown in motif IV together with the effects of the 7-substituents (*see* **Scheme 8**).

From the data discussed in the preceding section, it is apparent that the incorporation of 7-substituted 8-aza-7-deazapurine nucleosides in oligonucleotide duplexes leads to a significant increase of the duplex stability as indicated by the high Tm values. Already 7-halogeno substituents lead to a strong stabilization, 7-propynyl residues offer—in some cases—a further stability increase. The extraordinary base pair stability found for the 7-substituted 8-aza-7-deazapurines are not observed in the case of the 5-substituted pyrimidines carrying the same substituents *(48)*. The base pair stabilization shown in **Tables 2–5** results from an increased polarizability of the nucleobase, which will increase stacking interactions and the electronegative hydrophobic character of the 7-substituents protruding in the major groove of the DNA helix. As a result, water molecules will be expelled and the base stacking will be increased. The special high anti-conformation of the 8-aza-7-deazapurine nucleoside system and the increased proton donor activity of the 2-amino group induced by the electron-withdrawing 7-substituents are also of importance. Such a base pair stabilization is not observed for the **29a**–dT pair (*see* duplex **56 · 57**) indicating that the third H-bond is extremely weak in the case of the nonhalogenated compounds. The base pair motifs formed in those duplexes are displayed in **Schemes 7** and **8**.

Table 6 indicates that the replacement of the dA–dT by a **29c**–dT pair leads to a harmonization of the dA–dT vs. the dG–dC base pair stability *(68,69)*. Moreover, the **29c**–dT pair shows similar mismatch discrimination as that of dA–dT *(66)*. As a result, the duplex stability becomes independent of the base pair composition. This phenomenon can circumvent problems arising from the higher stability of dG–dC-rich oligonucleotide duplexes containing mismatches compared to perfectly matching ones being less stable owing to the high content of dA–dT base pairs. Thus, hybridization can be performed in a more accurate way by improving the data read out of hybridization experiments performed in solution or on arrays.

Table 4
T*m*-Values and Thermodynamic Data for Duplexes Containing iG$_d$ Derivatives 28a, c–d^a *(50,67)*

Duplex	T*m* [°C]	ΔT*m*^{b} [°C]	ΔH° [kcal/mol]	ΔS° [cal/mol K]	ΔG°$_{310}$ [kcal/mol]
5'-d(TAG GTC AAT ACT) (**30**) 3'-d(ATC CAG TTA TGA) (**31**)	51		−85	−237	−11.6
3'-d(TAiG iGTiC AAT AiCT) (**48**) 5'-d(ATiC iCAiG TTA TiGA) (**49**)	60	−	−94	−257	−14.7
3'-d(TA**28a 28a**T iC AAT A iC T) (**50**) 5'-d(AT iC iC A**28a** TTA T**28a**A) (**51**)	61	0.3	−96	−263	−14.6
3'-d(TA**28c 28c**T iC AAT A iC T) (**52**) 5'-d(AT iC iC A**28c** TTA T**28c**A) (**53**)	69	2.3	−106	−283	−17.6
3'-d(TA**28d 28d**T iC AAT A iC T) (**54**) 5'-d(AT iC iC A**28d** TTA T**28d**A) (**55**)	73	3.3	−119	−321	−20.0

^{a}Measured UV-spectrophotometrically at 260 nm in 1*M* NaCl, 100 m*M* MgCl$_2$, 60 m*M* Na-cacodylate, pH 7.0. The oligonucleotide concentration is 5 μ*M* + 5 μ*M*. d(iC), 5-methyl-2'-deoxyisocytidine; d(iG), 2'-deoxyisoguanosine.
^{b}*See* **Table 2**.

Table 5
T*m*-Values and Thermodynamic Data for Duplexes Containing 29a, c–e^a *(39,45)*

Duplex	T*m* [°C]	ΔT*m*^{b} [°C]	ΔH° [kcal/mol]	ΔS° [cal/mol K]	ΔG°$_{310}$ [kcal/mol]
5'-d(TAG GTC AAT ACT) (**30**) 3'-d(ATC CAG TTA TGA) (**31**)	47	−	−84	−236	−10.6
5'-d(TAG G T C **29a 29a**T A C T) (**56**) 3'-d(ATC C**29a** G T T **29a** T G A) (**57**)	51	1	−98.7	−278.6	−12.3
5'-d(TAG G T C **29c 29c**T A C T) (**58**) 3'-d(ATC C**29c** G T T **29c** T G A) (**59**)	67	5	−105.4	−285.0	−17.0
5'-d(TAG G T C **29d 29d**T A C T) (**60**) 3'-d(ATC C**29d** G T T **29d** T G A) (**61**)	66	4.8	−104.9	−284.7	−16.6
5'-d(TAG G T C **29e 29e**T A C T) (**62**) 3'-d(ATC C**29e** G T T **29e** T G A) (**63**)	68	5.3	−111	−300	−18.0

^{a,b}*See* **Table 2**.

4. Notes

1. From the second zone, compound **18** was obtained as a colorless foam (205 mg, 8%). Evaporation of the slow-migrating zone (petroleum ether/ethyl acetate 1:2) yielded the debrominated **19** (110 mg, 5%) *(54)*.
2. Apart from the N-9 compound, the debrominated N-8 isomer, compound **22**, was formed *(54)*.

Table 6
Comparison of T*m*-Values and Thermodynamic Data of Oligonucleotides Incorporating dA-dT, *29c*-dT or dG–dC Basepairs^a (66)

Duplex	Tm [°C]	ΔTm^b [°C]	ΔH° [kcal/mol]	ΔS° [cal/mol K]	ΔG°$_{310}$ [kcal/mol]
5'-d(TAG GTC **A**AT ACT) (**30**) 3'-d(ATC CAG **T**TA TGA) (**31**)	47	–	–84	–236	–10.6
5'-d(TAG GTC **29c**AT ACT) (**64**) 3'-d(ATC CAG **T**TA TGA) (**65**)	54	7	–100	–281	–12.9
5'-d(TAG GTC **G**AT ACT) (**66**) 3'-d(ATC CAG **C**TA TGA) (**67**)	53	6	–89	–248	–12.1
5'-d(TAG GTC **29c29c**T ACT) (**68**) 3'-d(ATC CAG **T TA** TGA) (**69**)	56	4.5	–91	–252	–13.4
5'-d(TAG GTC **GG**T ACT) (**70**) 3'-d(ATC CAG **CC**A TGA) (**71**)	57	5	–104	–290	–14.1

^{a,b}See **Table 2**.

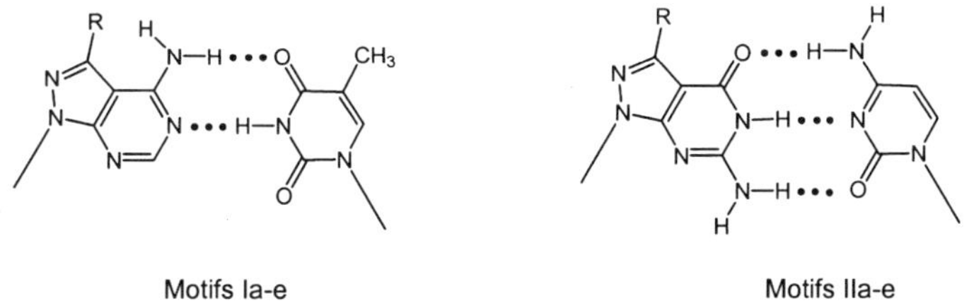

Motifs Ia-e   Motifs IIa-e

a: R = H, c: R = Br, d: R = I, e: R = Propynyl

Scheme. 7. Basepair motifs related to dA-dT and dG-dC.

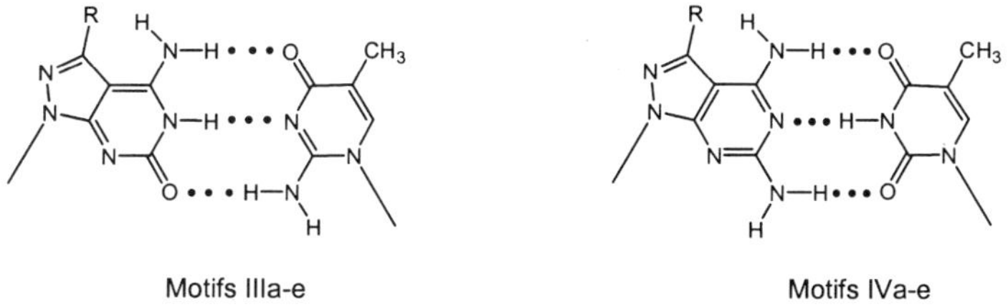

Motifs IIIa-e   Motifs IVa-e

a: R = H, c: R = Br, d: R = I, e: R = Propynyl

Scheme. 8. Basepair motifs related to iG$_d$ and n^{2}A$_d$-dT.

Acknowledgments

We thank Ms. Elisabeth Feiling for the DNA synthesis and Mrs. Monika Dubiel for the help in preparing the manuscript. Financial support by the Roche Diagnostics GmbH is gratefully acknowledged.

References

1. Suhadolnik, R. J. (1970) *Nucleoside Antibiotics.* Wiley, New York.
2. Rozenski, J., Crain, P. F., and McCloskey, J. A. (1999) The RNA Modification Database: 1999 update. *Nucleic Acids Res.* **27,** 196, 197.
3. Warren, R. A. J. (1980) Modified bases in bacteriophage DNAs. *Annu. Rev. Microbiol.* **34,** 137–158.
4. Martin, J. C. (1989) *Nucleotide Analogues as Antiviral Agents.* American Chemical Society, Washington, D.C.
5. Simons, C. (2001) *Nucleoside Mimetics: Their Chemistry and Biological Properties.* Gordon and Breach, Amsterdam.
6. Ramzaeva, N., Rosemeyer, H., Leonard, P., et al. (2000) Oligonucleotides functionalized by fluorescein and rhodamine dyes: *Michael* addition of methyl acrylate to 2'-deoxypseudouridine. *Helv. Chim. Acta* **83,** 1108–1126.
7. Prober, J. M., Trainor, G. L., Dam, R. J., et al. (1987) A system for rapid DNA sequencing using fluorescent chain-terminating dideoxynucleotides. *Science* **238,** 336–341.
8. Mizusawa, S., Nishimura, S., and Seela, F. (1986) Improvement of the dideoxy chain termination method of DNA sequencing by use of deoxy-7-deazaguanosine triphosphate in place of dGTP. *Nucleic Acids Res.* **14,** 1319–1324.
9. Seela, F. (2002) Base-modified nucleosides and oligonucleotides: synthesis and application. *Collection Symposium Series* **5,** 1–15.
10. Kool, E. T. (2002) Replacing the nucleobases in DNA with designer molecules. *Acc. Chem. Res.* **35,** 936–943.
11. Chadwick, D. J. and Cardew, G. (1997) *Oligonucleotides as Therapeutic Agents.* Wiley, Chichester, UK.
12. Uhlmann, E. and Peyman, A. (1990) Antisense oligonucleotides: a new therapeutic principle. *Chem. Rev.* **90,** 543–584.
13. Ross, J. (1997) *Nucleic Acid Hybridization: Essential Techniques.* Wiley, New York.
14. Seela, F. and Wei, C. (1997) Oligonucleotides containing consecutive 2'-deoxyisoguanosine residues: synthesis, duplexes with parallel chain orientation, and aggregation. *Helv. Chim. Acta* **80,** 73–85.
15. Schena, M. (2003) *Microarray Analysis.* Wiley, Hoboken, NJ.
16. Forman, J. E., Walton, I. D., Stern, D., Rava, R. P., and Trulson, M. O. (1998) Thermodynamics of duplex formation and mismatch discrimination on photolithographically synthesized oligonucleotide arrays. *J. Am. Chem. Soc.* **120,** 206–228.

17. Seela, F., Wei C., Melenewski, A., and Feiling, E. (1998) Parallel-stranded duplex DNA and self-assembled quartet structures formed by isoguanine and related bases. *Nucleosides & Nucleotides* **17,** 2045–2052.
18. Seela, F. and Kröschel, R. (2001) Quadruplex and pentaplex self-assemblies of oligonucleotides containing short runs of 8-aza-7-deaza-2'-deoxyisoguanosine or 2'-deoxyisoguanosine. *Bioconjugate Chem.* **12,** 1043–1050.
19. Seela, F., Wiglenda, T., Rosemeyer, H., Eickmeier, H., and Reuter, H. (2002) 7-Deaza-2'-deoxyxanthosine dihydrate forms water-filled nanotubes with C-H(((O hydrogen bonds. *Angew. Chem. Int. Ed. Engl.* **41,** 603–605.
20. Freier, S. M. and Altmann K.-H. (1997) The ups and downs of nucleic acid duplex stability: structure-stability studies on chemically-modified DNA: RNA duplexes. *Nucleic Acids Res.* **25,** 4429–4443.
21. Sági, J., Szemzö, A., Ébinger, K., et al. (1993) Base-modified oligodeoxynucleotides. I. Effect of 5-alkyl, 5-(1-alkenyl) and 5-(1-alkynyl) substitution of the pyrimidines on duplex stability and hydrophobicity. *Tetrahedron Lett.* **34,** 2191–2194.
22. Barnes, T. W., III, and Turner, D. H. (2001) Long-range cooperativity in molecular recognition of RNA by oligodeoxynucleotides with multiple C5-(1-propynyl) pyrimidines. *J. Am. Chem. Soc.* **123,** 4107–4118.
23. Barnes, T. W., III, and Turner, D. H. (2001) C5-(1-propynyl)-2'-deoxy-pyrimidines enhance mismatch penalties of DNA:RNA duplex formation. *Biochemistry* **40,** 12,738–12,745.
24. Wagner, R. W., Matteucci, M. D., Lewis, J. G., Gutierrez, A. J., Moulds, C., and Froehler, B. C. (1993) Antisense gene inhibition by oligonucleotides containing C-5 propyne pyrimidines. *Science* **260,** 1510–1513.
25. Froehler, B. C., Wadwani, S., Terhorst, T. J., and Gerrard, S. R. (1992) Oligodeoxynucleotides containing C-5 propyne analogs of 2'-deoxyuridine and 2'-deoxycytidine. *Tetrahedron Lett.* **33,** 5307–5310.
26. Gutierrez, A. J., Matteucci, M. D., Grant, D., Matsumura, S., Wagner, R. W., and Froehler, B. C. (1997) Antisense gene inhibition by C-5-substituted deoxyuridine-containing oligodeoxynucleotides. *Biochemistry* **36,** 743–748.
27. Ahmadian, M., Zhang, P., and Bergstrom, D. E. (1998) A comparative study of the thermal stability of oligodeoxyribonucleotides containing 5-substituted 2'-deoxyuridines. *Nucleic Acids Res.* **26,** 3127–3135.
28. Graham, D., Parkinson, J. A., and Brown, T. (1998) DNA duplexes stabilized by modified monomer residues: synthesis and stability. *J. Chem. Soc., Perkin Trans.* **1,** 1131–1138.
29. Armitage B. A. (2003) The impact of nucleic acid secondary structure on PNA hybridization. *Drug Discov. Today* **8,** 222–228.
30. Nielsen, P. E., Egholm, M., Berg, R. H., and Buchardt, O. (1991) Sequence-selective recognition of DNA by strand displacement with thymine-substituted polyamide. *Science* **254,** 1498–1500.
31. Nielsen P. E. and Haaima, G. (1997) Peptide nucleic acid (PNA): a DNA mimic with a pseudopeptide backbone. *Chem. Soc. Rev.* **26,** 73–78.

32. Uhlmann, E., Peyman, A., Breipohl, G., and Will, D. W. (1998) PNA: synthetic polyamide nucleic acids with unusual binding properties. *Angew. Chem. Int. Ed. Engl.* **37,** 2796–2823.
33. Petersen, M. and Wengel, J. (2003) LNA: a versatile tool for therapeutics and genomics. *J. Trends Biotechnol.* **21,** 74–81.
34. Seela, F. and Thomas, H. (1995) Duplex stabilization of DNA: oligonucleotides containing 7-substituted 7-deazaadenines. *Helv. Chim. Acta* **78,** 94–108.
35. Ramzaeva, N. and Seela, F. (1996) Duplex stability of 7-deazapurine DNA: oligonucleotides containing 7-bromo- or 7-iodo-7-deazaguanine. *Helv. Chim. Acta* **79,** 1549–1558.
36. Seela, F. and Chen, Y. (1996) Oligonucleotides containing 7- or 8-methyl-7-deazaguanine: steric requirements of major groove substituents on the DNA structure. *Chem. Commun.* 2263, 2264.
37. Seela, F. and Becher, G. (1998) Stabilisation of duplex DNA by 7-halogenated 8-aza-7-deazaguanines. *Chem. Commun.* 2017, 2018.
38. Seela, F. and Becher, G. (1999) Oligonucleotides containing pyrazolo(3,4-*d*]pyrimidines: the influence of 7-substituted 8-aza-7-deaza-2'-deoxyguanosines on the duplex structure and stability. *Helv. Chim. Acta* **82,** 1640–1655.
39. Seela, F. and Becher, G. (2001) Pyrazolo[3,4-*d*]pyrimidine nucleic acids: adjustment of dA-dT to dG-dC base pair stability. *Nucleic Acids Res.* **29,** 2069–2078.
40. Seela, F. and Kaiser, K. (1988) 8-Aza-7-deazaadenine N^8- and N^9-(β-D-2'-deoxyribofuranosides): building blocks for automated DNA synthesis and properties of oligodeoxyribonucleotides. *Helv. Chim. Acta* **71,** 1813–1823.
41. Seela, F., Becher, G., and Zulauf, M. (1999) 8-Aza-7-deazapurine DNA: synthesis and duplex stability of oligonucleotides containing 7-substituted bases. *Nucleosides Nucleotides* **18,** 1399, 1400.
42. Seela, F. and Becher, G. (2000) Synthesis, base pairing, and fluorescence properties of oligonucleotides containing 1*H*-pyrazolo[3,4-*d*]pyrimidin-6-amine (8-Aza-7-deazapurin-2-amine) as an analogue of purin-2-amine. *Helv. Chim. Acta* **83,** 928–942.
43. Sun, L. (2002) Pronucleotides and oligonucleotides of 7-halogenated 8-aza-7-deazaadenine 2'-deoxy-β-D-ribonucleosides. Diploma work, University of Osnabrueck.
44. Seela. F., Ramzaeva, N., and Zulauf M. (1997) Duplex stability of oligonucleotides containing 7-substituted 7-deaza- and 8-aza-7-deazapurine nucleosides. *Nucleosides Nucleotides* **16,** 963–966.
45. He, J. and Seela, F. (2002) Propynyl groups in duplex DNA: stability of base pairs incorporating 7-substituted 8-aza-7-deazapurines or 5-substituted pyrimidines. *Nucleic Acids Res.* **30,** 5485–5496.
46. Seela, F. and Driller, H. (1988) 8-Aza-7-deaza-2'-deoxyguanosine: phosphoramidite synthesis and properties of octanucleotides. *Helv. Chim. Acta* **71,** 1191–1198.
47. Seela, F. and Driller, H. (1989) Alternating d(G-C)$_3$ and d(C-G)$_3$ hexanucleotides containing 7-deaza-2'-deoxyguanosine or 8-aza-7-deaza-2'-deoxyguanosine in place of dG. *Nucleic Acids Res.* **17,** 901–910.

48. He, J. and Seela, F. (2002) 8-Aza-7-deazapurine-pyrimidine base pairs: the contribution of 2- and 7-substituents to the stability of duplex DNA. *Tetrahedron* **58**, 4535–4542.
49. Seela, F., Kröschel, R., and He, Y. (2001) Parallel DNA containing pyrazolo[3,4-d]pyrimidine analogues of isoguanine. *Nucleosides, Nucleotides, Nucleic Acids* **20**, 1283–1286.
50. Seela, F. and Kröschel, R. (2003) The base pairing properties of 8-aza-7-deaza-2'-deoxyisoguanosine and 7-halogenated derivatives in oligonucleotide duplexes with paralled and antiparallel chain orientation. *Nucleic Acids Res.* **31**, 7150–7158.
51. Seela, F. and Kaiser, K. (1986) Phosphoramidites of base-modified 2'-deoxyinosine isosteres and solid-phase synthesis of d(GCI*CGC) oligomers containing an ambiguous base. *Nucleic Acids Res.* **14**, 1825–1844.
52. Seela, F., Becher, G., and Chen, Y. (2000) Fluorescence properties and base pair stability of oligonucleotides containing 8-aza-7-deaza-2'-deoxyisoinosine or 2'-deoxyisoinosine. *Nucleosides, Nucleotides Nucleic Acids* **19**, 1581–1598.
53. Seela, F. and Zulauf, M. (1998) Synthesis of 7-alkynylated 8-aza-7-deaza-2'-deoxyadenosines *via* the Pd-catalysed cross-coupling reaction. *J. Chem. Soc., Perkin Trans.* **1**, 3233–3239.
54. Seela, F., Zulauf, M., and Becher, G. (1997) Unexpected dehalogenation of 3-bromopyrazolo[3,4-d]pyrimidine nucleosides during nucleobase-anion glycosylation. *Nucleosides & Nucleotides* **16**, 305–314.
55. Winkeler, H.-D. and Seela, F. (1983) Synthesis of 2-amino-7-(2'-deoxy-β-D-erythro-pentofuranosyl)-3,7-dihydro-4*H*-pyrrolo[2,3-*d*]pyrimidin-4-one: a new isostere of 2'-deoxyguanosine. *J. Org. Chem.* **48**, 3119–3122.
56. Hoffer, M. (1960) α-thymidin. *Chem. Ber.* **93**, 2777–2781.
57. Seela, F. and Becher, G. (1998) Synthesis of 7-halogenated 8-aza-7-deaza-2'-deoxyguanosines and related pyrazolo[3,4-d]pyrimidine 2'-deoxyribonucleosides. *Synthesis* 207–214.
58. Seela, F. and Zulauf, M. (1999) Synthesis of oligonucleotides containing pyrazolo[3,4-d]pyrimidines: the influence of 7-substituted 8-aza-7-deazaadenines on the duplex structure and stability. *J. Chem. Soc., Perkin Trans.* **1**, 479–488.
59. *Users' Manual of the DNA Synthesizer.* Applied Biosystems, Weiterstadt, Germany, p. 392.
60. Mc Dowell, J. A. and Turner, D. H. (1996) Investigation of the structural basis for thermodynamic stabilities of tandem GU mismatches: solution structure of (rGAG GU CUC)$_2$ by two-dimensional NMR and simulated annealing. *Biochemistry* **35**, 14,077–14,089.
61. Seela, F. and Steker, H. (1985) Facile synthesis of 2'-deoxyribofuranosides of allopurinol and 4-amino-1H-pyrazolo[3,4-d]pyrimidine via phase-transfer glycosylation. *Helv. Chim. Acta* **68**, 563–570.
62. Seela, F. and Steker, H. (1985) Synthesis of the β-D-deoxyribofuranoside of 6-amino-1H-pyrazolo[3,4-d]-pyrimidin-4(5H)-one: a new isoster of 2'-deoxyguanosine. *Heterocycles* **23**, 2521–2524.

63. Seela, F., Ramzaeva, N., and Becher, G. (1996) 7-deazapurine DNA: oligonucleotides containing 7-substituted 7-deaza-2'-deoxyguanosine and 8-aza-7-deaza-2'-deoxyguanosine. *Collect. Czech. Chem. Commun.* **61,** s258–s261.
64. Kazimierczuk, Z., Mertens, R., Kawczynski, W., and Seela, F. (1991) 2'-deoxyisoguanosine and base-modified analogues: chemical and photochemical synthesis. *Helv. Chim. Acta* **74,** 1742–1748.
65. Seela, F. and Driller, H. (1988) 8-aza-7-deaza-2',3'-dideoxyguanosine: deoxygenation of its 2'-deoxy-β-D-ribofuranoside. *Helv. Chim. Acta* **71,** 757–761.
66. Becher, G., He, J., and Seela, F. (2001) Major-groove-halogenated DNA: the effects of bromo and iodo substituents replacing H–C(7) of 8-aza-7-deazapurine-2,6-diamine or H–C(5) of uracil residues. *Helv. Chim. Acta* **84,** 1048–1065.
67. Seela, F., He, Y., and Wei, C. (1999) Parallel-stranded oligonucleotide duplexes containing 5-methylisocytosine-guanine and isoguanine-cytosine base pairs. *Tetrahedron* **55,** 9481–9500.
68. Nguyen, H.-K., Auffray, P., Asseline, U., Dupret, D., and Thuong, N. T. (1997) Modification of DNA duplexes to smooth their thermal stability independently of their base content for DNA sequencing by hybridization. *Nucleic Acids Res.* **25,** 3059–3065.
69. Seela, F. and He, Y. (2003) 6-Aza-2'-deoxyisocytidine: Synthesis, properties of oligonucleotides, and base-pair stability adjustment of DNA with parallel strand orientation. *J. Org. Chem.* **68,** 367–377.

11

Introduction of Hypermodified Nucleotides in RNA

Darrell R. Davis and Ashok C. Bajji

Summary

The anticodon domain of lysine transfer ribonucleic acid (tRNA) is a model system for investigation of the structural and biochemical effects of nucleoside posttranscriptional modification. To enable detailed study of the biophysical and structural effects of hypermodified nucleosides, methods have been developed to synthesize RNA oligonucleotides containing the modified nucleosides found in lysine tRNA. We describe in detail the synthesis of protected phosphoramidites of the nucleosides methylaminomethyl-2-thiouridine (mnm^5s^2U), methylcarboxymethyl-2-thiouridine (mcm^5s^2U), and 2-thiomethyl-N-6-carbamoylthreonyl-adenosine (ms^2t^6A). We also describe methods for using these nucleoside phosphoramidite reagents to synthesize RNA oligonucleotides with modified nucleosides incorporated at the specific sequence locations corresponding to their positions in the native lysine tRNAs.

Key Words: Hypermodified; nucleotides; tRNA; phosphoramidite; *tert*-butyl hydroperoxide; deprotection of oligoribonucleotides; labile modifications; DBU; purification of oligoribonucleotides; ion-exchange HPLC; gel filtration chromatography.

1. Introduction

The anticodon domain of transfer ribonucleic acids (tRNAs) is unique in the RNA world in that certain families of tRNA isoacceptors contain hypermodified nucleosides *(1,2)*. The wobble position and the purine nucleosides adjacent to the anticodon triplet at position 37 account for most of the observed chemical diversity. Modified uridine nucleosides of the s^2mx^5U family, for example, are unique to the tRNA wobble position, and the t^6A family of amino-acid-modified adenosines is likewise unique to position 37 in tRNA (*see* **Fig. 1**). Members of these two nucleoside groups are found throughout the three kingdoms. Nucleoside modifications within the anticodon domain affect codon–anticodon fidelity *(3,4)*, reading frame maintenance *(5)*, aminoacyl synthetase recognition *(6)*, and retroviral transcription initiation *(7)*. Studies of the

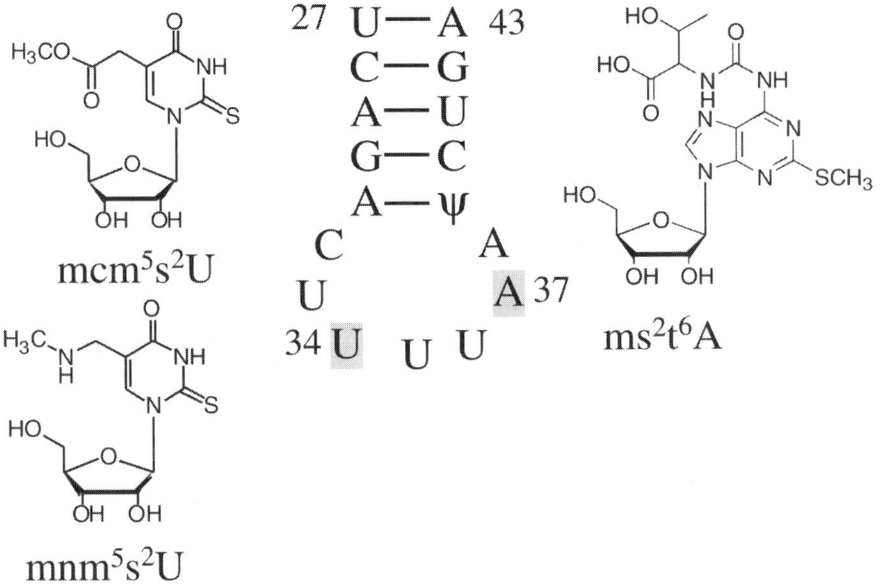

Fig. 1. The RNA sequence of the 17 nucleotide anticodon domain of human tRNA[Lys3] is shown. All RNAs were made with the naturally occurring ψ39 nucleoside. Human tRNA[Lys3] has mcm⁵s²U34 and ms²t⁶A37 and was made with an additional 3' dT. Natural *E. coli* tRNA[Lys] has the same primary sequence from residue 30 to 40 and contains mnm⁵s²U34 and t⁶A37 without the thiomethyl.

biophysical and structural properties of modified RNAs have been facilitated by the development of synthetic methods to site specifically incorporate nucleosides containing chemically sensitive groups *(8–11)*.

Automated RNA synthesis by the phosphoramidite method is a mature technology, and the protocols have been carefully optimized, allowing for the routine synthesis of RNA oligonucleotides. However, hypermodified nucleosides require numerous modifications to the standard conditions. Approaches have been developed that address the unique requirements of these interesting nucleosides but maintain the convenience provided by the automated phosphoramidite chemistry.

2. Materials

1. Automated oligonucleotide synthesizer (Applied Biosystems).
2. Ribonucleoside phosphoramidites and controlled pore glass (CPG) supports (Glen Research).
3. Flash chromatography (FC) equipment.
4. Silica gel 235–400 mesh.

5. DOWEX 50WX8-200 ion-exchange resin (Aldrich).
6. Resource Q high-pressure (performance) liquid chromatography (HPLC) resin (Pharmacia).
7. Buffer A: 20 mM NaOAc, pH 6.5, 20 mM LiClO$_4$, and 5% CH$_3$CN; Buffer B: 600 mM LiClO$_4$).
8. HPLC system with column heater.
9. SpeedVac (Savant) vacuum concentrator.
10. NAP-25 gel filtration prepacked columns (Pharmacia).
11. *tert*-butyl hydroperoxide (T-HYDRO) 70% in water (Aldrich).
12. 2-cyanoethyl-*N,N*-diisopropyl chlorophosphoramidite (Aldrich).

3. Methods

The methods describe (1) the synthesis of 17 or 18 nucleotide RNA oligonucleotides containing single modifications and combinations of hypermodified nucleosides, (2) deprotection and purification methods for 1–10-μmol scale RNA synthesis products, (3) synthesis of the protected phosphoramidite of mcm^{5}s^{2}U, (4) synthesis of the protected phosphoramidite of mnm^{5}s^{2}U, and (5) synthesis of the protected phosphoramidite of ms^{2}t^{6}A.

3.1. RNA Oligonucleotide Synthesis With Hypermodified Phosphoramidites

3.1.1. Solid Supports, Coupling, and Oxidation

The oligonucleotides were synthesized on an Applied Biosystems 394 oligoribonucleotide synthesizer on a 1-μmol scale using 0.05M acetonitrile (ACN) solutions of PAC amidites (PAC A, isopropyl PAC G, and acetyl C) from Glen Research. The labile hydroquinone-*o',o'*-diacetic acid linker was used with a dT residue at the 3' end of the oligonucleotides containing mcm^{5}s^{2}U *(12)*. This was necessary as the methyl ester is hydrolyzed under conditions required to release oligonucleotides from RNA linkers commercially available. For hypermodified RNAs that did not contain a methylester-modified nucleoside, the standard RNA solid support with the appropriate 3' ribonucleoside phosphoramidite was used. The concentrations of mcm^{5}s^{2}U, ms^{2}t^{6}A, and pseudouridine (ψ) amidites were 0.12M each (*see* **Note 1**). The phosphoramidites were coupled for 25 min. The typical I$_2$/H$_2$O oxidation was replaced by a protocol using 2 × 5 min oxidation cycles with 10% tBuOOH in ACN *(13)*.

3.1.2. Base Deprotection of RNAs Containing mnm^{5}s^{2}U and t^{6}A

The *Escherichia coli* tRNA^{Lys} anticodon contains both t^{6}A37 and mnm^{5}s^{2}U34. The deprotection conditions are described below and emphasize the unique challenges presented by the trifluoroacetyl-protected methylaminomethyl side chain. Briefly, base deprotection of the phosphate produces acrylonitrile, which must be removed prior to generating the secondary amine *(14)*. To accomplish this, the CPG-bound RNA sequences were transferred from the column to a screw-cap glass vial, to which was added 1 mL of 10% *tert*-butylamine in dry pyridine *(15)*. The solution was stirred at room temperature for 1 h. The supernatant was decanted and the residue dried under vacuum before treating with 3 mL of NH_4OH: ethanol (3:1 v/v) at room temperature for 1 h and then at 55°C for 4 h. The supernatant was decanted, the support material washed with an additional 1.0 mL of NH_4OH: ethanol and the combined washings were lyophilized on a SpeedVac concentrator.

3.1.3. Deprotection for RNAs Containing ms^2t^6A and mcm^5s^2U

The presence of ms^2t^6A37 and mcm^5s^2U34 in the anticodon of human $tRNA^{Lys,3}$ presents a set of challenges different from that of the *E. coli* tRNA. The methyl ester must be retained during the base treatment to remove the amide protection of the C, A, and G nucleosides, as well as retaining the carbamoyl functionality of ms^2t^6A. Furthermore, the ester of course must be retained, but the carboxylate protection of ms^2t^6A is removed, as well as the secondary *tert*-butyldimethylsilyl (TBS) hydroxyl protection. The CPG-bound RNA sequences were transferred from the column to a screw-cap glass vial, to which was added 1 mL of 5% 1,8-diazabicyclo[5.4.0]undec-7-ene (DBU) in methanol. The solution was stirred at room temperature for 2 h. The supernatant was decanted, the support was washed with an additional 1.0 mL of 5% DBU in methanol *(16)*, and the combined washings were lyophilized on a SpeedVac concentrator. The key elements of this protocol are the choice of protection for the carboxylate and the use of DBU in methanol.

3.1.4. Removal of Silyl Groups and RNA Purification

The dried material from either of the previously mentioned reactions (1 µmol scale) was dissolved in 1 mL neat $Et_3N \cdot 3HF$ (Aldrich) and stirred at room temperature for 9–12 h *(17,18)*. The reaction was quenched by adding 0.1 mL of H_2O, and the RNA was precipitated by adding 10 mL of n-butanol and allowing the solution to stand at –20°C for 6 h. After centrifugation, the pellet was dried under reduced pressure. The residue was dissolved in 1 mL of $1M$ tetrabutylammonium fluoride in tetrahydrofuran (THF) and stirred at room temperature for 12–18 h (*see* **Note 2**). To the solution was added 1 mL of H_2O and then it was loaded onto a NAP™-25 (Pharmacia) size exclusion column previously equilibrated with 5 column volumes of H_2O to remove the tetrabutylammonium counterions. The RNA was eluted with water and the fractions monitored by ultraviolet (UV) absorption. The RNA containing elute (as monitored by UV absorption at 260 nm) was lyophilized and purified by anion-exchange HPLC *(19)* (*see* **Fig. 2**). In our laboratory we have achieved excel-

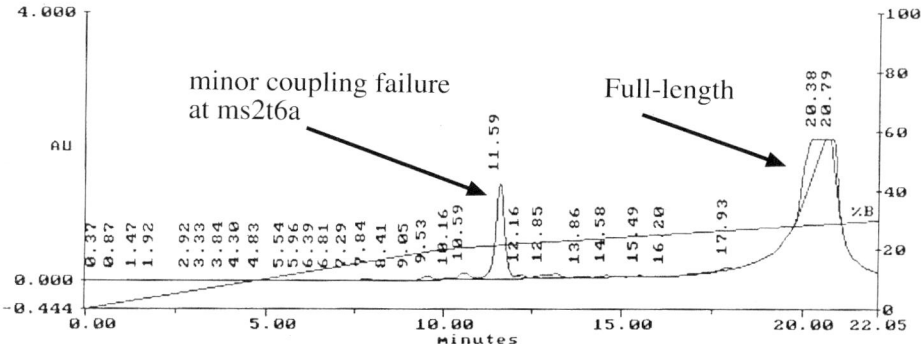

Fig. 2. Ion-exchange HPLC trace of fully modified human tRNA[Lys,3] after deprotection. Minor coupling failure owing to a low concentration of ms²t⁶A amidites elutes at 11.6 min. The HPLC analysis shows otherwise complete synthesis of full-length product and little degradation from the deprotection treatment.

lent results with a Pharmacia HR 10 column (1 cm × 15 cm) packed with Pharmacia Resource 15 Q material. The chromatography buffer is (A) 20 mM NaOAc, pH 6.5, containing 20 mM LiClO$_4$, and 5% ACN; buffer (B) 20 mM NaOAc, pH 6.5, containing 600 mM LiClO$_4$, and 5% ACN. The elution gradient is 0–15% B in 5 min, 15–55% B in 70 min, at a flow rate of 3 mL per min, 60°C. Typical elution times are 17 mers in 15 min and 31mers in 20 min. The fractions containing full-length material as judged by HPLC were collected, lyophilized, and then dialyzed against 2 × 1 L of deionized H$_2$O. The dialyzed material could then be lyophilized in preparation for further study.

3.2. Synthesis of the Protected Phosphoramidite of mcm⁵s²U

Both the 2-thio and 2-oxo nucleobases were synthesized following the protocol of Fissekis and Sweet *(20)*. We used the more common Vorbruggen coupling conditions to attach the silylated nucleobases to protected ribose with SnCl$_4$ Lewis acid catalysis *(21)*. The 2-oxo nucleoside phosphoramidite synthesis follows the protocol described for the 2-thio nucleoside, simply replacing the thiourea with urea, and therefore only the 2-thio is described in detail (*see* **Scheme 1**) *(22)*.

3.2.1. Methyl α-Formylsuccinate

To a suspension of 50 g, 0.925 mol freshly prepared sodium methoxide in dry ether at 4°C, a mixture of 135 g, 0.925 mol dimethyl succinate and 111.0 g, 1.85 mol methyl formate was added dropwise. The ice-cooled solution was stirred for 2 h and then allowed to warm to room temperature and stirred overnight. The brown residue was allowed to settle, and the ether was decanted.

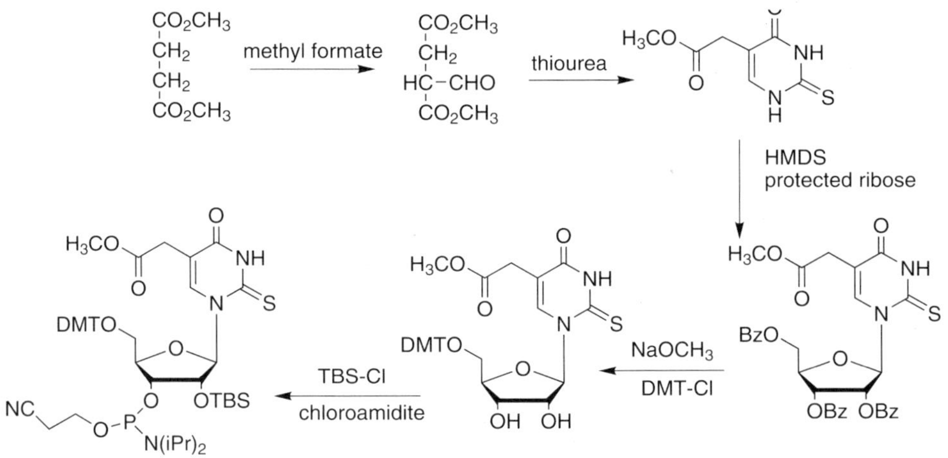

Scheme. 1. Synthesis of mcm^{5}s^{2}U amidite.

The residue was washed with cold petroleum ether then dissolved in 300 mL of cold 3*N* HCl plus a few drops of concentrated HCl. The product was extracted with ethyl acetate and the organic layer dried over sodium sulfate. The solvent was removed under reduced pressure and the pale yellow residue distilled under reduced pressure (95–105°C, approx 4 mm) to give 47.1 g (30%).

3.2.2. 5-Carbomethoxy-2-Thiouracil

Sodium metal (2.67 g, 116 mmol) was dissolved in 150 mL of methanol, then 8.83 g, 116 mmol thiourea was added followed by 20 g, 114 mmol methyl α-formylsuccinate. The solution was heated at reflux for 6 h, then cooled, and the solvent removed under reduced pressure. The yellow residue was triturated with 150 mL of 15% acetic acid and refrigerated for 4 h. The solid was collected by filtration and then washed with ice-cold ethanol to yield 11.68 g (51%).

3.2.3. 1-(2',3',5'-tri-O-Benzoyl-β-D-Ribofuranosyl)-5-Carbomethoxy-Methyluracil

1. A mixture of 6.0 g, 30 mmol 5-carbomethoxy-2-thiouracil, 90 mL hexamethyldisilizane, 3 mL chlorotrimethylsilane, and 30 mL 1,4-dioxane was heated at reflux overnight. Excess hexamethyldisilizane and dioxane were then removed under reduced pressure. The residue was dissolved in 30 mL of dry ACN, cooled on ice, and an ACN solution of 5.03 g, 19.4 mmol SnCl$_4$ added.
2. The solution was stirred at room temperature for 1.5 h, then 13.75 g, 27.3 mmol 1-*O*-acetyl-2,3,5-tri-*O*-benzoyl-β-D-ribofuranose was added as a solution in 25 mL of ACN. The solution was stirred for 2 h, then quenched with saturated sodium bicarbonate, diluted with 1,2-dichloroethane, and filtered through a Celite

pad. The organic layer was separated, dried over sodium sulfate, and the solvent removed under reduced pressure. The residue was purified by silica gel chromatography using a mixture of dichloromethane (DCM) and ethyl acetate (9:1) to give 7.26 g (42%).

3.2.4. 5-Carbomethoxymethyl-2-Thiouridine

The 4.95 g, 7.7 mmol of sugar-protected nucleoside was stirred in a solution of freshly prepared sodium methoxide (sodium metal, 0.27 g, 11.6 mmol) in 120 mL of dry methanol for 3 h at room temperature. The reaction mixture was then neutralized by adding DOWEX 50WX8-200 resin (strongly acidic), followed by filtration. The resin was washed with methanol (2 × 50 mL) and the solvent from the combined methanol fractions was removed under reduced pressure. The solid, product residue was triturated with petroleum ether and then crystallized from ethyl acetate to yield 1.92 g (75%) of the nucleoside as a colorless solid.

3.2.5. 5'-O-(4,4'-Dimethoxytrityl)-5-Carbomethoxymethyl-2-Thiouridine

To a solution of 1.00 g, 3.01 mmol 5-carbomethoxymethyl-2-thiouridine in 30 mL dry pyridine was added 1.22 g, 3.61 mmol 4,4'-dimethoxytritylchloride at room temperature. The reaction mixture was stirred at room temperature for 12 h. The solvent was evaporated under reduced pressure, and the brown residue was coevaporated with toluene (2 × 25 mL). The crude material was purified by silica gel chromatography using a mixture of DCM and methanol (9:1) to give a colorless solid (1.53 g, 80%).

3.2.6. 5'-O-(4,4'-Dimethoxytrityl)-2'-O-(tert-Butyldimethylsilyl)-5-Carbomethoxymethyl-2-Thiouridine

The 1.50 g, 2.36 mmol 5'-dimethoxytrityl (DMT) nucleoside was dissolved in dry pyridine (15 mL) under nitrogen atmosphere. To this was added 0.64 g, 9.45 mmol imidazole and 0.43 g, 2.84 mmol *tert*-butyldimethylsilyl chloride. The reaction mixture was stirred at room temperature for 10 h. The solvent was evaporated under reduced pressure and the residue was coevaporated with toluene (2 × 30 mL). The pale yellow residue was dissolved in 100 mL DCM and washed with 5% aqueous sodium hydrogen carbonate (2 × 50 mL) followed by 50 mL H_2O, dried over sodium sulfate, filtered, and the solvent then removed under reduced pressure to give a mixture of 2' and 3' isomers. The crude mixture of 2' and 3' isomers was separated by FC over silica gel using a mixture of DCM and ethyl acetate (9:1) to afford the higher R_f 2' isomer (0.71 g, 40%) and lower R_f 3'-isomer (0.69 g, 39%). To a methanolic solution of the 3' isomer was added a trace of triethylamine and the solution stirred at room temperature to

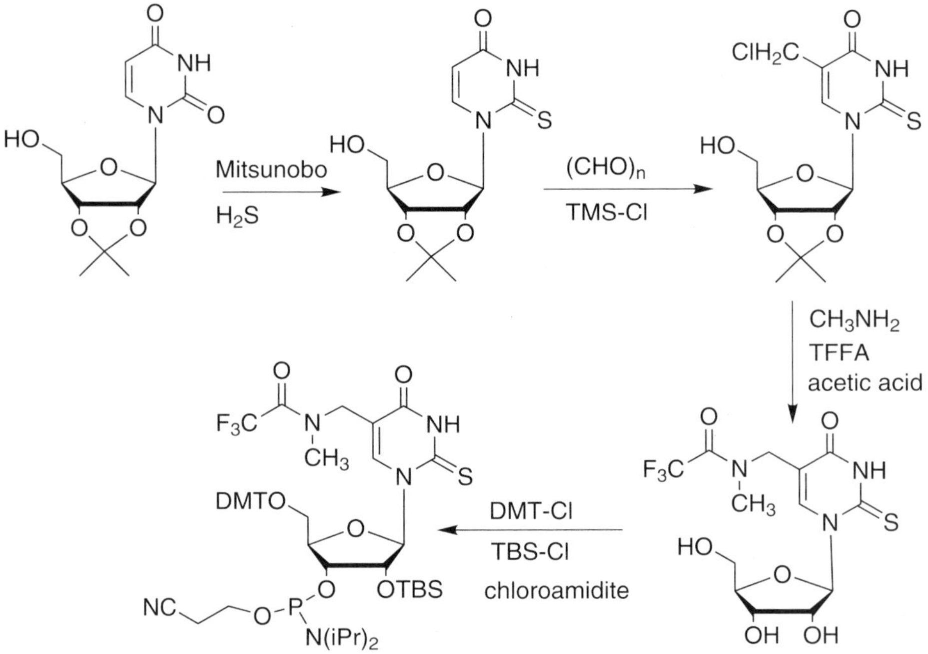

Scheme. 2. Synthesis of mnm^{5}s^{2}U amidite.

give a mixture of 2' and 3' isomers. The isomerization and chromatography were repeated two times to obtain an additional quantity of the 2' isomer for an overall yield of 1.22 g (69%).

3.2.7. 5'-O-(4,4'-Dimethoxytrityl)-2'-O-(tert-Butyldimethylsilyl)-5'-Carbomethoxymethyl-2-Thiouridine-3'-(Cyanoethyl-N,N-Diisopropylphosphoramidite)

The 2'-TBS nucleoside (0.25 g, 0.33 mmol) was dissolved in 15 mL dry THF under argon atmosphere. To this was added 8.1 mg, 0.06 mmol dimethylaminopurine (DMAP) and 114 µL, 0.66 mmol diisopropylethylamine (DIEA). To the above reaction mixture, 148 µL, 0.665 mmol 2-cyanoethyl-N,N-diisopropyl chlorophosphoramidite (Aldrich) was added while stirring. After 30 min a white solid formed, and the stirring was continued for an additional 3 h (*see* **Notes 3** and **4**). The reaction mixture was extracted with 50 mL ethyl acetate, washed with 5% sodium bicarbonate (25 mL), and finally with H$_2$O (2 × 25 mL). The organic layer was dried over sodium sulfate, filtered, and the solvent evaporated under reduced pressure. The residue was purified by FC over silica gel using DCM and ethyl acetate (9:1) to yield 0.23 g (75%). The material exists as a 1:1 ratio of stereoisomers to phosphorous, and two chemical shifts were observed for some of the ^{1}H nuclear magnetic resonance (NMR) resonances.

3.3. Synthesis of the Protected Phosphoramidite of mnm^{5}s^{2}U

The synthesis of the modified wobble nucleoside found in *E. coli* tRNA^{Lys} has been described from other laboratories *(23,24)*. We previously described the synthesis of the mnm^{5}s^{2}U nucleoside phosphoramidite *(25)*. In this section we describe the details of how the protected mnm^{5}s^{2}U nucleoside can be made from commercially available starting materials. 2',3'-*O*-isopropylidene-2-thiouridine was synthesized from 2',3'-*O*-isopropylidene-uridine (Aldrich) by the method of Malkiewicz et al. *(26)*; 5-methyl-aminomethyl-2-thiouridine was synthesized as described by Ikeda et al. *(24)*, and the secondary amino side chain was protected as the trifluoroacetate to give the *N*-protected nucleoside *(26)*. The 5' DMT,2'-TBS protected nucleoside was then synthesized in a straightforward fashion, following the general procedure described in **Subheading 3.2.6.** for mcm^{5}s^{2}U and, therefore, is not describes in detail. This method below describes the synthesis of the trifluoroacetyl protected nucleoside and then the final phosphoramidite product (*see* **Scheme 2**).

3.3.1. 2,5'-Anhydro-2',3'-O-Isopropylidene Uridine

2',3'-*O*-isopropylidene uridine (10 g, 34 mmol) was suspended in 90 mL of dry 1,4-dioxane along with 10.1 g, 39 mmol triphenylphosphine. Diisopropylazodicarboxylate (DIAD) (8.4 g, 42 mmol) was added slowly over 15 min, allowing the red color to dissipate. Sufficient DIAD should be added in total so that the red ultimately persists (*see* **Note 5**). The reaction was stirred for 3 h, then 10 mL of H$_2$O was added and the reaction heated for 30 min. The solvent was removed under reduced pressure then dried by coevaporation with toluene. The residue was resuspended in 100 mL of benzene and heated for 10 min, followed by cooling to room temperature for 1 h. The solid product was collected by filtration along with additional material collected by crystallization from the benzene solution to yield 8.9 g (89%).

3.3.2. 2',3'-O-Isopropylidene-2-Thiouridine

The 10 g, 37 mmol anhydro compound was dissolved in 200 mL of dry pyridine and frozen in a steel bomb precooled to –90°C (dry ice/ether/isopropanol). Condensed, liquid hydrogen sulfide (–90°C) was added to the pyridine solution and the bomb sealed at room temperature for 4 d. The hydrogen sulfide gas was vented slowly into a KOH/bleach solution and the pyridine removed under reduced pressure, followed by coevaporation with toluene to remove residual pyridine. The viscous residue was crystallized from hot chloroform to yield 9.2 g (83%) of pale yellow crystals.

3.3.3. 2',3'-O-Isopropylidene-5-Chloro-2-Thiouridine

1. The 6 g, 20 mmol 2-thionucleoside and 1.8 g of paraformaldehyde were heated at 60°C for 9 h in 60 mL of 0.5M triethylamine in water. The solvent was removed under reduced pressure, coevaporated with 2 × 50 mL of methanol and then dried *in vacuo*.
2. The crude hydroxymethyl nucleoside was dissolved in 100 mL of dry dioxane, and 5 equivalents of chlorotrimethylsilane were added and the mixture stirred at 60°C for 3 h. Solvent was removed under reduced pressure and then coevaporated with methanol. The resulting product was carried forward for the synthesis of the methylaminomethyl product (*see* **Note 6**).

3.3.4. 2',3'-O-Isopropylidene-5-Methylaminomethyl-2-Thiouridine

The 7.0g, 20 mmol chloromethyl nucleoside was dissolved in 80 mL of dry 1,4-dioxane and 80 mL of CH_3NH_2/methanol (1:1 v/v) chilled to –10°C in a stoppered flask. The solution was allowed to stand overnight at room temperature and then concentrated under reduced pressure and purified by silica gel chromatography with $CHCl_3$/methanol (5:1 v/v) to yield 4.8 g (60%).

3.3.5. 5-(N-Trifluoroacetyl)-Methylaminomethyl-2-Thiouridine

1. The 1.0 g, 2.9 mmol methylaminomethyl nucleoside was coevaporated from dry pyridine (3 × 5 mL) then dissolved in 15 mL of dry pyridine along with 1.6 g, 14.8 mmol chlorotrimethylsilane. The reaction was stirred for 2 h, then 2.0 g, 9.6 mmol trifluoroacetic anhydride was added and stirring continued for 12 h at room temperature. The reaction was quenched by adding cold, saturated sodium bicarbonate and stirring at 0°C for 4 h. The reaction was diluted with ethyl acetate and the organic layer washed with additional saturated sodium bicarbonate. The solvent was removed and the product purified by silica gel chromatography using chloroform/ethyl acetate/acetone (5:1:1) to yield 1.0 g (80%).
2. The 2',3'-acetonide was removed by heating the protected nucleoside at reflux in 25% acetic acid for 2 h. The solution was then evaporated to dryness and coevaporated with methanol to afford quantitative conversion to the *N*-protected nucleoside. This was then converted to the 5'-DMT, 2'-TBS nucleoside as described for mcm^{5}s^{2}U in **Subheadings 3.2.5.** and **3.2.6.**

3.3.6. 5(-O-(4,4'-DMT)-2'-O-(tert-Butyldimethylsilyl)-5-(N-Trifluoroacetyl)-Methylaminomethyl-2-Thiouridine-3'-(Cyanoethyl)N,N-Diisopropyl-Phosphoramidite)

The 0.60 g, 0.74 mmol 5'-DMT, 2'-TBS nucleoside was dissolved in 12 mL of dry THF under argon atmosphere. To this was added 0.018 g, 0.14 mmol DMAP, and 0.316 mL, 1.62 mmol DIEA. The solution was stirred while adding 0.328 mL, 1.48 mmol 2-cyanoethyl *N,N*-diisopropylphosphonamidic chloride. After 30 min a white precipitate formed, and the stirring was continued

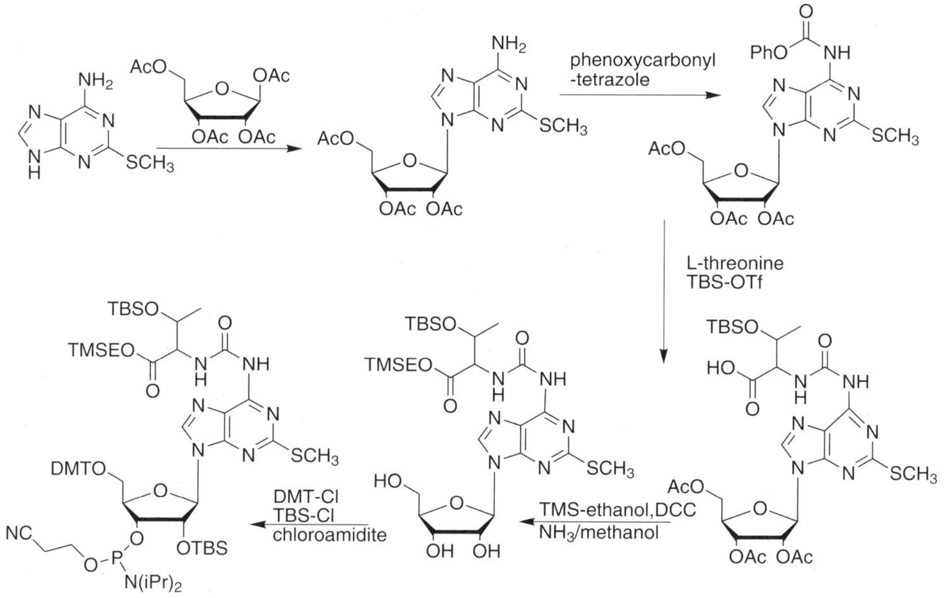

Scheme. 3. Synthesis of ms²t⁶A amidite.

for 3 h. The reaction was quenched by adding 100 mL of ethyl acetate and then extracting with saturated sodium bicarbonate, followed by saturated sodium chloride. The organic layer was separated, dried over anhydrous sodium sulfate, and the solvent was removed under reduced pressure. The residue was purified by FC on silica gel using DCM:ACN (18:1) to yield 0.47 g (64%) of a white foam (*see* **Note 7**). The material exists as a 1:1 ratio of stereoisomers to phosphorous, and two chemical shifts are observed for some of the NMR resonances: ^{31}P NMR (Acetone-d6) 155.79, 155.57.

3.4. Synthesis of the Protected Phosphoramidite of ms²t⁶A

The hypermodified nucleoside ms²t⁶A is found exclusively at position 37 of certain U36 containing tRNA isoacceptors from all three kingdoms of life *(2)*. Our original publication on this synthesis involved coupling 2-methylthio-adenosine with 1-*O*-acetyl-tri-*O*-benzoyl ribose followed by removal of the benzoyl groups and reprotection as the triacetate*(11,27)*. This was to address some problems we experienced with direct coupling to the tetra-acetyl ribose. However, on revisiting this issue, we came to the conclusion that this problem was artifactual, and therefore the tetra-acetyl ribose can be substituted for 1-*O*-acetyl, tri-*O*-benzoyl ribofuranose, eliminating the tedious deprotection/reprotection steps as shown in **Scheme 3**. However, we have chosen to leave the experimental section as originally described because the selective deprotection conditions are informative.

3.4.1. 2',3',5'-Tri-O-Benzoyl-2-Methylthioadenosine

To a suspension of 10.0 g, 55.18 mmol 2-methylthioadenine *(28,29)* and 33.40 g, 66.21 mmol 1-*O*-acetyl-2,3,5-tri-*O*-benzoyl-β-D-ribofuranose in 160 mL anhydrous nitromethane was added 10 mL $SnCl_4$ dropwise at 0°C. The temperature of the reaction mixture was slowly raised to room temperature and stirring continued for 16 h at room temperature. At the end of this period the solvent was evaporated, and the brown residue was triturated with a saturated solution of sodium bicarbonate and the solid collected by filtration. The crude product was purified by FC using a mixture of DCM and methanol (95:5) to give the product (16.58 g, 48%) as a pale yellow solid.

3.4.2. 2-Methylthioadenosine

To the 16.0 g, 25.73 mmol tri-*O*-benzoate in 150 mL methanol was added freshly prepared sodium methoxide (0.7 g of sodium in 5.0 mL of anhydrous methanol). The mixture was stirred at 25°C for 4 h and then neutralized with DOWEX 50 (H^+, 200 mesh), and filtered. The solvent was evaporated completely, and the crude product was then triturated with 5% ethyl acetate in petroleum ether and filtered. The product was washed with diethyl ether and dried to afford 7.4 g (92%) *(30)*. This product was sufficiently pure to use for the next step and was therefore carried forward without further purification.

3.4.3. 2',3',5'-Tri-O-Acetyl-2-Methylthioadenosine

A mixture of 6.0 g, 19.15 mmol 2-methylthioadenosine, 50 mL anhydrous pyridine, and 30 mL acetic anhydride was stirred at 0°C for 4 h. The excess reagent and pyridine were removed under reduced pressure and coevaporated with toluene (2 × 50 mL). The crude product was crystallized from a mixture of ethyl acetate and petroleum ether to afford 7.2 g (86%) of the triacetate as a colorless solid.

3.4.4. [9-(2',3',5'-Tri-O-Acetyl-β-D-Ribofuranosyl)-2-Methylthiopurin-6-yl]Phenylcarbamate

2-Methylthioadenosine triacetate (1.4 g, 3.18 mmol) was dissolved in 5 mL of 1,4-dioxane, and 1.82 g, 9.56 mmol phenoxycarbonyl tetrazole was added to the stirred solution *(31)*. The solvent was removed *in vacuo,* and then the suspension/evaporation step was repeated twice before adding 5 mL of 1,4-dioxane a final time. The thick slurry was heated at 35–40°C for 16 h. At the end of this period, the solvent was removed under reduced pressure, and the crude product was purified by FC using a mixture of DCM and ethyl acetate (8:2) to afford 1.32 g (74%) of the carbamate. The ^{1}H NMR spectrum showed that the product was relatively pure; however, minor peaks were seen at 6.8 and 9.3 ppm, indicating a persistent impurity that resisted multiple attempts to obtain absolutely clean material.

3.4.5. N-[[9-(2',3',5'-Tri-O-Acetyl-β-D-Ribofuranosyl)-2-Methylthiopurin-6-yl]Carbamoyl]-L-Threonine

To a 1.3 g, 2.32-mmol solution of the phenylcarbamate in 10 mL pyridine was added 0.83 g, 6.97 mmol L-threonine. The mixture was stirred at 35°C for 10 h, then the excess L-threonine was filtered off and washed with pyridine. The combined filtrate was evaporated to dryness, and the residue was coevaporated with toluene (3 × 10 mL). The pale yellow residue was purified by FC over silica gel using a mixture of DCM and methanol (7:3) to afford the coupled product (1.2 g, 88%) as a colorless solid.

3.4.6. N-[[9-(2',3',5'-Tri-O-Acetyl-β-D-Ribofuranosyl)-2-Methylthiopurin-6-yl]Carbamoyl]-O-Tert-Butyldimethylsilyl-L-Threonine

1. To a 0.8 g, 1.37-mmol solution of the threonine nucleoside in anhydrous DCM was added 0.69 g, 6.85 mmol triethylamine, followed by 1.04 g, 5.48 mmol *tert*-butyl-dimethylsilyl triflate at room temperature, and stirring was then continued for a further 6 h at room temperature. The reaction mixture was diluted with 60 mL DCM, washed with water, dried over sodium sulfate, filtered, and the solvent removed under reduced pressure.
2. The crude product was treated with 2*M* ammonia in 30 mL methanol for 5 min and then evaporated to dryness. The product was purified by FC over silica gel using a mixture of DCM and methanol (9:1) to give the TBS protected product (0.78 g, 82%) as a pale yellow foam.

3.4.7. N-[[9-(2',3',5'-Tri-O-Acetyl-β-D-Ribofuranosyl)-2-Methylthiopurin-6-yl]Carbamoyl]-O-tert-Butyldimethylsilyl-L-Threonine Trimethylsilylethyl Ester

A solution of 0.28 g, 1.34 mmol dicyclohexylcarbodiimide in anhydrous 2 mL DCM was added to a 0.78 g, 1.12-mmol mixture of the TBS protected nucleoside and 0.034 g, 0.28 mmol *N*,*N*-dimethylaminopyridine in anhydrous DCM under nitrogen atmosphere at 0°C. The reaction mixture was stirred for 10 min at 0°C, then 0.17 g, 1.45 mmol trimethylsilylethanol was added at 0°C. After the addition was complete, the reaction mixture was stirred at room temperature for 10 h, then the separated solid was filtered off and washed with 5 mL DCM. The combined filtrate was evaporated to dryness under reduced pressure, and the residue was triturated with ether, filtered, and the filtrate evaporated to dryness. The pale yellow product was purified by FC over silica gel using a mixture of DCM and methanol (95:5) to give 0.8 g (90%) of the carboxyl protected nucleoside as a colorless solid. It proved difficult to obtain analytically pure samples of this product, but the product could be successfully carried forward. The ^{1}H NMR spectrum indicated the compound was correctly synthesized, but thin-layer chromatography (TLC) and NMR analysis showed the presence of cyclohexylurea contamination with the same Rf as the product.

3.4.8. N-[[9-(β-D-Ribofuranosyl)-2-Methylthiopurin-6-yl]Carbamoyl]-O-tert-Butyldimethylsilyl-L-Threonine Trimethylsilylethyl Ester

The 1.3 g, 1.63 mmol acetyl protected nucleoside was dissolved in 2M ammonia in 25 mL methanol and stirred at room temperature for 3 h. The solvent was evaporated, and the residue was triturated with a mixture of ethyl acetate and petroleum ether (1:9) and filtered to afford the 2',3',5' hydroxyl nucleoside (1.00 g, 92%) as a colorless solid.

3.4.9. N-[[9-(5'-O-(4,4'-Dimethoxytrityl)-β-D-Ribofuranosyl)-2-Methylthiopurin-6-yl]Carbamoyl]-O-tert-Butyldimethylsilyl-L-Threonine Trimethylsilylethyl Ester

A 0.5 g, 0.74-mmol mixture of the free ribose nucleoside and 0.30 g, 0.89 mmol 4,4'-DMTrCl in 20 mL anhydrous pyridine were stirred at 25°C under argon atmosphere for 16 h. The solvent was evaporated under reduced pressure, and the residue was coevaporated with toluene (2 × 20 mL). The product was purified by FC over silica gel using a mixture of DCM and methanol (9:1) to afford the 5-DMT nucleoside (0.62 g, 89%) as a pale yellow solid.

3.4.10. N-[[9-(2'-O-tert-Butyldimethylsilyl-5'-O-(4,4'-Dimethoxytrityl)-β-D-Ribofuranosyl)-2-Methylthiopurin-6-yl]Carbamoyl]-O-tert-Butyldimethylsilyl-L-Threonine Trimethylsilylethyl Ester

1. The 0.50 g, 0.535 mmol 5'-DMT nucleoside was dissolved in 20 mL anhydrous pyridine under argon atmosphere. To this was added 0.15 g, 2.0 mmol imidazole and 0.10 g, 0.69 mmol *tert*-butyl-dimethylsilylchloride. The reaction mixture was stirred for 24 h at which time the starting material was converted to a nearly equimolar mixture of 2' and 3' TBS products. The solvent was removed under reduced pressure, and the residue was coevaporated with toluene (2 × 25 mL), then extracted with DCM. The organic solution was washed with 5% aqueous sodium bicarbonate, then with water, dried over sodium sulfate, filtered, and the solvent evaporated. The crude product was purified by FC over silica gel using a mixture of ethyl acetate and DCM (3:7) as the eluent to yield 0.18 g of the 2'-TBS nucleoside as a foam.
2. The 3'-TBS product was isomerized to an equimolar mixture of 2' and 3' TBS isomers by stirring in methanol with a trace of triethylamine. The isomerization

and chromatography was repeated twice to obtain additional 2'-TBS material, resulting in an overall yield of 0.41g (70%).

3.4.11. N-[[9-(2'-O-tert-Butyldimethylsilyl-3'-(2-Cyanoethyl-N,N-Diisopropylphosphoramidite)-5'-O-(4,4'-Dimethoxytrityl)-β-D-Ribofuranosyl)-2-Methylthiopurin-6-yl]Carbamoyl]-O-tert-Butyldimethylsilyl-L-Threonine Trimethylsilylethyl Ester

The 0.25 g, 0.23 mmol 5'-DMT, 2'-TBS nucleoside was dissolved in 10 mL dry THF under argon atmosphere. To this was added 5.59 mg, 0.046 mmol DMAP and 79 μL, 0.46 mmol DIEA. To this reaction mixture 102 μL, 0.46 mmol 2-cyanoethyl-N,N-diisopropyl chlorophosphoramidite was added while stirring. After 30 min a white solid formed, and the stirring was continued for an additional 3 h (*see* **Note 3**). The reaction mixture was extracted with 50 mL ethyl acetate, washed with 25 mL 5% sodium bicarbonate and finally with H_2O (2 × 25 mL). The organic layer was dried over sodium sulfate, filtered, and the solvent removed under reduced pressure. The residue was purified by FC over silica gel using DCM and ethyl acetate (6:1) to yield 0.22 g (75%). ^{31}P NMR (Me_2SO-$d6$): 151.0, 149.6. The upfield region of the ^1H NMR spectrum has considerable overlap owing to the presence of two diastereomers and the multiple protecting groups. However, the downfield region indicates the compound has the characteristic peaks at 9.21 and 10.0 for the amide protons on the threonyl side chain and the H8 proton at 8.41 ppm.

4. Notes

1. Nucleoside phosphoramidites were lyophilized from a frozen benzene solution, further dried under vacuum over P_2O_5 in a drying pistol, flushed with argon, and then dissolved in dry ACN prior to attaching the reagent bottle to the synthesizer.
2. For RNAs containing ms^2t^6A, the use of both $Et_3N·3HF$ and tetrabutylammonium fluoride (TBAF) provides complete deprotection of trimethylsilylethyl carboxylate protected RNAs. We found that the trimethylsilylethyl group is not removed with $Et_3N·3HF$ alone, and that TBAF alone gives incomplete deprotection of the secondary TBS protecting groups.
3. The phosphotitylation reactions are conveniently followed by taking a 50-μL aliquot of the reaction mixture, removing the solvent, oxidizing for 3 min with tBuOOH/toluene, followed by evaporation and then silica gel TLC. The amidite product is found at the origin, but the unreacted starting material is unaffected by the oxidation procedure. This protocol is very helpful because the starting material and products often migrate with identical Rf in the solvent systems we investigated.
4. Many solvents have been used for synthesizing nucleoside phosphoramidites. We found THF to be convenient when used with the very reactive chlorophosphoramidite reagent. The reaction is characterized by the production of white precipitate and is complete in 3 h at room temperature.

5. For making 2,5'-anhydrouridine, either DIAD or diethylazodicarboxylate (DEAD) works equally well. DEAD is no longer available from major chemical suppliers.
6. In the synthetic protocol for the mnm^{5}s^{2}U nucleoside, the transformation of the 5-hydroxymethyl to the 5-chloromethyl remains a problematic conversion. We have found the chlorotrimethylsilane procedure described to be superior to using HCl gas and more convenient *(24)*. Fortunately, the crude product could be carried forward because we never had good luck with isolating the chloro product in acceptable yield.
7. FC of the mnm^{5}s^{2}U phosphoramidite was surprisingly fickle. The combination of CH$_2$Cl$_2$:ethyl acetate, for example, is commonly used in our laboratory, yet for this phosphoramidite the recoveries were very poor. Adding a small amount of triethylamine (0.1–1%) was not helpful. The reported elution solvent of CH$_2$Cl$_2$:ACN resulted in excellent purification and recovery using standard silica gel FC.

Acknowledgments

The work was supported by grant GM55508 from the National Institutes of Health, and by NIH grants RR06262, RR13030, and CA42014, which support Core Facilities at the University of Utah.

References

1. Sprinzl, M., Horn, C., Brown, M., Ioudovitch, A., and Steinberg, S. (1998) Compilation of tRNA sequences and sequences of tRNA genes. *Nucleic Acids Res.* **26,** 148–153.
2. Limbach, P. A., Crain, P. F., and McCloskey, J. A. (1994) Summary: the modified nucleosides of RNA. *Nucleic Acids Res.* **22,** 2183–2196.
3. Yarian, C., Marszalek, M., Sochacka, E., et al. (2000) Modified nucleoside dependent Watson–Crick and wobble codon binding by tRNA^{Lys}UUU species. *Biochemistry* **39,** 13,390–13,395.
4. Von Ahsen, U., Green, R., Schroeder, R., and Noller, H. (1997) Identification of 2'-hydroxyl groups required for interaction of a tRNA anticodon stem-loop region with the ribosome. *RNA* **3,** 49–56.
5. Atkins, J. F., Herr, A. J., Massire, C., O'Connor, M., Ivanov, I., and Gesteland, R. F. (2000) Poking hole in the sanctity of the triplet code: inferences for framing. In *The Ribosome: Structure, Function, Antibiotics, and Cellular Interaction* (Garrett, R. A., Douthwaite, S. R., Liljas, A., Matheson, A. T., Moore, P. B., and Noller, H. F., eds.), ASM Press, Washington, D.C., pp. 369–383.
6. Tamura, K., Himeno, H., Asahara, H., Hasegawa, T., and Shimizu, M. (1992) In vitro study of *E. coli* tRNA^{Arg} and tRNA^{Lys} identity elements. *Nucleic Acids Res.* **20,** 2335–2339.
7. Isel, C., Westhof, E., Massire, C., Le Grice, S. F. J., Ehresmann, B., Ehresmann, C., and Marquet, R. (1999) Structural basis for the specificity of the initiation of HIV-1 reverse transcription. *EMBO J.* **18,** 1038–1048.

8. Boudou, V., Langridge, J., van Aerschot, A., et al. (2000) Synthesis of the anticodon hairpin tRNAfMet containing N-{[9-(b-D-ribofuranosyl)-9H-purin-6-yl]carbamoyl}-L-threonine (=N^6-{{[(1S,2R)-1-carboxy-2-hydroxylpropyl] amino}carbonyl}adenosine, t^6A. *Helvet. Chim. Acta* **83**, 152–161.
9. Stuart, J. W., Gdaniec, Z., Guenther, R. H., et al. (2000) Functional anticodon architecture of human tRNA^{Lys,3} includes disruption of intraloop hydrogen bonding by the naturally occurring amino acid modification t^6A. *Biochemistry* **39**, 13,396–13,404.
10. Sundaram, M., Durant, P. C., and Davis, D. R. (2000) Hypermodified nucleosides in the anticodon of tRNALys stabilize a canonical U-turn structure. *Biochemistry* **39**, 12,575–12,584.
11. Bajji, A. C., Sundaram, M., Myszka, D. G., and Davis, D. R. (2002) An RNA complex of the HIV-1 A-loop and tRNA^{Lys,3} is stabilized by nucleoside modifications. *J. Am. Chem. Soc.* **124**, 14,302–14,303.
12. Pon, R. T. and Yu, S. (1997) Hydroquinone-O,O'-diacetic acid as a more labile replacement for succinic acid linkers in solid-phase oligonucleotide synthesis. *Tetrahedron Lett.* **38**, 3327–3330.
13. Jager, A. and Engels, J. (1984) Synthesis of deoxynucleoside methylphonates via a phophonamidite approach. *Tetrahedron Lett.* **25**, 1437–1440.
14. Griffey, R. H., Monia, B. P., Cummins, L. L., et al. (1996) 2'-O-aminopropyl ribonucleotides: a zwitterionic modification that enhances the exonuclease resistance and biological activity of antisense oligonucleotides. *J. Med. Chem.* **39**, 5100–5109.
15. Sinha, N. D., Beirnat, J., McManus, J., and Koster, H. (1984) Polymer support oligonucleotide synthesis XVIII: use of b-cyanoethyl-N,N-dialkylamino-N-morpholino phosphoramidite of dexoynucleosides for the synthesis of DNA fragments simplifying deprotection and isolation of the final product. *Nucleic Acids Res.* **12**, 4539–4557.
16. Shah, K., Wu, H., and Rana, T. M. (1994) Synthesis of uridine phosphoramidite analogs: reagents for site-specific incorporation of photoreactive sites into RNA sequences. *Bioconj. Chem.* **5**, 508–512.
17. Gasparutto, D., Livache, T., Bazin, H., et al. (1992) Chemical synthesis of a biologically active natural tRNA with its minor bases. *Nucleic Acids Res.* **20**, 5159–5166.
18. Wincott, F. E. and Usman, N. (1994) 2'-(Trimethylsilyl)ethoxymethyl protection of the 2'-hydroxyl group in oligoribonucleotide synthesis. *Tetrahedron Lett.* **35**, 6827–6830.
19. Sproat, B., Colonna, F., Mulla, B., et al. (1995) An efficient method for the isolation and purification of oligoribonucleotides. *Nucleosides Nucleotides* **14**, 255–273.
20. Fissekis, J. D. and Sweet, F. (1970) Synthesis of 5-carboxymethyluridine: a nucleoside from transfer ribonucleic acid. *Biochemistry* **9**, 3136–3142.
21. Vorbruggen, H. and Strehlke, P. (1973) Eine Einfache Synthese von 2-Thiopyrimidin-nucleosiden. *Chem. Ber.* **106**, 3039–3061.

22. Bajji, A. and Davis, D. R. (2000) Synthesis and biophysical characterization of tRNA^{Lys,3} anticodon stem-loop RNAs containing the mcm^{5}s^{2}U nucleoside. *Org. Lett.* **2**, 3865–3868.
23. Vorbruggen, H. and Krolikiewicz, K. (1980) Synthesis of 5-methylaminomethyl-2-thiouridine, a rare nucleoside from t-RNA. *Liebigs. Ann. Chem.* 1438–1447.
24. Ikeda, K., Tanaka, S., and Mizuno, Y. (1975) Syntheses of potential antimetabolites. XX syntheses of 5-carbomethoxymethyl- and -methylaminomethy-2-thiouridine (The "first letters" of some anticodons and closely related nucleosides from uridine. *Chem. Pharm. Bull.* **23**, 2958–2964.
25. Sundaram, M., Crain, P. F., and Davis, D. R. (2000) Synthesis and characterization of the native anticodon domain of *E. coli* tRNA^{Lys}: simultaneous incorporation of modified nucleosides mnm^{5}s^{2}U, t^{6}A, and pseudouridine using phosphoramidite chemistry. *J. Org. Chem.* **65**, 5609–5614.
26. Malkiewicz, A. J., Nawrot, B., and Sochacka, E. (1987) Transformation of some tRNA "wobble uridines" to their 2-thioanalogues. *Z. Naturforsch.* **42b**, 360–366.
27. Bajji, A. and Davis, D. R. (2002) Synthesis of the tRNA^{Lys,3} anticodon stem-loop domain containing the hypermodified ms^{2}t^{6}A nucleoside. *J. Org. Chem.* **67,** 5352–5358.
28. Saxena, N. K. and Bhakuni, D. S. (1979) Synthesis of 9-a-L-rhamnopyranosyl-2-alkylthioadenines. *Indian J. Chem.* **18B,** 348–351.
29. Taylor, E. C., Vogl, O., and Cheng, C. C. (1958) Studies in purine chemistry: II, a facile synthesis of 2-substituted adenines. *J. Am. Chem. Soc.* **81,** 2442–2248.
30. Kikugawa, K., Suehiro, H., Yanase, R., and Aoki, A. (1977) Platelet aggregation inhibitors: IX, chemical transformation of adenosine into 2-thioadenosine derivatives. *Chem. Pharm. Bull.* **25,** 1959–1969.
31. Adamiak, R. W. and Stawinski, J. (1977) A highly effective route to N,N'-disubstituted ureas under mild conditions: an application to the synthesis of tRNA anticodon loop fragments containing ureidonucleosides. *Tetrahedron Lett.* **22, 1935,** 1936.

12

Chemical Methods for Peptide–Oligonucleotide Conjugate Synthesis

Dmitry A. Stetsenko and Michael J. Gait

Summary

Methods of peptide–oligonucleotide conjugate synthesis are presented that may be useful in the study of cell targeting and delivery of oligonucleotides and their analogs. The first method involves total stepwise solid-phase synthesis on a single support. The second involves preparation of oligonucleotides containing 2'-aldehydes and subsequent chemoselective ligation with functionalized peptides to form hydrazine, oxime, or thiazolidine linkages.

Key Words: Oligonucleotide; peptide; conjugate; aldehyde; solid-phase synthesis; chemoselective ligation; hydrazone; hydrazine; oxime; thiazolidine.

1. Introduction

Peptide conjugates of oligonucleotides and their analogs are being studied extensively in attempts to improve cell-specific targeting and cellular delivery for antisense and other applications involving inhibition of gene expression *(1–4)*. Numerous methods of chemical synthesis of peptide–oligonucleotide conjugates have been developed *(2,5–6)* that are divided into two main synthetic approaches: total stepwise solid-phase synthesis and solution-phase coupling of independently prepared peptide and oligonucleotide fragments—chemoselective ligation. No single route is likely to be adequate for all applications, and thus we have selected one robust method for each main approach.

In total stepwise synthesis, the same solid support is used for both peptide and oligonucleotide assemblies. Peptide synthesis may be followed by oligonucleotide synthesis *(7)*, or vice versa *(8)*, or a branched linker is attached to the support permitting independent growth of both peptide and oligonucleotide chains *(9)*. The poor compatibility of peptide and oligonucleotide synthesis

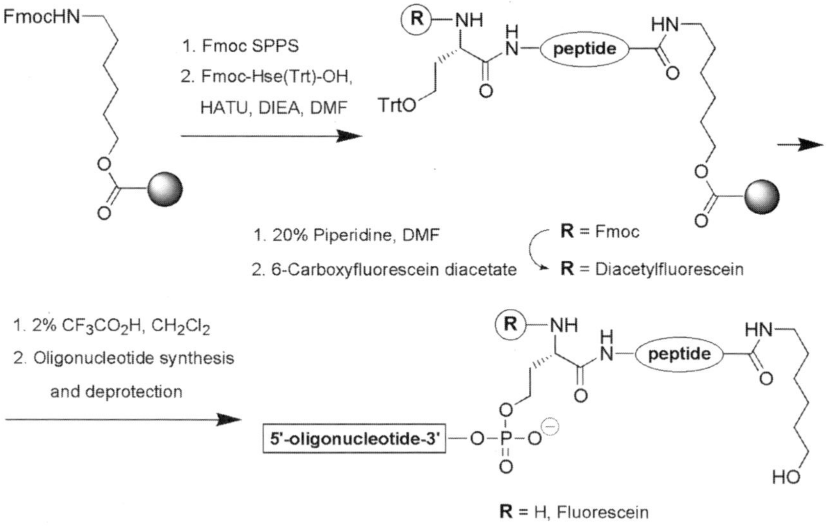

Scheme. 1. Total stepwise synthesis of oligonucleotide-(3'-N)-peptide conjugates on a homoserine functionalized support.

chemistries has presented considerable difficulties, and up to now only a few routes show some promise *(10)*. A novel and facile method of synthesis of oligonucleotide 3'-conjugates was recently described, including some short peptides that made use of a ω-aminoalkyl succinate/L-homoserine combination linker (**Scheme 1**) *(11)*. We extended this method toward the total stepwise solid-phase synthesis of longer peptides conjugated to antisense oligonucleotides and their 2'-O-methyl analogs *(12)*. Peptides are assembled first on solid support by use of O-(7-azobenzotriazol-1-yl)-N,N,N',N',-tetramethyluronium hexafluorophosphate (HATU)/diisopropylethylamine (DIEA) *in situ* activation protocol *(13)* and 9-fluorenylmethoxycarbonyl (Fmoc) chemistry *(14)*. An optional fluorescein or other label can be introduced after incorporation of the homoserine linker. After removal of the trityl group, the peptide-loaded support is subjected to standard oligonucleotide chain assembly by the phosphoramidite method *(15)*. After ammonia deprotection, products are purified by high-pressure (performance) liquid chromatography (HPLC).

Fragment coupling of peptides to oligonucleotides has proved valuable in the case of longer peptides or for those containing certain amino acids, such as arginine, which at present cannot be fully introduced by a stepwise solid-phase route. Ligation products do not require additional deprotection steps and may therefore be used directly in biological assays after purification. Oligonucleotide and peptide fragments are assembled separately followed by a solution-phase chemoselective ligation mediated by mutually reactive groups introduced during solid-phase assembly or postsynthetically. Popular linkages include disulfide *(16)*, thioether *(17)*, and amide *(18,19)*. We have also described a native ligation method towards peptide–oligonucleotide conjugation *(20)*. Another promising approach involves aqueous reaction at pH near or below neutral of a weakly basic nucleophile with an electrophile, such as an aldehyde. Conjugation is specific because of the lower pK_a than in the side chain amino groups of Lys and Arg and is also effective in the presence of organic cosolvents or denaturing agents. Suitable nucleophiles include an *O*-alkyl hydroxylamine to form an oxime *(21–23)* or an alkyl- or arylhydrazine, acyl- or sulfonylhydrazide, or carbazate to form a hydrazone *(24,25)*. A hydrazone linkage appears to be less stable at low pH but may be stabilized by reduction to a hydrazine. In addition, 1,2-aminothiols such as N-terminal cysteinyl peptides react with aldehydes to form thiazolidines, which are stable over a wide pH range *(22)*. All such conjugation reactions are best conducted at a slightly acidic pH *(26)*. We described recently a phosphoramidite reagent for incorporation of 1,2-diol groups into oligonucleotides using 2'-modification (**Scheme 2**) *(27)*. A 2'-aldehyde is generated by treatment with periodate and can then be conjugated through use of any of the three previously mentioned chemistries (**Scheme 3**).

We showed the highly efficient conjugation single or multiple peptides to oligodeoxynucleotides and their 2'-*O*-methyl analogs *(28)*. An advantage of this method is that conjugates bind to complementary ribonucleic acid (RNA) with unimpaired, and sometimes enhanced, binding strength.

2. Materials

2.1. Reagents

1. *Solid supports*: For oligonucleotide synthesis, use standard 0.2- or 1-µ*M* columns prepacked with an appropriate support (Applied Biosystems, Glen Research, or Cruachem). For stepwise solid-phase peptide–oligonucleotide conjugate synthesis, use aminomethyl-PS200 (Amersham Biosciences), ArgoPore™LL (Sigma) or 500Å LCAA-CPG (Pierce). LCAA-CPG is the support of choice for synthesis of oligonucleotides and analogs up to 50 nucleotides long, but it is recommended only for stepwise synthesis of short peptides (<12 amino acids). Aminomethyl-PS200 is better suited for peptide synthesis and continuous-flow machine assemblies (*see* **Note 1**). Oligonucleotide synthesis on PS200 is satisfactory with good to moderate final product yields. ArgoPore resin results in better peptide synthesis performance, but oligonucleotide assemblies sometimes provide lower yields than PS200, although with higher product purity. For peptide amide synthesis by Fmoc/*tert*-butyl scheme use Rink amide NovaGel™HL (Novabiochem) or a polyethyleneglycol-polysterene copolymer support with peptide amide linker (PAL-PEG-PS) (Applied Biosystems).
2. *Fmoc amino acids*: Use standard Fmoc amino acids for regular peptide synthesis (Novabiochem, Bachem, Applied Biosystems, or any other speciality supplier).

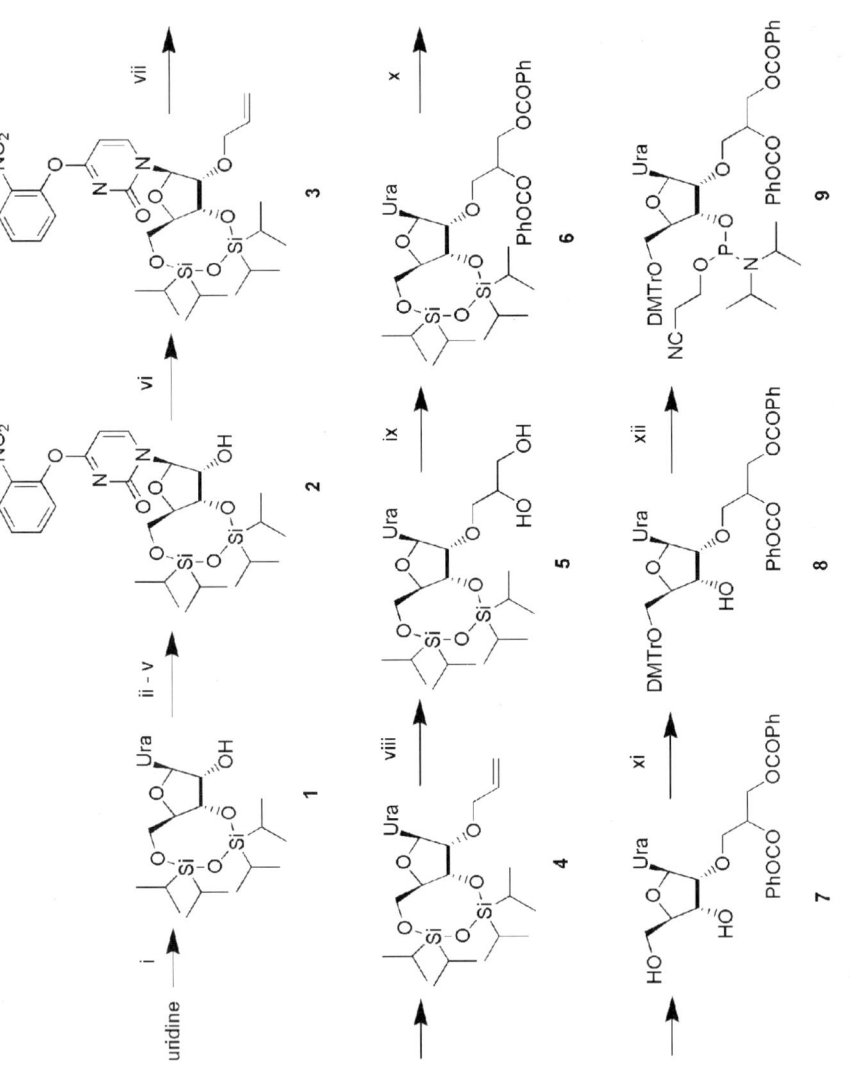

Scheme 2. (i) TIPSCl₂, pyridine; (ii) Me₃SiCl, Et₃N, pyridine; (iii) 2-mesitylenesulfonyl chloride, DMAP, Et₃N, CH₂Cl₂; (iv) 2-nitrophenol, DABCO; (v) p-TsOH, CH₂Cl₂; (vi) allylOCO₂Me, Pd₂dba₃, Ph₃P, THF; (vii) 2-nitrobenzaldoxime, TMG, MeCN; (viii) NMMO, OsO₄, THF; (ix) PhCOCN, Et₃N, MeCN; (x) TBAF, THF; (xi) DMTrCl, pyridine; (xii) 2-cyanoethoxy-N,N-diisopropylaminochlorophosphine, DIEA, CH₂Cl₂.

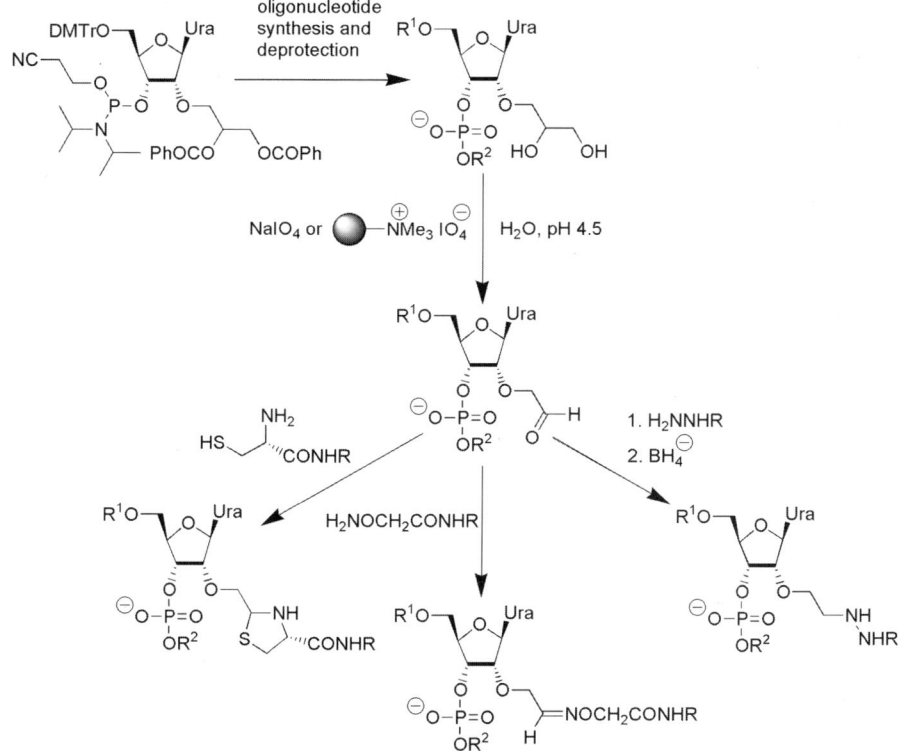

Scheme 3. Synthesis of peptide-oligonucleotide conjugates by a fragment coupling approach. R, peptide residue, R_1 and R_2, deprotected oligonucleotide chains; P, polymer support.

Fmoc amino acid monomers lacking side-chain protection (Gly, Ala, Pro, Val, Leu, Ile, Phe, Gln, Sar) are compatible with stepwise solid-phase peptide–oligonucleotide conjugate synthesis (Novabiochem, Millipore, Advanced ChemTech, or any other supplier). Nonstandard Fmoc monomers for stepwise solid-phase peptide–oligonucleotide conjugate synthesis require base-labile side-chain protecting groups: Lys(Tfa) (Senn Chemicals or Novabiochem), His(Boc) (Bachem), Asp(Dmab), and Glu(Dmab) (Novabiochem), Trp(Boc) (all major companies). Other special monomers are: Fmoc-Tyr(Clt)-OH and Fmoc-asparagine pentafluorophenyl ester (Novabiochem) and Fmoc-Hse(Trt)-OH (Senn Chemicals or Novabiochem).

3. *Oligonucleotide synthesis monomers*: Use β-cyanoethyl 2'-deoxy, 2'-O-methyl or other phosphoramidites from any major supplier (Applied Biosystems, Glen Research, or Cruachem).

4. *Ligation monomers*: Four major companies now provide good selections: Boc-aminooxyacetic acid (Fluka, Senn Chemicals, or Novabiochem), Fmoc-aminooxyacetic acid (Senn Chemicals), Tris-Boc-hydrazinoacetic acid (Novabiochem), Fmoc-hydrazinobenzoic acid (Bachem), and pentafluorophenyl *S*-benzylthiosuccinate (Glen Research).
5. *Solid-supported reagents*: Polymer-bound periodate and borohydride reagents (polystyrylmethyl trimethylammonium metaperiodate and borohydride) (Novabiochem or Aldrich).
6. *Solvents*: Distil dimethyl formamide (DMF) (BDH) *in vacuo* prior to use, but discard the first fraction (approx 5% v/v). Reflux CH_2Cl_2 over CaH_2 for 8 h before distillation. Use other high-grade solvents as supplied.
7. *Other reagents*: Piperidine (Romil), DIEA (Aldrich 99.5%), HATU (Applied Biosystems) (*see* **Note 2**), HOAt (Millipore), 6-carboxyfluorescein diacetate (Sigma or Berry & Associates), Beaucage sulfurizing reagent (Glen Research), Fmoc-aminohexanol (Fluka), benzoyl cyanide (Aldrich), and other fine chemicals supplied by any major company.
8. *Capping solutions for peptide synthesis*:
   a. Reagent 1: 20% acetic anhydride in tetrahydrofuran (THF).
   b. Reagent 2: 20% *N*-methyl imidazole (NMI), 40% 2,4,6-collidine in THF (*see* **Note 3**).
9. *Matrix-assisted laser desorption/ionization time-of-flight (MALDI-TOF)*: For low-molecular-weight compounds use either 2,5-dihydroxybenzoic acid (10 mg/mL^{-1} in MeOH) or 2,4,6-trihydroxyacetophenone (40 mg/mL^{-1} in 50% aqueous MeCN).

2.3. Special Equipment

1. *Automated peptide synthesis*: A continuous-flow Pioneer (Applied Biosystems) peptide synthesizer capable of a 50–200-μM scale synthesis by Fmoc/*tert*-butyl chemistry was used but other peptide synthesizers may also be appropriate.
2. *Automated oligonucleotide synthesis*: Use a commercial deoxyribonucleic acid (DNA)/RNA synthesizer suitable for 0.2–1-μM scale oligonucleotide assemblies (e.g., Applied Biosystems 394).
3. *MALDI-TOF mass spectroscopy*: Use a commercial MALDI-TOF workstation (e.g., Applied Biosystems Voyager DE).
4. *Nuclear magnetic resonance (NMR) spectroscopy*: Use a 300-MHz NMR spectrometer (e.g., Bruker DRX 300) to record one-dimensional ^{1}H-, ^{13}C-, or ^{31}P-NMR spectra using tetramethylsilane as an internal standard or 85% aqueous H_3PO_4 as an external standard.
5. *Ultraviolet (UV)-visible spectroscopy*: Use a dual-band UV-Vis spectrophotometer (e.g., Lambda 40, PerkinElmer).

3. Methods

3.1. Stepwise Solid-Phase Peptide–Oligonucleotide Conjugate Synthesis

3.1.1. Functionalization of Solid Support

3.1.1.1. Preparation of 6-N-(Fluorenylmethoxycarbonyl) Aminohexyl Succinate Linker *(11)*

1. Suspend 3.73 g, 11 mmol 6-N-Fmoc-aminohexanol, 1.49 g, 14.85 mmol succinic anhydride, and 134 mg, 1.1 mmol dimethylaminopurine (DMAP) in 75 mL of 1,2-dichloroethane, add 5 mL of pyridine, and reflux the mixture for 24 h.
2. Transfer the solution to a separation funnel, wash successively 3× with 50 mL 0.5M NaHSO$_4$, 50 mL H$_2$O, and 30 mL brine, and dry over anhydrous Na$_2$SO$_4$.
3. Remove the solvent *in vacuo,* and purify the compound by flash chromatography on a Kieselgel 60 column eluted with 1% MeOH in CHCl$_3$. Pool the appropriate fractions, evaporate, and crystallize the residue from EtOAc-hexane (2:1 v/v) mixture. Yield 4.31 g (89%) of the title compound as white crystals. Thin-layer chromatography (TLC) (5% MeOH in CHCl$_3$): R_f 0.46. MALDI-TOF MS: [M+H]$^+$ 439.85 (observed), 440.51 (calculated). ^{1}H-NMR (DMSO-d_6, δ, ppm): 1.26 (m, 4H, CH$_2$), 1.37 (t, 2H, J = 6.1 Hz, CH$_2$), 1.53 (t, 2H, J = 6.4 Hz, CH$_2$), 2.47 (s, 4H, succinyl CH$_2$), 2.94 (t, 2H, J = 6.9 Hz, CH$_2$), 3.98 (t, 2H, J = 6.4 Hz, CH$_2$), 4.19 (t, 1H, J = 6.9 Hz, 9-fluorene CH), 4.28 (d, 2H, J = 6.8 Hz, fluorenyl CH$_2$), 7.31 (t, 2H, J = 7.4 Hz, Fmoc aromatics), 7.40 (t, 2H, J = 7.3 Hz, Fmoc aromatics), 7.67 (d, 2H, J = 7.3 Hz, Fmoc aromatics), 7.87 (d, 2H, J = 7.4 Hz, Fmoc aromatics).

3.1.1.2. Sarcosine Functionalization of Amino-Modified Solid Support (Aminomethyl–ArgoPore LL As an Example)

1. Dissolve 78 mg, 0.25 mmol Fmoc-Sar-OH and 91 mg, 0.24 mmol HATU in 5 mL DMF, add 44 µL, 0.25 mmol DIEA, and shake or stir the mixture for 5 min. Add (43 µL, 0.25 mmol) DIEA, and pour the yellow solution onto the resin (0.5 g, approx 0.4 mmol, 0.2–0.6 mmol g^{-1} amino groups).
2. Keep the slurry for 1 h with occasional swirling, filter the solution into a sintered glass filter, and wash the beads 5× with 5 mL DMF, 3× with 5 mL MeOH, and 3× with 5 mL diethyl ether.
3. Check the Fmoc group loading by weighing about 5 mg of the resin into a 1.5-mL Eppendorf tube, treating it with 1 mL of 20% piperidine in DMF (v/v) for 15 min, withdrawing a 50-µL aliquot, adjusting the volume by DMF to 1 mL, and reading the absorbance value at 301 nm.
4. Treat the resin on the sinter with equal volumes (2.0 mL of each) of capping solutions for 15 min, wash successively 5× with 5 mL DMF, 3× with 5 mL MeOH, and 3× with 5 mL diethyl ether, and dry *in vacuo*. Yield approx 0.51 g, at loading 253 µM/g.

3.1.1.3. 6-N-(9-Fluorenylmethoxycarbonyl)Aminohexyl Succinate Attachment

1. Treat 0.5 g of the resin from the above stage on a sintered glass filter with about 10 mL 20% piperidine in DMF (v/v) for 10 min. Wash thoroughly 10× with 5 mL DMF.
2. Simultaneously dissolve 29 mg, 0.07 mmol 6-N-Fmoc-aminohexyl succinate and 24 mg, 0.07 mmol HATU in 2 mL DMF, add 11.4 µL, 0.07 mmol DIEA, and shake for 5 min. Add (11.4 µL 0.07 mmol) DIEA and pour the solution over the beads.
3. Wait for about 45 min, wash the filter 5× with 5 mL DMF, 3× with 5 mL MeOH, and 3× with 5 mL diethyl ether, and dry *in vacuo*. Check Fmoc loading and cap the resin using the procedures described. Yield approx 0.5 g, loading 108 µmol g^{-1}. This resin is ready for use in either manual or machine-assisted peptide synthesis.

3.1.2. Solid-Phase Synthesis of a Peptide Fragment

1. Subject the resin (200–250 mg) to automated peptide assembly *(14)* on a 50-µmol scale. Use HATU/DIEA-mediated double couplings with four to five equivalents of Fmoc-amino acids and allow 30 min for each coupling. Use Fmoc-amino acids that do not require side-chain protection or that have base-labile protecting groups removable by ammonia treatment, for example, Lys(Tfa) *(11)*, His(Boc) *(29)*, or Asp(Dmab) *(12)*, the carboxylic group of which is protected by a base-labile ester group (Dmab) *(30)*. Omit acetic anhydride capping steps for the automated synthesis (except before and after all manual coupling steps). Instead, use as a capping agent HATU/DIEA-mediated coupling of 4-chlorophenoxyacetic acid (four to five equivalents) in each third amino acid position (except before manual stage).
2. Manual Fmoc-asparagine pentafluorophenyl ester coupling is recommended for Asn incorporation to minimize dehydration of the side-chain amide. In this case remove the column from the synthesizer with the last Fmoc group still attached. Cap the support with the acetylation reagent, wash with THF, and dry *in vacuo*. Check the Fmoc loading, and remove the Fmoc group by flowing through with approx 5–10 mL 20% piperidine in DMF for 10 min, and wash thoroughly 10× with 5 mL DMF. Add 120 mg, 0.23 mmol predissolved Fmoc-asparagine pentafluorophenyl ester and 6 mg, 0.05 mmol HOAt in 2 mL DMF. After 4 h, drain, wash 2× with 5 mL DMF, and repeat the coupling for another 4 h (or leave overnight). Wash 5× with 5 mL DMF, 3× with 5 mL MeOH, and 3× with 5 mL diethyl ether, and dry *in vacuo*. Check the Fmoc loading and then cap the resin by acetylation as usual and resume the automated synthesis.
3. Fmoc-Tyr(Clt)-OH has its phenolic group transiently protected by a mild acid labile 2-chlorotrityl group *(31)*, which has to be removed after the end of assembly—but before the last Hse(Trt) residue attachment—and the tyrosine residue reprotected by an isobutyryl group. For Tyr reprotection, remove the column from the synthesizer with the last Fmoc group still attached, wash 3× with 5 mL MeOH and 3× with 5 mL CH$_2$Cl$_2$. Cleave the 2-chlorotrityl group by flowing

through 10 mL 2% CF_3CO_2H in CH_2Cl_2 (v/v) until the yellow color of the dripping solution is no longer visible (approx 5 min); wash 3× with 5 mL MeOH and 3× with 5 mL CH_2Cl_2. Cap the resin with 5 mL of 10% isobutyric anhydride, 10% NMI, 20% 2,4,6-collidine in THF for 1 h; wash 3× with 5 mL MeOH, 5× with 5 mL DMF, and resume the automated synthesis.

4. For Hse incorporation, remove the column from the synthesizer with the last Fmoc group still attached, cap the resin by acetylation, dry *in vacuo,* and check the Fmoc loading. Remove the Fmoc group as described and wash thoroughly 10× with 5 mL DMF. Dissolve 73 mg, 0.13 mmol Fmoc-Hse(Trt)-OH, and 45 mg, 0.12 mmol HATU in 1 mL DMF, add 22 µL, 0.13 mmol DIEA, shake for 5 min, add more DIEA (44 µL, 0.25 mmol), and pour the solution over the resin. Wait for 3 h, then drain, wash 2× with 5 mL DMF, and repeat the coupling once more. After a further 3 h, wash the resin 5× with 5 mL DMF, 3× with 5 mL MeOH, and 3× with 5 mL diethyl ether, and dry *in vacuo.* Check the Fmoc loading and then cap the resin by acetylation. Homoserine incorporation can also be accomplished in a machine-assisted way using the conditions for normal Fmoc-amino acid coupling.

5. To label the peptide with fluorescein, deprotect the last Fmoc group and wash thoroughly 10× with 5 mL DMF. Dissolve 50 mg, 0.1 mmol 6-carboxyfluorescein diacetate and 39 mg, 0.1 mmol HATU in 1 mL DMF, add 19 µL, 0.1 mmol DIEA, shake for 5 min, add more (19 µL, 0.1 mmol) DIEA, and pour over the resin. Wait for 2 h, wash with DMF (2 × 5 mL), and repeat the coupling. After a further 2 h, drain the resin, wash 2× with 5 mL DMF, 2× with 5 mL MeOH, and 2× with 5 mL CH_2Cl_2, and cap the resin by acetylation. Wash 3× with 5 mL MeOH and 3× with 5 mL CH_2Cl_2.

6. To remove the Trt group from the last Hse residue, flow through 10 mL of 2% CF_3CO_2H in CH_2Cl_2 (v/v) until the yellow color of the dripping solution is no longer visible (approx 5 min), wash 3 × with 5 mL MeOH , and 3× with 5 mL diethyl ether, and dry *in vacuo.* The resin is now ready for automated oligonucleotide synthesis.

7. Check the integrity of the peptide by treating a 5-mg portion of the resin in a screw-capped tube or vial with 0.75 mL 30% aqueous NH_3 and 0.25 mL EtOH at 55°C for 14 h. Decant the supernatant, wash the support with 50% aqueous EtOH, half-evaporate the combined solution by a stream of nitrogen or argon, filter by centrifugation in a Spin-X tube (Costar), and evaporate to dryness by use of a SpeedVac concentrator. Dissolve the peptide in 10% aqueous MeCN and analyze by RP-HPLC and MALDI-TOF mass spectroscopy (*see* **Subheading 3.2.1.**).

3.1.3. Solid-Phase Synthesis of an Oligonucleotide Fragment, Cleavage and Purification of a Conjugate

1. Assemble oligonucleotides, for example, those containing 2'-deoxynucleotides, 2'-*O*-methylribonucleotides, or locked nucleic acids, on a suitable DNA/RNA synthesizer by the β-cyanoethyl phosphoramidite method *(15)* following

manufacturer's recommendations. Fill a 1-µmol All-Fit DNA synthesis column with an appropriate amount of peptidyl resin calculated from the last Fmoc group measurement. Install it on the synthesizer and carry out a standard DMTr-off protocol (use of 2'-O-methyl protocol is recommended with 95-s DMTr deprotection and 925-s coupling time). Monitor DMTr stepwise coupling yields if possible. For phosphorothioate synthesis, use a 30-s treatment with Beaucage's reagent (1 g in 100 mL MeCN) instead of normal iodine oxidation.

2. After the end of assembly, detach the column, wash it twice with 10 mL MeCN by connection to a syringe and dry *in vacuo*. Transfer the support into a screw-capped pressure-proof tube or vial, add 0.75 mL 30% aqueous ammonia and 0.25 mL EtOH, shake and deprotect at 55°C for 6–16 h. After cooling, evaporate most of the solution under a stream of nitrogen or argon. Add 0.25 mL H_2O, transfer the residue into a Spin-X centrifuge tube and centrifugate for 5 min at 11,600g. Wash the filter with 0.25 mL H_2O, and evaporate the combined washings on a SpeedVac vacuum concentrator.

3. Reconstitute the crude product in water and purify by HPLC. For oligonucleotide and conjugate RP-HPLC purification use a Phenomenex Bondclone 10 C_{18} column (3.9 × 300 mm). Buffer A, 0.1M triethylammonium acetate, pH 7.0; buffer B, MeCN, gradient of B 0–60% (45 min), flow rate 1 mL/min. Pool the appropriate fractions, reevaporate twice with water, and lyophilize. For oligonucleotide and conjugate IE-HPLC purification use a NucleoPac PA-100 (Dionex) (9 × 250 mm) column, gradient of sodium perchlorate (1–400 mM) in a buffer of 20 mM Tris-HCl, pH 6.8, 25% formamide, and 280-nm wavelength detection. Pool the appropriate fractions, desalt on a NAP-10 column, and lyophilize. Dissolve the product in water, and quantify the oligonucleotide or conjugate by measuring the UV absorbance at 260 nm. Store the solution frozen.

4. Check the molecular mass of the product by running MALDI-TOF mass spectra. For oligonucleotides and peptide–oligonucleotide conjugates use a freshly prepared 1:1 v/v mixture of 2,6-dihydroxyacetophenone (40 mg/mL in MeOH) and 80 mg/mL aqueous diammonium hydrogen citrate as a matrix and 1:1 v/v of matrix:sample.

3.2. Peptide–Oligonucleotide Conjugate Synthesis by a Chemoselective Ligation

3.2.1. N-Terminally Modified Peptide Fragment Synthesis

Assemble a peptide sequence on a suitable synthesizer by standard Fmoc/*tert*-butyl chemistry starting from a compatible amide support (Rink or PAL) on a 50–100-µM scale, and deprotect the last Fmoc group. For thiazolidine ligation, furnish the sequence with an N-terminal Cys residue using a standard monomer, for example, Fmoc-Cys(Trt)-OH and automated protocol *(26)*, and proceed with trifluoroacetic acid (TFA) deprotection. For oxime and hydrazone ligation, detach the column and couple the appropriate monomer (Fmoc- or Boc-aminooxyacetic acid *[26]*, Tris-Boc-hydrazinoacetic acid *[24]*, or Fmoc-hydrazinobenzoic acid *[25]*) manually, for example, using a carbodiimide activation (*see* **Note 4**). Repeat the coupling if necessary. The exact amount of the reagent, number of steps, and deprotection conditions depend on the monomer used, scale of the synthesis, and initial resin loading. The following is an example describing N-terminal hydrazide incorporation *(28)*.

1. Acylate the Fmoc-deprotected peptidyl resin by treating with 4.5 equivalents pentafluorophenyl *S*-benzylthiosuccinate *(20)* and 1 equivalent HOAt in 2.5 mL DMF for 4 h at 20°C. Wash 3× with 5 mL DMF and drain.
2. Treat the peptidyl support with $0.5M$ hydrazine hydrate solution in 3 mL 1,4-dioxane for 2 h at 20°C, filter, wash 3× with 5 mL DMF, 2× with 5 mL MeOH, and 2× with 5 mL diethyl ether, and dry *in vacuo*.
3. Cleave the peptide from the resin and deprotect by addition of 5 mL TFA–phenol-1,2-ethanedithiol-water (925:25:25:25 v/v/v/v). Transfer the supernatant into a 50-mL polypropylene tube, wash the support twice with 2-mL portions of TFA, and blow the TFA away under a stream of nitrogen delivered via a glass Pasteur pipet to the bottom of the tube. When the volume is down to approx 1 mL, add 45 mL of cold (–20°C) diethyl ether to precipitate the peptide, vortex for 1 min, centrifuge at 2500*g* for 5 min, and discard the ethereal layer. Repeat the ether washes three times. After the last wash, dry the peptide pellet in air for 5 min and then *in vacuo*.
4. Analyze the crude peptide by HPLC and purify (if necessary). For analytical RP-HPLC of peptides use dual wavelength (215 and 230 nm) UV detection and a 5 µ*M* Phenomenex C_{18} column (4.6 × 250 mm), flow rate 1.5 mL/min. For preparative peptide HPLC use a 5 µ*M* Vydac RP-C_8 column (25 × 300 mm) and dual wavelength (215 and 230 nm) UV detection, flow rate 10 mL/min^{-1}. Buffer A, 0.1% aqueous TFA; buffer B, 10% A in MeCN (v/v). Use a gradient of B from 10 to 90% in 30 min for the HPLC analysis and purification of most peptides. Pool the appropriate fractions and lyophilize. Store at –20°C.
5. Check the molecular mass of the peptide by running MALDI-TOF spectra. Use either 10 mg/mL α-cyano-4-hydroxycinnamic acid in MeCN–H_2O–3% aqueous TFA (6:3:1 v/v/v) or 10 mg/mL sinapinic acid in MeCN–H_2O–3% aqueous TFA (5:4:1 v/v/v) as a matrix, matrix:sample ratio 2:1 v/v.

3.2.2. Preparation of 2'-Diol Phosphoramidite

The following method is adapted from our previous work *(27)*. The diol phosphoramidite is prepared by a multistep synthesis starting from 3',5'-*O*-(1,1,3,3-tetraisopropyldisiloxane-1,3-diyl)uridine *(32)* using modified O^4–protection *(33)*, 2'-allylation *(34)*, benzoylation *(35)*, 3',5'-desilylation *(36)*, 5'-dimethoxytritylation *(37)*, and phosphitylation *(38)* procedures.

3.2.2.1. Preparation of 3',5'-*O*-(1,1,3,3-Tetraisopropyldisiloxane-1,3-Diyl)Uridine (Compound **1**)

Coevaporate 3.66 g, 15 mmol uridine 2× with 20 mL dry pyridine, dissolve in 50 mL dry pyridine, and add 5.0 g, 15.8 mmol 1,3-dichloro-1,1,3,3-tetraisopropyldisiloxane via syringe to the septum-sealed flask. Stir the mixture at 20°C until TLC (MeOH–CH$_2$Cl$_2$ 2:3 v/v) shows complete disappearance of the starting material (approx 3 h). Dilute the mixture with 300 mL EtOAc, transfer it to a 1-L separating funnel, wash 2× with 150 mL H$_2$O, 10% aqueous NaHCO$_3$ (2 × 150 mL), and 125 mL brine, dry over anhydrous Na$_2$SO$_4$, and evaporate to dryness. Coevaporate the residue 3× with 50 mL toluene, and purify the residue by chromatography on a silica gel column eluted with EtOAc–CHCl$_3$ (1:4, then 1:2 v/v). Combine the appropriate fractions, evaporate, coevaporate 3× with 50 mL CHCl$_3$ and 3× with 50 mL CH$_2$Cl$_2$, and dry *in vacuo*. Yield 6.88 g (94%) as a white foam. R_f 0.55 (EtOH–CHCl$_3$ 1:9 v/v). ^{1}H-HMR (CDCl$_3$, δ, ppm): 9.34 (br s, 1H, NH), 7.74 (d, 1H, H-6, $J_{5,6}$ = 8.1 Hz), 5.74 (s, 1H, H-1'), 5.70 (d, 1H, H-5), 4.34 (dd, 1H, H-3', $J_{3',4'}$ = 4.7 Hz, $J_{3',2'}$ = 8.8 Hz), 4.21 (m, 2H, H-4', CH_2b), 4.14 (d, 1H, H-2', $J_{2',3'}$ = 8.8 Hz), 4.00 (dd, 1H, CH_2a), 3.44 (s, 1H, 2'-OH), 1.00–1.10 (m, 28H, Pr^i).

3.2.2.2. Preparation of 3',5'-*O*-(1,1,3,3-Tetraisopropyldisiloxane-1,3-Diyl)-*O*^{4}–2-Nitrophenyluridine (Compound 2)

Coevaporate 3',5'-*O*-(1,1,3,3-tetraisopropyldisiloxane-1,3-diyl)uridine (4.33 g, 8.9 mmol) 3× with 10 mL dry pyridine and dissolve in 45 mL dry CH$_2$Cl$_2$. To the magnetically stirred solution add 8.9 mL, 63.6 mmol Et$_3$N and 4.45 mL, 35.1 mmol Me$_3$SiCl under argon atmosphere. Control the reaction by TLC (EtOAc–hexane 1:2 v/v, the product's R_f 0.45). After approx 20 min transfer the reaction mixture to a separating funnel and wash 2× with 50 mL 100 mL 10% aqueous NaHCO$_3$ and H$_2$O, dry the organic phase over anhydrous Na$_2$SO$_4$, and remove the solvent *in vacuo*. Coevaporate the oily residue 2× with 20 mL toluene, and dissolve in 45 mL CH$_2$Cl$_2$. To the magnetically stirred solution add 6.23 mL, 44.5 mmol Et$_3$N, 2.89 g, 13.2 mmol 2-mesitylenesulfonyl chloride, and 0.26 g, 2.2 mmol DMAP under argon atmosphere. Control the reaction by TLC (EtOAc–hexane 1:2 v/v, the product's R_f 0.7). After approx 30 min, add 0.20 g, 1.8 mmol 1,4-diazabicyclo[2.2.2]octane, and 2.49 g, 17.8 mmol 2-nitrophenol. Control the reaction by TLC (EtOAc–hexane 1:2 v/v, the product's R_f 0.6). After approx 1 h dilute the mixture with 45 mL CHCl$_3$, transfer to a separating funnel, and wash 2× with 80 mL 50 mL 10% aqueous NaHCO$_3$ and H$_2$O, dry the organic phase over anhydrous Na$_2$SO$_4$, and evaporate to dryness *in vacuo*. Dissolve the oil in 25 mL CH$_2$Cl$_2$ with magnetic stirring, add 3.38 g, 17.8 mmol p-TsOHxH$_2$O in 25 mL 1,4-dioxane, and stir for 2 min, then quench the reaction by adding 2.67 mL, 19.1 mmol Et$_3$N. Dilute the reaction mixture with 100 mL CHCl$_3$, transfer to a separating funnel, and wash 2× with 50 mL 100 mL 10% aqueous NaHCO$_3$ and 2× with 50 mL H$_2$O, dry the organic phase over anhydrous Na$_2$SO$_4$, and evaporate to dryness *in vacuo*. Purify the residual oil by chromatography on a silica gel column eluted with a gradient of EtOAc in benzene from 0 to 25% v/v. Yield 3.53 g (65%). R_f 0.12 (EtOAc–hexane 1:2 v/v).

3.2.2.3. Preparation of 3',5'-O-(1,1,3,3-Tetraisopropyldisiloxane-1,3-Diyl)-2'-O-Allyl-O^4-2-nitrophenyluridine (Compound 3)

Coevaporate 3.53 g, 5.8 mmol 3',5'-O-(1,1,3,3-tetraisopropyldisiloxane-1,3-diyl)-O^4-2-nitrophenyluridine 3× with 10 mL dry pyridine, dissolve in 17 mL dry THF, add 0.68 mL, 11.6 mmol allylmethylcarbonate. To the magnetically stirred solution add 53 mg, 0.06 mmol Tris(dibenzylideneacetone)dipalladium and 60 mg, 0.23 mmol triphenylphosphine in 12 mL dry THF under argon at 20°C. Reflux the reaction mixture with condenser for approx 30 min. Control the reaction by TLC (EtOAc–hexane 1:2 v/v). Remove the solvent *in vacuo*, and purify the oily residue by chromatography on a silica gel column eluted with a gradient of EtOAc in benzene from 0 to 40% v/v. Yield 3.37 g (90%). R_f 0.6 (EtOAc–hexane 1:2 v/v). MALDI-TOF: [M+H]$^+$ 647.8 (calculated), 647.2 (found).

3.2.2.4. Preparation of 3',5'-O-(1,1,3,3-Tetraisopropyldisiloxane-1,3-Diyl)-2'-O-Allyluridine (Compound 4) *(33)*

Dissolve 3.45 g, 20.8 mmol 2-nitrobenzaldoxime and 2.35 g, 18.7 mmol 1,1,3,3-tetramethylguanidine in 20 mL dry MeCN. To the stirred solution add 3.37 g, 5.2 mmol 3',5'-O-(1,1,3,3-tetraisopropyldisiloxane-1,3-diyl)-2'-O-allyl-O^4-2-nitrophenyluridine, and continue stirring for approx 3 h at 20°C. Monitor the reaction by TLC (EtOAc–hexane 1:1 v/v). Remove the solvent *in vacuo*, dissolve the rest in 150 mL EtOAc, wash 5× with 50 mL H$_2$O, dry the organic phase under anhydrous Na$_2$SO$_4$, and evaporate *in vacuo*. Purify the compound by chromatography on a silica gel column eluted with a gradient of EtOAc in benzene from 0 to 25% v/v. Yield 2.32 g (85%). ^{1}H-HMR (CDCl$_3$, δ, ppm): 9.35 (br d, 1H, NH), 7.92 (d, 1H, H6, $J_{5,6}$ = 8.1Hz), 5.95 (m, 1H, CH =), 5.76 (m, 1H, H1') 5.69 (m, 1H, H5), 5.40 (dd, 1H, = CH2a, $J_{= CH2a, = CH2b}$ = 1.8Hz), 5.20 (dd, 1H, = CH2b), 4.39 (m, 2H, CH2), 4.27–4.00 (m, 5H, H2', H3', H4', H5'), 1.12–0.95 (m, 28H, Pr^i).

3.2.2.5. Preparation of 3',5'-O-(1,1,3,3-Tetraisopropyldisiloxane-1,3-Diyl)-2'-O-(2,3-Dihydroxypropyl)Uridine (Compound 5)

Dissolve 2.1 g, 4 mmol 3',5'-O-(1,1,3,3-tetraisopropyldisiloxane-1,3-diyl)-2'-O-allyluridine in dry 45 mL THF. Add 0.65 g, 4.8 mmol N-methylmorpholine-N-oxide in 20 mL H$_2$O and 0.04 mg, 0.16 mmol OsO$_4$ in 5 mL THF. Stir the solution for 4 h at 20°C and for 2 d at 4°C. When TLC shows complete conversion, quench the reaction by addition of saturated aqueous Na$_2$S$_2$O$_3$, and dilute with 20 mL CHCl$_3$. Wash the mixture with 20 mL 10% aqueous NaHCO$_3$ and 20 mL H$_2$O. Dry the organic layer over anhydrous Na$_2$SO$_4$ and evaporate *in vacuo*. Purify the residue by chromatography on silica gel column eluted with a gradient of EtOH in $CHCl_3$ (0 to 6%). Yield 2.25 g (90%) as a white solid. R_F 0.4 (EtOH–$CHCl_3$ 1:9 v/v). ^{1}H-HMR ($CDCl_3$, δ, ppm): 9.80 (br d, 1H, NH), 7.88 (d, 1H, H6, $J_{5,6}$ = 8.1Hz), 5.74 (d, 1H, H1'), 5.72 (d, 1H, H5), 4.27–3.74 (m, 11H, H2', H3', H4', H5', CH_2, CH(OH), -CH_2OH, CH(OH), CH_2OH), 1.26–0.98 (m, 28H, Pr^i).

3.2.2.6. PREPARATION OF 3',5'-O-(1,1,3,3-TETRAISOPROPYLDISILOXANE-1,3-DIYL)-2'-O-(2,3-DIBENZOYLOXY)PROPYLURIDINE (COMPOUND 6)

Coevaporate 2.25 g, 4.0 mmol 3',5'-O-(1,1,3,3-tetraisopropyldisiloxane-1,3-diyl)-2'-O-(2,3-dihydroxypropyl)uridine 3× with 10 mL dry pyridine, dissolve in 40 mL dry MeCN, and add 1.23 mL, 8.8 mmol Et_3N and 1.06 g, 8.0 mmol benzoyl cyanide with stirring at 20°C. Control the reaction by TLC (EtOH–$CHCl_3$ 1:9 v/v). After approx 30 min, transfer the reaction mixture to a separating funnel, wash with 50 mL 10% aqueous $NaHCO_3$ and 2× with 50 mL H_2O, dry the organic phase over anhydrous Na_2SO_4 and evaporate *in vacuo*. Purify the product by chromatography on a silica gel column eluted with a gradient of EtOAc in benzene from 0 to 40% v/v. Yield 2.62 g (85%) as a white solid. R_F 0.8 (EtOH–$CHCl_3$ 1:9 v/v). ^{1}H-HMR ($CDCl_3$, δ, ppm): 8.78 (br d, 1H, NH), 8.1 (m, 4H, *o*-Ph) 7.91 (d, 1H, H6), 7.6 (m, 2H, *p*-Ph), 7.45 (m, 4H, *m*-Ph), 5.78 (m, 1H, H1'), 5.71 (m, 2H, H5, CHOBz), 4.78–4.70 (m, 3H, H3', H5'), 4.30–4.17 (m, 2H, CH_2OBz), 4.02–3.77 (m, 4H, CH_2, H2', H4'), 1.15–0.97 (m, 28H, Pr^i).

3.2.2.7. PREPARATION OF 2'-O-(2,3-DIBENZOYLOXYPROPYL)URIDINE (COMPOUND 7)

Dissolve 2.63 g, 3.42 mmol 3',5'-O-(1,1,3,3-tetraisopropyldisiloxane-1,3-diyl)-2'-O-(2,3-dibenzoyl-oxypropyl)uridine in 6 mL dry THF, and add 1*M* TBAF solution in 6 mL THF while stirring at 20°C. Control the reaction by TLC (EtOH–$CHCl_3$ 1:9 v/v). After approx 15 min evaporate the mixture to dryness and purify the compound by chromatography on a silica gel column eluted with a gradient of EtOH in $CHCl_3$ (0 to 15%). Yield 1.53 g (85%). R_f 0.25 (EtOH–$CHCl_3$ 1:9 v/v). ^{1}H-HMR (acetone-d_6, δ, ppm): 9.79 (br d, 1H, NH), 8.10 (m, 1H, H6, $J_{5,6}$ = 8.1 Hz), 8.05–8.01 (m, 4H, *o*-Ph), 7.62–7.60 (m, 2H, *p*-Ph), 7.50–7.47 (m, 4H, *m*-Ph), 6.00 (d, 1H, H1', $J_{1',2'}$ = 7.00 Hz), 5.65 (m, 1H, CHOBz, $J_{CH2a,CHOBz}$ = 6.88 Hz), 5.54 (m, 1H, H5), 4.79 (td, 1H, CH_2a, $J_{CH2a,CH2b}$ = 11.3 Hz), 4.67 (m, 1H, CH_2b-, $J_{CH2b,CH}$ = 2.94 Hz), 4.42 (k_b, 1H, H3', $J_{3',4'}$ = 6.50 Hz), 4.26 (k, 1H, CH_{2a}OBz, $J_{CH,CH2aOBz}$ = 6.88 Hz, $J_{CH2aOBz,CH2bOBz}$ = 12.0 Hz), 4.23 (m, 1H, H2', $J_{2',3'}$ = 7.00 Hz), 4.20 (k, 1H, CH_{2b}OBz, $J_{CH,CH2bOBz}$ = 2.94 Hz), 4.02 (ddd, 1H, H4', $J_{4',5'a}$ = 6.04 Hz), 3.87 (t, 1H, H5'a, $J_{5'a,5'b}$ = 12.35 Hz), 3.81 (t, 1H, H5'b, $J_{4',5'a}$ = 1.01 Hz), 3.45 (m, 2H, 3'-OH, 5'-OH). ^{13}C-HMR (acetone-d_6, δ, ppm): 166.38 (2'-O-[2,3-{OBz}$_2$Pr], C = O), 164.7 (2'-O-[2,3-{OBz}$_2$Pr], C = O), 164.0 (C-4), 151.39 (C-2), 141.20 (C-6), 134.05 (2'-O-[2,3-{OBz}$_2$Pr], C-4), 130.98 (2'-O-[2,3-{OBz}$_2$Pr], C-2,

C-6), 130.85 (2'-O-[2,3-{OBz}$_2$Pr], C-2, C-6), 130.39 (2'-O-[2,3-{OBz}$_2$Pr], C-1), 130.33 (2'-O-[2,3-{OBz}$_2$Pr], C-1, 2'-O-[2,3-{OBz}$_2$Pr], C-3, C-5), 129.41 (2'-O-[2,3-{OBz}$_2$Pr], C-3, C-5), 102.30 (C-5'), 88.55 (C-2'), 85.89 (C-1'), 85.66 (C-4'), 83.95 (C-3'), 72.00 (2'-O-[2,3-{OBz}$_2$*Pr*], C-2), 69.86 (2'-O-[2,3-{OBz}$_2$*Pr*], C-1), 64.06 (2'-O-[2,3-{OBz}$_2$*Pr*], C-3), 59.37 (C-5').

3.2.2.8. Preparation of 5'-O-(4,4'-Dimethoxytrityl)-2'-O-(2,3-Dibenzoyloxypropyl)Uridine (Compound 8)

Coevaporate 1.49 g, 2.84 mmol 2'-O-(2,3-dibenzoyloxypropyl)uridine 3× with 10 mL dry pyridine, dissolve in 14 mL dry pyridine, and add 1.44 g, 4.26 mmol 4,4'-dimethoxytrityl chloride in one portion with stirring. Monitor the reaction by TLC (EtOH–CHCl$_3$ 1:19 v/v). After approx 2 h, quench the reaction with 5 mL MeOH, half-evaporate *in vacuo*, dilute by 60 mL CHCl$_3$, wash with 50 mL 10% aqueous NaHCO$_3$ and 2× with 50 mL H$_2$O, dry the organic phase over anhydrous Na$_2$SO$_4$ and evaporate *in vacuo*. Purify the product by chromatography on a silica gel column eluted with a gradient of MeOH in CHCl$_3$ containing 0.5% v/v Et$_3$N from 1 to 5% v/v. Yield 2.24 g (95%) as a white solid. R$_F$ 0.3 (EtOH–CHCl$_3$ 1:19 v/v). ^{1}H-HMR (CDCl$_3$, δ, ppm): 9.79 (br d, 1H, NH), 8.1–8.05, 7.65–7.25, 6.95–6.85 (m, 19H, Ph, 4,4'-DMTr, H6), 5.92 (d, 1H, H1'), 5.75 (m, 1H, C*H*OBz), 5.3 (d, 1H, H5), 4.8 (dd, 1H, C*H*$_{2a}$OBz), 4.72 (m, 1H, 3'-OH), 4.62 (m, 1H, C*H*$_{2b}$OBz), 4.50 (m, 1H, H3'), 4.40 (m, 1H, C*H*$_{2a}$), 4.10–4.00 (m, 3H, C*H*$_{2b}$, H-2', H-4'), 3.85 (s, 6H, C*H*$_3$O), 3.62 (m, 1H, H-5'a), 3.55 (m, 1H, H-5'b). ^{13}C-HMR (CDCl$_3$, δ, ppm): 166.38 (2'-O-[2,3-{OBz}$_2$Pr], C = O), 164.7 (2'-O-[2,3-{OBz}$_2$Pr], C = O), 164.0 (C-4), 151.39 (C-2), 141.20 (C-6), 134.05 (2'-O-[2,3-{OBz}$_2$Pr], C-4), 130.98 (2'-O-[2,3-{OBz}$_2$Pr], C-2, C-6), 130.85 (2'-O-[2,3-{OBz}$_2$Pr], C-2, C-6), 130.39 (2'-O-[2,3-{OBz}$_2$Pr], C-1), 130.33 (2'-O-[2,3-{OBz}$_2$Pr], C-1, 2'-O-[2,3-{OBz}$_2$Pr], C-3, C-5), 129.41 (2'-O-[2,3-{OBz}$_2$Pr], C-3, C-5), 102.30 (C-5'), 88.55 (C-2'), 85.89 (C-1'), 85.66 (C-4'), 83.95 (C-3'), 72.00 (2'-O-[2,3-{OBz}$_2$*Pr*], C-2), 69.86 (2'-O-[2,3-{OBz}$_2$*Pr*], C-1), 64.06 (2'-O-[2,3-{OBz}$_2$*Pr*], C-3), 59.37 (C-5').

3.2.2.9. Preparation of 5'-O-(4,4'-Dimethoxytrityl)-2'-O-(2,3-Dibenzoyloxypropyl)Uridine β-Cyanoethyl-N,N-Diisopropyl-Phosphoramidite (Compound 9)

Coevaporate 0.41 g, 0.5 mmol 5'-O-(4,4'-dimethoxytrityl)-2'-O-(2,3-dibenzoyloxypropyl)uridine 3× with 2 mL dry MeCN, dissolve in 1.5 mL dry CH$_2$Cl$_2$, and add 0.39 mL, 2.3 mmol DIEA and 0.24 mL, 1.0 mmol β-cyanoethoxy-N,N-diisopropylaminochlorophosphine under argon. Monitor the reaction by TLC (CH$_2$Cl$_2$–Et$_3$N 98:2 v/v). Stir for approx 50 min, then dilute with 20 mL CH$_2$Cl$_2$, and wash with 15 mL 10% aqueous NaHCO$_3$ and 15 mL brine, dry over anhydrous Na$_2$SO$_4$, and evaporate *in vacuo*. Triturate the foam with cold hexane, keep for 18–24 h at 5°C, and decant and dry the solid *in vacuo*. Yield 0.41 g (80%) as a white solid. R_f 0.4 (CH_2Cl_2–Et_3N 98:2 v/v). MALDI-TOF: [M+H]$^+$ 1029.1 (calculated), 1029.0 (found). ^{31}P NMR ($CDCl_3$, δ, ppm): 151.55, 150.73, 151.44, 150.57 (2d, mixture of diastereomers). The phosphoramidite is stable when stored at –20°C for at least 12 mo.

3.2.3. Solid-Phase Synthesis of "Protected" 2'-Aldehyde Oligonucleotides

1. Dry down the 2'-diol phosphoramidite overnight *in vacuo* over KOH. Dissolve under argon in one of spare synthesizer bottles to 0.15M in anhydrous MeCN.
2. Assemble the oligonucleotide sequence on a synthesizer by the β-cyanoethyl phosphoramidite method *(15)* on a 1-μmol scale in DMTr-off mode. Incorporate the diol unit into any desired position within the oligonucleotide sequence. An extended coupling time of 10–15 min is recommended. A coupling yield of >96% for the diol phosphoramidite should be expected.
3. After the end of the synthesis, remove the column, and carry out all the necessary steps as described in **Subheading 3.1.3**.
4. Analyze and purify the oligonucleotide by HPLC. For 2'-modified oligonucleotide analysis use of RP-HPLC in ion pair mode on a DIAKS-130-CETYL column (4 × 250 mm) is recommended. Buffer A, 5% aqueous MeCN (v/v), 2 mM tetrabutylammonium dihydrogen phosphate, 48 mM KH_2PO_4, pH 7.0; buffer B, 40% aqueous MeCN (v/v), 2 mM tetrabutylammonium dihydrogen phosphate, 48 mM KH_2PO_4, pH 7, flow rate 1 mL/min^{-1} at 45°C. Use a logarithmic gradient of 0–45.9% B for 1 min, 45.9–49.2% B for 1 min), 49.2–53.6% B for 3 min, 53.6–56.9% B for 5 min, 56.9–60.2 B for 10 min, 60.2–62.1% B for 10 min, and 62.1–63.5% B for 10 min. For RP- and IE-HPLC purification conditions *see* **Subheading 3.1.3**.
5. Check the molecular mass of the oligonucleotides by MALDI-TOF spectrometry as described in **Subheading 3.1.3**.

3.2.4. Solution-Phase Chemoselective Ligation of 2'-Aldehyde Oligonucleotides and N-Terminally Modified Peptides

A 2'-aldehyde oligonucleotide can be generated from its 2'-diol precursor either by $NaIO_4$ or solid-supported periodate treatment. Conjugation of the thus obtained 2'-aldehyde may be effected through use of any of the three alternative chemistries, thiazolidine, oxime, or hydrazone. For greater stability, the hydrazone linkage may be reduced to a hydrazine by a solid-supported borohydride reduction (*see* **Note 5**).

1. *Preparation of 2'-aldehyde oligonucleotides*: Treat a solution of 2'-diol oligonucleotide (5 A_{260} U, approx 43 nmol) in 50 μL of 0.2M acetate buffer, pH 4.5, with a few granules of polymer-supported periodate. After 1 h of shaking, decant off the solution, and wash the support 2× with 20 mL H_2O. Evaporate the solution *in vacuo*. Alternatively, add 5 μL of 0.1M $NaIO_4$ to a solution of 2'-diol

oligonucleotide (5 A_{260} U, approx 43 nmol) in 50 µL 0.2M acetate buffer, pH 4.5. After 30 min, evaporate the reaction mixture *in vacuo*. The latter method is not recommended for subsequent thiazolidine ligation because of Cys oxidation.

2. *Thiazolidine ligation*: Add a solution of N-terminal Cys peptide (50–70 nmol) in a mixture of 0.4M acetate buffer, pH 4.5, and MeCN or NMP (3:7–7:3 v/v, 50 µL) to the oligonucleotide from the above stage. Incubate the reaction mixture at 37°C for 0.3–18 h, then evaporate *in vacuo* and purify by polyacrylamide gel electrophoresis (PAGE) or HPLC as appropriate.
3. *Oxime ligation*: To the oligonucleotide from **step 1**, add a solution of aminooxyacetyl peptide (60–100 nmol) in 50 µL of 0.4M acetate buffer, pH 4.7, and dimethyl sulfoxide (DMSO) (3:7–7:3 v/v). Incubate the reaction mixture at 37°C for 0.3–5 h and purify by PAGE or HPLC.
4. *Hydrazone ligation*: To the oligonucleotide from **step 1**, add a solution of hydrazine (hydrazinoacetyl or hydrazinobenzoyl) or hydrazide peptide (60–100 nmol) in 50 µL of 0.4M acetate buffer, pH 4.7, and DMSO (3:7–7:3 v/v). Incubate the reaction mixture at 37°C for 0.3–5 h, then add a few granules of polymer-supported borohydride reagent. Incubate the reaction mixture for 30 min more and purify by PAGE or HPLC.
5. For PAGE purification, use denaturing gel electrophoresis of oligonucleotides and conjugates in 15% PAGE containing 2M urea in Tris-borate buffer (50 mM Tris-HCl, 50 mM boric acid, 1 mM ethylenediaminetetraacetic acid, pH 7.5). Cut the bands, elute the product with 0.5M LiClO$_4$, desalt using a Microcon® YM-3 tube (Millipore) with a cut-off limit of 3000, or by use a NAP-10 column, and lyophilize. Use of PAGE purification is recommended for most peptide conjugates (*see* **Note 6**).
6. Check the molecular mass of the conjugates by MALDI-TOF spectrometry as described in **Subheading 3.1.3**.

4. Notes

1. The batch of PS200 support used has a small bead size of approx 30 µM. For easier handling, we recommend use of Omnifit columns (150 mm) and 25 µM disposable frits for all the functionalization steps.
2. We used HATU/DIEA/DMF *(13,14)* activation protocol exclusively for stepwise solid-phase synthesis of peptide–oligonucleotide conjugates and for most manual incorporations as well. However, other peptide coupling reagents, for example, HBTU, TBTU, and PyBOP *(14)* may be used with equal success to assemble normal peptides for a chemoselective ligation.
3. Our attempts to apply an on-machine acetic anhydride capping step via a predissolved single-bottle capping mixture, for example, Ac$_2$O–pyridine 1:1 v/v, 10 min, during Fmoc/*tert*-butyl peptide synthesis were not successful, whereas the use of separate capping reagents works well only for manual synthesis. Thus, we switched to the milder HATU-mediated chemistry described for automated synthesis.

4. So far the best peptide synthesis results were obtained with pentafluorophenyl S-benzylthiosuccinate/hydrazine treatment for N-terminal hydrazide incorporation and, most recently, with Tris-Boc-hydrazinoacetic acid/HATU/DIEA/DMF and Fmoc-hydrazinobenzoic acid/N,N'-diisopropylcarbodiimide/CH_2Cl_2 coupling for hydrazine incorporation (data not yet published). Use of Boc-aminooxyacetic acid in our hands led to multiple products after deprotection on RP-HPLC.
5. Hydrazone ligation/borohydride treatment two-step protocol is not compatible with oligonucleotides or peptides labeled with fluorescein owing to its reduction.
6. From the point of stability of the linkages, the most stable is the hydrazine bond obtained after borohydride reduction, the next being oxime, which is, in turn, more stable than thiazolidine *(28)*.

References

1. Gait, M. J. (2003) Peptide-mediated cellular delivery of antisense oligonucleotides and their analogues. *Cell. Mol. Life Sci.* **60,** 844–853.
2. Tung, C.-H. and Stein, S. (2000) Preparation and applications of peptide–oligonucleotide conjugates. *Bioconjugate Chem.* **11,** 605–618.
3. Fischer, P. M., Krausz, E., and Lane, D. V. (2001) Cellular delivery of impermeable effector molecules in the form of conjugates with peptides capable of mediating membrane translocation. *Bioconjugate Chem.* **12,** 825–841.
4. Langel, Ü. (2002) *Cell-Penetrating Peptides: Processes and Applications.* CRC Press, Boca Raton, LA, USA.
5. Stetsenko, D. A., Arzumanov, A. A., Korshun, V. A., and Gait, M. J. (2000) Peptide–oligonucleotide conjugates as enhanced antisense agents. *Mol. Biol. (Rus.)* **34,** 998–1006.
6. Zubin, E. M., Romanova, E. A., and Oretskaya, T. S. (2002) Modern methods for the synthesis of peptide–oligonucleotide conjugates. *Russ. Chem. Rev.* **71,** 239–264.
7. Robles, J., Pedroso, E., and Grandas, A. (1994) Stepwise solid-phase synthesis of the nucleopeptide Phac-Phe-Val-Ser ($p^{3'}$ACT)-Gly-OH. *J. Org. Chem.* **59,** 2482–2486.
8. Bergmann, F. and Bannwarth, W. (1995) Solid phase synthesis of directly linked peptide–oligodeoxynucleotide hybrids using standard synthesis protocols. *Tetrahedron Lett.* **36,** 1839–1842.
9. Antopolsky, M. and Azhayev, A. (1999) Stepwise solid-phase synthesis of peptide–oligonucleotide conjugates on new solid supports. *Helv. Chim. Acta* **82,** 2130–2140.
10. Antopolsky, M. and Azhayev, A. (2000) Stepwise solid-phase synthesis of peptide–oligonucleotide phosphorothioate conjugates employing Fmoc peptide chemistry. *Tetrahedron Lett.* **41,** 9113–9117.
11. Stetsenko, D. A. and Gait, M. J. (2001) A convenient solid-phase method for synthesis of 3'-conjugates of oligonucleotides. *Bioconjugate Chem.* **12,** 576–586.

12. Stetsenko, D. A., Malakhov, A. D., and Gait, M. J. (2002) Total stepwise solid-phase synthesis of oligonucleotide-(3'-N)-peptide conjugates. *Org. Lett.* **4,** 3259–3262.
13. Carpino, L. A., El-Faham, A., Minor, C. A., and Albericio, F. (1994) Advantageous applications of azabenzotriazole (triazolopyridine)-based coupling reagents to solid-phase peptide synthesis. *J. Chem. Soc. Chem. Commun.* 201, 202.
14. Chan, W. C. and White, P. D. (2000) *Fmoc Solid Phase Peptide Synthesis: A Practical Approach.* Oxford University Press, Oxford, U.K.
15. Brown, T. and Brown, D. J. S. (1991) Modern machine-aided methods of oligodeoxyribonucleotide synthesis. In *Oligonucleotides and Analogues: A Practical Approach* (F. Eckstein, ed.). Oxford University Press, Oxford, UK, pp. 1–24.
16. Antopolsky, M., Azhayeva, E., Tengvall, U., et al. (1999) Peptide–oligonucleotide phosphorothioate conjugates with membrane translocation and nuclear localization properties. *Bioconjugate Chem.* **10,** 598–606.
17. Harrison, J. G. and Balasubramanian, S. (1998) Synthesis and hybridization analysis of a small library of peptide–oligonucleotide conjugates. *Nucleic Acids Res.* **26,** 3136–3145.
18. McMinn, D. L. and Greenberg, M. M. (1999) Convergent solution-phase synthesis of a nucleopeptide using a protected oligonucleotide. *Bioorg. Med. Chem. Lett.* **9,** 547–550.
19. Kachalova, A. V., Stetsenko, D. A., Romanova, E. A., Tashlitsky, V. N., Gait, M. J., and Oretskaya, T. S. (2002) A new and efficient method for synthesis of 5'-conjugates of oligonucleotides through amide-bond formation on solid phase. *Helv. Chim. Acta* **85,** 2409–2416.
20. Stetsenko, D. A. and Gait, M. J. (2000) Efficient conjugation of peptides to oligonucleotides by "native ligation." *J. Org. Chem.* **65,** 4900–4908.
21. Cebon, B., Lambert, J. N., Leung, D., et al. (2000) New DNA modification strategies involving oxime formation. *Austr. J. Chem.* **53,** 333–340.
22. Forget, D., Boturyn, D., Defrancq, E., Lhomme, J., and Dumy, P. (2001) Highly efficient synthesis of peptide–oligonucleotide conjugates: chemoselective oxime and thiazolidine formation. *Chemistry* **7,** 3976–3984.
23. Forget, D., Renaudet, O., Boturyn, D., Defrancq, E., and Dumy, P. (2001) 3'-oligonucleotide conjugation via chemoselective oxime bond formation. *Tetrahedron Lett.* **42,** 9171–9174.
24. Olivier, N., Olivier, C., Gouyette, C., Huynh-Dinh, T., Gras-Masse, H., and Melnyk, O. (2002) Synthesis of oligonucleotide-peptide conjugates using hydrazone chemical ligation. *Tetrahedron Lett.* **43,** 997–999.
25. Spetzler, J. C. and Tam, J. P. (1995) Unprotected peptides as building blocks for branched peptides and peptide dendrimers. *Int. J. Peptide Protein Res.* **45,** 78–85.
26. Shao, J. and Tam, J. P. (1995) Unprotected peptides as building blocks for the synthesis of peptide dendrimers with oxime, hydrazone, and thiazolidine linkages. *J. Am. Chem. Soc.* **117,** 3893–3899.
27. Kachalova, A. V., Zatsepin, T. S., Romanova, E. A., Stetsenko, D. A., Gait, M. J., and Oretskaya, T. S. (2000) Synthesis of modified nucleotide building blocks

containing electrophilic groups in the 2'-position. *Nucleosides Nucleotides Nucleic Acids* **19,** 1693–1707.
28. Zatsepin, T. S., Stetsenko, D. A., Arzumanov, A. A., Romanova, E. A., Gait, M. J., and Oretskaya, T. S. (2002) Synthesis of peptide–oligonucleotide conjugates with single and multiple peptides attached to 2'-aldehyde through thiazolidine, oxime and hydrazine linkages. *Bioconjugate Chem.* **13,** 822–830.
29. Stetsenko, D. A., Williams, D., and Gait, M. J. (2001) Synthesis of peptide–oligonucleotide conjugates: application to basic peptides. *Nucl. Acids Res.* **Suppl. 1,** 153, 154.
30. Chan, W. C., Bycroft, B. W., Evans, D. J., and White, P. D. (1995) A novel 4-aminobenzyl ester-based carboxy-protecting group for synthesis of atypical peptides by Fmoc-Bu^t solid-phase chemistry. *J. Chem. Soc. Chem. Commun.* 2209, 2210.
31. Athanassopoulos, P., Barlos, K., Gatos, D., Hatzi, O., and Tzavara, C. (1995) Application of 2-chlorotrityl chloride in convergent peptide synthesis. *Tetrahedron Lett.* **36,** 5645–5648.
32. Chui, H. M.-P., Meroueh, M., Scaringe, S. A., and Chow, C. S. (2002) Synthesis of a 3-methyluridine phosphoramidite to investigate the role of methylation in a ribosomal RNA hairpin. *Bioorg. Med. Chem.* **10,** 325–332.
33. Sproat, B. S. and Lamond A. I. (1991) 2'-O-Methyloligoribonucleotides: synthesis and applications. In *Oligonucleotides and Analogues: A Practical Approach* (F. Eckstein, ed.). IRL Press: New York, pp. 49–86.
34. Sproat, B. S., Iribarren, A. I., Garcia, R. G., and Beijer, B. (1991) New synthetic routes to synthons suitable for 2'-O-allyloligoribonucleotide assembly. *Nucl. Acids Res.* **19,** 733–738.
35. Holy, A. and Soucek, M. (1971) Benzoyl cyanide; a new benzoylating agent in nucleoside and nucleotide chemistry. *Tetrahedron Lett.* **12,** 185–188.
36. Matsuda, A., Yasuoka, J., Sasaki, T., and Ueda, T. (1991) Improved synthesis of 1-(2-azido-2-deoxy-β-D-arabinofuranosyl)cytosine (cytarazid) and thymine, inhibitory spectrum of cytarazid on the growth of various human tumor cells in vitro. *J. Med. Chem.* **34,** 999–1002.
37. Jones, R. A. (1984) Preparation of protected deoxyribonucleotides. In *Oligonucleotide Synthesis: A Practical Approach* (M. J. Gait, ed.). IRL Press, Washington, DC, pp. 23–34.
38. Beaucage, S. L. (1993) Oligodeoxyribonucleotide synthesis: phosphoramidite approach. In *Protocols for Oligonucleotides and Analogs: Synthesis and Properties* (S. Agrawal, ed.). Humana Press, Totowa, NJ, pp. 33–63.

13

Biotin-Labeled Oligonucleotides With Extraordinarily Long Tethering Arms

A. Michael Morocho, Valeri Karamyshev, Olga Shcherbinina, and Nikolai Polushin

Summary

We describe synthesis of four novel biotin phosphoramidites with tethering arms ranging from 20 to 74 atoms in length. One of these phosphoramidites is a uridine derivative with a biotin moiety attached through the 2'-position. The biotin phosphoramidites were synthetized based on robust and efficient methoxyoxalamido (MOX) and succinimido (SUC) precursor strategies from MOX/SUC precursors containing a secondary hydroxyl. They are highly stable in solution (coupling efficiency remains equally high for at least 2 wk after phosphoramidite installation on the synthesizer) and are ideal for the streamlined production of biotin-labeled oligonucleotides. Protocols for the synthesis of biotinylated primers and purification of biotinylated sequencing products by means of streptavidin-coated magnetic beads are also presented.

Key Words: Biotin-labeled oligonucleotides; biotin phosphoramidites; tethering arm; precursor strategy.

1. Introduction

Biotin-labeled oligonucleotides are widely used for deoxyribonucleic acid (DNA)/ribonucleic acid (RNA) detection/isolation owing to the extremely high affinity of the biotin–streptavidin interaction ($K_d = 10^{-15}M$) *(1)*. There are several methods for the attachment of biotin moieties to oligonucleotides, but the easiest and most efficient is through the use of biotin phosphoramidites on the DNA/RNA synthesizer. Quite a handful of such phosphoramidites is commercially available *(2)*, yet none of these have a tethering arm exceeding 20 atoms.

For some applications, especially those executed on a solid support, it is beneficial to have a biotin moiety attached through a considerably longer noncharged tethering arm. For example, it has been shown that tether lengths of at least 40 atoms for surface-bound oligonucleotides is crucial for optimum hybridization to its target *(3)*, and unfavorable steric interactions are reduced by positioning the bound oligonucleotide away from an otherwise crowded surface *(4)*. Obviously, it is possible to make longer linkers by sequentially incorporating spacers to the end of the oligonucleotide prior to adding a terminal biotin. However, this approach builds up charge within the linker and results in adverse intermolecular repulsion to other bound linkers and target DNA. For these reasons the use of biotin phosphoramidites with exceptionally long uncharged tethers should be favorable.

In this chapter we describe the synthesis of biotin phosphoramidites with extraordinarily long tethers and biotinylated oligonucleotides derived from them. Owing to space limitations we describe the synthesis of only four novel biotin phosphoramidites with tethering arms ranging from 20 to 74 atoms in length. One of these phosphoramidites is a uridine derivative with a biotin moiety attached through the 2'-position. The biotin phosphoramidites were synthesized based on robust and efficient methoxyoxalamido (MOX) and succinimido (SUC) precursor strategies from MOX/SUC precursors containing a secondary hydroxyl *(5–7)* . They are highly stable in solution because coupling efficiency remains equally high for at least 2 wk after phosphoramidite installation on the synthesizer, and they are ideal for the streamlined production of biotin-labeled oligonucleotides. All phosphoramidites have an *O*-dimethoxytrityl (DMT) group on the biotin moiety for coupling efficiency determination or reverse-phase purification and for increased solubility in acetonitrile (ACN). We believe the use of long uncharged tethers will increase the efficiency of many biotin-based nucleic acid processing events staged at the solid-phase surface.

2. Materials

1. The following laboratory-grade solvents are required: chloroform, methanol, triethylamine (TEA), ethanol, and formamide.
2. The following anhydrous solvents are required: pyridine, dioxane, ACN, dichloromethane (DCM), toluene, ethyl acetate, ether, and pentane.
3. The following high purity reagents are required: (+)-biotin, 4-nitrophenol; 1,3-dicyclohexylcarbodiimide (DCC), 4,7,10-trioxa-1,13-tridecanediamine, dimethyl oxalate, 4-dimethyl-aminopyridine, 4,4'-dimethoxytrityl chloride (DMTrCl), Silica Gel 60, Na_2SO_4, 1,8-diazabicyclo(5.4.0)undec-7-ene (DBU), *trans*-4-aminocyclohexanol hydrochloride, tetrazole; 2-cyanoethyl *N,N,N',N'*-tetraisopropylphosphorodiamidite, $NaHCO_3$, NaCl, ethanolamine, ammonium acetate, and dextran blue.
4. *Trans*-4-MOX-1-cyclohexanol (synthesize as described in *(7)* or purchase from Fidelity Systems, Gaithersburg, MD; www.fidelitysystems.com).

5. 5'-dimethoxytrityl-2'-succinimido-2'-deoxyuridine (Fidelity Systems, Gaithersburg, MD; www.fidelitysystems.com).
6. ASM-800 DNA synthesizer (Biosset, Novosibirsk, Russia) or similar 0.05–1-μmol synthesis scale DNA synthesizer.
7. Regular DNA monomers: dA-CE phosphoramidite, dC^{Ac}-CE phosphoramidite, dG-CE phosphoramidite, dG-CE phosphoramidite (Transgenomic, Wayne, PA).
8. Ancillary reagents for DNA synthesis (Transgenomic, Wayne, PA).
9. 0.25M Ethylthio tetrazole (ETT) in CH_3CN (Transgenomic, Wayne, PA).
10. 100-nmol scale synthetic columns (Biosset, Novosibirsk, Russia).
11. G-25 NAP-10 column (Amersham Pharmacia Biotech AB, Uppsala, Sweden).
12. Equipment and reagents for polyacrylamide gel electrophoresis (PAGE).
13. Dynabeads M-280 streptavidin (Dynal Biotech, Oslo, Norway).

3. Methods
3.1. DMT-Bt-Arm17-Ach Phosphoramidite (Compound 5)

Synthesis of DMT-Bt-Arm17-Ach phosphoramidite is outlined in **Scheme 1**. This is the simplest of the presently described biotin phosphoramidites. It can be introduced only at the 5'-end and provides an uncharged tether of 23 atoms in length.

3.1.1. Biotynyl-13-Methoxyoxalyl-4,7,10-Trioxa-Tridecanediamide (Compound 2)

3.1.1.1. (+)-BIOTIN-4-NITROPHENYL ESTER

1. Add anhydrous pyridine (400 mL) to the mixture of (+)-biotin (12.2 g, 50 mmol) and 4-nitrophenol (7.65 g, 55 mmol). Heat the suspension to 85–95°C, and stir until almost completely solubilized. Stop heating and carefully, under moderate vacuum (10–15 mm Hg), concentrate the solution to approx 300 mL. During concentration the solution cools down and (+)-biotin partly crystallizes. This step is needed to remove any possible traces of water.
2. Heat the suspension again to solubilize the precipitated (+)-biotin. Quickly add DCC (10.3 g, 50 mmol), and let the reaction mixture stir at room temperature for 2 h. Heat the reaction mixture to 85–95°C, and add the second portion of DCC (2.06 g, 10 mmol). Let the reaction mixture stir at room temperature for an additional hour, concentrate to approx 150 mL, and leave it stirring overnight.
3. Filter the reaction mixture, and wash the urea crystals with cold dioxane (2 × 50 mL). Concentrate the crude solution of (+)-biotin-4-nitrophenyl ester to approx 150 mL.

3.1.1.2. 1-BIOTINYL-4,7,10-TRIOXA-13-TRIDECANEAMINE

1. Add the crude biotin-4-nitrophenyl ester solution from **Subheading 3.1.1.1.** portionwise over a 1-h period to a stirring solution of 4,7,10-trioxa-1,13-

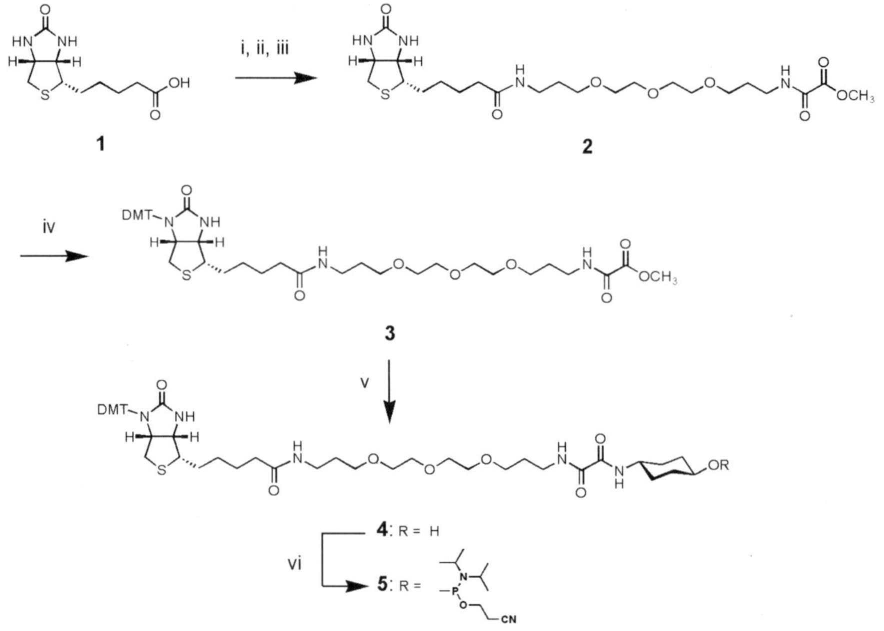

Scheme 1. Reaction scheme for the preparation of DMT-Bt-Arm17-Ach phosphoramidite. Reagents: **(i)** 4-nitrophenol and DCC in pridine; **(ii)** 4,7,10-trioxa-1,13-tridecanediamine in ACN; **(iii)** dimethyl oxalate and TEA in methanol; **(iv)** DMTrCl and 4-dimethyl-aminopyridine in pyridine; **(v)** *trans*-4-aminocyclohexanol hydrochloride, DBU and pyridine in methanol; **(vi)** 2-cyanoethyl *N,N,N',N'*-tetraisopropylphosphorodiamidite and tetrazole in DCM.

tridecanediamine (43.8 mL, 200 mmol) (*see* **Note 1**) in ACN (200 mL). Stir the reaction mixture for an additional 15 min, and then add 1.6. L ether in approx 200 mL portions with stirring. Let the reaction mixture stir for a few more hours. Filter the reaction mixture, wash the precipitate with ether (2 × 200 mL), and dry under vacuum.

3.1.1.3. 1-Biotynyl-13-Methoxyoxalyl-4,7,10-Trioxa-Tridecanediamide

1. Dissolve the solid from **Subheading 3.1.1.2.** in MeOH (200 mL) containing TEA (13.9 mL, 100 mmol). Add the solution portionwise to a stirring solution of dimethyl oxalate (17.7 g, 150 mmol) in MeOH (100 mL) over a 1-h period. Continue stirring for another hour.
2. Concentrate the reaction mixture to dryness. Raise the bath temperature to approx 60°C, and continue to concentrate for 30–40 min. This will remove the majority of nonreacted dimethyl oxalate (*see* **Note 2**).
3. Dissolve the residue in DCM (200 mL) containing 3% of MeOH and fractionate by flash column chromatography on silica gel (approx 600 mL) using increasing

proportions of MeOH (3–12%) in chloroform as eluent. Combine and evaporate to dryness the product-containing fractions to obtain white hydroscopic solid (13.8 g, 52%). R_f = 0.51 (chloroform:methanol, 8:2); ^{1}H nuclear magnetic resonance (NMR) (dimethyl sulfoxide [DMSO]-d$_6$) δ 8.85–8.95 (br. t, 1H, NHCOCO), 7.70–7.78 (t, 1H, CONH, Bt), 6.40–6.46 (s, 1H, N*H*CONH, Bt), 6.33–6.40 (s, 1H, NHCON*H*, Bt), 4.26–4.35 (m, 1H, CH, Bt), 4.08–4.17 (m, 1H, CH, Bt), 3.73–3.80 (s, 3H, COCOOCH$_3$), 3.42–3.57 (m, 8H, 4×CH$_2$), 3.35–3.42 (q, 4H, 2×CH$_2$), 3.12–3.23 (q, 2H, CH$_2$), 3.02–3.12 (m, 3H, CH (Bt) + CH$_2$), 2.76–2.86 (d/d, 1H, 1/2CH$_2$, Bt), 2.51–2.62 (d, 1H, 1/2CH$_2$, Bt), 2.00–2.08 (t, 2H, CH$_2$CO, Bt), 1.20–1.75 (overlapping multiplets, 10H, 3×CH$_2$ (Bt) + 2×CH$_2$); electrospray ionization mass spectrometry (ESI-MS) *m/z* 533.4 (M + H$^+$); calculated for C$_{23}$H$_{40}$N$_4$O$_8$S: 532.6.

3.1.2. 1-(N-4,4'-Dimethoxytrityl-Biotynyl)-13-Methoxyoxalyl-4,7,10-Trioxa-Tridecanediamide (Compound 3)

1. Dissolve compound **2** (10.65 g, 20 mmol) and 4-dimethyl-aminopyridine (0.24 g, 2 mmol) in anhydrous pyridine (150 mL). Concentrate the solution to approx 50 mL to remove any traces of water. To the solution add DMTrCl (10.16 g, 30 mmol), and stir the reaction mixture for 72 h.
2. Dilute the reaction mixture with chloroform (500 mL) and extract with saturated saline solution (500 mL). Separate the organic layer, dry it (Na$_2$SO$_4$), filter, and remove solvent *in vacuo*. Dissolve the remaining oil in toluene (100 mL), and concentrate the solution to dryness to remove traces of pyridine. Redissolve the oil in chloroform (50 mL) containing 0.1% of TEA and fractionate by flash column chromatography on silica gel (approx 500 mL) using increasing proportions of MeOH (1–10%) in chloroform (should contain 0.1% of TEA) as eluent. Combine and evaporate to dryness the product-containing fractions to obtain a yellowish solid (15.2 g, 89%). R_f = 0.64 (chloroform:methanol, 8:2); ESI-MS *m/z* 835.8 (M + H$^+$), 936.5 (M + Et$_3$NH$^+$); calculated for C$_{44}$H$_{58}$N$_4$O$_{10}$S: 835.0.

3.1.3. DMT-Bt-Arm17-Ach Alcohol (Compound 4)

1. Add MeOH (30 mL), pyridine (30 mL), and DBU (8.97 mL, 60 mmol) to *trans*-4-aminocyclohexanol hydrochloride (7.58 g, 50 mmol). Stir the mixture until the solid is completely dissolved. To the solution add compound **3** (8.35 g, 10 mmol) and stir for 1 h.
2. Dilute the reaction mixture with chloroform (500 mL), and extract with saturated saline solution (500 mL). Separate the organic layer, dry it (Na$_2$SO$_4$), filter, and concentrate to dryness. To the residue add toluene (100 mL), and concentrate the mixture to dryness again to remove traces of pyridine. Redissolve the oil in chloroform (50 mL) containing 0.1% of TEA, and fractionate by flash column chromatography on silica gel (approx 400 mL) using increasing proportions of MeOH (0–10%) in chloroform (should contain 0.1% of TEA) as eluent. Combine and evaporate to dryness the product-containing fractions to obtain a yellowish solid (7.89 g, 86%). R_f = 0.50 (chloroform:methanol, 8:2); ^{1}H NMR (DMSO-d$_6$)

δ 8.66–8.75 (t, 1H, NHCOCO), 8.42–8.49 (d, 1H, COCONH, ACH), 7.70–7.78 (t, 1H, CONH, Bt), 7.15–7.34 (m, 5H, DMT), 7.00–7.10 (t, 4H, DMT), 6.83–6.90 (d, 4H, DMT), 6.70–6.76 (s, 1H, NHCON-DMT, Bt), 4.53–4.58 (d, 1H, OH, ACH), 4.23–4.38 (m, 2H, 2×CH, Bt), 3.71–3.78 (s, 6H, OCH$_3$, DMT), 3.42–3.57 (m, 9H, 4×CH$_2$ + CH), 3.33–3.42 (m, 5H, 2×CH$_2$ + CH), 3.02–3.22 (m, 5H, CH (Bt) + 2×CH$_2$), 2.15–2.25 (d, 2H, CH$_2$, Bt), 1.98–2.09 (t, 2H, CH$_2$CO, Bt), 1.10–1.86 (overlapping multiplets, 16H, 3×CH$_2$ (Bt) + 4×CH$_2$ (ACH) + 2×CH$_2$); ESI-MS m/z 918.8 (M + H$^+$), 1020.0 (M + Et$_3$NH$^+$); calculated for C$_{49}$H$_{67}$N$_5$O$_{10}$S: 918.1.

3.1.4. DMT-Bt-Arm17-Ach Phosphoramidite (Compound 5)

1. To the mixture of dry compound **4** (4.59 g, 5 mmol) and dry tetrazole (0.35 g, 5 mmol) placed under blanket of Ar add sequentially DCM (50 mL) and 2-cyanoethyl N,N,N',N'-tetraisopropylphosphorodiamidite (2.47 mL, 7.5 mmol). Stir the reaction mixture for 3 h. Follow the progress of the reaction by thin-layer chromatography (TLC) in pyridine/ethyl acetate (2:8, v/v) mixture containing 1% of TEA: The spot of the starting material stays close to the bottom, and the product has R_f of 0.34. In the course of the reaction the starting solids are disappearing and the crystals of triethylammonium salt are typically formed.
2. Dilute the reaction mixture with DCM (200 mL) containing 2% of TEA, and extract with 10% NaHCO$_3$ solution (250 mL). Separate the organic layer, dry it (Na$_2$SO$_4$), filter, and concentrate to approx 30 mL. Precipitate the residue into ether/pentane mixture (1:1, v/v, 500 mL). Let the oil settle down and decant the supernatant. The precipitation at this point mainly removes H-phosphonate of the phosphitylating reagent. Redissolve the oil in DCM (30 mL) containing 2% of TEA and fractionate by flash column chromatography on silica gel (approx 300 mL) using increasing proportions of pyridine (5, 10, 20%) in ethyl acetate (should contain 2% of TEA) as eluent (500 mL of each portion). The product should not stay on the column longer than 1 h. Combine and evaporate to dryness the product-containing fractions. To completely remove pyridine, coevaporate the product with toluene (100 mL). Reconstitute the oil in DCM (30 mL) and precipitate into cold pentane (600 mL). Centrifuge, decant the supernatant, and dry the residue to obtain white solid (3.44 g, 60%). ^{31}P NMR (DMSO-d$_6$) δ 144.75; ESI-MS m/z 1140.9 (M + Na$^+$), 1220.1 (M + Et$_3$NH$^+$); calculated for C$_{58}$H$_{84}$N$_7$O$_{11}$PS: 1118.4.

3.2. DMT-Bt-Arm34-Ach Phosphoramidite (Compound 8)

Synthesis of DMT-Bt-Arm34-Ach phosphoramidite is outlined in **Scheme 2**. This biotin phosphoramidite can be introduced only at the 5'-end and provides an uncharged tether of 40 atoms in length.

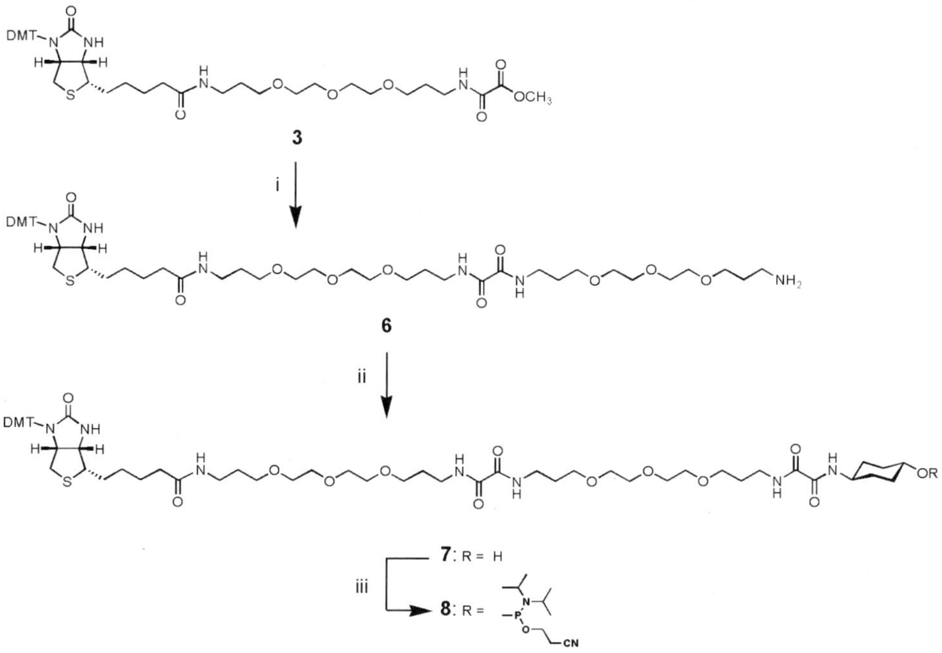

Scheme 2. Reaction scheme for the preparation of DMT-Bt-Arm34-Ach phosphoramidite. Reagents: (i) 4,7,10-trioxa-1,13-tridecanediamine in methanol; (ii) trans-4-MOX-1-cyclohexanol and DBU in methanol; (iii) 2-cyanoethyl N,N,N',N'-tetraisopropylphosphorodiamidite and tetrazole in DCM.

3.2.1. DMT-Bt-Arm34-Ach Alcohol (Compound 7)

3.2.1.1. COMPOUND 6

1. Dissolve compound 3 (8.35 g, 10 mmol) in MeOH (100 mL), and add the solution dropwise to a stirring solution of 4,7,10-trioxa-1,13-tridecanediamine (10.9 mL, 50 mmol) in MeOH (20 mL) over a 1-h period (see **Note 3**). Stir the reaction mixture for an additional 15 min, and then concentrate to dryness.
2. Dissolve the residue in chloroform (300 mL) and extract with saturated saline solution (300 mL). Separate the organic layer, dry it (Na_2SO_4), filter, and concentrate to dryness. Reconstitute the residual oil in toluene (60 mL), and precipitate the solution into pentane/ether mixture (3:1 v/v) (600 mL). Let the oil settle down and decant the supernatant. Repeat the precipitation and dry the product *in vacuo* to obtain the yellowish oil.

3.2.1.2. COMPOUND 7

1. Dissolve the oil from **Subheading 3.2.1.1.** in MeOH (100 mL) and add the solution to *trans*-4-MOX-1-cyclohexanol (2.41 g, 12 mmol). Add DBU (1.5 mL, 10 mmol) and stir overnight. Check completion of the reaction by TLC: In methanol/chloroform (1:9 v/v) the product has slightly lower mobility than *trans*-4-MOX-1-cyclohexanol.
2. Concentrate the reaction mixture to dryness, reconstitute in chloroform (300 mL), and extract with saturated saline solution (300 mL). Separate the organic layer, dry it (Na_2SO_4), filter, and concentrate to approx 80 mL. Fractionate the mixture by flash column chromatography on silica gel (approx 300 mL) using increasing proportions of MeOH (2–10%) in chloroform (should contain 0.1% of TEA) as eluent. Combine and evaporate to dryness the product-containing fractions to obtain a white solid (10.14 g, 85%). ^{1}H NMR (DMSO-d_6) δ 8.65–8.75 (t, 3H, NHCOCONH + NHCOCO), 8.42–8.49 (d, 1H, COCONH, ACH), 7.68–7.76 (t, 1H, CONH, Bt), 7.15–7.34 (m, 5H, DMT), 7.00–7.10 (t, 4H, DMT), 6.81–6.88 (d, 4H, DMT), 6.70–6.76 (s, 1H, NHCON-DMT, Bt), 4.53–4.58 (d, 1H, OH, ACH), 4.23–4.38 (m, 2H, 2×CH, Bt), 3.71–3.78 (s, 6H, OCH$_3$, DMT), 3.42–3.57 (m, 17H, 8×CH$_2$ + CH), 3.33–3.42 (m, 9H, 4×CH$_2$ + CH), 3.02–3.23 (m, 9H, CH (Bt) + 4×CH$_2$), 2.15–2.25 (d, 2H, CH$_2$, Bt), 1.98–2.09 (t, 2H, CH$_2$CO, Bt), 1.10–1.86 (overlapping multiplets, 22H, 3×CH$_2$ (Bt) + 4×CH$_2$ (ACH) + 4×CH$_2$); ESI-MS m/z 1193.2 (M + H$^+$), 1294.2 (M + Et$_3$NH$^+$); calculated for $C_{61}H_{89}N_7O_{15}S$: 1192.4.

3.2.2. DMT-Bt-Arm34-Ach Phosphoramidite (Compound 8)

Prepare compound **8** from compound **7** exactly as compound **5** (*see* **Subheading 3.1.4.**). Compound **8** is a white solid (66%). ^{31}P NMR (DMSO-d_6) δ 144.75; ESI-MS m/z 1494.3 (M + Et$_3$NH$^+$); calculated for $C_{70}H_{106}N_9O_{16}PS$: 1392.7.

3.3. DMT-Bt-Arm68-Ach Phosphoramidite (Compound 12)

Synthesis of DMT-Bt-Arm68-Ach phosphoramidite is outlined in **Scheme 3**. This is the longest of the presently described biotin phosphoramidites. It can be introduced only at the 5'-end and provides an uncharged tether of 74 atoms in length.

3.3.1. MOX-Arm17-Ach Alcohol (Compound 9)

1. Dissolve *trans*-4-MOX-1-cyclohexanol (2.01 g, 10 mmol) in methanol/DCM (25 mL) mixture (1:1, v/v), and add the solution dropwise to a stirring solution of 4,7,10-trioxa-1,13-tridecanediamine (8.76 mL, 40 mmol) in MeOH (50 mL) over a 30-min period. A partial precipitation of the product occurs during this time. Stir the reaction mixture for an additional 30 min, and complete the precipitation

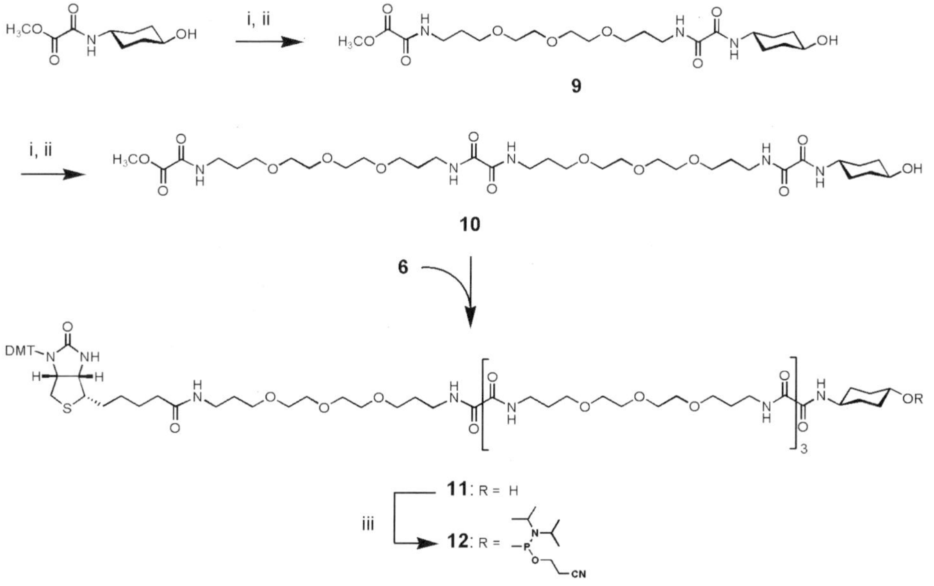

Biotin-Labeled Oligonucleotides

Scheme 3. Reaction scheme for the preparation of DMT-Bt-Arm68-Ach phosphoramidite. Reagents: **(i)** 4,7,10-trioxa-1,13-tridecanediamine in methanol/DCM; **(ii)** dimethyl oxalate and TEA in methanol; **(iii)** 2-cyanoethyl N,N,N',N'-tetraisopropylphosphorodiamidite and tetrazole in DCM.

by addition of diethyl ether (300 mL). Cool the reaction mixture to 4°C, and incubate at this temperature for at least 2 h. Filter the precipitate, wash it with diethyl ether (2 × 30 mL), and dry.

2. Add the dried amino alcohol portionwise to a stirring solution of dimethyl oxalate (40 mmol) and TEA (20 mmol) in methanol (50 mL) over a 2-h period. Stir the reaction mixture for another 30 min, concentrate it to dryness, and remove the excessive dimethyl oxalate by sublimation at approx 5 mmHg and approx 70°C. Reconstitute the residue in chloroform (50 mL) and fractionate the mixture by flash column chromatography on silica gel (approx 300 mL) using increasing proportions of MeOH (0–5%) in chloroform as eluent. Combine and evaporate to dryness the product-containing fractions to obtain white solid (3.95 g, 83%). $R_f = 0.37$ (chloroform:methanol, 9:1); ^{1}H NMR (DMSO-d_6) δ 8.83–8.93 (t, 1H, COCONH, MOX), 8.65–8.73 (t, 1H, NHCOCO), 8.38–8.46 (d, 1H, COCONH, ACH), 4.53–4.58 (d, 1H, OH, ACH), 3.73–3.78 (s, 3H, OCH$_3$, MOX), 3.43–3.57 (m, 9H, 4×CH$_2$ + CH), 3.25–3.43 (m, 5H, 2×CH$_2$ + CH), 3.10–3.24 (q, 8H, 2×CH$_2$), 1.58–1.84 (overlapping multiplets, 8H, + 2×CH$_2$ (ACH) + 2×CH$_2$), 1.30–1.49 (multiplet, 2H, CH$_2$, ACH), 1.07–1.26 (multiplet, 2H, CH$_2$, ACH); ESI-MS m/z 476.3 (M + H$^+$), 498.3 (M + Na$^+$); calculated for $C_{21}H_{37}N_3O_9$: 475.5.

3.3.2. MOX-Arm34-Ach Alcohol (Compound 10)

Prepare compound **10** from compound **9** exactly as compound **9** (*see* **Subheading 3.3.1.**). Compound **10** is a white solid (80%). $R_f = 0.33$ (chloroform: methanol, 9:1); ^{1}H NMR (DMSO-d_6) δ 8.83–8.93 (t, 1H, COCONH, MOX), 8.65–8.75 (t, 3H, NHCOCONH + NHCOCO), 8.42–8.49 (d, 1H, COCONH, ACH), 4.53–4.58 (d, 1H, OH, ACH), 3.73–3.78 (s, 3H, OCH$_3$, MOX), 3.43–3.57 (m, 17H, 8×CH$_2$ + CH), 3.25–3.43 (m, 9H, 4×CH$_2$ + CH), 3.10–3.24 (q, 8H, 4×CH$_2$), 1.60–1.85 (overlapping multiplets, 12H, + 2×CH$_2$ (ACH) + 4×CH$_2$), 1.33–1.50 (multiplet, 2H, CH$_2$, ACH), 1.10–1.28 (multiplet, 2H, CH$_2$, ACH); ESI-MS *m/z* 750.6 (M + H$^+$), 772.6 (M + Na$^+$); calculated for C$_{33}$H$_{59}$N$_5$O$_{14}$: 749.8.

3.3.3. DMT-Bt-Arm68-Ach Alcohol (Compound 11)

3.3.3.1. Compound 6

Prepare compound **6** as described previously (*see* **Subheading 3.2.1.1.**) proceeding from 2 mmol of compound **3**.

3.3.3.2. Compound 11

1. Dissolve the oil from **Subheading 3.3.3.1.** in MeOH (15 mL) and add the solution to the solution of compound **10** (1.65 g, 2.2 mmol) in methanol/DCM (10 mL). Add DBU (0.15 mL of 1 mmol), and let the reaction mixture stir overnight at room temperature. Check completion of the reaction by TLC: In methanol/ethyl acetate (2:8 v/v) the product (R_f 0.28) has lower mobility than compound **10**.
2. Concentrate the reaction mixture to dryness, reconstitute in chloroform (100 mL), and extract with saturated saline solution (100 mL). Separate the organic layer, dry it (Na$_2$SO$_4$), filter, and concentrate to dryness. Reconstitute the residue in methanol/ethyl acetate mixture (1: v/v) (50 mL) containing 0.1% of TEA. Fractionate the mixture by flash column chromatography on silica gel (approx 300 mL) using increasing proportions of MeOH (10–20%) in ethyl acetate (should contain 0.1% of TEA) as eluent. Combine and evaporate to dryness the product-containing fractions to obtain a white solid (2.37 g, 68%). If the product still has impurities recrystallize it from ethyl acetate. ^{1}H NMR (DMSO-d_6) δ 8.65–8.75 (t, 7H, 3×NHCOCONH + NHCOCO), 8.40–8.48 (d, 1H, COCONH, ACH), 7.68–7.76 (t, 1H, CONH, Bt), 7.15–7.34 (m, 5H, DMT), 7.00–7.10 (t, 4H, DMT), 6.81–6.88 (d, 4H, DMT), 6.70–6.76 (s, 1H, NHCON-DMT, Bt), 4.52–4.55 (d, 1H, OH, ACH), 4.23–4.38 (m, 2H, 2×CH, Bt), 3.71–3.78 (s, 6H, OCH$_3$, DMT), 3.42–3.57 (m, 33H, 16×CH$_2$ + CH), 3.33–3.42 (m, 17H, 8×CH$_2$ + CH), 3.02–3.23 (m, 17H, CH (Bt) + 8×CH$_2$), 2.15–2.25 (d, 2H, CH$_2$, Bt), 1.98–2.09 (t, 2H, CH$_2$CO, Bt), 1.10–1.86 (overlapping multiplets, 30H, 3×CH$_2$ (Bt) + 4×CH$_2$ (ACH) + 8×CH$_2$); ESI-MS *m/z* 1439.4 (M - DMT + H$^+$), 1461.4 (M - DMT + Na$^+$), 1741.4 (M + H$^+$), 1764.5 (M + Na$^+$); calculated for C$_{85}$H$_{133}$N$_{11}$O$_{25}$S: 1741.1.

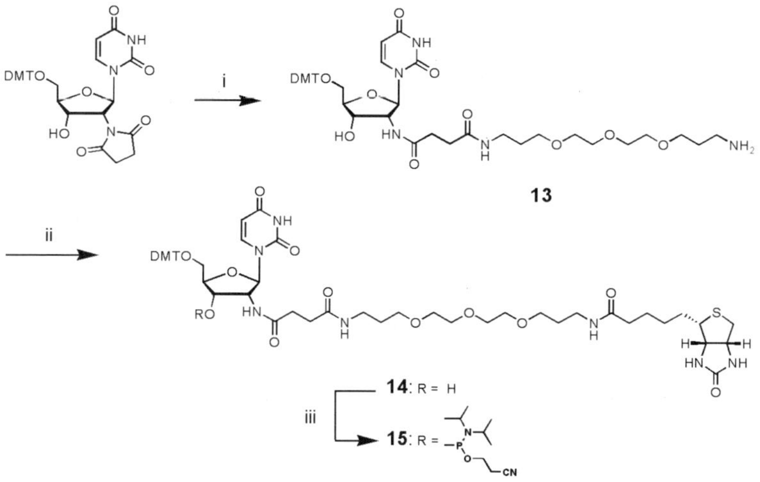

Scheme 4. Reaction scheme for the preparation of 5'-DMT-2'-Bt-Arm20-2'-dU phosphoramidite. Reagents: **(i)** 4,7,10-trioxa-1,13-tridecanediamine in methanol; **(ii)** (+) biotin-4-nitrophenol ester in pyridine; **(iii)** 2-cyanoethyl N,N,N',N'-tetraisopropylphosphorodiamidite and tetrazole in DCM.

3.3.4. DMT-Bt-Arm68-Ach Phosphoramidite (Compound 12)

Prepare compound **12** from compound **11** exactly as compound **5** (*see* **Subheading 3.1.4.**), except that during flash chromatography use a fourth portion of eluent (40% pyridine in ethyl acetate containing 2% of TEA). Compound **12** is a white solid (79%). ^{31}P NMR (DMSO-d_6) δ 144.77; ESI-MS *m/z* 2042.6 (M + Et$_3$NH$^+$); calculated for C$_{94}$H$_{150}$N$_{13}$O2$_{16}$PS: 1941.3.

3.4. 5'-DMT-2'-Bt-Arm20-2'-dU Phosphoramidite (Compound 15)

Synthesis of 5'-DMT-2'-Bt-Arm20-2'-dU phosphoramidite is outlined in **Scheme 4**. This biotin phosphoramidite allows internal and multiple incorporations of biotin moiety and provides an uncharged tether of 20 atoms in length.

3.4.1. 5'-DMT-2'-Bt-Arm20-2'-dU (Compound 14)

3.4.1.1. COMPOUND **13**

1. Dissolve 5'-dimethoxytrityl-2'-succinimido-2'-deoxyuridine (1.88 g, 3 mmol) in MeOH (30 mL), and add the solution to solution of 4,7,10-trioxa-1,13-tridecanediamine (3.28 mL, 15 mmol) in MeOH (20 mL). Stir the reaction mixture for 48 h. Follow the course of the reaction by TLC in chloroform/methanol (9:1 v/v): The starting nucleoside has R_f of 0.38; the product has very low mobility in this system.

2. Concentrate the reaction mixture to dryness. Dissolve the residue in chloroform (300 mL), and extract with saturated saline solution (300 mL). Separate the organic layer, dry it (Na_2SO_4), filter, and concentrate to dryness. Reconstitute the residual oil in DCM (30 mL), and precipitate the solution into ether mixture (600 mL). Filter the precipitate, wash it with ether, and dry the product *in vacuo* to obtain the yellowish solid.

3.4.1.2. (+)-Biotin-4-Nitrophenyl Ester

Prepare (+)-biotin-4-nitrophenyl ester as described previously (*see* **Subheading 3.1.1.1.**) proceeding from 4 mmol of (+)-biotin.

3.4.1.3. Compound 14

1. To the crude compound **13** add sequentially pyridine (20 mL) and the solution of (+)-biotin-4-nitrophenyl ester, and stir the reaction mixture for 2 h. Concentrate the reaction mixture to dryness, and coevaporate twice with toluene to remove residual pyridine.
2. Reconstitute the residue in 5% MeOH in chloroform (50 mL) containing 0.1% of TEA, and fractionate the mixture by flash column chromatography on silica gel (approx 300 mL) using increasing proportions of MeOH (5–12%) in chloroform (should contain 0.1% of TEA) as eluent. Combine and evaporate to dryness the product-containing fractions to obtain 1.93 g of 60% yellowish solid. $R_f = 0.30$ (chloroform:methanol, 8:2); ^{1}H NMR (DMSO-d_6) δ 11.28–11.34 (s, 1H, HN, U), 7.92–7.99 (d, 1H, 2'-NH), 7.79–7.86 (t, 1H, CONH), 7.70–7.86 (t, 1H, NHCO), 7.58–7.66 (d, 1H, H6, U), 7.18–7.25 (m, 9H, Ar, DMT), 6.84–6.95 (d, 4H, Ar, DMT), 6.40–6.46 (s, 1H, N*H*CONH, Bt), 6.33–6.40 (s, 1H, NHCON*H*, Bt), 5.82–5.88 (d, 1H, 1'-H), 5.65–5.70 (d, 1H, 3'-OH), 5.35–5.24 (d/d, 1H, H5, U), 4.55–4.68 (q, 1H, 2'-H), 4.25–4.333 (m, 1H, CH, Bt), 4.08–4.18 (m, 2H, CH (Bt) + 3'-H), 3.96–4.03 (m, 1H, 4'-H), 3.70–3.78 (s, 3H, OCH$_3$, DMT), 3.42–3.57 (m, 8H, 4×CH$_2$), 3.30–3.42 (m, 4H, 2×CH$_2$), 3.13–3.30 (m, 2H, 5'- CH$_2$), 3.00–3.30 (m, 5H, CH (Bt) + 2×CH$_2$), 2.76–2.86 (d/d, 1H, 1/2CH$_2$, Bt), 2.52–2.60 (d, 1H, 1/2CH$_2$, Bt), 2.10–2.47 4H, 2×CH$_2$, Suc), 2.00–2.08 (t, 2H, CH$_2$CO, Bt), 1.22–1.67 (overlapping multiplets, 10H, 3×CH$_2$ (Bt) + 2×CH$_2$); ESI-MS *m/z* 1074.9 (M + H$^+$), 1096.9 (M + Na$^+$), 1112.8 (M + K$^+$), 1176.0 (M + Et3NH$^+$); calculated for $C_{54}H_{71}N_7O_{14}S$: 1074.2.

3.4.2. 5'-DMT-2'-Bt-Arm20-2'-dU Phosphoramidite (Compound 15)

1. To the mixture of dry compound **14** (1.07 g of 1 mmol) and dry tetrazole (70 mg, 1 mmol) placed under a blanket of Ar add sequentially DCM (30 mL) and 2-cyanoethyl *N',N'*-tetraisopropylphosphorodiamidite (495 μL, 1.5 mmol). Stir the reaction mixture for 12 h. Follow the progress of the reaction by TLC in pyridine/ethyl acetate (1:1 v/v) mixture: The product has higher R_f (0.36) than the one of the starting material (0.27). In the course of reaction the starting solids are disappearing and the crystals of triethylammonium salt are typically formed.

2. Dilute the reaction mixture with DCM (200 mL) containing 2% of TEA and extract with 10% NaHCO$_3$ solution (250 mL). Separate the organic layer, dry it (Na$_2$SO$_4$), filter and concentrate to approx 20 mL. Precipitate the residue into ether/pentane mixture (2:1 v/v) (300 mL). Let the oily precipitate settle down and decant the supernatant. Redissolve the oil in 30 mL DCM containing 2% of TEA and fractionate by flash column chromatography on silica gel (approx 150 mL) using increasing proportions of pyridine (20, 30, 40, 50, 60, 65%) in ethyl acetate (should contain 2% of TEA) as eluent (150 mL of each portion). The product should not stay on the column longer than 1 h. Combine and evaporate to dryness the product-containing fractions. To completely remove pyridine, coevaporate the product with toluene (100 mL). Reconstitute the residue in DCM (20 mL), and precipitate into ether/pentane mixture (1:1 v/v) (300 mL). Centrifuge, decant the supernatant, and dry the residue to obtain a white solid (854 mg, 67%). ^{31}P NMR (DMSO-d$_6$) δ 146.92, 148.78; ESI-MS m/z 1274.8 (M + H$^+$), 1297.0 (M + Na$^+$), 1313.0 (M + K$^+$) 1376.2 (M + Et$_3$NH$^+$); calculated for C$_{63}$H$_{88}$N$_9$O$_{15}$PS: 1274.5.

3.5. Synthesis of Biotin-Labeled Oligonucleotide

3.5.1. Coupling of Biotin Phosphoramidites on the DNA Synthesizer

For automated solid-phase synthesis of biotin-labeled oligonucleotides we use Biosset ASM-700/800 DNA synthesizers (Novosibirsk, Russia). However, any other DNA synthesizer with at least one extra position for phosphoramidites could be used. Follow the manufacturer's instructions for the machine and reagent preparation. We prefer to use dC^{Ac}-phosphoramidite instead of the Bz-protected one because this allows deprotection with ethanolamine *(8) (see* **Table 1** for the coupling conditions).

We typically remove the last trityl on the synthesizer. The appearance of an intense orange color at the last detritylation step indicates the efficiency of the synthesis. Leave the DMT group on if a biotin-labeled oligonucleotide is purified by means of RP-HPLC or RP cartridge.

3.5.2. Deprotection of Biotin-Labeled Oligonucleotide

In general, use the deprotection method that suits your settings the best or which is dictated by oligonucleotide composition (base sensitive nucleotides, etc.). The tethers described above are stable toward ammonium hydroxide (ammonium hydroxide/40% aqueous methylamine, 1:1 v/v) or primary aliphatic amines. Avoid the use of aqueous NaOH or other alkaline solutions for cleavage/deprotection because the oxalamide group is not very stable toward these reagents. Our method of choice for deprotection of biotin-labeled oligonucleotide is ethanolamine treatment.

Table 1
Coupling Conditions

Phosphoramidite	Solvent	Concentration	Coupling time/min	
			0.45M Tetr.	0.25M ETT
DMT-Bt-Arm17-Ach	ACN	0.1M	10	3
DMT-Bt-Arm34-Ach	ACN	0.1M	10	3
DMT-Bt-Arm68-Ach	ACN/DCM, 1:1	0.1M	15	5
5'-DMT-2'-Bt-Arm20-2'-dU	ACN	0.1M	—	15

3.5.2.1 DEPROTECTION WITH ETHANOLAMINE

1. Dry the synthetic columns (50–200-nmol scale) under medium vacuum for 10–15 min.
2. Transfer the controlled pore glass (CPG) support from the column to a 1.7-mL centrifuge tube.
3. Add ethanolamine (40 µL), seal the tube, and finger vortex it so that all the CPG is covered with ethanolamine.
4. Incubate for 20–25 min at 70°C.
5. Dilute the reaction mixture to 1 mL with water, and desalt on a G-25 NAP-10 column following the manufacturer's protocol.

3.5.3. Purification of Biotin-Labeled Oligonucleotide

Generally, the method of purification is dictated by the desired purity of the oligonucleotide that, in turn, is dictated by the intended application. In our experience, preparative PAGE is a reliable and cost-effective procedure for the purification of up to 1µmol of material. **Fig. 1** shows an example of biotin-labeled PAGE-purified oligonucleotides.

3.6. Purification of DNA Sequencing Reactions

Background noise often compromises the results of DNA sequencing especially in the case of direct sequencing of genomic DNA (up to 8 Mb in size). Purification of biotinylated sequencing products by means of streptavidin-coated magnetic beads dramatically reduces the background noise and substantially increases the quality of sequencing. In our hands biotinylated primers with tether lengths of 23 and 40 atoms give the best results for capture/release of the sequencing products when Dynabeads M-280 Streptavidin are used.

Synthesize the biotin-labeled primer/Fimer (*see* **Note 4**) as described in **Subheading 3.5.** using biotin phosphoramidites DMT-Bt-Arm17-Ach (*see* **Subheading 3.1.**) or DMT-Bt-Arm34-Ach (*see* **Subheading 3.2.**). For DNA

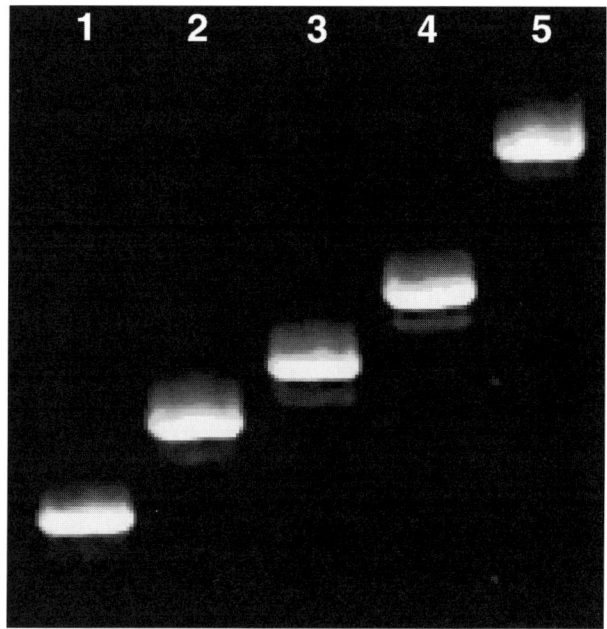

Fig. 1. Analytical PAGE of a 21-nt oligonucleotide (5'-GTA-ATA-CGA-CTC-ACT-ATA-GGG-3') labeled with fluorescein at the 3'-end and with biotin at the 5'-end. For biotin labeling phosphoramidites DMT-Bt-Arm17-Ach *(lane 2)*, 5'-DMT-2'-Bt-Arm20-2'-dU *(lane 3)*, DMT-Bt-Arm34-Ach (lane 4) and DMT-Bt-Arm68-Ach (lane 5) were used. *Lane 1*: nonbiotinylated precursor.

sequencing use standard protocols. For direct sequencing of genomic DNA use protocol described in Chapter 18 of this book.

1. Add 10 μL of Dynabeads M-280 Streptavidin and 10 μL of a 3M ammonium acetate solution to each reaction vessel.
2. Incubate for 30 min with gentle mixing.
3. Wash once with 60 μL of 1M ammonium acetate solution and twice with 60 μL of 70% aqueous EtOH.
4. Elute the sequencing fragments by resuspending in 5 μL of a loading buffer (85% formamide, 5 mM ethylenediaminetetraacetic acid, pH 8.0, (10 mg/mL dextran blue) at 95°C for 5 min with mixing.

4. Notes

1. The practical advantage of our synthetic approach lies in its amenability to structural versatility by using an arsenal of distinctly different diamines. It is possible to tailor the synthesis to suit practically any requirement in chemical or physical

properties of the tethering arms. In this chapter we use 4,7,10-trioxa-1,13-tridecanediamine because we believe it is the most suitable in terms of flexibility and biological applicability.
2. Dimethyl oxalate is relatively inexpensive and always used in excess to suppress dimer formation. However, if needed, the excess reagent could easily be recovered from the reaction mixture by sublimation under reduced pressure and medium heating on the rotary evaporator. Alternatively, a Kugelrohr apparatus may be used to avoid possible blockage as the oxalate condenses within the recovery flask.
3. In the preparations where 4,7,10-trioxa-1,13-tridecanediamine reacts with a MOX/SUC precursor, the diamine is always used in significant excess to suppress dimer formation. However, owing to substantial differences in solubility between the diamine and the formed amino alcohol in ether, 4,7,10-trioxa-1,13-tridecanediamine could easily be separated and recycled, if needed, to reduce costs.
4. A Fimer is a primer with at least one proprietary chemical modification that inhibits unwanted reactions during cycle sequencing (e.g., primer–dimer artifacts and nonspecific polymerase chain reaction). For further reading, *see* Chapter 18 of this book.

References

1. Green, N. M. and Toms, E. J. (1973) The properties of subunits of avidin coupled to sepharose. *Biochem. J.* **133,** 687–700.
2. Glen Research. (2002) Catalog. 67, 68.
3. Shchepinov, M. S., Case-Green, S. C., and Southern, E. M. (1997) Steric factors influencing hybridisation of nucleic acids to oligonucleotide arrays. *Nucleic Acids Res.* **25,** 1155–1161.
4. Southern, E. M., Mir, K., and Shchepinov, M.S. (1999) Molecular interaction on microarrays. *Nat. Genet.* **21(Suppl. 1),** 5–9.
5. Polushin, N., Malykh, A., Malykh, O., et al. (2001) 2'-modified oligonucleotides from methoxyoxalamido and succinimido precursors: synthesis, properties and applications. *Nucleosides Nucleotides Nucleic Acids* **20,** 507–514.
6. Polushin, N. (2001) The MOX/SUC precursor strategies: robust ways to construct functionalized oligonucleotides. *Nucleosides Nucleotides Nucleic Acids* **20,** 973–976.
7. Polushin, N. (2000) The precursor strategy: terminus methoxyoxalamido modifiers for single and multiple functionalization of oligodeoxyribonucleotides. *Nucleic Acids Res.* **28,** 3125–3133.
8. Polushin, N., Morocho, A. M., Chen, B. C., and Cohen, J. S. (1994) On the rapid deprotection of synthetic oligonucleotides and analogs. *Nucleic Acids Res.* **22,** 639–645.

14

Universal Labeling Chemistry for Nucleic Acid Detection on DNA Chips

Ali Laayoun, Eloy Bernal-Méndez, Isabelle Sothier, Mitsuharu Kotera, and Alain Troesch

Summary

An efficient strategy for nucleic acid labeling and analysis on deoxyribonucleic acid (DNA) chips has been developed. This approach, which combines the fragmentation and the labeling steps, is based on the reactivity of the phosphates of DNA and ribonucleic acid (RNA) fragments and is using reporter molecules bearing a bromomethyl- or aryldiazomethane-reactive group. In this chapter, we describe the preparation of the reactive label and protocols for efficient labeling of any nucleic acid sequence, DNA or RNA, prior to their hybridization, detection, and analysis on DNA chips.

Key Words: DNA chip; labeling; cleavage; bromomethyl; diazomethyl.

1. Introduction

Deoxyribonucleic acid (DNA) chips (or DNA microarrays), which are miniaturized arrays of oligonucleotide probes *(1–3)*, provide an outstanding and rapid means in biomedical research and in vitro diagnostics for nucleic acid analysis by hybridization. Because the technology relies on the hybridization of labeled nucleic acid targets to large sets of probes *(4)*, the control of the preanalytical steps is fundamental for obtaining a specific and sensitive detection of the target.

Detection is usually achieved through the labeling of the nucleic acid with a convenient reporter molecule and the use of an appropriate optical detection. Labeling of the target sequence is crucial and usually achieved by enzymatic incorporation of the label during the amplification step *(5)*. Chemical approaches for postamplification labeling *(6–8)* have been described as well. However, these procedures lead to the modifications of the nucleobases, with a significant risk of alteration of base pairing capability and specificity, which are the critical parameters for nucleic acid hybridization-based analysis, particularly on high-density DNA chips.

In our laboratory, we have been working on the development of a new and selective chemistry for nucleic acid labeling on their phosphates to preserve the hybridization efficiency and specificity *(9,10)*. In this chapter, we describe the latest developments of these labeling procedures and the results obtained with high-density DNA chip (GeneChip®, Affymetrix) analysis.

1.1. Sample Processing for Identification on GeneChip

1.1.1. Cleavage

GeneChip hybridization-based analysis is generally performed with amplified DNA or ribonucleic acid (RNA) fragments, which generate an enrichment of the sequence to be analyzed. To achieve a good signal uniformity and hybridization specificity on high-density probe arrays, labeled nucleic acids have to be fragmented *(1)*. We have developed two different chemical cleavage protocols, for DNA and RNA (**Fig. 1**). RNA cleavage is performed by incubation in a buffer containing imidazole and manganese chloride *(11)*. DNA is cleaved by discrete acidic depurination and β-elimination, leading to 5'-phosphate fragments *(12)*.

1.1.2. Phosphate Labeling

Compared to the nucleic bases, the phosphates are relatively inert to chemical attack and have been long discarded as possible alkylation sites. The approach developed in our laboratory takes advantage of the combination of labeling and cleavage conditions that improves the detection sensitivity and hybridization specificity as compared to the other existing labeling methods.

1.1.3. Labeling During Cleavage

The labeling during cleavage (LDC) approach was first used to label 3'-phosphate of cleaved RNA fragments with a reporter molecule containing bromomethyl group, 5-(bromomethyl)fluorescein (5-BMF) shown in **Fig. 2**. RNA transcripts of a region of *Mycobacterium tuberculosis* 16S ribosomal ribonucleic acid (rRNA) were cleaved using a buffer containing metal ion and imidazole (**Fig. 1**) *(11)*, whereas 5-BMF reacted with the 3'-phosphates formed. Compared to biological methods, this technique was efficient for RNA labeling and preserved the hybridization specificity *(9,10)*. However, it showed poor labeling yields on single- and double-stranded DNA fragments *(13)*. Subsequently, a new class of reporter molecules based on diazo chemistry (**Fig. 2**) that selectively and efficiently react on the phosphates of any nucleic acid

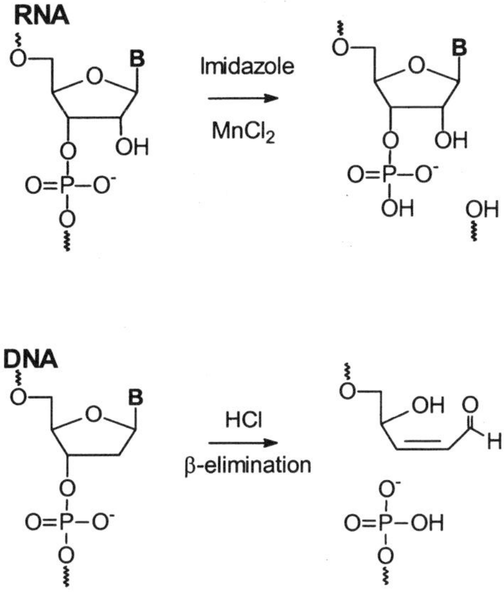

Universal Labeling Chemistry

Fig. 1. Chemical cleavage of RNA and DNA.

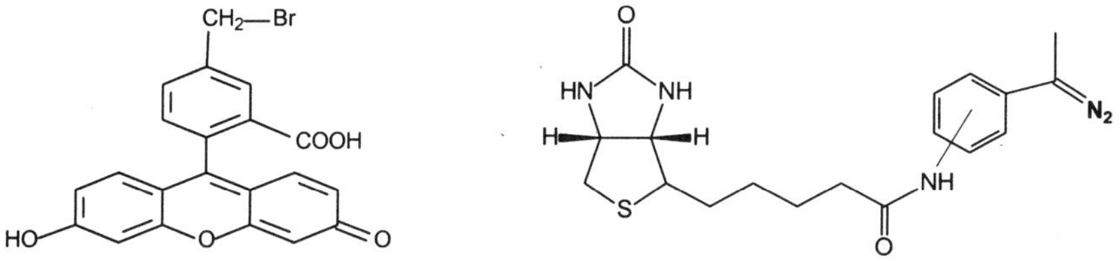

5-BMF

ortho-, para- and *meta*-BioPMDAM

Fig. 2. Structure of 5-(bromomethyl)fluorescein (5-BMF) and diazo–biotin derivatives: *ortho-, para-* and *meta*-BioPMDAM.

sequence was explored. These molecules were designed to bear an aromatic ring, which modulates the reactivity and stabilizes the diazo function, and biotin as detectable moiety. The *meta*-substituted label, *meta*-biotinylphenylmethyldiazomethyl (*m*-BioPMDAM) (**Figs. 2** and **3**) that exhibited the best reactivity/stability profile was used to label RNA and DNA under LDC conditions. The acidic treatment was used for depurination and cleavage of DNA *(12)* (**Fig. 1**).

In all cases, labeled fragments were hybridized to the *Mycobacterium* DNA chip *(14)* and detected to evaluate the signal intensities and base-calling percentage that measures the percentage of homology between the experimentally

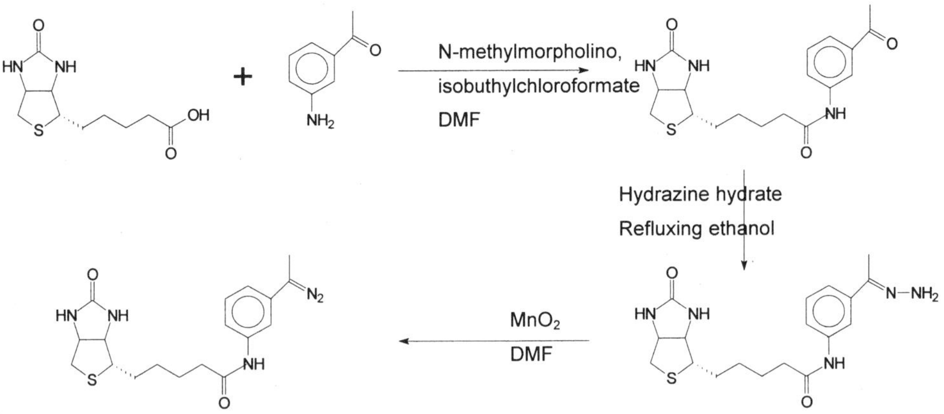

Fig. 3. Synthesis of *m*-BioPMDAM.

derived sequence and the reference sequence tiled on the array. **Table 1** highlights the performances of LDC, with 5-(bromomethyl)fluorescein (5-BMF) and *m*-BioPMDAM labels, on RNA transcripts and DNA amplicons. In LDC procedure, the diazo–biotin selectively and efficiently reacted on the phosphates of RNA and DNA, whereas 5-BMF is more appropriate for LDC of RNA.

1.1.4. Labeling Plus Cleavage

In the labeling plus cleavage (LPC) procedure, used to label DNA and RNA, the labeling was achieved prior to the cleavage. The label, 5-BMF or *m*-BioPMDAM, was added to nucleic acid sample (DNA amplicons or RNA transcripts of a region of *M. tuberculosis* 16S rRNA) without any prior purification step. The cleavage of DNA into smaller fragments was then achieved under acidic conditions. After eliminating the excess of the label, nucleic acid fragments were hybridized onto *Mycobacterium* DNA chip, scanned, and data analyzed (**Table 1**). In LPC protocol, the highly reactive diazo function reacts on internucleotidic phosphates of RNA and of double-stranded DNA. The signal intensities were strong and the signal–background ratios high, all significantly better than with 5-BMF.

1.1.5. LDC vs LPC

LDC using 5-BMF chemistry is a straightforward and efficient labeling procedure we recommend for detection and analysis of RNA targets on DNA chips. Its labeling performances are higher than those of enzymatic incorporation of the label.

Table 1
Labeling Results of Amplified DNA and RNA Targets From a 16S rRNA *M. Tuberculosis* Locus (202 nt)

Labeling procedure			BC (%)	Chip results[a]	
				Median intensity (rfu)	Signal background
RNA	LDC[b]	*m*-BioPMDAM	97.3	20968	43.0
		5-BMF	95.7	11058	31.0
	LPC[c]	*m*-BioPMDAM	96.8	13402	30.0
		5-BMF	90.8	9312	13.0
DNA	LDC	*m*-BioPMDAM	99.5	2538	4.5
		5-BMF	62.7	157	1.5
	LPC	*m*-BioPMDAM	100	4700	10.7
		5-BMF	0	18	0.2

[a]Chip results are given in terms of base call (BC) percentage (percent of homology between the experimentally derived sequence and the reference sequence tiled on the array), median signal intensity (relative fluorescence unit), and signal–background ratios.
[b]LDC, labeling during cleavage.
[c]LPC, labeling plus cleavage.

With diazo-based LDC (*m*-BioPMDAM), labeling intensities are also very high compared to existing enzymatic and chemical labeling procedures. Any nucleic acid target, DNA or RNA, can be labeled with this technology regardless of its sequence, size, or base composition.

LPC procedure using *m*-BioPMDAM is also a rapid, straightforward, and universal labeling process of DNA and RNA. It allows an efficient labeling of nucleic acid targets without any effect on the basepairing capability and hybridization specificity. Unlike 5-BMF, with *m*-BioPMDAM labeling and fragmentation steps can be achieved separately. Thus, diazo labeling can be used for other detection systems that do not require target fragmentation before hybridization, for example, low-density probes detection formats.

2. Materials

1. 5-BMF (Molecular Probes, Leiden, The Netherlands).
2. *m*-BioPMDAM was prepared according to the protocol described in the **Subheading 3.1.** and stored at 4°C as a 0.1*M* solution in dimethyl sulfoxide (DMSO). For this synthesis, all the reagents were purchased from Acros Organics (Geel, Belgium), except ethanol and manganese (IV) oxide (Merck KgaA, Darmstadt, Germany).
3. Fast Start *Taq* DNA polymerase (Roche, Mannheim, Germany).

4. Set of 2'-deoxyadenosine 5'-triphosphate, 2'-deoxycytidine 5'-triphosphate, 2'-deoxyguanosine 5'-triphosphate, 2'-deoxythymidine 5'-triphosphate (Promega, Madison, WI).
5. GeneAmp PCR system 9700 (Applied Biosystems, Foster City, CA).
6. MEGAscript T7 kit (Ambion, Austin, TX).
7. Imidazole for molecular biology ± 99.5% (Sigma-Aldrich, Saint Quentin Fallavier, France).
8. Manganese II chloride ($MnCl_2$) (Sigma-Aldrich, Saint Quentin Fallavier, France).
9. HCl $5N$ (VWR, Zaventem, Belgium).
10. Water deoxyribonuclease, ribonuclease, and protease free (Acros, Geel, Belgium).
11. QIAvac 6S columns (Qiagen, Venlo, The Netherlands).
12. Sodium chloride ≥99.5% (Sigma-Aldrich, Saint Quentin Fallavier, France).
13. Sodium dihydrogen phosphate monohydrate ≥99.5% (Sigma-Aldrich, Saint Quentin Fallavier, France).
14. Ethylenediaminetetraacetic acid (EDTA) 99.5% (Sigma-Aldrich, Saint Quentin Fallavier, France).
15. Triton X-100 for molecular biology (Sigma-Aldrich, Saint Quentin Fallavier, France).
16. Betaine monohydrate 99+% (Acros, Geel, Belgium).
17. Dodecyl trimethyl ammonium bromide 99% (DTAB) (Sigma-Aldrich, Saint Quentin Fallavier, France).
18. Tris (hydroxymethyl) aminomethane (Merck, Darmstadt, Germany).
19. Tween-20 molecular biology grade (Calbiochem, San Diego, CA).
20. Hybridization buffer: $0.9M$ NaCl, 60 mM NaH_2PO_4, pH 7.4, 6 mM EDTA, 0.05% Triton X-100, $3M$ betaine, 5 mM DTAB.
21. Bovine serum albumin (BSA) (Invitrogen, Carlsbad, CA).
22. Streptavidin/R-phycoerythrin conjugate (RPE) (Dako, Carpinteria, CA).
23. *Mycobacterium* DNA chip (BMX-Myco) were provided by Affymetrix (Santa Clara, CA).
24. Genechip Fluidics Station 400 (Affymetrix, Santa Clara, CA).
25. GeneArray scanner (Agilent, Palo Alto, CA).

3. Methods
3.1. m-BioPMDAM Label Synthesis
3.1.1. Synthesis of m-*Biotinylacetophenone (**Fig. 3**)*

1. D-biotin (10.0 g, 41 mmol) is dissolved in anhydrous dimethyl formamide (DMF) (200 mL) at 30–35°C, under argon. The solution is cooled down on ice and *N*-methylmorpholine (5.9 mL, 53.7 mmol) and isobutyl-chloroformiate (8.4 mL, 66 mmol) are added.
2. After 30 min, 3-aminoacetophenone (8.24 g, 61 mmol) is added. The solution is stirred at room temperature for 2 h, and the solvent is evaporated under vacuum.
3. The residue is taken in MeOH (30 mL) and precipitated with 10 vol of H_2O. The product is then filtrated and washed with water, CH_2Cl_2, and diethyl ether. The

resulting product is dried under vacuum overnight; 13.6 g (92%) is obtained as a light yellow powder (see **Note 1**).

F 145°C.– IR (KBr): 3280, 2931, 2857, 1691, 1590, 1540, 1487, 1434, 1298, 1266 cm^{-1}.– ^{1}H RMN (DMSO-d$_6$, 300 MHz) δ 1.3–1.7 (m, 6H), 2.33 (t, J = 8 Hz, 2H), 2.55 (s, 3H), 2.58 (d, J = 12 Hz, 1H), 2.83 (dd, J = 12 and 5 Hz, 1H), 3.13 (m, 1H), 4.15 (m, 1H), 4.31 (m, 1H), 6.34 (s, 1H), 6.41 (s, 1H), 7.44 (t, J = 8 Hz, 1H), 7.64 (d, J = 8 Hz, 1H), 7.85 (d, J = 8 Hz, 1H), 8.17 (s, 1H), 10.05 (s, 1H).–MS (FAB/glycérol), m/z: 362 [M+H]$^+$.

3.1.2. Synthesis of m-Biotinylhydrazone (**Fig. 3**)

The 37.6 mmol acetophenone is reacted with hydrazine monohydrate (5.6 mL, 116 mmol) for 2 h in refluxing ethanol (100 mL). After cooling down on an ice bath, the white precipitate is filtrated and washed with water and diethyl ether. Overnight drying under vacuum gives the product in 62% yield, as a white solid.

F 185°C.–IR (KBr): 3298, 2931, 2857, 1698, 1665, 1626, 1541, 1494, 1470, 1446, 1330, 1265 cm^{-1}. –^{1}H RMN (DMSO-d$_6$, 300 MHz) δ 1.3–1.7 (m, 6H), 1.98 (s, 3H), 2.26 (t, J = 8 Hz, 2H), 2.56 (d, J = 12 Hz, 1H), 2.81 (dd, J = 12 and 5 Hz, 1H), 3.11 (m, 1H), 4.13 (m, 1H), 4.29 (m, 1H), 6.39 (s, 3H), 6.42 (s, 1H), 7.22 (m, 2H), 7.50 (d, J = 8 Hz, 1H), 7.84 (s, 1H), 9.82 (s, 1H).–MS (FAB/glycérol), m/z: 376 [M+H]$^+$.

3.1.3. Synthesis of m-BioPMDAM (**Fig. 3**)

1. The 11.5 mmol hydrazone is dissolved in anhydrous DMF (60 mL) plus 3 mL of dry methanol and made to react with MnO$_2$ (15 g, 172 mmol). After 30 min stirring at room temperature under argon, the solution is filtered on a sintered glass funnel containing Celite and 3Å molecular sieve powder (0.5 cm, eight each).
2. The solution is concentrated to dryness and the product is washed with diethyl ether until a fine powder is obtained. m-BioPMDAM (3.85g, 90 %) is obtained as a fuchsia-colored powder (see **Note 2**). Ongoing stability studies showed that m-BioPMDAM is stable at least 12 mo at 4°C as a powder and as a 0.1M solution in DMSO.

   F 160°C.–IR (KBr): 3278, 2935, 2859, 2038, 1704, 1666, 1605, 1577, 1536, 1458, 1430, 1263 cm^{-1}.–^{1}H RMN (DMSO-d$_6$, 300 MHz) δ 1.3–1.7 (m, 6H), 2.11 (s, 3H), 2.28 (t, J = 8 Hz, 2H), 2.57 (d, J = 12 Hz, 1H), 2.81 (dd, J = 12 and 5 Hz, 1H), 3.11 (m, 1H), 4.13 (m, 1H), 4.29 (m, 1H), 6.33 (s, 1H), 6.41 (s, 1H), 6.60 (m, 1H), 7.25 (m, 3H), 9.84 (s, 1H).

3.2. Target Preparation

1. Preparation and amplification of the *Mtb* 16S rRNA locus were carried out from freshly grown colonies, according to the procedure described by Troesch et al. *(14)*. Polymerase chain reaction (PCR) amplification was carried out in a 50-µL

reaction volume using the Fast Start *Taq* DNA polymerase (Roche Molecular Biochemicals), 200 μM of each deoxynucleotide triphosphate (Promega), and 0.3 μM of primers.
2. PCR was performed in a PerkinElmer 9700 thermal cycler with an initial denaturation step at 94°C for 5 min and cycling conditions of 94°C for 30 s, 68°C for 30 s, 72°C for 45 s for 35 cycles, and 72°C for 7 min for the last cycle. PCR products were analyzed by agarose gel electrophoresis.
3. The promoter-tagged PCR amplicons were used for generating single-stranded RNA targets by in vitro transcription. Each 20-μL reaction mixture contained 8 μL of PCR product (approx 50 ng). Transcription was carried out using the in vitro transcription kit MEGAscript (Ambion). The reaction was performed at 37°C for 2 h and in vitro transcribed RNA was analyzed by agarose gel electrophoresis.

3.3. Labeling

3.3.1. Labeling of RNA

1. Labeling reactions were carried out in 100 μL reaction volume. In LDC procedure, 5 μL of in vitro transcripts were incubated at 60°C for 10 min in labeling buffer 1 containing 30 mM imidazole, 5 mM MnCl$_2$, 2 mM *m*-BioPMDAM, or at 60°C for 30 min in labeling buffer 2 containing 6 mM imidazole, 60 mM MnCl$_2$, 2 mM 5-BMF.
2. For LPC, conditions were the same except that labeling and cleavage were done sequentially: 10 min at 60°C for labeling with 2 mM 5-BMF followed by 10 min at 60°C for cleavage with 6 mM imidazole, 60 mM MnCl$_2$, or 10 min at 60°C for labeling with 2 mM *m*-BioPMDAM, followed by 10 min at 60°C for cleavage with 30 mM imidazole, 5 mM MnCl$_2$.

3.3.2. Labeling of DNA

Labeling of DNA was also carried out in 100 μL reaction volume.

1. For LDC labeling of DNA, 10 μL of PCR amplicons were incubated at 95°C for 10 min in 3 mM HCl, 10 mM *m*-BioPMDAM, or in 10 mM 5-BMF.
2. For LPC protocol, labeling and fragmentation were done sequentially: 10 min at 95°C for labeling with 10 mM *m*-BioPMDAM or 10 mM 5-BMF, followed by 10 min at 95°C for cleavage in 3 mM HCl (*see* **Note 3**).
3. For all labeling protocols, unreacted label was removed prior to the hybridization step by means of silica membrane purification, using 6S Qiavac columns (Qiagen), according to the manufacturer's instructions (*see* **Note 4**).

3.4. Probe Array Hybridization and Analysis

3.4.1. Hybridization

1. 100 μL of the labeled and purified nucleic acid fragments were added to 400 μL of hybridization buffer, denatured at 95°C for 10 min, and hybridized onto the *Mycobacterium* DNA chip *(14)*.

2. After 30 min of incubation at 45°C in hybridization buffer, the chip was washed twice in 0.45M NaCl, 30 mM NaH$_2$PO$_4$, 3 mM EDTA, pH 7.4, 0.005% Triton X-100 at 30°C.
3. For biotinylated fragments, the DNA chip was then stained for 10 min with a solution made of 600 µL of 0.05M Tris, pH 7.0, 0.5M NaCl, 0.02% Tween-20, 500 µg/mL BSA, and 6 µL of streptavidin/RPE conjugate (300 µg/mL; Dako) (*see* **Note 5**).
4. The final washing step in 0.9M NaCl, 0.06M NaH$_2$PO$_4$, 6 mM EDTA, pH 7.4, 0.05% Triton X-100, is achieved prior to the analysis.

3.4.2. Analysis

The fluorescent signal emitted by the target bound to the DNA chip was detected by a GeneArray scanner (Agilent) (*see* **Note 6**), at a pixel resolution of 3 µM. Probe array cell intensities, nucleotide base-calls, and reports were generated by functions available on the GeneChip software (Affymetrix, Santa Clara, CA).

4. Notes

1. Precipitation has to be carried out gently and under stirring to obtain the pure product as a fine powder.
2. The reagent has to be stored under inert atmosphere.
3. The labeling reagent (5-BMF or *m*-BioPMDAM) should be added at the end to avoid hydrolysis of the label prior to labeling reaction.
4. Add binding buffer (PN) (750 µL) and mix. Apply the samples to the wells of the QIAquick strips. Switch on the vacuum source. After all liquid has been pulled through, switch off the vacuum. Wash wells of strips by adding washing buffer (PE) (1 mL) to each well and switch on the vacuum source. After buffer PE in all wells has been drawn through, apply maximum vacuum for an additional 2 min to dry the membrane. Switch off the vacuum and replace waste with the collection microtube rack. To elute add elution buffer (EB) (120 µL) to the center of each well, let stand 1 min, and switch on the vacuum source for 1 min. Switch off the vacuum source and ventilate QIAvac slowly.
5. Store Streptavidin/RPE at 4°C in the dark.
6. Scanning of the chip has to be performed immediately after hybridization.

Acknowledgments

We thank C. Tora and L. Menou for their assistance in the development of methods described in this chapter. E.B.M. acknowledges a Marie Curie Industry Host Fellowship (Quality of Life Program).

References

1. Chee, M., Yang, R., Hubell, E., et al. (1996) Accessing genetic information with high-density DNA arrays. *Science* **274,** 610–614.
2. Ramsay, G. (1998) DNA chips: state-of-the-art. *Nat. Biotechnol.* **16,** 40–44.

3. Pirrung, M. C. (2002) How to make a DNA Chip. *Angew. Chem. Int Ed. Engl.* **41,** 1276–1289.
4. McGall, G. H and Fidanza, J. A. (2001) Photolithographic synthesis of high-density oligonucleotides arrays. In *DNA Arrays* (Rampal, J. B., ed.), Humana, Totowa, NJ, pp. 71–101.
5. Zhu, Z., Chao, J., Yu, H, and Waggoner, A. S. (1994) Directly labeled DNA probes using fluorescent nucleotides with different length linkers. *Nucleic Acids Res.* **22,** 3418–3422.
6. Kessler, C. (1995) Methods for nonradioactive labeling of nucleic acids. In *Nonisotopic Probing, Blotting, And Sequencing* (Kricka, L. J., ed.), Academic Press, San Diego, CA, pp. 41–109.
7. Hermanson, G. T. (1996) Nucleic acid and oligonucleotides modification and conjugation. In *Bioconjugate Techniques* (Hermanson, G. T., ed.), Academic Press, San Diego, CA, pp. 639–671.
8. Van Gijlswijk, R. P. M., Talman, E. G., Peekel, I. et al. (2002) Use of horseradish peroxidase- and fluorescein-modified cisplatin derivatives for simultaneous labeling of nucleic acids and proteins. *Clin. Chem.* **48,** 1352–1359.
9. Laayoun, A. (2002) Process for labeling a ribonucleic acid, and labeled RNA fragments which are obtained thereby. *US Patent* 6,376,179.
10. Monnot, V., Tora, C., Lopez, S., Menou, L., and Laayoun, A. (2001) Labeling during cleavage (LDC), a new labeling approach for RNA. *Nucleosides Nucleotides Nucleic Acids* **20,** 1177–1179.
11. Breslow, R. and Xu, R. (1993) Recognition and catalysis in nucleic acid chemistry. *Proc. Natl. Acad. Sci. USA* **90,** 1201–1207.
12. Browne, K. A. (2002) Metal ion-catalyzed nucleic acid alkylation and fragmentation. *J. Am. Chem. Soc.* **124,** 7950–7962.
13. Lhomme, J., Constant, J.-F., and Demeunynck, M. (2000) Abasic DNA structure, reactivity, and recognition. *Biopolymers* **52,** 65–83.
14. Troesch, A., Nguyen, H., Miyada, C. G., et al. (1999) *Mycobacterium* species identification and rifampin resistance testing with high-density DNA probe arrays. *J. Clin. Microbiol.* **37,** 49–55.

15

Postsynthetic Functionalization of Triple Helix-Forming Oligonucleotides

Alexandre S. Boutorine and Jian-Sheng Sun

Summary
The design of molecules that recognize specific sequence on the deoxyribonucleic acid (DNA) double helix would provide interesting tools to interfere with DNA information processing at an early stage of gene expression. This chapter describes in detail the protocol of conjugation between terminally phosphorylated oligonucleotides and chemically or biologically active ligands possessing electrophilic or nucleophilic functional groups. The synthetic procedure includes chemical activation of oligonucleotide terminal phosphate and introduction in this way of a nucleophilic or electrophilic group (such as amino or carboxyl groups) into oligonucleotide terminus using aliphatic amino group of a ligand or a linker. The attachment of a topoisomerase inhibitor camptothecin to a triple helix-forming oligonucleotide is taken as an example of such synthesis. The described method has general interest because any functional ligand containing a primary or secondary amino group or aliphatic carboxyl group could be attached to the terminal phosphate of an oligonucleotide in a similar way.

Key Words: Oligonucleotides; terminal phosphate; chemical activation; conjugate; ligand; linker; covalent attachment; synthesis; triple helix; topoisomerase inhibitor.

1. Introduction
The fast progress in deciphering genomic information of an increasing number of organisms (including human genome) together with the development of functional genomics is providing the basic knowledge required to use gene-specific reagents for both basic understanding of cell physiology and therapeutic developments. The field of chemical genomics has the ambitious goal of designing molecules that could act selectively on every single gene or gene product in a cell and in vivo. The design of molecules that recognize specific sequence on the deoxyribonucleic acid (DNA) double helix would provide

interesting tools to interfere with DNA information processing at an early stage of gene expression. The ability to specifically manipulate gene expression has a wide range of applications in experimental biology and in gene-based biotechnology and therapeutics. It is an important challenge in biological and biomedical sciences (reviewed in **ref. *1***).

Oligonucleotides that can form complexes with DNA by triple helix formation (antigene strategy), represent a promising class of synthetic molecules that can interfere with gene expression and gene modification. They are potential therapeutic agents *(2–4)*. During the past 15 yr, triple helix-forming oligonucleotides (TFOs) have been extensively modified especially in their backbone. The high sequence specificity and high affinity of TFOs have been exploited to downregulate or upregulate transcription by binding to their target oligopyrimidine·oligopurine sequences, or to induce directed mutagenesis and to promote homologous recombination, as well, regarding direct modification on genomic DNA at selected gene loci. TFOs are therefore powerful tools for gene-specific chemistry. The main achievements in the field have been recently reviewed (*see* **ref. 5** for review and the references therein).

The functionalization of TFOs refers to their covalent conjugation to chemically or biologically active ligands. First, it may be small molecules interacting with target DNA; the conjugation serves to direct their action to the defined target double-stranded DNA sequences or to stabilize triplexes (DNA-cleaving agents, antibiotics, enzyme inhibitors, intercalators, positively charged groups). Second, it may be sequence-specific molecules as peptides or oligocarboxamide minor groove binders; the aim is to increase recognition sequences and to avoid their limitation to only polypurine tracts. Third, certain attached molecules could facilitate cellular penetration of the oligonucleotides.

An interesting application using TFO-ligand conjugates was described recently *(6–8)*. It is a sequence-specific double-stranded DNA cleavage by recruiting of the intracellular enzyme topoisomerase. Topoisomerases are involved in the control of DNA topology. They unwind DNA molecules by cleaving one (topoisomerase I) or two (topoisomerase II) DNA strands, forming cleaved complex, and then quickly reseal the break. In the presence of inhibitors, such as the antitumor alkaloid camptothecin (CPT), the cleaved complex was trapped leading to DNA fragmentation and later cell apoptosis. Thus topoisomerase inhibitor can convert topoisomerases I and II into site-specific nucleases. To direct the action of these new nucleases to a specific gene or a group of genes involved in a determined biological process, such as signal transduction pathway and cell apoptosis, covalent attachment of CPT to TFO was done. These kinds of conjugates could be useful for chemical genomics and for potential therapeutic developments.

Triple Helix-Forming Oligonucleotides

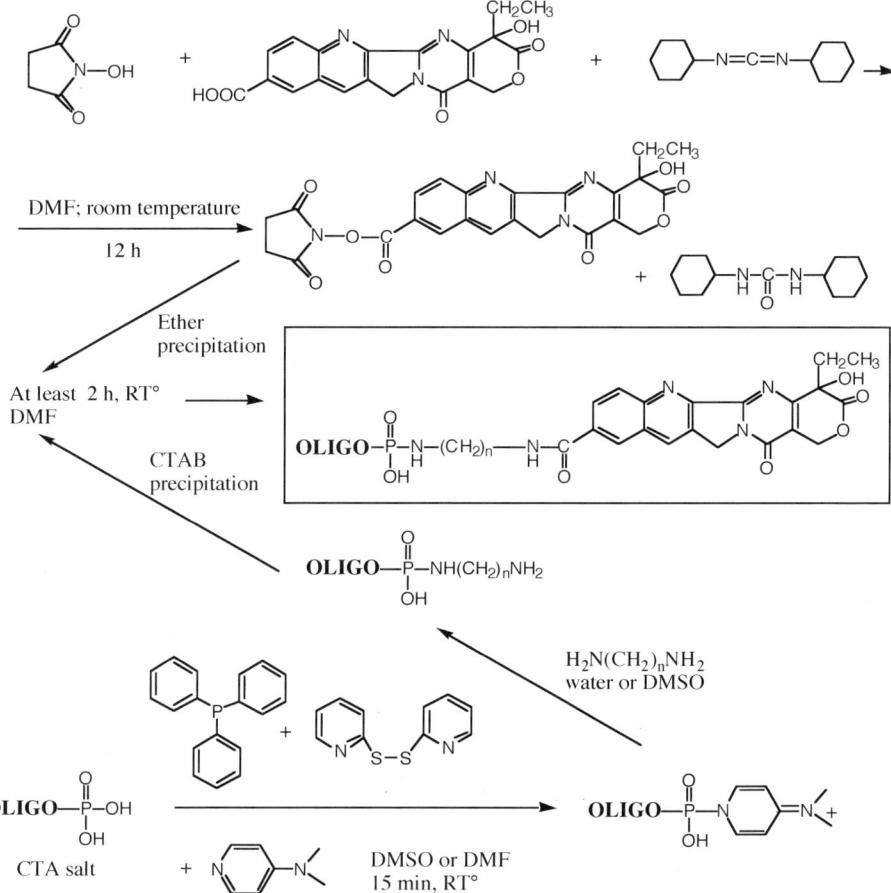

Fig. 1. Schema of synthesis of oligonucleotide–CPT conjugates.

This chapter describes in detail the protocol of making such TFO–CPT conjugates via chemical activation of oligonucleotide terminal phosphate and the introduction in this way of a nucleophilic group into the oligonucleotide terminus *(9–13)* (**Fig. 1**) It must be noted that the described method has general interest: any functional ligand containing a primary or secondary amino group or aliphatic carboxyl group could be attached to the terminal phosphate of the oligonucleotide in a similar way.

2. Materials

1. Oligonucleotides at least eight nucleotides long possessing a 3'- or 5'-phosphate group (*see* **Note 1**).

2. Bidistilled or deionized water.
3. 1.5- and 2-mL Eppendorf vials.
4. Triphenylphosphine (PPh_3).
5. Dipyridyl disulfide (PyS_2).
6. Dimethylaminopyridine (DMAP).
7. 10-CPT.
8. Dicyclohexylcarbodiimide.
9. N-hydroxysuccinimide.
10. Dimethyl sulfoxide (DMSO) or dimethyl formamide (DMF), quality anhydrous, no more than 0.001% of water.
11. Aliphatic diamine $H_2N(CH_2)_nNH_2$, where $n \geq 2$.
12. 8% (w/v) Water solution of hexadecyltrimethylammonium bromide (CTAB).
13. Freshly distilled dry triethylamine (TEA).
14. 3% Lithium perchlorate solution in acetone.
15. $0.2M$ Sodium acetate in ethanol.
16. Acetone.
17. Ethanol.
18. Diethyl ether.
19. TBE buffer: $0.089M$ Tris-hydroxymethylaminomethane, $0.089M$ boric acid, $0.001M$ ethylenediaminetetraacetic acid, pH 8.3.
20. 20% solution of acrylamide/N,N'-methylene bis-acrylamide (19:1) in TBE buffer.
21. Fluorescent screen covered with polyethylene or polypropylene film, for example, a Kieselgel F254 thin-layer chromatography plate from Merck.
22. Solution for high-pressure (performance) liquid chromatography (HPLC) purification (5% and 40% acetonitrile [ACN] in $0.025M$ ammonium acetate).

3. Methods

The chemical methods are based on the activation of oligonucleotide terminal phosphate and the attachment of a nucleophilic group to it *(9–13)* (**Fig. 1**). Because the synthesis is done on a microscale and no heavy chemical equipment is necessary, all the steps of synthesis could be conducted in 1.5- or 2-mL Eppendorf vials or small glass tubes. To confirm attachment of CPT to oligonucleotide, a denaturing gel electrophoresis system is necessary. The product may be detected by an ultraviolet (UV) shadowing method (as a dark band migrating slower than the band of control—the initial oligonucleotide—when the gel slab is placed on a fluorescent screen and irradiated by a UV lamp). In the case of CPT, this band may also be detected by blue fluorescence when irradiated by UV light at 365 nm. For final purification of the product, a HPLC system supplied with a C-18 reverse-phase column (0.46 × 25 cm or larger) is needed.

3.1. Precipitation of Oligonucleotide by CTAB

This procedure transforms oligonucleotide (usually nonsoluble in organic solvents) into the form of a CTAB salt that is soluble in polar aprotic organic solvents, such as DMSO, DMF, and pyridine.

1. Dissolve 100–800 µg in 50–200 µL of H_2O. Add at least 5 vol of $0.2M$ sodium acetate in ethanol. Keep the tube 10–20 min at –20°C to allow formation of a good coagulated pellet (*see* **Note 2**). Spin the tube at 15,000g, wash the pellet with ethanol to completely remove the traces of salt and dry the pellet. This procedure is suitable when there is at least 200 µg or more of oligonucleotides that are slightly contaminated by salts soluble in ethanol.
   For lower quantities of oligonucleotide 10 vol of 3% lithium perchlorate solution in acetone must be used for precipitation followed by agitation, incubation 3–5 min at room temperature (better in cold), centrifugation, washing of the pellet by acetone, and drying. This precipitation is rapid and quantitative even for very short oligonucleotides, but contaminating salts have a better chance to coprecipitate with the oligonucleotide.
2. Dissolve the pellet in a minimal volume (25–50 µL) of deionized H_2O, add 2 µL of 8% (w/v) solution of CTAB in water, and vortex the tube. A white coagulated pellet forms. Centrifuge the tube 2 min at 15,000g until precipitation of the pellet. Add again 1 µL of 8% CTAB solution and shake slightly. If any new coagulation appears, centrifuge the tube, and add a new portion (1 µL) of CTAB. Repeat the procedure until the precipitate stops forming after addition of a new portion (*see* **Note 3**). Then remove the supernatant (*see* **Note 4**).
3. Dissolve the pellet in water-free ethanol (200–300 µL, heat, and vortex to accelerate dissolution), and dry in a vacuum drier at least 30 min. Water must be very well removed and the following reaction must be carried out in water-free solutions.

3.2. Activation of the Oligonucleotide Terminal Phosphate and Attachment of Diamine Linker

1. Prepare solution of 2–5 mg of aliphatic diamine in 50 µL of H_2O (*see* **Note 5**). Better results were obtained for higher amines (10–12 methylene groups) when the diamine was dissolved in a 25–30% alcohol–water mixture. If the amino linker is commercially available in the form of HCl, trifluoroacetic acid (TFA), and other salts, 3–5 µL of TEA must be added to transform the amine into a free base form.
2. Dissolve dry oligonucleotide CTA salt in 50 µL of water-free DMSO or DMF, add 5 mg of DMAP, agitate until dissolution, add 6.6 mg (30 µmol) of dipyridyldisulfide in 25 µL of DMSO and 7.9 mg (30 µmol) of triphenylphosphine in 25 µL of DMSO (dissolution of the chemicals can be accelerated by heating at 50–70°). Agitate the mixture thoroughly, and incubate 15 min at room temperature.

3. Add 10 vol of cold 3% lithium perchlorate solution in acetone and immediately centrifuge 2 min at 15,000g. Remove supernatant acetone and quickly rinse the pellet with cold acetone, but do not dry it.
4. Immediately after removal of the acetone dissolve the pellet in the aliphatic diamine solution and incubate at room temperature. For simple amines 20 min of incubation is enough; for complicated organic amines incubate at least 3 h (*see* **Note 6**).
5. Because the yield of the linker attachment is usually close to quantitative, no additional purification is necessary. However, to remove excess of diamine, it is necessary to precipitate the conjugate twice by ethanol/sodium acetate as it is described in **Subheading 3.1.1.**
6. For attachment of other primary amines *see* **Notes 5** and **6**.

3.3. Transformation of the Oligonucleotide Linker Conjugate Into DMF Solution of CTAB Salt

Dissolve the pellet of oligonucleotide amino linker conjugate in 25–50 µL of H$_2$O and repeat exactly the same procedure of CTAB precipitation as it is described in **Subheadings 3.1.2.** and **3.1.3.** Dissolve the CTAB salt of the conjugate in 50–100 µL of dry DMF.

3.4. Activation of Carboxyl Group of 10-Carboxycamptothecin

1. *Method 1*: Dissolve 3.9 mg (10 µmol) of the 10-carboxycaptothecin in 80 µL of dry DMF. Add 1.5 mg (13 µmol) of *N*-hydroxysuccinimide to the solution. Start the reaction on addition of 2.7 mg (13 µmol) of dicyclohexylcarbodiimide in 20 µL of dry DMF. Incubate the reaction mixture overnight at room temperature. After 12 h of incubation remove the pellet of dicyclohexylurea by centrifugation and rinse it with 20 µL of DMF. Add the combined DMF fractions by 20 µL portions to 1.5 mL of diethyl ether and centrifuge. Wash the precipitated *N*-hydroxysuccineimide ester of CPT twice with 500 µL of ether, dry *in vacuo*, and dissolve in 50–100 µL of dry DMF.
2. *Method 2*: Dissolve 1 mg (2.5 µmol) of CPT in 50 µL of DMF. Add 6.6 mg (30 µmol) of dipyridyl disulfide in 25 µL of DMF and 7.9 mg (30 µmol) of triphenylphosphine in 25 µL of DMF. After 10 min of incubation at room temperature the mixture should be directly added to the DMF solution of amino linker-modified oligonucleotide together with 1 µL of TEA (*see* **Note 7**).

3.5. Attachment of CPT to Amino-Modified Oligonucleotide

To 50–100 µL of DMF solution of oligonucleotide linker conjugate add either a solution of 1 mg of CPT *N*-hydroxysuccinimide ester (**Subheading 3.4., step 1**) in DMF or a solution of 1 mg of activated CPT (**Subheading 3.4., step 2** together with 1 µL of TEA). Mix the solutions and incubate at least 2 h (preferable overnight) at room temperature. Precipitate the product twice, using the acetone/lithium perchlorate procedure the first time, and the second time using the ethanol/acetate procedure, as it was described in **Subheading 3.1.1.** The usual yield of the reaction is about 85–95%.

3.6. Analysis and Purification of the Conjugate

Several methods could be used for detection of conjugate, its characterization, and purification. In this chapter we shall not go into the detail of each method because it is not an aim of this publication.

1. *Denaturing gel electrophoresis*: Apply 3–5 μg of the product in 7*M* urea—standard TBE buffer in the well of a 20% denaturing polyacrylamide gel slab, and run electrophoresis until the bromophenol blue passes more than two-thirds of the slab. Formation of the product could be detected by a UV-shadowing method; the product is visible on UV-fluorescent screen as a retarded (2–3 mm) band compared to the control—untreated oligonucleotide with amino linker. This band shows blue fluorescence when irradiated by UV light at 365 nm. For purification of the product, excise the fluorescent band, and extract the product by soaking the gel slice in water or in 0.2*M* Tris-HCl and 0.1% sodium dodecyl sulfate, then isolate the conjugate by precipitation with ethanol/acetate as described previously (*see* **Subheading 3.1.2.**).
2. *HPLC*: Owing to different hydrophobicity of the conjugate compared to the initial oligonucleotide with amino linker, the product may be separated by HPLC on a reversed-phase column. In our practice to purify up to 1 mg of the product we use a semipreparative C-18 X-Terra column (Waters, 7 μ*M*, 300 × 7.8 mm) and a 5–40% ACN linear gradient in 0.02*M* ammonium acetate, flow rate 2 mL/min. The retention times are 6–8 min for the starting oligonucleotide and 1–2 min more for the conjugate. Detection at 260 nm usually gives very large peaks that are difficult to operate. In order to identify a conjugate we use detection at 376 nm (maximum of CPT absorption). The isolated material is then lyophilized several times using a SpeedVac centrifuge lyophilizer to remove ammonium acetate, and then it is characterized by mass spectrometry.
3. *UV-visible spectrophotometry*: Record the electronic spectrum of the product on the UV-visible spectrophotometer in the wavelength interval between 220 and 500 nm. Two absorption maxima will be observed: one at 260 nm mainly corresponding to those of an oligonucleotide and the second at 376 nm corresponding to those of CPT (33,300 mol^{-1}cm^{-1}). Their molar ratio is about 1:1.

4. Notes

1. If the commercial oligonucleotide is not phosphorylated, 5'-terminal phosphate may be introduced by a simple preparative enzymatic phosphorylation using adenosine triphosphate and T-4 polynucleotide kinase. The method is described in an article by Grimm et al. *(13)*.
2. Sometimes, especially when the oligonucleotide is contaminated by salts or when its concentration is low, no evident coagulation is observed. Even after 20–30 min at –20°C the mixture is homogeneously turbid. In this case just centrifuge

the tube at 15,000g for 5 min, but be careful, because the pellet distributes on the tube walls and is not visible. Usually repetitive resuspension of the pellet in a small water volume (rinse tube walls well) and a second precipitation improves the quality of coagulation.
3. Titration of the oligonucleotide by CTAB is used because it is very important to add the molar quantity of CTAB equivalent to the molar quantity of oligonucleotide phosphates (for example, 17 molecules of CTAB/1 molecule of phosphorylated 16-mer oligonucleotide). Otherwise, CTAB salt of oligonucleotide will dissolve in excess of CTAB solution.
4. The usual yield of this procedure is 90–99% depending on the length and structure of oligonucleotides. Thymidylate-containing oligonucleotides usually precipitate worse and form a glass-like or oil-like pellet. In this case it is quite difficult to remove the supernatant without taking the oligonucleotide. You can visually control the presence of oligonucleotide in the supernatant by precipitation of it with acetone/lithium perchlorate solution. If you have difficulties in separating the pellet and the supernatant, it is possible to dry all the mixture without removal of the supernatant. Oligonucleotide in supernatant can be precipitated then by 3% $LiClO_4$ in acetone and used again.
5. Any chemically and biologically active molecule soluble in water and possessing an aliphatic amino group could be attached in this way to the terminal phosphate of oligonucleotide. The yield of the reaction depends on several structural factors, such as hydrophobicity, chemical nature of the ligand, accessibility of the amino group, and so forth. It must be noted that in many cases the ligands with a primary amino group are in the form of HCl or TFA salts, and to deprotonate them in the reaction mixture, they must be added to activated oligonucleotide together with 3–5 µL of TEA.
6. When you attach special linkers or ligands that contain amino groups and that are not soluble in water, avoid **step 3**, and add the ligand directly to the activation mixture in 50–100 µL of dry DMSO or DMF together with 3–5 µL of TEA, if necessary. Thus, the coupling of aromatic amines is possible on direct addition of 1–2 mg of amine to the activation mixture in organic solutions after 15 min of activation, but they need much longer incubation times compared to aliphatic amines. For example, conjugates of oligonucleotide with aniline (100% yield after 1 d) and *para*-aminobenzoic acid (80% yield after 2 d) were obtained. However, the rate and yield of reaction depend on the nature of the amine and aromatic ring of the molecule.

    In the case of terminal phosphate activation, the drawback of direct addition without removal of activating agents is a possible attachment of two ligands to the same terminal phosphate of oligonucleotide *(14,15)*.

    The exhaustive review of methods of conjugate synthesis using Mukaiyama reagents is published in *(13)*.
7. The second method is more rapid and technically simple, but it gives more secondary products owing to the presence of an oxidizing-reducing couple throughout the synthesis.

References

1. Sun, J. S. and Hélène, C. (2003) Oligonucleotides and derivatives as gene-specific control agents. *Nucleosides Nucleotides Nucleic Acids* **22**, 489–505.
2. Le Doan, T., Perrouault, L., Praseuth, D., et al. (1987) Sequence-specific recognition, photocrosslinking and cleavage of the DNA double helix by an oligo-[α]-thymidilate covalently linked to an azidoproflavine derivative. *Nucl. Acids Res.* **15**, 7749–7760.
3. Moser, H. E. and Dervan, P. B. (1987) Sequence-specific cleavage of double helical DNA by triple helix formation. *Science* **238**, 645–650.
4. François, J. C., Saison-Behmoaras, T., and Hélène, C. (1988) Sequence-specific recognition of the major groove of DNA by oligodeoxynucleotides via triple helix formation. Footprinting studies. *Nucleic Acids Res.* **16**, 11,431–11,440.
5. Giovannangeli, C. and Hélène, C. (2000) Triplex-forming molecules for modulation of DNA information processing. *Curr. Opin. Mol. Ther.* **2**, 288–296.
6. Arimondo, P. B., Bailly, C., Boutorine, A., Sun, J. S., Garestier, T., and Hélène, C. (1999) Targeting topoisomerase I cleavage to specific sequences of DNA by triple helix-forming oligonucleotide conjugates. A comparison between a rebeccamycin derivative and camptothecin. *C R Acad. Sci. Ser. III* **322**, 785–790.
7. Arimondo, P. B., Moreau, P., Boutorine, A., et al. (2000) Recognition and cleavage of DNA by rebeccamycin- or benzopyridoquinoxaline conjugated of triple helix-forming oligonucleotides. *Bioorg. Med. Chem.* **8**, 777–784.
8. Arimondo, P. B., Boutorine, A., Baldeyrou, B., et al. (2001) Design and optimization of camptothecin conjugates of triple helix-forming oligonucleotides for sequence-specific DNA cleavage by topoisomerase I. *J. Biol. Chem.* **277**, 3132–3140.
9. Godovikova, T. S., Zarytova, V. F., and Khalimskaya, L. M. (1986) Reactive phosphamidates of mono- and dinucleotides. *Bioorg. Khim.* **12**, 475–481.
10. Godovikova, T. S., Zarytova, V. F., Maltseva, T. V., and Khalimskaya, L. M. (1989) Reactive oligonucleotide derivatives with a zwitter-ionic terminal phosphate group for affinity reagents and probe construction. *Bioorg. Khim.* **15**, 1246–1252.
11. Knorre, D. G., Alekseyev, P. V., Gerassimova, Y. V., Silnikov, V. N., Maksakova, G. A., and Godovikova, T. S. (1998) Intraduplex photocrosslinking of p-azidoaniline residue and amino acid side chains linked to the complementary oligonucleotides via a new phosphorylating intermediate formed in the Mukaiyama system. *Nucleosides Nucleotides* **17**, 397–410.
12. Boutorine, A. S., Le Doan, T., Battioni, J. P., Mansuy, D., Dupré, D., and Hélène, C. (1990) Rapid routes of synthesis of chemically reactive and highly radioactively labeled α- and β-oligonucleotide derivatives for in vivo studies. *Bioconjug. Chem.* **1**, 350–356.
13. Grimm, G. N., Boutorine, A. S., and Hélène, C. (2000) Rapid routes of synthesis of oligonucleotide conjugates from nonprotected oligonucleotides and ligands possessing different nucleophilic or electrophilic functional groups. *Nucleosides Nucleotides Nucleic Acids* **19**, 1943–1965.

14. Kostenko, E., Dobrikov, M., Pyshnyi, D., et al. (2001). 5'-bis-pyrenylated oligonucleotides displaying excimer fluorescence provide sensitive probes of RNA sequence and structure. *Nucleic Acids Res.* **29,** 3611–3620.
15. Sinyakov, A. N., Ryabinin, V. A., Grimm, G. N., and Boutorine, A. S. (2001) Stabilization of DNA triple helices using conjugates of oligonucleotides and synthetic ligands. *Mol. Biol. Engl. Tr.* **35,** 251–260.

16

Fluorescence-Based On-Line Detection as an Analytical Tool in RNA Electrophoresis

Ute Scheffer and Michael Göbel

Summary

In this chapter, we present methods for denaturing and native gel electrophoresis of nucleic acids based on fluorescently labeled probes and automatic signal detection by a deoxyribonucleic acid sequencer. Specific examples are given for the determination of ribonucleic acid (RNA) fragmentation patterns, of (deoxy)ribozyme kinetics, and of direct or competitive gel-shift assays. These methods can replace widely used radioisotope-based protocols, for example, for secondary structure mapping of RNA and for the characterization of nucleic acid ligand interactions.

Key Words: Band-shift; cleavage of RNA; Cy5; dye; electrophoresis; fragmentation of RNA; gel shift; fluorescence; mobility shift; nonradioactive; ribozyme kinetics; sequencer.

1. Introduction

Various methods of molecular biology to analyze ribonucleic acid (RNA) are based on gel electrophoresis. Most of these procedures are traditionally performed by using radioactively labeled RNA probes and a standard electrophoresis chamber. However, some important drawbacks come along with radioactivity: the short half-lives of the radioisotopes (usually ^{32}P or ^{35}S), the increasing costs for waste disposal and radiation protection, and the time-consuming detection by autoradiography and densitometry or phosphorimaging. In this chapter we describe a fast, sensitive, and quantitative method for the analysis of RNA cleavage or of RNA ligand interactions using stable, fluorescently labeled RNA probes and an ALFexpress™ automatic deoxyribonucleic (DNA) sequencer. During electrophoresis, Cy5-labeled RNA products are detected as they pass through a laser beam (633 nm) at the bottom of the gel. Cy5 is a carbocyanine dye with an absorption maximum at 643 nm and λ_{em} = 667 nm. It withstands a wide variety of solvents and incubation conditions. Laser excitation of the dye label takes place simultaneously in the 40 lanes of the gel. Thus a great number of probes can be analyzed in parallel. The emitted fluorescence is measured by a linear array of photodiodes with a detection limit similar to that of modern laser fluorescence scanners *(1)*. Detector signals are recorded by the system's computer and analyzed by the AlleleLinks™ software package. Compared to a simple electrophoresis chamber, the sequencer offers superior band resolution owing to the long migration distance (26 cm) and the controlled temperature of the gel, typically 55°C. Gel temperatures ranging from 4–20°C, necessary for gel-shift assays, can be achieved by addition of a cryostat.

In the last several years RNA cleavage by ribozymes and functional mimics of ribonucleases has received considerably attention, especially regarding potential applications in molecular biology and drug design. Up to now radioactive methods are predominantly in use to screen and characterize RNA degrading catalysts. In this chapter, detailed protocols are given for applications of Cy5-labeled probes to analyze RNA fragmentation patterns *(2–4)* and to determine kinetic parameters of an RNA-cleaving DNA enzyme *(5)*. Furthermore, this technique will be applicable for the secondary and tertiary structure mapping of RNA.

The gel-shift assay, also known as band-shift assay, or electrophoretic mobility shift assay, is a powerful tool for the qualitative and quantitative analysis of protein–nucleic acid or ligand–nucleic acid interactions. This method is based on the differences in electrophoretic mobilities of free and protein-bound RNA *(6)*. The structural diversity of RNA motifs and the absence of a cellular repair mechanism for RNA make RNA-binding ligands that selectively disrupt RNA–protein interactions attractive drug targets. The identification of such compounds is an interesting challenge for modern medicinal chemistry. However, owing to their low molecular weight, a significant change of RNA mobilities is rarely observed. Competitive gel-shift assays may be used instead. A small ligand replaces an RNA-binding protein or peptide, thus abolishing the highly detectable effect of the direct gel-shift experiment.

A well-characterized example of an RNA binding protein specifically recognizing a defined region of RNA is the HIV-1 Tat protein interacting with the transactivation response element (TAR) (reviewed in **ref. 7**). Here we describe band-shift assays with a Cy5-labeled TAR model (31-mer RNA sequence, spanning nt 18–44 of HIV-1 TAR RNA) and two Tat model peptides—a 36-mer Tat peptide *(8)* and an 11-mer Tat-polyethylene glycol (PEG) conjugate *(9)*.

2. Materials
2.1. Analysis of RNA Cleavage and Measurement of Ribozyme Kinetics
2.1.1. Electrophoresis

1. ALFexpress DNA sequencer with standard gel cassette (Amersham Biosciences, Uppsala, Sweden).
2. 16% Polyacrylamide gel (prepared from a 40% ready-made stock solution of acrylamide and *bis*-acrylamide in a ratio of 19:1) containing $7M$ urea, 1X Tris-borate-EDTA (TBE) buffer.
3. 10% Ammonium persulfate (APS) in diethyl pyrocarbonate (DEPC)-treated water.
4. N,N,N',N'-tetramethylethylenediamine (TEMED).
5. 10X TBE: $1M$ Tris, $0.8M$ boric acid, 10 mM ethylenediaminetetraacetic acid (EDTA), filtrated through a 0.45 µm filter (Schleicher & Schüll, ME25).
6. Running buffer: 0.5X TBE; dilute the 10X stock solution.
7. Loading buffer (× 2): 5mg/mL dextran blue 2000 in formamide (Amersham Biosciences, Uppsala, Sweden).

2.1.2. Construction of Labeled Probes

1. Cy5-labeled oligonucleotides were obtained from BioSpring GmbH (Frankfurt, Germany).
2. The deoxyribozyme was prepared by standard phosphoramidite chemistry on a 381A DNA synthesizer (Applied Biosystems).

2.1.3. Purification of Oligonucleotides

1. Gel electrophoresis apparatus (Biometra, Göttingen; Germany).
2. 16% Polyacrylamide gel (prepared from a 40% ready-made stock solution of acrylamide and *bis*-acrylamide in a ratio of 19:1) containing $7M$ urea, 1X TBE buffer.
3. Running buffer: 0.5X TBE; dilute the 10X stock solution with DEPC-treated water.
4. Loading buffer (×2): 80% formamide, 10 mM EDTA, 0.1% bromophenol blue.
5. Elution buffer: 500 mM ammonium acetate, 0.1% sodium dodecyl sulfate (SDS), 2 mM EDTA.
6. Shaker mixer 5432 (Eppendorf, Hamburg, Germany).
7. Quantum Prep Freeze'N Squeeze spin columns (Bio-Rad, Munich, Germany).
8. Ultrafiltration units Microcon YM3 (Millipore, Germany).
9. NAP-10 columns (Amersham Biosciences, Uppsala, Sweden).
10. SpeedVac, Thermo Savant (Thermo Life Sciences, Dreieich, Germany).

2.1.4. Base Hydrolysis

1. Thermoblock TB1 (Biometra, Göttingen, Germany).
2. Hydrolysis buffer (×2): 200 mM sodium carbonate solution.
3. Stop solution: $1M$ acetic acid.

2.1.5. Ribonuclease (RNase) T1 Treatment

1. Thermoblock (Biometra, Göttingen, Germany).
2. RNase T1, 1000 U/μL, (MBI Fermentas), diluted to 1 U/μL.
3. RNase T1 incubation buffer (×5): 50 mM Tris-HCl, pH 8.0, 0.1 mM EDTA, 1.5M NaCl.

2.1.6. Measurement of Kinetic Parameters

1. Tris-MgCl$_2$ buffer (×5): 250 mM Tris-HCl, pH 8.0, 50 mM MgCl$_2$, 0.05% SDS.
2. ALF gel loading buffer (×2): 5mg/mL dextran blue 2000 in formamide.

Electropherograms were analyzed using the AlleleLinks 1.01 software package (Amersham Biosciences, Uppsala, Sweden).

2.2. Primer Extension

See Chapter 19 of this book.

2.3. Determination of RNA Ligand Interactions by Gel-Shift Experiments

2.3.1. Electrophoresis

1. ALFexpress DNA sequencer (Amersham Biosciences, Uppsala, Sweden) equipped with Cooler Kit and a short gel cassette.
2. Water bath RC6 (Lauda Dr. R. Wobster GmbH & CO. KG, Lauda-Königshofen, Germany).
3. Acrylamide stock solution: 40% ready-made stock solution of acrylamide and *bis*-acrylamide in a ratio of 29:1.
4. 50% Glycerol.
5. 10% APS in DEPC-treated water.
6. TEMED.
7. 10X TB: 1M Tris, 0.8M boric acid, filtrated through a 0.45 μM filter (Schleicher & Schüll, ME25).
8. 10% Triton X-100: 10% Triton X-100 (w/v) in DEPC-treated water.
9. 0.5X TB running buffer: 0.5X TB, 0.2% glycerol, 0.01% Triton X-100.
10. ALF loading buffer: 10 mg/mL dextran blue 2000, 50% glycerol.

2.3.2. Probes

Tat 36-mer was purchased from Interactiva (Ulm, Germany); Tat 11-mer was prepared by standard Fmoc chemistry and purified by RP18 high-pressure (performance) liquid chromatography; Cy5-labeled oligonucleotides were obtained from Biospring (Frankfurt, Germany).

2.3.2.1. Bioconjugate Preparation

1. Reaction buffer: 40 mM Tris-Hcl, pH 7.0, 100 mM KCl, and 5 mM MgCl$_2$.
2. 10 mM (50 μg/μL) Methoxy-PEG-maleimide, MW 5000, (N-Mal-5000, ethoxy-PEG) in reaction buffer. Available from Shearwater Polymers, Inc.

3. NaBH$_4$, acetone, and 2N HCl.
4. Sephadex G50 (Amersham Biosciences, Uppsala, Sweden).

2.3.3. PEG Staining

1. Fixing solution: 5% glutaaraldehyde.
2. Staining solutions: 0.1M perchloric acid; 5% barium chloride solution; 0.1N iodine solution (ready-to-use solution, Merck).
3. Orbital shaker: Polymax 1040 (Heidolph).
4. Gel documentation system (Biostep, Jahnsdorf, Germany).

2.3.4. Gel-Shift Assay

1. 10X Tris-KCl buffer: 500 mM Tris-HCl, pH 7.0, 200 mM KCl.
2. 10 mg/mL transfer RNA (tRNA) in DEPC-treated water.

3. Methods
3.1. Inactivation of RNases

All glassware should be baked at 180°C for 6 h. Plasticware, tubes, and most solutions have to be treated with DEPC. Add DEPC to solutions at a concentration of 0.1% (1 mL DEPC/L), shake vigorously, incubate overnight, and autoclave in liquid cycle (20 min, 121°C, 2.1 bar). Solutions not compatible with DEPC treatment are prepared by mixing molecular biology-grade powdered reagents in DEPC-treated ultrapure water.

3.1.1. Purification of Oligonucleotides

1. Oligonucleotides are purified on denaturing polyacrylamide gel electrophoresis (PAGE) (16% monomer, 7M urea) to remove undesirable shorter oligonucleotides (*see* **Note 1**). Construct a chamber (100 × 100 × 0.6 mm in size) by separating the two glass plates with spacer strips down the edges of the plates, then seal the edges and the bottom to form a liquid-tight box (*see* **Note 2**). Take 10 mL of the acrylamide gel solution, start the reaction with 100 µL APS and 10 µL TEMED, and cast the gel. Prepare 0.5 L of 0.5X TBE running buffer by diluting the 10X stock solution with DEPC-treated water. Prerun the gel for 1 h at 120 V. Mix appropriate amounts of nucleic acids (approx 20 µg nucleic acid per 13-mm broad lane) and gel loading buffer. Denature the sample by heating at 90°C for 3 min, then rapidly cool on ice. Flush the wells of the gel with fresh running buffer and load the gel. Run at constant voltage (200 V) and stop electrophoresis when the dye has migrated an appropriate distance.
2. After gel electrophoresis the bands of interest are excised (*see* **Note 3**), the gel fragments transferred to a nuclease-free tube, minced, and submerged with 500–1000 µL elution buffer (500 mM ammonium acetate, 0.1% SDS, 2 mM EDTA). The gel fragments are then incubated under vigorous shaking overnight at room temperature (Mixer 5432, Eppendorf, Hamburg, Germany).

3. To remove the gel fragments, the supernatant is aliquoted onto Quantum Prep Freeze'N Squeeze spin columns (Bio-Rad, Munich, Germany) and spun at 13,000g for 3 min at room temperature.
4. After transferring the oligonucleotide solution from the collecting tube to ultrafiltration units, the oligonucleotides are concentrated by ultrafiltration (Microcon YM3, Millipore). The retentate is filled with DEPC-treated water to give a final volume of 1 mL and desalted on a NAP-10 column. Collect 4 fractions of 500 µL each. The pooled fractions 1–3 are lyophilized to dryness and the pellet is dissolved in DEPC-treated water to give a concentration of approx 0.5 µg/µL.

3.2. Analysis of RNA Cleavage

3.2.1. Electrophoresis

Assemble the standard cassette, following the ALFexpress manual. Prepare 60 mL of the acrylamide gel solution, and filtrate the gel solution through a 0.45-µm filter membrane. Start the reaction by adding 450 µL APS and 45 µL TEMED, and cast the gel (see **Note 4**). Prepare 2 L of 0.5X TBE running buffer by diluting the 10X stock solution. Prior to electrophoresis add 1 vol of loading buffer (5 mg/mL dextran blue 2000 in formamide) to each sample (0.1–0.5 pmol Cy5-labeled nucleic acid per lane). Flush the wells of the gel with fresh running buffer, and load 10 µL of each sample on the gel. The running conditions are: 1500 V (maximum), 60 mA (maximum), 30 W (constant), 55°C, 2-s sampling interval. For 40–50-mer RNAs a running time of 350 min is sufficient.

3.2.1.1. CONSTRUCTION OF LABELED PROBES

To avoid very short fragments complicating the analysis and to achieve best resolution of all possible cleavage products, a DNA spacer of 10 deoxythymidines was placed between the Cy5 dye and the RNA part. Separation between substrate and the longest degradation product is improved by attachment of four to six additional deoxynucleotides (T4 or T6) at the 3'-end.

3.2.1.2. BASE HYDROLYSIS

Partial alkaline hydrolysis of RNA is used to prepare base ladders to analyze cleavage patterns and to identify cleavage sites. To 0.5 pmol Cy5-labeled oligonucleotide pipet 4 µL 2X hydrolysis buffer, and fill up with DEPC-treated water to a final volume of 8 µL. Incubate for 12 min at 90°C, and stop the reaction by adding 2 µL 1M HOAc (see **Note 5**). Add 1 vol ALF gel loading buffer, and analyze the fragments on a denaturing 16% polyacrylamide gel. Electrophoresis conditions are described in **Subheading 3.2.1.**

3.2.2. RNase T1 Treatment

RNase T1 cleaves single-stranded RNA after guanosine residues forming 2',3'-cyclic phosphates. Partial digestion of Cy5-labeled RNA with this enzyme thus generates a ladder of G-specific fragments. In a final volume of 10 µL, mix 0.5 pmol Cy5-labeled RNA with 1 U RNase T_1 in RNase T_1 1X incubation buffer. Incubate for 12 min at 37°C, stop the reaction by adding 10 µL of ALF loading buffer, and load 10 µL on the gel (a denaturing 16% polyacrylamide gel, electrophoresis conditions as described in **Subheading 3.2.1.**).

3.2.2.1. Measurement of (Deoxy)Ribozyme Kinetics

The determination of k_{cat} and K_m from multiple-turnover kinetics is a key experiment to characterize the function of ribozymes and other RNA cleaving catalysts. The substrate concentration should be in the range of the expected K_m, typically 0.1–100 nM. A 10-fold lower concentration as for the least substrate concentration is chosen for the ribozyme. Quantitative comparison between samples requires the addition of an internal standard to correct the variable amplification of different photodetectors of the sequencer. Prepare premixed solutions for ribozyme, substrate, and internal standard in Tris-$MgCl_2$ 1X buffer (*see* **Note 6**). The internal standard should be included in the substrate solution. Start the reactions by mixing appropriate amounts of catalyst and substrate solutions, and incubate the reaction mixtures at 37°C. Stop the reactions by adding 1 vol of ALF-loading buffer at various times (time frame is typically 1–30 min) covering the first 10–15% of the reaction. Load 10 µL on a denaturing 16% polyacrylamide gel; electrophoresis conditions are described in **Subheading 3.2.1.**

3.2.2.2. Analysis

1. Cleavage patterns: Cleavage sites induced by RNA degrading agents can be easily identified by comparison with the appropriate base- and G-ladder (*see* **Fig. 1**).
2. RNA Cleavage: Electropherograms are integrated using the AlleleLinks 1.01 software. The percentage of degraded RNA is then calculated from the ratio of integrals for product and (product + substrate).
3. Kinetic Parameters: Integrals of RNA cleavage products are calibrated by the integral of the internal standard (*see* **Fig. 2**). For each substrate concentration, plot the amount of cleaved RNA vs time and calculate k_{obs} from the slope of a best-fit line. A modified Eadie–Hofstee plot of k_{obs} vs k_{obs}/(substrate concentration) is used to determine k_{cat} from the y intercept and K_m from the negative slope of a best-fit line.

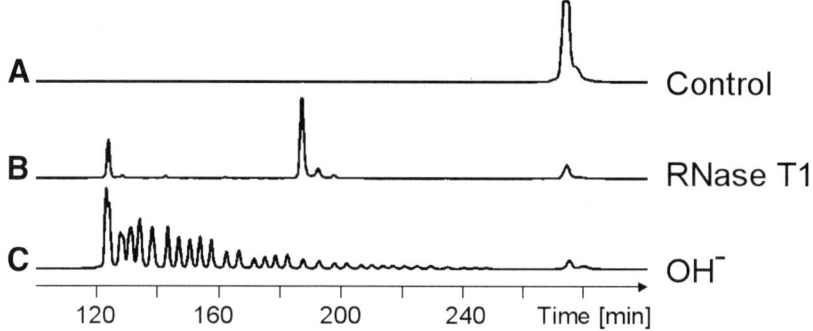

Fig. 1. Base- and G-ladder of Cy5-labeled TAR RNA substrate. **A** contains only untreated Cy5-labeled TAR RNA. An RNase T1 digest is shown in **B**, and a base hydrolysis is shown in **C**. Sequence of Cy5-labeled TAR RNA (5'-3'): Cy5-T_{10}-GGCCAGAUCUGAGCCUGGGAGCUCUCUGGCC-T_6.

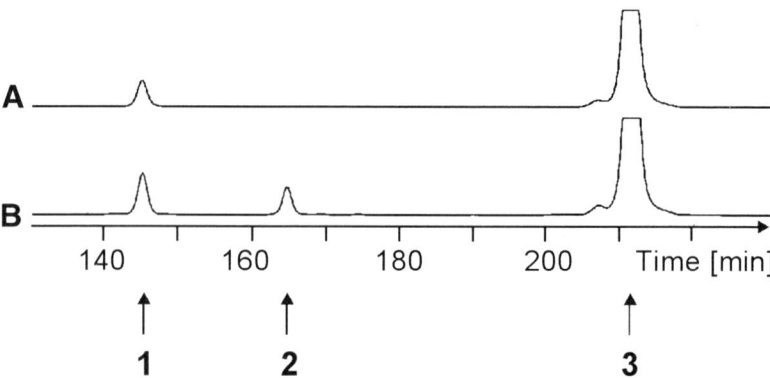

Fig. 2. Cleavage of Cy5-labeled RNA substrate by a deoxyribozyme. In **A** the untreated Cy5-labeled RNA substrate (**3**) together with the Cy5-labeled internal control (**1**) is shown. After incubation with the deoxyribozyme, a specific cleavage product (**2**) is formed in **B**. Sequence of the Cy5-labeled RNA substrate (5'-3'): Cy5-d(CTAGCCGACTGCCGATCTC)r(G→CUGAC)d(TGAC) and sequence of the deoxyribozyme (5'-3'):TCAGTCAGGGCTAGCTACAACGAGAGATCG→: cleavage site.

3.3. Determination of RNA Ligand Interactions by Gel-Shift Experiments

3.3.1. Electrophoresis

Assemble the short gel cassette, following the ALFexpress manual. Prepare 40 mL of the appropriate acrylamide gel solution (8 or 12% monomer, 0.1% Triton X-100, 0.2% glycerol), and filtrate the gel solution through 0.45 µm filter membrane. Start the reaction by adding 350 μL APS and 35 μL TEMED and cast the gel. Prepare 2 L of 0.5X TB running buffer. Add 100 mL of 10X TB stock solution, 8 mL 50% glycerol, and 2 mL 10% Triton X-100 solution to approx 800 mL ultrapure H_2O. Bring to the final volume with ultrapure water. Mix appropriate amounts of nucleic acids (approx 0.5–2 ng nucleic acid per lane) and ALF gel loading buffer ×5 and load the gel. The running conditions are: 1200 V (maximum), 40 mA (maximum), 35 W (constant), 16°C, 2-s sampling interval, 250 min running time.

3.3.1.1. BIOCONJUGATE PREPARATION

The cysteine SH-residue of the Tat 11-mer peptide is attached to methoxy–PEG–maleimide by nucleophilic addition. Prior to synthesis of peptide–maleimide conjugates the SH-groups of the peptide must be reduced by $NaBH_4$. Degas all buffers used in the procedure in a SpeedVac (2 min). Carefully add $NaBH_4$ powder (approx 5 mg) to a solution of 1 mg of 0.63 μmol Tat peptide in 125 μL reaction buffer. After a few minutes, add 10 μL acetone to stop the reaction. Adjust the pH of the solution to about 7.0 (pH optimum for the specific reaction of maleimides with sulfhydryl groups) with 15 μL 2N HCl (*see* **Note 7**; pH after addition of acetone: >8.0). Add 125 μL of the 10 mM N-Mal-5000 methoxy–PEG solution (6.3 mg, 1.25 μmol) dissolved in reaction buffer (molar ratio Tat 11-mer:N-Mal-5000 methoxy–PEG: 1:2). Incubate the reaction mixture for 2 h at room temperature. Desalt on a Sephadex G-50 column. Determine the peptide concentration via Bradford assay, and analyze the PEG-conjugated product on 6%/16.5% SDS PAGE (Schägger and Jagow) following a double-staining procedure. At first a PEG-specific staining *(10)* is carried out. Place the gel in a plastic tray, and fix it in 50 mL 5% glutaraldehyde for 15 min at room temperature. Replace the fixing solution with 50 mL 0.1M perchloric acid, and incubate for another 15 min at room temperature. Remove the perchloric acid solution, and add 12.5 mL 5% barium chloride solution and 5 mL 0.1N iodine solution. PEG bands appear within a few minutes, and a photo is taken of the stained gel. After destaining the gel in water, the protein part can be visualized by Coomassie staining (*see* **Note 8**) or silver staining. Again a photo is taken, the two images are overlaid in the computer thus allowing to identify the doubly stained band of the conjugate.

3.3.1.2. GEL-SHIFT ASSAY

This technique is often used to investigate interactions of proteins with nucleic acids. Existing methods for direct or competitive gel-shift assays have been adapted for the ALFexpress and Cy5-labeled RNA probes. In the given example, the inhibition of the Tat–TAR complex by neomycin or streptomycin is studied in competitive experiments *(11)*. To achieve a sufficient resolution

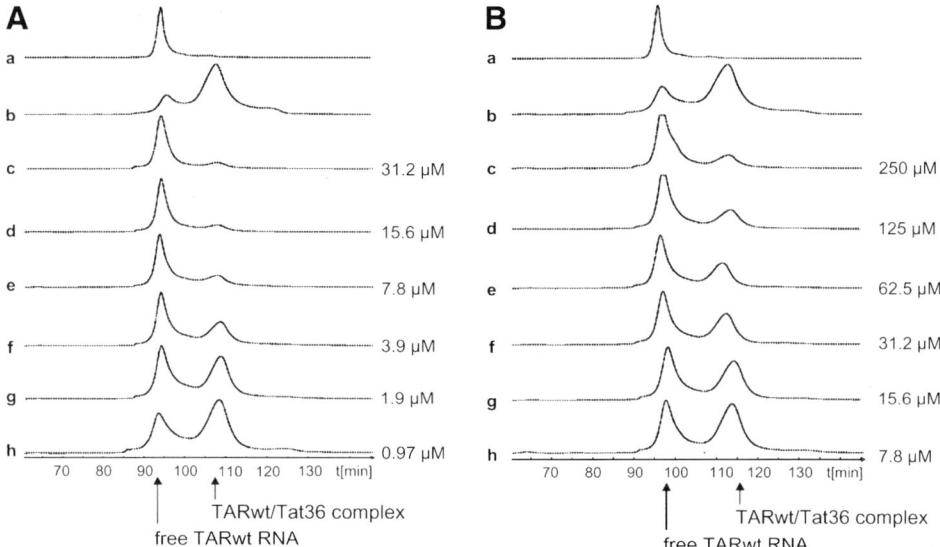

Fig. 3. Competition assay with neomycin (**A**) and streptomycin (**B**). Unbound Cy5-labeled TAR RNA is applied in **A** and Cy5-labeled TAR RNA complexed with Tat36 in **B**. In *lanes c–h* Cy5-labeled TAR RNA/Tat36 complexes are incubated with the indicated concentrations of competitor.

of free and complexed RNA, a synthetic 36-mer Tat peptide (Tat36; *see* **Fig. 3**) is used. Alternatively the Tat–PEG conjugate (*see* **Subheading 3.3.1.1.**) can be applied. Set up a negative control containing only Cy5-labeled TAR RNA and nonspecific tRNA, a positive control containing labeled TAR RNA, tRNA and Tat model peptide (36-mer or Tat–PEG conjugate), and the test sample. The latter contains Cy5-labeled TAR RNA, tRNA, Tat model peptide, and the low-molecular-weight RNA ligand. Prepare a mixture of Cy5-labeled TAR RNA, tRNA, and Tat model peptide in 2X Tris-KCl buffer. Add a solution of the RNA ligand (neomycin or streptomycin) to 5 µL of this mixture, and fill up to a final volume of 10 µL with DEPC-treated water. The final assay concentrations are 60 nm for Cy5-labeled TAR RNA and 600 nm for Tat36 or Tat–PEG conjugate, respectively. Incubate for 10 min at room temperature. Add 2 µL gel loading buffer, vortex the tube for 10 s to mix, and briefly centrifuge in a microcentrifuge. Clean sample wells with fresh running buffer and load the gel. Electrophoresis conditions are described in **Subheading 3.3.1.**

4. Notes

1. All oligonucleotide probes intended to be analyzed on the sequencer should be purified by denaturing PAGE. Other purification methods are not recommended

because the high sensitivity of the sequencer uncovers even very low background values caused by imperfect separation of undesired shorter oligonucleotides.
2. Rinse all plasticware and laboratory equipment that is in contact with the RNA probe and cannot be autoclaved with RNase Away™ (Roth, Karlsruhe, Germany).
3. Cy5-labeled probes appear as blue bands. Unlabeled oligonucleotides can be visualized by the ultraviolet-shadowing method.
4. Allow the gel to polymerize for a minimum of 2 h.
5. Incubation time for ribonucleic acids up to 10 mers, 35 min; for 40–50mers, 12 min; and for 100 mers, 2 min.
6. The SDS included in this buffer prevents the material from sticking to the reaction tubes.
7. Add $2N$ HCl as 5-µL aliquots; after each addition of HCl check the pH by spotting 1 µL of reaction mix onto universal indicator paper.
8. Optional: Add a ScotchBrite® sponge to the destaining solution; this hastens the destaining time and lowers the background.

References

1. Filée, P., Delmarcelle, M., Thamm, I., and Bernard, J. (2001) Use of an ALFexpress DNA sequencer to analyze protein–nucleic acid interactions by band shift assay. *Biotechniques* **30,** 1044–1051.
2. Scarso, A., Scheffer, U., Göbel, M., et al. (2002) A peptide template as an allosteric supramolecular catalyst for the cleavage of phosphate esters. *Proc. Natl. Acad. Sci.* **99,** 5144–5149.
3. Schmidt, C., Welz, R., and Müller, S. (2000) RNA double cleavage by a hairpin-derived twin ribozyme. *Nucleic Acids Res.* **28,** 886–894.
4. Glaesner, W., Merkl, R., Schmidt, S., Cech, D., and Fritz, H. J. (1992) Fast quantitative assay of sequence-specific endonuclease activity based on DNA sequencer technology. *Biol. Chem. Hoppe Seyler* **373,** 1223–1225.
5. Santoro, S.W. and Joyce, G. F. (1997) A general purpose RNA-cleaving DNA enzyme. *Proc. Natl. Acad. Sci.* **94,** 4262–4266.
6. Carey, J. (1991) Gel retardation. *Methods Enzymol.* **208,** 102–117.
7. Karn, J. (1999) Tackling Tat. *J. Mol. Biol.* **293,** 235–254.
8. Churcher, M. J., Lamont, C., Hamy, F., et al. (1993) High affinity binding of TAR RNA by the human immunodeficiency virus type-1 tat protein requires base pairs in the RNA stem and amino acid residues flanking the basic region. *J. Mol. Biol.* **230,** 90–110.
9. Wang, J., Huang, S. Y., Choudhury, I., Leibowitz, M. J., and Stein, S. (1995) Use of polyethylene glycol–peptide conjugate in a competition gel-shift assay for screening potential antagonists of HIV-1 tat protein binding to TAR RNA. *Anal. Biochem.* **232,** 238–242.
10. Kurfürst, M. K. (1992) Detection and molecular weight determination of polyethylene glycol-modified Hirudin by staining after sodium dodecyl sulfate–polyacrylamide gel electrophoresis. *Anal. Biochem.* **200,** 244–248.

11. Mei, H. Y., Galan, A. A., Halim, N. S., et al. (1995) Inhibition of an HIV-1 tat-derived peptide binding to TAR RNA by aminoglycoside antibiotics. *Bioorg. Med. Chem. Lett.* **5,** 2755–2760.

17

Design and Optimization of Molecular Beacon Real-Time Polymerase Chain Reaction Assays

Jacqueline A. M. Vet and Salvatore A. E. Marras

Summary

During the last few years, several innovative technologies have become available for performing sensitive and accurate genetic analyses. These techniques use fluorescent detection strategies in combination with nucleic acid amplification protocols. Most commonly used is the real-time polymerase chain reaction (PCR). To achieve the maximum potential of a real-time PCR assay, several parameters must be evaluated and optimized independently. This chapter describes the different steps necessary for establishing a molecular beacon real-time PCR assay: (1) target design, (2) primer design, (3) optimization of the amplification reaction conditions using SYBR Green, (4) molecular beacon design, and (5) molecular beacon synthesis and characterization. The last section provides an example of a multiplex quantitative real-time PCR.

Key Words: Real-time PCR; molecular beacon; SYBR Green; fluorescence; quantification of nucleic acids; SNP detection.

1. Introduction

The polymerase chain reaction (PCR), first described by Mullis and Saiki in 1985 *(1)*, has made it possible to detect rare target nucleic acid sequences isolated from cell, tissue, or blood samples. Real-time PCR is a powerful improvement on the basic PCR technique *(2)*. The use of fluorescent detection strategies in combination with appropriate instrumentation enables accurate quantification of nucleic acids. Quantification of nucleic acids is achieved by measuring the increase in fluorescence during the exponential phase of PCR. The point at which the fluorescence rises significantly above background is called the threshold cycle (Ct) (**Fig. 1A**). There is an inverse linear relationship between the log of the starting amount of template and the corresponding

From: *Methods in Molecular Biology, vol. 288: Oligonucleotide Synthesis: Methods and Applications*
Edited by: P. Herdewijn © Humana Press Inc., Totowa, NJ

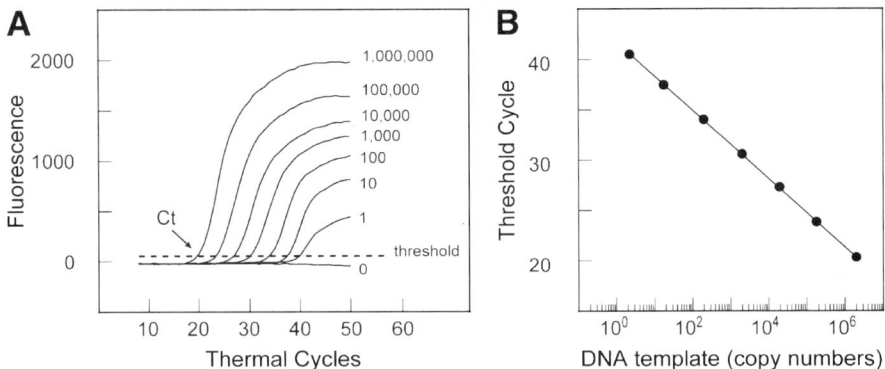

Fig. 1. (**A**) The threshold cycle (Ct) is the cycle at which the fluorescence rises significantly above the background. The fluorescence increases as the molecular beacons bind to the amplification products that accumulate during each successive cycle. In the early cycles of amplification, the change in fluorescence is usually undetectable, but at some point during amplification, the accumulation of amplified DNA results in a detectable change in the fluorescence of the reaction mixture. The Ct number decreases as the number of target molecules initially present in a reaction increases. (**B**) The standard curve can be used to determine the starting amount of an unknown template, based on its Ct. Given known starting amounts of the target, a standard curve can be constructed by plotting the log of the starting amount vs the Ct. The Ct is inversely proportional to the logarithm of the number of target molecules initially present. In real-time PCR quantitative results can be obtained over a wide dynamic range of target concentrations.

Ct-value during real-time PCR (**Fig. 1B**). Compared with end point quantification methods, real-time amplification assays offer reproducible results and have a much wider dynamic range. In addition, the use of fluorescent agents and probes that only generate a fluorescence signal on binding to their target enables real-time amplification assays to be carried out in sealed tubes, eliminating the risk of carryover contamination. Different techniques are available to monitor real-time amplification. The amplification process can be monitored using nonspecific double-stranded deoxyribonucleic acid (DNA) binding dyes or specific fluorescent hybridization probes. Dyes, such as SYBR Green, produce enhanced fluorescence signals on intercalating with double-stranded DNA complexes *(3)*. Although double-stranded DNA binding dyes are a simple, fast, and inexpensive method to monitor the amplification of a template, their major disadvantage is that the dye binds nonspecifically to all double-stranded DNA complexes, such that primer–dimers and nonspecific amplification products cannot be distinguished from the intended amplification product. The use of fluorescent hybridization probes enhances the overall

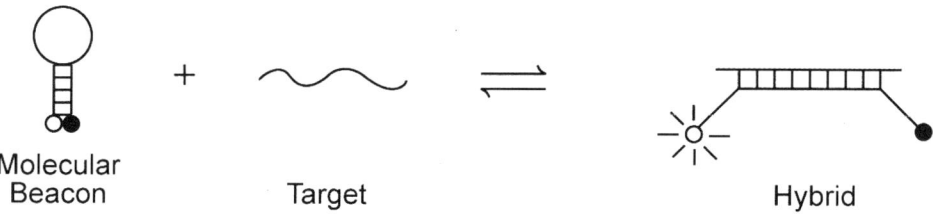

Fig. 2. Principle of operation of molecular beacons. Molecular beacons are single-stranded oligonucleotides that possess a hairpin structure. Free molecular beacons are nonfluorescent because the hairpin stem keeps the fluorophore in close proximity to the quencher. When the probe sequence in the hairpin loop hybridizes to its target, forming a rigid double helix, a conformational change occurs that removes the quencher from the vicinity of the fluorophore, thereby restoring fluorescence.

assay sensitivity by eliminating background signals owing to the synthesis of nonspecific amplification products. The hybridization probe is designed to be complementary to a target sequence within the expected amplification product, so that the probes do not bind to nonspecific amplification products, thus enhancing the specificity of the assay. The different fluorescent probe-based techniques include adjacent probes *(4)*, 5' nuclease probes *(5)*, molecular beacons *(6),* and duplex scorpion primers *(7)*. This chapter describes the use of molecular beacons in conjunction with PCR. Molecular beacons (**Fig. 2**) hybridize at the annealing temperature to the amplification products and do not interfere with primer annealing and extension. Molecular beacon real-time PCR assays are simple, fast, sensitive, accurate, allow a high-throughput format, and enable the detection of a series of different agents in the same assay tube *(8)*. Molecular beacon real-time PCR assays have been used in numerous studies to detect genomic DNA sequences, single nucleotide polymorphisms (SNPs), messenger ribonucleic acid (mRNA) expression levels, and pathogens *(9–18)*. The development and application of molecular beacons was recently reviewed by Vet et al. *(19)*, and more information is available at www.molecular-beacons.org.

To achieve the maximum potential of real-time PCR, several parameters must be evaluated and optimized independently. The amplification conditions, the molecular beacon design, and synthesis affect the efficiency and accuracy of real-time PCR. When determining which conditions to optimize, the ultimate assay goal must be considered. When the goal is to screen for a specific sequence, for example, in SNP detection assays or in assays to identify bacterial species with conserved DNA regions, both the design of the primers and molecular beacon is limited to the region that contains the specific sequence. In those cases, the selection of primers depends on the molecular beacon target sequence, that is, the primers should not overlap with the molecular beacon sequence. As a result, the choice of primers is limited. When the choice of primers and molecular beacon is not limited to a restricted sequence region, for example, in assays that determine the mRNA expression of a gene, the primers should be designed and tested prior to developing the molecular beacon. After optimization of the PCR assay, a molecular beacon can be designed that will detect the amplified template. We have outlined the different steps necessary for establishing a reliable real-time PCR assay. These steps include: (1) target design, (2) primer design, (3) optimization of the amplification reaction conditions using SYBR Green, (4) molecular beacon design, and (5) molecular beacon synthesis and characterization. The last section provides an example of a multiplex quantitative real-time PCR.

2. Materials

1. 5 U/µL AmpliTaq Gold DNA polymerase (Applied Biosystems, CA) (*see* **Note 1**).
2. 10X concentrated PCR buffer (supplied with DNA polymerase, containing 100 mM Tris-HCl, pH 8.3, 500 mM KCl, 0.01% (w/v) gelatin, without $MgCl_2$).
3. 25 mM $MgCl_2$ (Applied Biosystems, CA).
4. Deoxynucleotide-triphosphate (dNTP) mixture: 25 mM dATP, 25 mM dCTP, 25 mM dGTP, 25 mM dTTP (Promega, WI).
5. TE buffer: 1 mM EDTA, 10 mM Tris-HCl, pH 8.0.
6. PCR primers (dissolved in TE). Stock solutions of 100 µM should be kept for longtime storage at –20°C. Work solutions (diluted in Milli-Q water) should be kept at –20°C.
7. 50X SYBR Green for DNA in Milli-Q water. SYBR Green is available as a 10,000X stock in dimethyl sulfoxide (DMSO) (Molecular Probes, OR); 200X SYBR Green solutions in DMSO can be kept for longtime storage protected from light at –20°C. Work solutions of 50X SYBR Green in Milli-Q water are prepared fresh weekly and should be kept protected from light at –20°C prior to use.
8. Molecular beacon (dissolved in TE). Stock solutions of 100 µM should be kept for longtime storage protected from light at –20°C. Work solutions (diluted in TE at, for example, 5 µM) should be kept protected from light at –20°C for 1 mo.
9. Oligonucleotide target that is perfectly complementary to the probe sequence of the molecular beacon (but not complementary to the arm sequences). Store work solutions (diluted in Milli-Q water at 25 µM) at –20°C.
10. Spectrofluorometric thermal cycler. There are several platforms available for performing real-time PCR analyses (**Table 1**). These platforms range from ultrarapid, air-heated thermal cyclers, where PCR is performed in glass capillaries, to tube-based and microtiter plate-based systems for high-throughput assays.

Table 1
Specification of the Different Available Spectrofluorometric Thermal Cyclers

Company	Model	Fluorophore choices	Multiplex capabilities	Sample capacity	Remarks
Applied Biosystems	PRISM® 7000 Sequence Detection System	FAM, TET, TMR and Texas Red	Up to four targets	96 wells	High throughput
Applied Biosystems	PRISM® 7700 and 7900HT Sequence Detection System	FAM, TET, Rhodamine-6G, TMR and Texas Red	Up to five targets (*see* **Note 9**)	96 wells or 384 wells	High throughput
Bio-Rad	iCycleriQ™ real-time PCR Detection System	FAM, HEX, Texas Red and Cy5	Up to four targets	96 wells	High throughput, temperature gradient block (*see* **Note 2**)
Cepheid	SmartCycler	FAM, Cy3, Texas Red and Cy5	Up to four targets	16 units	Rapid thermal cycles, each unit is independently programmable (*see* **Note 2**)
MJ Research	Opticon	FAM and TMR	Up to two targets	96 wells	High throughput, temperature gradient block (*see* **Note 2**)
Roche Applied Science	LightCycler System	FAM, ROX and Cy5.5	Up to three targets (*see* **Note 9**)	32 wells	Rapid thermal cycles
Stratagene	Mx3000P and Mx4000™ Multiplex Q-PCR System	FAM, TMR, Texas Red and Cy5	Up to four targets	96 wells	High throughput

3. Methods

3.1. Target Design

A successful real-time PCR requires efficient amplification of the template. Both the source of the template *(20)* and the sequence of the primers and template affect this efficiency. Significant secondary structures of the template may hinder the primers or molecular beacon from annealing and prevent complete product extension by the DNA polymerase. It is therefore recommended to design primer pairs that amplify a target region of 75 to 250 basepairs. Besides a more efficient amplification of the target sequences, these shorter amplification products produce higher fluorescent signals because molecular beacons are better able to compete with the complementary strands for binding to the target strands. To avoid selecting a molecular beacon target sequence within a region with strong secondary structures, the target sequence can be analyzed using a DNA folding program, such as the DNA mfold server *(21)* (available at http://www.bioinfo.rpi.edu/applications/mfold).

3.2. Primer Design

The goal is to design primers with a melting temperature (Tm) higher than the Tm of any of the predicted template secondary structures. This ensures that the majority of possible secondary structures have been unfolded before the primer-annealing step. The following parameters are important when designing primers:

1. Design primers with a 50–60% GC content.
2. Maintain a Tm between 50–65°C.
3. Eliminate strong secondary structures (more than two basepairs).
4. Avoid repeats of Gs or Cs longer than three bases.
5. Check the sequence of the primers against each other to ensure that there are no 3' complementarities (avoids primer–dimers).
6. Place Gs and Cs on ends of the primers.
7. Avoid false priming by verifying the primers specificity using sites such as the Basic Local Alignment Search Tool (available at http://www.ncbi.nml.nih.gov/blast/).

We recommend designing a few primer sets for each individual PCR and testing these sets using SYBR Green as a reporter molecule.

3.3. Optimization Using SYBR Green Real-Time PCR

For optimizing the PCR and monitoring the amplification of nonspecific amplification products, different concentrations of $MgCl_2$, different PCR primer sets in varying concentrations, and different annealing temperatures should be tested.

Real-Time PCR Assays

1. Prepare 50-μL reactions that contain 1X PCR buffer, 2.5 U AmpliTaq Gold DNA polymerase, 250 μM of each dNTP, 1X SYBR Green, and variable concentrations of MgCl$_2$ (3–6 mM, in 0.5 mM steps), PCR primers (100 nM, 250 nM, 500 nM, and 1000 nM).
2. For each amplification condition, prepare two reactions. To one reaction, add approx 100,000 copies of template molecules to test the target amplification efficiency. To monitor the formation of nonspecific amplification products, Milli-Q water (instead of template) is added to the other reaction.
3. Program the thermal cycler to heat the reaction mixture at 95°C for 10 min to activate the AmpliTaq Gold DNA polymerase, followed by 40 amplification cycles. Each cycle consists of denaturation at 95°C for 30 s, primer annealing at the temperature determined by the primer selection software for 30 s (*see* **Note 2**), and primer extension at 72°C for 30 s.
4. Monitor the fluorescence of SYBR Green during the annealing step of the amplification reaction. Refer to the instrument's manual to enable the appropriate channel or emission filter for monitoring the fluorescence of SYBR Green. If no specific SYBR Green option is available on the instrument, monitor the fluorescence using the same channel or emission filter chosen to monitor the fluorescence of fluorescein (FAM). FAM and SYBR Green have roughly the same emission spectrum.

3.3.1. Data Interpretation

1. Analyze the acquired fluorescence data with the appropriate settings for background and baseline subtraction as indicated in the instrument's manual.
2. Determine the Ct for each reaction condition for both the template and no target control.
3. A reaction initiated with 100,000 copies of the template will give a Ct in the range of 21 to 25. The different conditions can be compared to determine the reaction with earliest Ct and strongest fluorescence signal. The reaction with no template added will indicate the presence of nonspecific amplification products. If the Ct for nonspecific amplification products is 30 or lower, a new primer pair should be chosen. The formation of the nonspecific amplification products will most likely affect the amplification of the template. If the threshold of nonspecific amplification products is between 30 and 40, a new primer pair should be chosen if the assay is designed to amplify template molecules as low as 100 copies. If the assay is designed for the amplification of template molecules as low as one copy, then it is preferable to have Cts of nonspecific amplification products to be higher than 40, or, even better, have no nonspecific amplification products.
4. The efficiency of the PCR can be determined by performing a serial dilution experiment, using 100,000, 10,000, 1000, 100, and 10 copies of template to initiate the PCR, and constructing a standard curve. Refer to the thermal cycler manual to automatically determine the Ct for the standards and to calculate a standard curve. A slope of –3.322 represents a PCR efficiency of 100% (*see* **Note 3**).

The correlation coefficient should be at least 0.995. (For multiplex quantitative PCR guidelines, *see* **Note 4**.)

After careful selection of the best primer pair and other optimized reaction conditions, a suitable molecular beacon can be designed.

3.4. Molecular Beacon Design

To successfully monitor PCR reactions, molecular beacons should be designed that are able to hybridize to their targets at the annealing temperature of the PCR, whereas the free molecular beacons should stay closed and be nonfluorescent at this temperature. This can be ensured by appropriately choosing the length of the probe sequence and the length of the arm sequences. The length of the probe sequence should be such that the molecular beacon will dissociate from its target at temperatures above the annealing temperature of the PCR. The Tm of the probe target-hybrid can be predicted using the "percent-GC" rule, which is available in most primer and hybridization probe design programs. The prediction should be made for the probe sequence alone before adding the stem sequences. In practice, the length of the probe sequence is usually between 15 and 30 nucleotides. The Tm of the probe-target hybrid should not be above 72°C to avoid interference of the DNA polymerase. Molecular beacons are very versatile. The general design, just outlined, can be altered to fit specific applications.

3.4.1. Molecular Beacons for Standard Target Detection

For standard target detection, in which the molecular beacon target sequence is not limited to a specific sequence region of the amplification product, a probe length of 22–30 nucleotides can be used, and the Tm of the probe-target hybrid should be 7–10°C above the annealing temperature of the PCR. Thermal denaturation profiles can be performed to confirm the theoretical prediction of the Tm (*see* **Subheading 3.6.2.**). Longer loop sequences ensure that probe-target hybrids that contain mismatches are stable at the annealing temperature of the PCR. This can be of importance to detect, for example, a retroviral subtype that contains one or two nucleotide substitutions *(18)* (*see* **Note 5**).

3.4.2. Allele-Discriminating Molecular Beacons

To ensure that a molecular beacon can discriminate single-nucleotide variations at the detection temperature, the following criteria should be met: In the presence of perfectly complementary targets, the molecular beacons must form a stable probe-target hybrid; and in the presence of mismatched targets, the molecular beacons must remain closed. Therefore:

1. Select a probe sequence that will dissociate from its target at temperatures 5–8°C higher that the annealing temperature of the PCR. The length of the probe sequence usually falls in the range between 15 and 21 nucleotides.

2. Measure the fluorescence of solutions of molecular beacons in the presence of each kind of target as a function of temperature, to determine the window of discrimination, which is the range of temperature in which perfectly complementary probe targets can form and in which mismatched probe-target hybrids cannot form (see **Subheading 3.6.2.**).

3.4.3. Selecting the Stem Sequence

After selecting a probe sequence, two complementary arm sequences should be added on either side of the probe sequence. To ensure that the molecular beacon remains closed in the absence of a target sequence, the stem should be stable at the annealing temperature of the PCR. Because the stem is formed by an intramolecular hybridization event, its Tm cannot be predicted by the percent-GC rule or nearest neighbor method used in the primer and hybridization probe software. Instead, a DNA folding program (such as the DNA mfold server *[21]*, available at http://www.bioinfo.rpi.edu/applications/mfold) should be used to predict its Tm. Usually the stem is 5–7 nucleotides long (see **Note 6**). In general, GC-rich stems that are five basepairs in length will melt between 55 and 60°C, GC-rich stems six basepairs in length will melt between 60 and 65°C, and GC-rich stems seven basepairs in length will melt between 65 and 70°C. Longer stems enhance the specificity of molecular beacons *(22)*. The folding program also predicts if the chosen molecular beacon sequence will form an unwanted secondary structure (see **Note 7**). Molecular beacons can also be designed with the help of a software package called Beacon Builder, which is available from Premier Biosoft International (see **Note 8**).

3.4.4. Choice of Fluorophores

Molecular beacons can be labeled with a wide range of fluorophores, and all fluorophores can be efficiently quenched by the same quencher, such as DABCYL, BHQ-1, and BHQ-2 *(8,23)*. The choice of the fluorophore is therefore mainly dependent on the apparatus used to perform the real-time PCR assay. **Table 1** provides information on what fluorophores can be used with the different spectrofluorometric thermal cyclers. For each spectrofluorometric thermal cycler, the fluorophores are listed that have minimum emission overlap and allow reliable multiplex detection assays to be carried out. However, this list only provides a small subset of the fluorophores that can be used. In the manual provided with each thermal cycler, a detailed list is available of fluorophores that are suitable for the selected light and emission filter installed on the thermal cycler (see **Note 9**).

3.5. Molecular Beacon Synthesis

Molecular beacons can be obtained from a large number of oligonucleotide synthesis companies (a list is available at www.molecular-beacons.org). These companies specialize in the synthesis, purification, and characterization of molecular beacons. It is recommended to order simultaneously an oligonucleotide target that is complementary to the probe (loop) sequence of the molecular beacon, which is used for additional characterization experiments of the molecular beacon (*see* **Subheadings 3.6.1.** and **3.6.2.**).

If a DNA synthesizer and high-pressure (performance) liquid chromatography (HPCL) apparatus are available, molecular beacon probes can be prepared in a one- or two-step synthesis process, followed by one purification step. A rising number of fluorophores and quenchers have become available as phosphoramidite derivatives or linked to controlled pore glass (CPG) columns (Glen Research, VA; Amersham, NJ; Biosearch Technologies, CA). This allows the synthesis of a complete molecular beacon in a single synthesis step on a DNA synthesizer. In most cases, no changes in the DNA synthesis protocol have to be made to incorporate the fluorophores and quenchers; however, we suggest referring to the instruction manuals that are supplied with the fluorophores and quenchers for special coupling and postsynthesis protocols.

In case a fluorophore or quencher is not available for direct incorporation during the DNA synthesis, chemically reactive fluorophore derivatives can be introduced in a post-DNA-synthesis step. In a first DNA synthesis step, the quencher DABCYL, BHQ-1, or BHQ-2, linked to a CPG column (Glen Research, VA; Biosearch Technologies, CA) is introduced at the 3'-terminal position and an amino or sulfhydryl phosphoramidite (Glen Research, VA) is incorporated at the 5'-terminal position of the oligonucleotides. In the second post-DNA-synthesis step, a succinimidyl ester or an iodoacetamide (or maleimide) derivative of a fluorophore (Molecular Probes, OR) is coupled to the 5'-amino or sulfhydryl moiety, respectively. Protocols describing the coupling of fluorophore derivatives to oligonucleotides can be downloaded at www.molecular-beacons.org.

All molecular beacons are purified by HPCL through a C-18 reverse-phase column, using a linear elution gradient of 20 to 70% buffer B (0.1M triethylammonium acetate [TEAA] in 75% acetonitrile, pH 6.5) in buffer A (0.1M TEAA, pH 6.5) that forms over 25 min at a flow rate of 1 mL/min (refer to www.molecular-beacons.org for typical chromatograms.) After the purification, the molecular beacons are precipitated with 1/10 vol 3M sodium acetate, pH 5.2, and 2.5 vol 100% ethanol. The molecular beacons are dissolved at a concentration of 100 µM in TE buffer and stored at –20°C.

3.6. Characterization of Molecular Beacons

The molecular beacon synthesis and purification is very crucial for the production of high-quality probes and, subsequently, for accurate and reliable real-time PCR experiments. Weak fluorescence signals during the real-time PCR

Real-Time PCR Assays

assay are often caused by poor signal-to-background ratios of the molecular beacons. These poor signal-to-background ratios are caused by the presence of uncoupled fluorophores in the preparation or by the presence of oligonucleotides containing a fluorophore but lacking a quencher. To determine the purity of a molecular beacon preparation, one can measure the extent to which its fluorescence increases on binding to its target. Preferably, the molecular beacons should have signal-to-background ratios above 20.

3.6.1. Signal-to-Background Ratio

The signal-to-background ratio of a molecular beacon can be determined using the same spectrofluorometric thermal cycler used for the real-time PCR assay. All measurements are taken at the same temperature as optimized for the annealing of the primers during the PCR, and the appropriate excitation and emission source settings are selected for the fluorophore that is used as a label for the molecular beacon.

1. Prepare one 25-μL reaction that contains 1X PCR buffer and the concentration of $MgCl_2$ that will be used in the PCR assay.
2. Determine the fluorescence signal of this solution (F_{buffer}).
3. Add 1 μL of a 5 μM molecular beacon solution to the 25 μL solution, and measure the new level of fluorescence (F_{closed}). Ensure that the fluorescence level exceeds that of the F_{buffer}. In case the fluorescence level is the same as F_{buffer}, add an additional 1 μL of a 5 μM molecular beacon solution to the solution and repeat the measurement.
4. Add 1 μL of 25 μM of an oligonucleotide target whose sequence is perfectly complementary to the probe sequence of the molecular beacon. Allow the molecular beacon to hybridize to the oligonucleotide target (about 1 min), and measure the fluorescence of the solution (F_{open}).
5. Calculate the signal-to-background ratio as $(F_{open}-F_{buffer})/(F_{closed}-F_{buffer})$.

3.6.2. Thermal Denaturation Profiles

To determine the window of discrimination, the fluorescence of the molecular beacon as a function of temperature is measured in the absence of target, in the presence of perfectly complementary target oligonucleotides, and in the presence of the mismatch target oligonucleotides:

1. For each molecular beacon, prepare three tubes containing 200 nM molecular beacon in 1X PCR buffer and the concentration of $MgCl_2$ that will be used in the PCR assay.
2. Add to one of the tubes a fivefold molar excess of an oligonucleotide that is perfectly complementary to the molecular beacon probe sequence, add a fivefold excess of an oligonucleotide that contains the mismatched target sequence to the other tube, and add only buffer to the third tube.

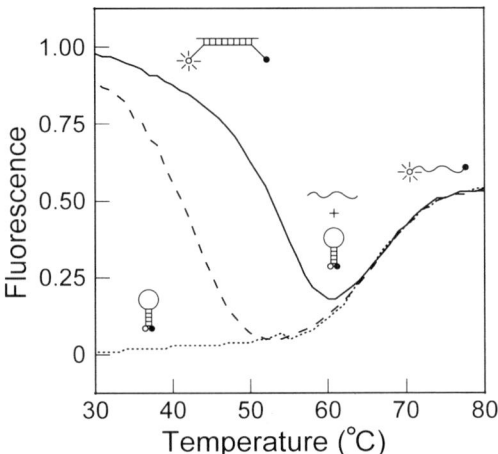

Fig. 3. Thermal denaturation profiles of molecular beacons in the presence of either wild-type target (continuous line), mutant target (dashed line), or no target (dotted line). A diagram indicates the state of the molecular beacon over the thermal denaturation profiles. Mismatched hybrids denature 10 to 12°C below the T_m of perfectly matched hybrids. In this example, an annealing temperature of 50°C will be optimal to ensure that the molecular beacon generates a high fluorescence signal in the presence of its perfect complementary target and that no fluorescence signals are generated in the presence of a mutant target.

3. Determine the fluorescence of each solution as a function of temperature, using a spectrofluorometric thermal cycler. Decrease the temperature of the solutions from 80°C to 30°C in 1°C steps, with each step lasting 1 min, while monitoring fluorescence during each step.

Figure 3 shows an example of a thermal denaturation profile. Select the appropriate annealing temperature, that is, the temperature at which only the perfectly complementary probe-target hybrids are formed.

3.7. Real-Time PCR

After characterization of the molecular beacon, the last step in the optimization of the real-time PCR is to determine the optimal concentration of the molecular beacon. In general, the amount of molecular beacon to be used is about the same as the amount of primers, so that both will be in excess to the expected amount of amplification products. It is suggested to test three molecular beacon concentrations: half the amount of the primers, the same amount as the primers, and twice as much. Use the reaction conditions as determined with the SYBR Green PCR, and test the performance of each molecular beacon concentration with different target concentrations (e.g., 100,000, 10,000, 1000, 100, and 10 molecules). Compare the Ct-values of the different amounts of template molecules for the three molecular beacon concentrations. If the Ct values are equal at all template amounts for the three molecular beacon concentrations, choose the lowest concentration of the molecular beacon used. Otherwise, choose the molecular beacon concentration that results in the lowest Ct value for the different amount of target molecules.

3.8. Multiplex Detection of Four Pathogenic Retroviruses Using Molecular Beacons

Four primer sets and four molecular beacons were designed according to the procedures described above. To detect HIV-1, HIV-2, HTLV-I, and HTLV-II equally well in the same assay, primer sets and molecular beacons were chosen to be compatible. The primers generated short amplification products (100–130 basepairs) and were chosen so that their target sequences would be unique to each retrovirus, highly conserved, present in most clinical subtypes, and not found in the human genome. The assay described here is adapted from Vet et al., 1999 *(18)*. The concentration of each primer set was adjusted so that the efficiency of amplification of each of the four types of amplification products was approximately equal during the exponential phase of the reaction. The length of each of the probe sequences (25–33 nucleotides) was selected so that probe-target hybrids were likely to form even if the retroviral target sequence was from a subtype that contained one or two nucleotide substitutions. The arm sequences were 6 nucleotides long, to ensure that the Tm of the stem would be above 65°C. The molecular beacons designed to detect HIV-1, HIV-2, and HTLV-I were labeled with, respectively, FAM, HEX, and Texas Red, and used DABCYL as a quencher. The molecular beacon to detect HTLV-II was labeled with Cy5 and used BHQ-2 as a quencher. These fluorophores were chosen to minimize crosstalk of the emission fluorescence between the filters available on the Bio-Rad iCycler IQ system that was used for this assay.

3.8.1. PCR Assays

Each 50-µL reaction contained the relevant template DNA, 1.00 µM of each HIV-1 primer, 0.25 µM of each HIV-2 primer, 0.50 µM of each HTLV-I primer, 0.25 µM of each HTLV-II primer, 0.10 µM of HIV-1–FAM, 0.20 µM HIV-2–HEX, 0.10 µM HTLV-I–Texas red, 0.10 µM HTLV-II–Cy5, 250 µM of each dNTP, 2.5 U AmpliTaq Gold DNA polymerase, 3 mM MgCl$_2$ in 1X PCR buffer. After activating the DNA polymerase by incubation for 10 min at 95°C, 40 cycles of amplification (94°C for 30 s, 55°C for 30 s, and 72°C for 30 s) were carried out in a 96-well spectrofluorometric thermal cycler (Bio-Rad iCycler IQ) and fluorescence was monitored during every thermal cycle at the 55°C annealing step.

To determine how well individual retroviruses can be distinguished from one another in this multiplex format, four assays were carried out in parallel, each initiated with 100,000 molecules of a different retroviral DNA. The results are shown in **Fig. 4**. The intensity of the fluorescence of each of the four molecular beacons (normalized on a scale from 0 to 1 to aid in their comparison) is plotted as a function of the number of thermal cycles completed. The only significant fluorescence that appeared in the course of the amplification reactions carried out in each assay tube was fluorescence from the molecular beacon that was complementary to the sequence of the expected amplification product. The color of the fluorescence identified the retroviral DNA that was originally added to the reaction mixture. No significant fluorescence developed in a control assay that did not contain any template DNA. These results demonstrate that each molecular beacon binds only to its intended target amplification product and that there is no crosstalk between the emission filters.

4. Notes

1. The use of hot-start *Taq* DNA polymerase is recommended because it often results in higher quality real-time PCR assays compared to the use of conventional *Taq* DNA polymerase. The major advantage of hot-start *Taq* DNA polymerases is that primer extension cannot occur during the preparation of the master mix. This greatly minimizes the extension of nonspecific primer annealing events that might occur prior to starting the PCR reaction. As a result, they increase the amplification yield of low copy numbers of template and result in more efficient multiplex amplification assays.
2. Some spectrofluorometric thermal cyclers have the possibility to program a temperature gradient that enables the user to optimize the annealing temperature at the same reaction plate with the different $MgCl_2$ and primer concentrations. The temperature gradient is also a very convenient option for optimizing SNP detection assays.
3. PCR efficiency can be calculated using the following formula:

    $$E = e^{\ln 10/-s} - 1$$

    where E = efficiency, s = slope of the standard curve,
    e = 2.718 (approx) and $\ln 10$ = 2.303 (approx)

4. Multiplex quantitative assays require both precise and accurate quantification of multiple gene targets in one assay tube and should therefore be carefully optimized. PCR efficiencies of the different amplification reactions should be maximized and equalized. Maximizing the efficiency of a reaction allows accurate quantification over a wider range of starting template concentrations and improves the reproducibility of replicate samples. It is equally important that the individual reactions have similar efficiencies because any difference in individual efficiencies will be amplified when the two reactions are combined. Always test for crossreactivity of the individual reaction components by comparing a set of wells containing one reaction alone to a set of wells containing all ingredients for

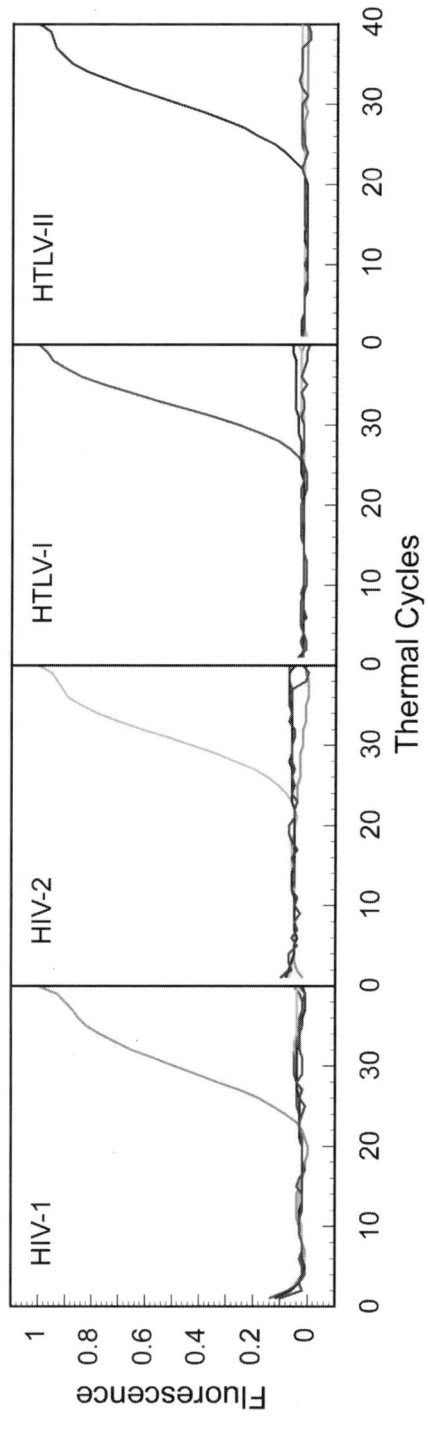

Fig. 4. Real-time detection of four retroviral DNAs in a multiplex format. Four assays were carried out in sealed tubes, each initiated with 100,000 molecules of different retroviral DNA. Each reaction contained four sets of PCR primers specific for unique HIV-1, HIV-2, HTLV-I, and HTLV-II nucleotide sequences and four molecular beacons, each specific for one of the four amplification products and labeled with a differently colored fluorophore. Fluorescent signals from the FAM-labeled molecular beacon (HIV-1 specific), from the HEX-labeled molecular beacon (HIV-2 specific), from the Texas Red-labeled molecular beacon (HTLV-I specific), and from the Cy5-labeled molecular beacon (HTLV-II specific), are plotted in green, yellow, red, and blue, respectively.

the multiplex PCR. If there is no crossreactivity, the Ct value of template X when amplified alone will be identical to the Ct value of template X in the presence of the other components.

5. Use a complementary oligonucleotide target that possesses the relevant nucleotide substitution(s) to perform a thermal denaturation profile as described in **Subheading 3.6.2.** Determine the temperature at which the mismatch hybrids are stable and use this as the assay temperature.
6. It has been observed that nucleotides can quench the fluorescence of fluorophores *(23)*. Guanosine exhibits the highest degree of quenching. Therefore, it is recommended not to use the guanosine nucleotide at the 5'-end next to the fluorophore.
7. It is important that the conformation of the free molecular beacons is the intended hairpin structure, rather than other structures that either do not place the fluorophore in the immediate vicinity of the quencher, or that form longer stems than intended. The former will cause high background signals, and the latter will make the molecular beacon sluggish in binding to its target. If the alternative structure results from the choice of the stem sequence, the identity of the stem sequence can be altered. If, on the other hand, the alternative structures arise from the identity of the probe sequence, the frame of the probe can be moved along the target sequence to obtain a probe sequence that is not self-complementary. Small stems within the probe's hairpin loop that are two to three nucleotides long do not adversely affect the performance of molecular beacons.
8. This software for designing primers and molecular beacons is convenient to use when the DNA fragment to work with is rather large. With small fragments often no suitable primers or probe sequences are found.
9. In spectrofluorometric thermal cyclers possessing a monochromatic light source, such as a laser or light emitting diode, wavelength-shifting molecular beacons can expand the number of targets that can be detected in one assay tube *(24)*. Wavelength-shifting molecular beacons contain a "harvester" fluorophore that, when hybridized to an amplification product, transfers the absorbed energy from the light source to a second "emitter" fluorophore that fluoresces in the desired color. However, when not hybridized to an amplification product, the energy from the harvester fluorophore is directly transferred to a quencher, and the wavelength-shifting molecular beacon remains dark.

Acknowledgments

We thank Henk Blom, Brenda Van der Rijt, Sanjay Tyagi, and Fred Russell Kramer for their advice and assistance. Furthermore, we thank Arno Pol and Benjamin Gold for a critical reading of this manuscript. This work was supported by National Institutes of Health grants HL-43521 and GM-070357-01.

References

1. Saiki, R. K., Scharf, S., Faloona, F., et al. (1985) Enzymatic amplification of beta-globin genomic sequences and restriction site analysis for diagnosis of sickle cell anemia. *Science* **230**, 1350–1354.
2. Higuchi, R., Fockler, C., Dollinger, G., and Watson, R. (1993) Kinetic PCR analysis: real-time monitoring of DNA amplification reactions. *Biotechnology (NY)* **11**, 1026–1030.
3. Morrison, T. B., Weis, J. J., and Wittwer, C. T. (1998) Quantification of low-copy transcripts by continuous SYBR Green I monitoring during amplification. *Biotechniques* **24**, 954–962.
4. Wittwer, C. T., Herrmann, M. G., Moss, A. A., and Rasmussen, R. P. (1997) Continuous fluorescence monitoring of rapid cycle DNA amplification. *Biotechniques* **22**, 130, 131.
5. Holland, P. M., Abramson, R. D., Watson, R., and Gelfand, D. H. (1991) Detection of specific polymerase chain reaction product by utilizingusing the 5'-3' exonuclease activity of Thermus aquaticus*Thermus aquaticus* DNA polymerase. *Proc. Natl. Acad. Sci. USA* **88**, 7276–7280.
6. Tyagi, S. and Kramer, F. R. (1996) Molecular beacons: probes that fluoresce uponon hybridization. *Nat. Biotechnol.* **14**, 303–308.
7. Solinas, A., Brown, L. J., McKeen, C., et al. (2001) Duplex Scorpion primers in SNP analysis and FRET applications. *Nucleic Acids Res.* **29**, E96.
8. Tyagi, S., Bratu, D. P., and Kramer, F. R. (1998) Multicolor molecular beacons for allele discrimination. *Nat. Biotechnol.* **16**, 49–53.
9. El-Hajj, H. H., Marras, S. A., Tyagi, S., Kramer, F. R., and Alland, D. (2001) Detection of rifampin resistance in *Mycobacterium tuberculosis* in a single tube with molecular beacons. *J. Clin. Microbiol.* **39**, 4131–4137.
10. Giesendorf, B. A., Vet, J. A., Tyagi, S., Mensink, E. J., Trijbels, F. J., and Blom, H. J. (1998) Molecular beacons: a new approach for semiautomated mutation analysis. *Clin. Chem.* **44**, 482–486.
11. Kostrikis, L. G., Tyagi, S., Mhlanga, M. M., Ho, D. D., and Kramer, F. R. (1998) Molecular beacons: spectral genotyping of human alleles. *Science* **279**, 1228, 1229.
12. Manganelli, R., Dubnau, E., Tyagi, S., Kramer, F. R., and Smith, I. (1999) Differential expression of 10 sigma factor genes in *Mycobacterium tuberculosis*. *Mol. Microbiol.* **31**, 715–724.
13. Marras, S. A., Kramer, F. R., and Tyagi, S. (1999) Multiplex detection of single-nucleotide variations using molecular beacons. *Genet. Anal.* **14**, 151–156.
14. Park, S., Wong, M., Marras, S. A., et al. (2000) Rapid identification of *Candida dubliniensis* using a species-specific molecular beacon. *J. Clin. Microbiol.* **38**, 2829–2836.
15. Piatek, A. S., Tyagi, S., Pol, A. C., et al. (1998) Molecular beacon sequence analysis for detecting drug resistance in *Mycobacterium tuberculosis*. *Nat. Biotechnol.* **16**, 359–363.
16. Szuhai, K., Ouweland, J., Dirks, R., et al. (2001) Simultaneous A8344G heteroplasmy and mitochondrial DNA copy number quantification in myoclonus

epilepsy and ragged-red fibers (MERRF) syndrome by a multiplex molecular beacon based real-time fluorescence PCR. *Nucleic Acids Res.* **29,** E13.
17. Täpp, I., Malmberg, L., Rennel, E., Wik, M., and Syvänen, A. C. (2000) Homogeneous scoring of single-nucleotide polymorphisms: comparison of the 5'-nuclease TaqMan assay and molecular beacon probes. *Biotechniques* **28,** 732–738.
18. Vet, J. A., Majithia, A. R., Marras, S. A., et al. (1999) Multiplex detection of four pathogenic retroviruses using molecular beacons. *Proc. Natl. Acad. Sci. USA* **96,** 6394–6399.
19. Vet, J. A., Van der Rijt, B. J., and Blom, H. J. (2002) Molecular beacons: colorful analysis of nucleic acids. *Expert Rev. Mol. Diagn.* **2,** 77–86.
20. Wilson, I. G. (1997) Inhibition and facilitation of nucleic acid amplification. *Appl. Environ. Microbiol.* **63,** 3741–3751.
21. Zuker, M. (2003) Mfold web server for nucleic acid folding and hybridization prediction. *Nucleic Acids Res.* **31,** 3406–3415.
22. Bonnet, G., Tyagi, S., Libchaber, A., and Kramer, F. R. (1999) Thermodynamic basis of the enhanced specificity of structured DNA probes. *Proc. Natl. Acad. Sci. USA* **96,** 6171–6176.
23. Marras, S. A., Kramer, F. R., and Tyagi, S. (2002) Efficiencies of fluorescence resonance energy transfer and contact-mediated quenching in oligonucleotide probes. *Nucleic Acids Res.* **30,** E122.
24. Tyagi, S., Marras, S. A., and Kramer, F. R. (2000) Wavelength-shifting molecular beacons. *Nat. Biotechnol.* **18,** 1191–1196.

18

High-Throughput Production of Optimized Primers (Fimers) for Whole-Genome Direct Sequencing

Nikolai Polushin, Andrei Malykh, A. Michael Morocho, Alexei Slesarev, and Sergei Kozyavkin

Summary

Fimers are specifically modified primers selected to inhibit nonspecific interactions occurring in cycle sequencing. They are postsynthetically derived from 2'-methoxyoxalamido or 2'-succinimido precursor oligonucleotides by treatment with appropriate small molecular weight modifiers (a primary aliphatic amine or hydroxide anion). We describe design, synthesis, and purification of fimers, and their use in protocols for direct sequencing of genomic deoxyribonucleic acid (DNA). Protocol for isolation of microbial genomic DNA is also reported.

Key Words: Primer; cycle sequencing; nonspecific events; genomic DNA; modified primers; precursor strategy; modifiers; fimers.

1. Introduction

Efficiency of primer-based amplification reaction depends on the specificity of primer hybridization. Ideally, under the elevated temperatures used in a typical amplification technique, such as cycle sequencing, the primer should hybridize only to the target sequence. However, there is a relatively narrow range of conditions (temperature, ionic concentrations, and denaturing agents) for specific annealing of an oligonucleotide to its complementary target. Outside this range a number of nonspecific interactions might take place leading to different side products. Two particularly adverse nonspecific events occurring in the linear cycle-sequencing reaction are primer–dimer extension and nonspecific polymerase chain reaction (PCR). These artifacts progress exponentially because the products of one round of replication become templates for subsequent rounds. Therefore, even if the side products are formed in negli-

Fig. 1. PCR and primer–dimer artifacts occurring during directed sequencing of BAC DNA using regular primers. Gel image was generated on an ABI 377 automated sequencer. Human BAC B1092P21 DNA was sequenced directly with 96 primers using the 100-cycle protocol. Filled arrows indicate lanes with exponentially amplified side products. Unfilled arrows designate lanes with amplified primer–dimers.

gible quantities in the beginning, their exponential growth may bring them to high concentrations relative to the concentration of linearly growing Sanger fragments, thus compromising the quality of sequencing chromatograms (**Fig. 1**).

Another type of artifact occurs when the primers cannot effectively discriminate between their "perfect" and "nonperfect" binding sites. Poor discrimination results in polymorphous "unreadable" chromatograms. The outcome of cycle sequencing could also be compromised if a primer has two segments that are complementary to each other. In this case an internal hairpin structure might form, preventing the primer from annealing to its target. This problem often

High-Throughput Production of Fimers

takes place when one cannot select an ideal primer in the particular sequence window.

In relatively simple cycle-sequencing protocols (25–50 cycles, plasmid or bacterial artificial chromosome [BAC] templates) nonspecific interactions could generally be eliminated by careful choice of primer sequences and by precise adjustment of reaction conditions. However, the discrimination between specific and nonspecific hybridization becomes a great challenge on sequencing directly off genomic deoxyribonucleic acid (DNA) (up to 8 MB in size). Such complex templates typically have a large number of potential nonspecific priming sites, as well as perfect and/or nonperfect repeats. The large size of genomic DNA also dictates the use of longer primers (25–35 nt) and a higher number (>100) of replication rounds, further aggravating problems associated with nonspecific interactions. To successively implement the whole-genome direct sequencing protocols, primers themselves must be modified in such a way that their ability to prime DNA synthesis would be preserved, but the potency to serve as a template for DNA polymerase would be eliminated or, at least, inhibited. It is also critical to increase the specificity of primer annealing.

Over recent years we, among others *(1)*, have been involved in the development of modified primers that would satisfy the aforementioned requirements *(2,3)*. In a search for a favorable modification we employed methoxyoxalamido (MOX) and succinimido (SUC) precursor strategies, schematically shown in **Fig. 2** *(4,5)*. In these techniques the precursor oligonucleotides containing 2'-MOX/SUC substituted nucleosides are first assembled. The modified oligonucleotides are then postsynthetically reacted with appropriate small molecular weight modifiers (a primary aliphatic amine or hydroxide anion) and deprotected to form the final 2'-modified oligonucleotides, defined here as fimers. The precursor strategies allow one to easily and cost-effectively generate a large number of fimers that could be screened for particular applications. Over 30 different modifications were evaluated in advanced sequencing protocols. The ones that we found the most suitable are presented in **Fig. 2B**.

There are three key characteristics that make fimers superior to unmodified or 2'-*O*-methyl modified *(1)* primers *(3)*. First, we found that primer extension reaction is completely inhibited when DNA polymerase approaches the modified nucleoside within the template strand. Thus, even a single, appropriately placed, modification renders fimers practically nonreplicative, preventing exponential growth of any side products. Second, compared to unmodified oligonucleotides, fimers have an improved annealing specificity that also contributes to their ability to suppress nonspecific events. Third, fimers have a decreased melting temperature (2–3°C per modification) that becomes beneficial when primers longer than 25 nucleotides are to be used.

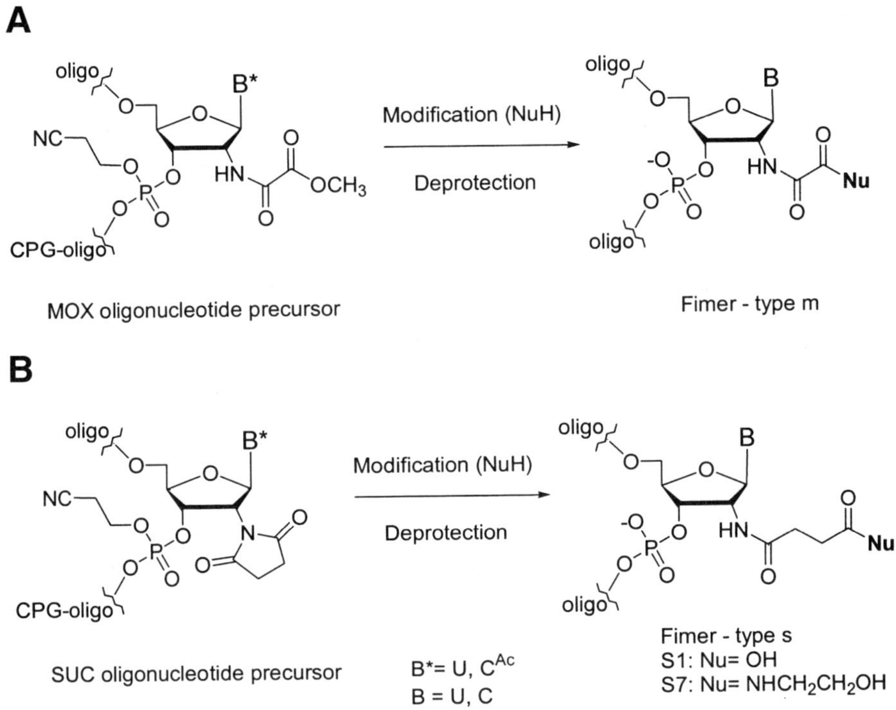

Fig. 2. Schematic representation of the 2'-MOX (**A**) and 2'-SUC (**B**) precursor strategies.

Figure 3 gives an example of fimer usage to increase yield and prevent PCR and primer–dimer formation in sequencing reactions. Placing the modification close to the 3'-end prevents both exponentially growing artifacts and does not compromise the efficiency of linear amplification. The optimal effect was found when the modification was made on the –5 position. **Figure 4** shows how fimer specificity was tuned by incorporating multiple modifications throughout the length of priming sequence. In this case a 40-mer unmodified primer cannot discriminate between perfect and nonperfect synthetic templates. Sequencing with such a primer results in a polymorphous chromatogram (**Fig. 4A**). However, by using a fimer having s1 type modifications at six different positions, we obtained the unambiguous sequencing chromatogram (**Fig. 4B**).

In this chapter we describe design, synthesis, and application of fimers selected for use in protocols for direct sequencing of genomic DNA. These modified primers are routinely used in our sequencing projects. The robustness of fimer technology was recently illustrated in the completion of the 1,694,969-nt sequence of the GC-rich *Methanopyrus kandleri* genome using a direct whole-genome sequencing approach *(6)*.

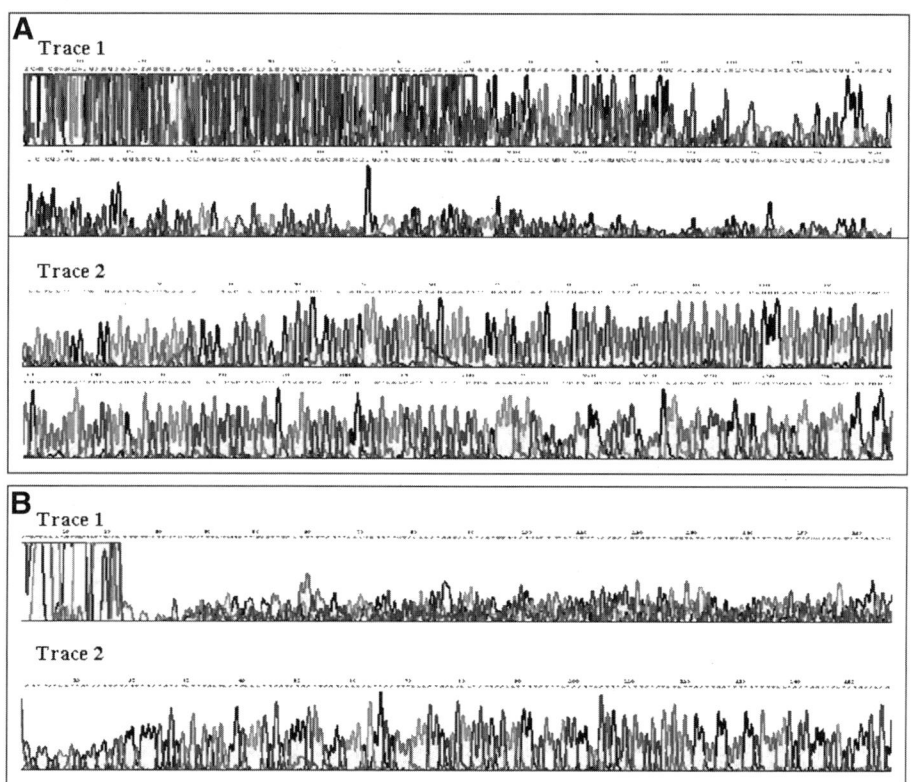

Fig. 3. Uses of fimers to increase yield and prevent PCR (**A**) and primer–dimer (**B**) formation in DNA sequencing reactions. Human BAC DNA (**A**) was sequenced using primer 5'-CCAGGATAGGAGGTATTTATG-3' (*Trace 1*) and the corresponding fimer having s1-type modification on the -5 position (*Trace 2*). *E. coli* genomic DNA (**B**) was sequenced with primer 5'-GATTTCGCGGGTGGCACCGTGGTGCA-3' (*Trace 1*) and the corresponding fimer having s7-type modification on the -7 and -15 positions (*Trace 2*).

2. Materials

2.1. Fimers Synthesis

1. ASM-800 DNA synthesizer (Biosset, Novosibirsk, Russia) or similar 50–200-nmol synthesis scale DNA synthesizer.
2. Regular DNA monomers: dA-CE phosphoramidite, dC^{Ac}-CE phosphoramidite, dG-CE phosphoramidite, dG-CE phosphoramidite (Transgenomic, Wayne, PA).
3. 2'-SUC monomers: 2'-SUC-2'-deoxyuridine and N^4-acetyl-2'-SUC-2'-deoxycytidine phosphoramidites (Fidelity Systems, Gaithersburg, MD).
4. Solvents: acetonitrile (ACN) (anhydrous), dichloromethane (VWR Scientific, Bridgeport, NJ).

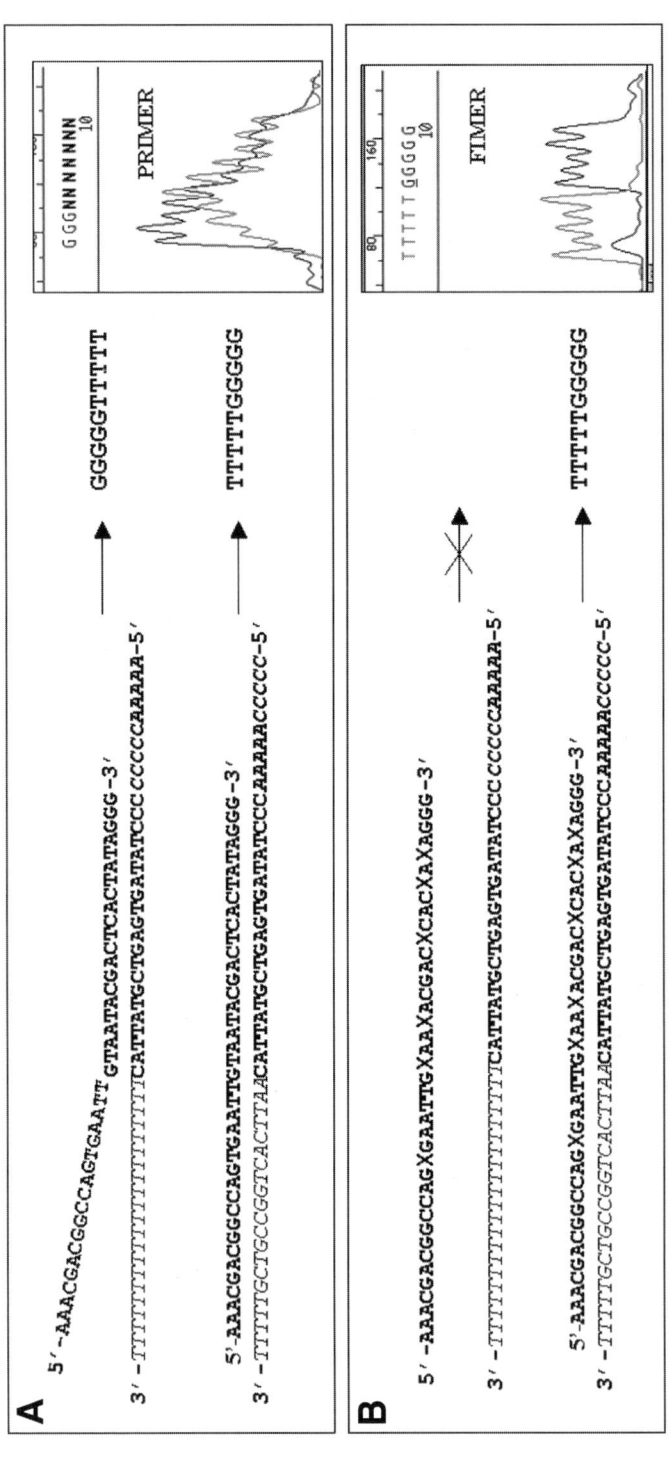

Fig. 4. Tuning fimer specificity by multiple SUC (s7-type) modifications. (**A**) 20 nucleotides at the 3′ end of the 40-mer primer have two priming sites in synthetic templates (bottom strands in **A, B**). The primer anneals to both templates to generate an ambiguous chromatogram. (**B**) Xs in the fimer sequence represent 2′-SUC modified nucleosides. Such fimer anneals only to one template to produce an unambiguous read.

5. Ancillary reagents for DNA synthesis (Transgenomic, Wayne, PA).
6. 0.25M Ethylthio tetrazole (ETT) in CH_3CN (Transgenomic, Wayne, PA).
7. 100-nmol-scale synthetic columns (Biosset, Novosibirsk, Russia).
8. 0.2-mL automation-friendly skirted 96-well PCR plate (Marsh Bio Products, Rochester, NY).
9. Easy Peel heat sealing foil (ABgene House, Epsom, UK).
10. Combi Thermo sealer (Advanced Biotechnologies, Epsom, UK).
11. 0.2M Aqueous NaOH.
12. Ethanolamine, redistilled (Aldrich, Milwaukee, WI).
13. Superfine Sephadex G-25 (Sigma, St. Louis, MO).
14. MultiScreen 45 µL column loader (Millipore, Bedford, MA).
15. MultiScreen-HV 96-well filtration plate, clear, styrene, 0.45 µm Durapore polyvinylidene difluoride (Millipore, Bedford, MA).
16. A 96-well microtitration plate, polystyrene: V-bottom, 220 µL/well (Marsh Bio Products, Rochester, NY).

2.2. Microbial DNA Isolation

1. Shaker–incubator.
2. Agarose gel electrophoresis equipment.
3. Reagents: mutanolysin, lysozyme, ribonuclease A (RNase A), proteinase K, Brij-58, deoxycholate, N-lauryl sarcosine, ammonium acetate, Tris-HCl, NaCl, Tris-saturated phenol (Sigma, St. Louis, MO).
4. Resuspension buffer (10 mM Tris-HCl, pH 7.6, 1M NaCl).
5. Lysis buffer (6 mM Tris-HCl, pH 7.6, 1M NaCl, 100 mM ethylenediaminetetraacetic acid (EDTA), 0.5% Brij-58, 0.2% deoxycholate, 0.5% N-lauryl sarcosine, 25 mg/mL lysozyme, 140 U/mL mutanolysin, 80 µg/mL RNase A. Dissolve Brij-58 at 55°C, and filter the solution.
6. Polypropylene 50-mL centrifuge tube (VWR Scientific, Bridgeport, NJ).

2.3. Direct Whole-Genome Sequencing

1. ABI 377 DNA sequencer (Applied BioSystems, Foster City, CA).
2. Thermocycler (MJ Research, Reno, NV).
3. Centrifuge with rotor for microtiter plates.
4. Vacuum centrifuge with rotor for mictotiter plates.
5. Big Dye Terminator Cycle Sequencing Ready Reaction kit (Applied Biosystems, Foster City, CA).
6. ThermoFidelase 2 (Fidelity Systems, Gaithersburg, MD).
7. Fimers (*see* **Subheading 3.2.** or custom order from Fidelity Systems).
8. 7-deaza-deoxyguanosine 5'-triphosphate (dGTP) (Roche, Indianapolis, IN).
9. 96- or 384-well polyethylene or polypropylene plates (MJ Research, Reno, NV).
10. Multichannel pipetters.
11. Superfine Sephadex G-25 (Sigma, St. Louis, MO).
12. Loading buffer: 85% formamide, 5 mM EDTA, pH 8.0, 10 mg/mL dextran blue.

3. Methods

The methods described below outline: (1) the general guidelines for fimer design, (2) fimer synthesis, (3) purification of fimer for cycle sequencing, (4) isolation of microbial genomic DNA, and (5) direct sequencing off genomic DNA.

3.1. General Guidelines for Fimer Design

Fimers should effectively direct sequencing reactions if the modification occurs at position 3 or farther from the 3'-end. For the optimal effect we recommend placing the modified nucleotide at either one of positions –5 through –7. This gives enough flexibility to use either one of the two commercially available U and C monomers. If neither of two nucleotides can be accommodated, move the priming frame accordingly. For fimers 20–29 nucleotides in length only one modification is necessary. For longer (30–40 nucleotides) fimers, place the second modification 10–15 nucleotides before the 5'-end. Contact Fidelity Systems for more details on fimer design.

3.2. Synthesis of Fimers

3.2.1. Assembling of 2'-SUC Precursor Oligonucleotides on the DNA Synthesizer

For automated solid-phase synthesis of 2'-SUC precursor oligonucleotides, we use Biosset ASM-700/800 DNA synthesizers (Novosibirsk, Russia). However, any other DNA synthesizer with at least one extra position for phosphoramidites could be used. Follow the manufacturer's instructions for the machine and reagent preparation. Use of dC^{Ac}-phosphoramidite instead of the Bz-protected version is required to prevent N^4-side product formation during derivatization/deprotection with ethanolamine *(7)*. Dilute 2'-SUC-2'-deoxyuridine and N^4-acetyl-2'-SUC-2'-deoxycytidine phosphoramidites in anhydrous ACN to $0.1M$ concentration. To ensure efficient (>97%) incorporation of modified phosphoramidites, use the manufacturer's recommended ribonucleic acid (RNA) synthetic cycles (10–15 min coupling time) and $0.25M$ ETT in ACN as activator. Use standard DNA synthetic cycles for assembling the unmodified portion of the fimer. For convenience ETT could be used instead of tetrazole throughout the whole synthesis. Typically a 50–200-nmol scale synthesis provides enough fimer quantity for any single sequencing project.

3.2.2. Fimers With s1-Type Modification

The 2'-SUC group is prone to hydrolysis. Thus, for use in a production setting where controlled pore glass (CPG)-bound precursor oligonucleotides are exposed to humidity for long periods of time (>12 h) before processing, and for the fimers that are not polyacrylamide gel electrophoresis (PAGE) purified, s1-type modification is preferable to implement. Modification of s1 type is a modification of choice in streamlined high-throughput fimer production for direct whole-genome sequencing.

1. Dry the synthetic columns (50–200-nmol scale) under medium vacuum for 10–15 min.
2. Transfer the CPG support from each column into the corresponding well of an assigned 96-well PCR plate (200 µL/well) (*see* **Notes 1** and **2**).
3. Add to each well 40 µL of 0.2M aqueous NaOH. Seal the plate with temperature-resistant sealing foil, vortex thoroughly, and centrifuge at 910g for 5 min.
4. Incubate for 15 min at 70°C.
5. Peel off the sealing foil, add to each well 100 µL of a freshly prepared 30% aqueous ethanolamine and seal the plate again. Vortex thoroughly.
6. Incubate for 20–25 min at 70°C.
7. Vortex the plate thoroughly, centrifuge at 910g for 5 min, and peel off the sealing foil.
8. *See* **Subheading 3.3.2.** and **Note 3**.

3.2.3. Fimers With s7-Type Modification

If the 2'-SUC precursor oligonucleotides are to be derivatized shortly after the synthesis, it is somewhat easier to generate fimers of s7 type modification because derivatization and deprotection are combined by treatment with ethanolamine (EA).

1. Dry the synthetic columns (50–200-nmol scale) under medium vacuum for 10–15 min.
2. Transfer the CPG support from each column into the corresponding well of an assigned 96-well PCR plate (200 µL/well) (*see* **Notes 1** and **2**).
3. Add to each well 140 µL of a freshly prepared 30% aqueous ethanolamine, seal the plate with temperature-resistant sealing foil, and vortex thoroughly.
4. Incubate for 25 min at 70°C.
5. Vortex the plate thoroughly, centrifuge at 910g for 5 min, and peel off the sealing foil.
6. Go to **Subheading 3.3.2.** (*see* **Note 3**).

3.3. Purification of Fimers

Generally, if there are no problems with automated synthesis, simple desalting is sufficient to prepare fimers of a quality adequate for cycle sequencing. More rigorous purification (PAGE or anion-exchange high-pressure [performance] liquid chromatography) is recommend only when fimers are used repeatedly or are longer than 40 nucleotides.

3.3.1. Preparation of a 96-Well Desalting Plate

1. Load a MultiScreen-HV 96-well filtration plate with superfine Sephadex G-25 using a MultiScreen 45 µL column loader in accordance with the manufacturer's instructions.

2. Add to each well 300 µL of dH$_2$O and let Sephadex soak for at least 2 h.
3. Place the MultiScreen plate with Sephadex on the top of a 96-well microtitration plate, and centrifuge at 910g for 5 min. There is no need to use a clean 96-well microtitration plate at this point because the only purpose of this operation is to remove excess water.

In high-throughput production settings it is more convenient to stock 96-well desalting plates. For storage do not remove water. Store the plates covered with lids and sealed with Parafilm in a sealed plastic bag at 4°C.

3.3.2. Desalting of Fimers

1. Place a 96-well desalting plate (*see* **Subheading 3.3.1.**) on the top of a clean 96-well microtitration plate (220 µL/well).
2. Load 18 µL of each crude fimer solution from the mother plate (*see* **Subheading 3.2.2** or **3.2.3**) onto Sephadex in the corresponding wells of the desalting plate. Try not to disturb the Sephadex layer.
3. Centrifuge at 910g for 5 min. Remove the desalting plate. Visually check sample volumes in the 96-well microtitration plate. The volumes should not differ considerably.
4. Add to each well 100 µL of dH$_2$O to make stock fimer solutions.
5. Measure fimer concentrations in three to five randomly chosen wells. The concentration should be in the range of 0.6–3.0 A$_{260}$/mL.
6. Store the plate of stock fimer solutions sealed with a sealing film at 4°C.

3.4. Microbial Genomic DNA Isolation

For genomic DNA isolation a Qiagen (Valencia, CA) genomic DNA isolation kit can be used. The protocol described as follows, however, works more robustly in our hands.

1. Inoculate 200 mL of corresponding media (Luria-Bertani, MRS, etc.) with bacterial cells. Grow for approx 16–17 h (A$_{600}$ should be around 0.4; if more—use fewer cells or increase the volume).
2. Harvest bacteria by centrifugation at 5000–7000g for 15 min at 4°C.
3. Wash once with 200 mL of cold resuspension buffer.
4. Resuspend bacteria in 5 mL of resuspension buffer supplemented with 25 mg/mL lysozyme. Add 5 mL of lysis buffer. Incubate 1 h at 37°C (some viscosity will be seen already after 30 min; do not shake and handle with care).
5. Add 10 mL of proteinase K solution (1 mg/mL), mix by gentle swirling. Incubate overnight (12–14 h) at 50°C.
6. Add 5 mg of proteinase K powder, mix gently, incubate for an additional 3 h at 50°C. (Do *not* expect to see transparent solution, although it is possible).
7. Extract at least twice with 20 mL of Tris-saturated phenol–chloroform, that is, until you see no interphase and the water phase is crystal clear, and then extract once with 20 mL of chloroform. Mix DNA solution and solvents by gentle inver-

sion of the tubes for approx 1 min (until it becomes homogeneous). For extraction use polypropylene centrifuge tubes. To separate phases centrifuge at 5000g for 20 min at room temperature. Use clean glass pipets for transferring water phase and solvents. The final volume of the DNA solution should be around 15 mL.
8. Treat with RNase for 1 h at 37°C. The final concentration of RNase should be 40 µg/mL (*see* **Note 4**).
9. Extract once with phenol–chloroform and once with chloroform.
10. Add 3 mL of 10*M* ammonium acetate solution. Precipitate with 40–50 mL of EtOH overnight at –20°C, and wash twice with 15 mL of 70% EtOH.
11. Dry DNA pellet and dissolve in 5 mL of 10 m*M* Tris-HCl, pH 8.0.
12. Measure A_{260} and check the quality by agarose gel.

3.5. Whole-Genome Direct Sequencing

The following protocol for direct genomic DNA sequencing was successfully tested in thousands of reactions with at least 50 different microorganisms.

3.5.1. Cycle-Sequencing Protocol

The amount of each component needed for one sequencing reaction is given in **Table 1**. One can run the reaction in larger volumes (15 and 20 µL), but we do not recommend decreasing the Big Dyes concentration. Dilution of Big Dyes results in shorter reads and lower signal.

Because the volume of ThermoFidelase 2 (*see* **Note 5**) per reaction is very small, a single mixture for at least 3 sequencing reactions should be prepared. The reactions can be done in individual 0.2-mL PCR tubes, 8-tube strips, or 48-, 96-, or 384-well plates. The amount of genomic DNA used per reaction for each particular microorganism may vary and depends on the size of the genome. Our most sensitive protocol (400 cycles) allows sequencing from as low as 300 ng of *Escherichia coli* genomic DNA (4.6 MB) with fimer designed for a single copy gene. For high-quality reads of up to 900 bp we use 3 µg DNA and run the reaction for 200 cycles. With universal fimer designed for 16 S recombinant DNA we sequenced from as low as 100 ng of *E. coli* DNA. This fimer can be used for bacteria typing.

To compensate for pipetting errors we recommend adding 10% more reagents. An example of a 96-reaction setup is given in **Table 2**. This example shows the calculation based on a 10-µL vol reaction.

1. Label the tube that is going to be used for reaction setup.
2. Add the desired amount of template to this tube based on calculated DNA concentration.
3. Add ThermoFidelase 2. Mix by pipetting.
4. Add Dye Terminator Ready Reaction mix.
5. Add a calculated amount of water.

Table 1
1-Reaction Setup

Reagent	Quantity per reaction
Genomic DNA	0.1–5 μg
ThermoFidelase 2	0.1 μL
1 mM 7-deaza-dGTP	0.1 μL
Dye Terminator Mix	1.0 μL
Fimer	4.0 μL
dH$_2$O	q.s.
Total volume	10.0 μL

Table 2
96-Reaction Setup

Reagent	Quantity
Genomic DNA (1μg/rxn)	106 μg
ThermoFidelase 2	10.6 μL
1 mM 7-deaza-dGTP	10.6 μL
Dye Terminator Mix	424.0 μL
dH$_2$O	q.s.
Volume	848.0 μL

6. Mix well by pipetting. Total volume of premix should be equal to 9 μL × (number of reactions).
7. Dispense 9 μL per tube/well.
8. Add fimers (1 μL of each stock solution, *see* **Subheading 3.3.2.**) using a multichannel pipet. Mix by pipetting.
9. Close tubes or seal plate.
10. Spin tubes or plate briefly.
11. Place tubes or plate in a thermocycler.
12. Use the following cycling conditions:
    a. Heat plate at 95°C for 2 min.
    b. Then heat at 95°C for 5 s (denaturation).
    c. Heat at 55°C for 20 s (annealing).
    d. Heat at 60°C for 2 min (extension).
    e. 200 cycles total, then 4°C (if 400 cycles, use 1 min extension).
    f. Add 10 μL 20 mM EDTA to each well.

3.5.2. Reaction Cleanup Using Millipore MultiScreen 96-Well Filtration Plates

Prepare Millipore MultiScreen 96-well filtration plates with Sephadex G-50 superfine as described in **Subheading 3.3.1.**

1. Place a 96-well desalting plate (*see* **Subheading 3.3.1.**) on the top of a clean 96-well microtitration plate (220-µL/well).
2. Transfer the entire 20 µL of each reaction mixture onto Sephadex of a corresponding well of the desalting plate. Try not to disturb the Sephadex layer.
3. Centrifuge at 910g for 5 min. Remove the desalting plate. Visually check sample volumes in the 96-well microtitration plate. The volumes should not differ considerably.
4. Dry samples for 20–30 min in a vacuum centrifuge.
5. Resuspend each pellet in 2 µL of loading buffer (85% formamide, 5 mM EDTA, pH 8.0, 10 mg/mL dextran blue).
6. Load 1.5 µL of each sample per lane on sequencing gel.
7. Run sequencing gel on ABI 377 DNA sequencer as recommended by the manufacturer.

4. Notes

1. If only a few fimers are needed for the project, the derivatization/deprotection could be done in 1.7-mL centrifuge tubes. In this case we found it more convenient to use neat ethanolamine instead of an aqueous solution. After deprotection dilute the reaction mixture with water to 1 mL and desalt on a NAP-10 Sephadex G-25 column (Amersham Pharmacia Biotech AB, Uppsala, Sweden) following the manufacturer's instructions.
2. Biosset's columns exactly fit into wells of a 96-well PCR plate. This significantly simplifies fimer production and decreases the possibility of mistakes.
3. The mother plate with crude fimer solution could be stored at –70°C for at least 3 mo.
4. If at this point agarose gel shows no indication of RNA presence omit **steps 8** and **9**. For unclear reasons one RNase treatment is not enough for some strains.
5. ThermoFidelase 2 is a thermostable DNA unlinking enzyme (type IB DNA topoisomerase). During initial heating of the reaction mixtures it removes topological linkage between two strands of circular DNA (plasmid, BAC, bacterial genomic DNA). The result is the increased availability of targets for primer annealing and higher yield of cycle sequencing reactions.

Acknowledgments

This work was supported in part by grants from the Department of Energy and the National Institutes of Health (to S. K.).

References

1. Stump, M.D., Cherry, J. L., and Weiss, R. B. (1999) The use of modified primers to eliminate cycle sequencing artifacts. *Nucleic Acids Res.* **27,** 4642–4648.
2. Polushin, N., Malykh, O., Pavlov, N., Malykh, A., and Kozyavkin, S. (1999) Chemically modified primers for advanced DNA sequencing. *Microb. Comp. Genom.* **4,** 107, P-0483.
3. Polushin, N. Malykh, A., Malykh, O., et al. (2001) 2'-Modified oligonucleotides from methoxyoxalamido and succinimido precursors: synthesis, properties and applications. *Nucleosides Nucleotides Nucleic Acids* **20,** 507–514.
4. Polushin, N. N. (1996) Synthesis of functionally modified oligonucleotides from methoxyoxalamido precursors. *Tetrahedron Lett* **37,** 3231–3234.
5. Polushin, N. N. (1999) Methoxyoxalamido and succinimido precursors for nucleophile addition to nucleosides, nucleotides, and oligonucleotides. *US Patent 5902879*.
6. Slesarev, A. I., Mezhevaya, K. V., Makarova, K. S., et al. (2002) The complete genome of hyperthermophile Methanopyrus kandleri AV19 and monophyly of archaeal methanogens. *Proc. Natl. Acad. Sci. USA* **99,** 4644–4649.
7. Polushin, N., Morocho, A. M., Chen, B. C., and Cohen, J. S. (1994) On the rapid deprotection of synthetic oligonucleotides and analogs. *Nucleic Acids Res.* **22,** 639–645.

19

Nonenzymatic Template-Directed RNA Synthesis

Marcus Hey and Michael Göbel

Summary
The aim of this chapter is to provide reliable protocols for the study of nonenzymatic oligomerization reactions of ribonucleic acid and its analogs. Traditional radioactive labels are replaced by fluorescent dyes. Product analysis is either based on reversed-phase high-pressure (performance) liquid chromatography or on gel electrophoresis with on-line detection applying a deoxyribonucleic acid sequencer. Three examples of primer extensions are given, one of which demonstrates how the reaction of several primers may be monitored simultaneously.

Key Words: Acridine; Cy5; dye; electrophoresis; fluorescence; HPLC; nonradioactive; oligonucleotide; oligomerization; prebiotic; primer; RNA; sequencer; self-replication; template.

1. Introduction

How did life emerge from nonliving matter? This fascinating topic has led to many speculations. The scientific challenge is to transform vague assumptions into clear questions that may be answered unequivocally by experiments. The ribonucleic acid (RNA) world hypothesis presumes the existence of an early stage of life characterized by a multiple use of RNA as genetic material and as catalyst for all necessary metabolic steps. The problem of how nucleic acids may be copied in the absence of proteins has induced a wealth of creative chemical and biochemical studies *(1–3)*.

After the discovery of catalytic RNAs by Cech *(4)* and Altman *(5)*, several impressive examples for template-directed extensions of RNA primers by ribozymes have been reported by Szostak et al. *(6–8)*, Bartel et al. *(9,10)*, and Joyce et al. *(11,12)*. None of them, however, includes self-replication. More simplified replicating systems may have existed before ribozymes, and currently alternative approaches to copy short oligonucleotides are under extensive investigation.

An option to achieve the replication of an existing nucleic acid is the nonenzymatic template-directed ligation of RNA or RNA analogs *(13,14)*. To answer the question why ribose and no other sugar was chosen by nature to build up the backbone of nucleic acids, Eschenmoser and coworkers systematically investigated structural alternatives of RNA *(15–17)*. Among these analogs, the most promising candidate with respect to duplex stability, fidelity of base recognition, and potential of prebiotic formation is pyranosyl-RNA (p-RNA), the ribopyranosyl isomer of natural RNA *(18)*. In experiments for nonenzymatic replication, it was shown that p-RNA is able to undergo template directed ligation via the 2',3'-cyclic phosphate. It was possible to link two p-RNA homoC tetramer-2'-phosphates by incubating with a p-RNA homoG octamer and a carbodiimide as condensing agent *(18)*. Later Eschenmoser succeeded in copying the template pr($4'$GGGCGGGC$^{2'}$) into its complement by ligation of two pr($4'$GCCC)-2',3'-cyclic phosphates *(19)*. Furthermore, they produced longer oligomers from the tetramer pr($4'$ATCG)-2',3'-cyclic phosphate. The partially self-complementary sequences led to long chain aggregates thus facilitating the formation of covalent bonds *(20)*. Each ligation step selectively resulted in 2',4'-phosphodiester bonds. Similar attempts with natural RNAs were not that efficient and formed unnatural 2',5'-linked RNA almost exclusively *(19)*.

In recent experiments with threofuranosyl nucleic acid, a tetrose analog of RNA *(21)*, Eschenmoser demonstrated an improved efficiency of ligation when adenine is substituted by 2,6-diaminopurine ("D" stands for the corresponding nucleoside). This is in accordance with a higher stability of duplexes containing D/T basepairs instead of A/T *(22)*.

The first successful self-replication of nucleic acids was reported by von Kiedrowski and colleagues *(23)*. Using the deoxyribonucleic acid (DNA) hexamer $^{3'}$GGCGCC$^{5'}$ as a template, they were able to ligate the trimers $^{5'}$CCGp$^{3'}$ and $^{5'}$CGG$^{3'}$ in the presence of carbodiimide. The product is identical with the template, thus leading to an autocatalytic process in accordance with theoretical models ("square root law of autocatalysis") *(24; see also* **ref. 25***)*.

In these experiments short oligomers were used as substrates and a single phosphodiester bond was formed in the ligation step. In contrast, prebiotic RNA synthesis probably started from monomers, and multiple phosphodiester bonds had to be formed. To enable self-reproduction and evolution, a source of monomeric building blocks should have been available. Their possible prebiotic origin is discussed elsewhere *(15,26)*. Furthermore, a mechanism for the assembly and copying of chains must have existed as well.

The pioneer who started the investigation of template-directed nonenzymatic RNA oligomerization is L. E. Orgel *(27)*. His group dominated the field for years and made important experimental contributions such as nucleotide activation by 2-methyl-phosphorimidazolides. These building blocks form 3',5'-linked RNA chains preferentially. Orgel and coworkers also introduced an experimental setup combining primer and template in a single DNA hairpin. After 5'-labeling with ^{32}P and incubation with monomers, primer extension at the 3'-end was analyzed by gel electrophoresis and autoradiography *(28)*. Because primer and template are attached covalently, experiments are possible at very low template concentrations. Aggregation phenomena, for example, of guanosine-rich strands are thus avoided.

Göbel and coworkers introduced a nonradioactive approach based on acridine-labeled oligonucleotide primers and high-pressure (performance) liquid chromatography (HPLC) analysis *(29)*. The dye has a threefold function. It stabilizes the primer–template duplexes and allows a sensitive quantification of the labeled chains by ultraviolet (UV) or fluorescence detectors. In addition, the lipophilicity of the dye retards the primer and its elongation products on reverse-phase columns. The nonlabeled template, the monomers, and buffer components are all eluted in front *(29,30)* (*see* **Scheme 1** and **Fig. 1**).

Both analytical setups lead to comparable results. The condensation of primer and activated mononucleotide preferentially forms 3',5'-bonds. Reaction rates increase with the temperature (0–37°C) and the concentration of imidazolides. The oligomerization with 2-MeImpG on a homoC-template is 10 times more efficient with primers carrying ribo- instead of deoxyribonucleotides at the 3'-end *(28)*. Chain extension runs best when the primer–template duplex adopts an A-type conformation *(29,30)*. For optimal rates and yields, therefore, primer or template (or both) should be RNA. Templates rich in guanosine tend to aggregate in the presence of alkali metal ions. In consequence, primer extension with 2-MeImpC on a homoG-template is inefficient at high sodium or potassium concentrations. These salts, on the other hand, have no beneficial effects. Omitting them allows one to copy G rich templates in fair yield *(29)*.

Another limitation of nonenzymatic RNA oligomerizations results from the low stability of A/U basepairs. Replacing adenine by 2,6-diaminopurine, as previously mentioned, increases the number of hydrogen bonds and leads to the stronger basepair D/U or D/T, respectively. Orgel studied the incorporation of 2-MeImpU on templates containing either A or D. Göbel investigated the complementary case, the template controlled primer extension by 2-MeImpA or 2-MeImpD. In both experiments, yields are largely improved by replacing A with D *(31,32)*.

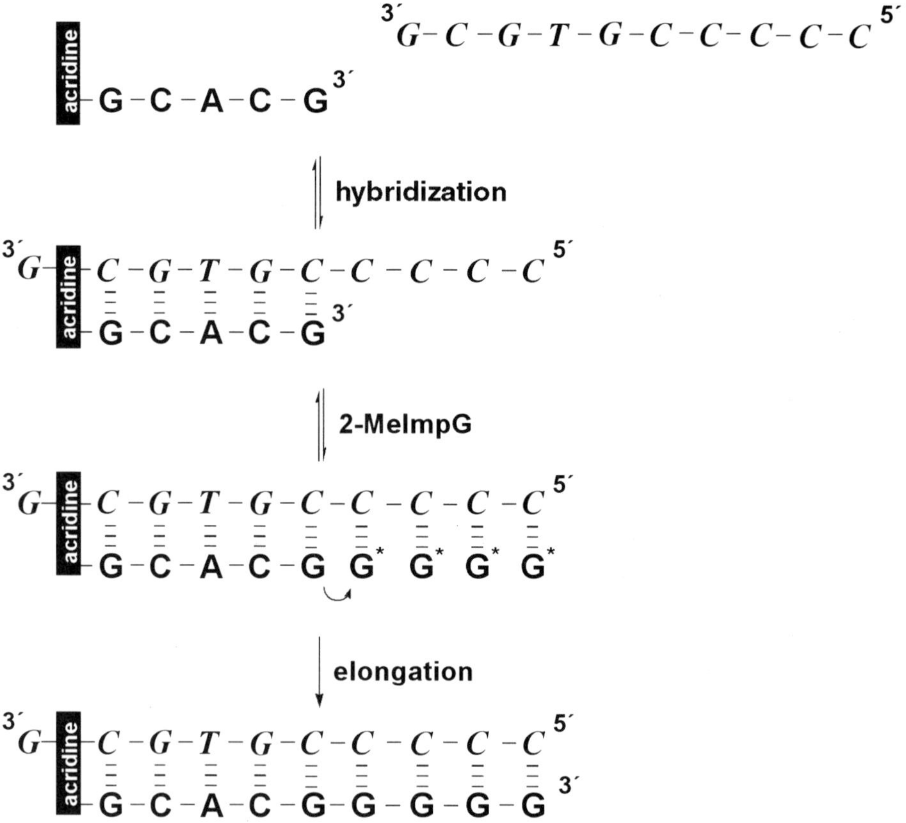

Scheme 1

```
                              3´G-C-G-T-G-C-C-C-C-C 5´
       [acridine]-G-C-A-C-G 3´
                              ↓ hybridization
   3´G-[acridine]-C-G-T-G-C-C-C-C-C 5´
                  ‖ ‖ ‖ ‖ ‖
                  G-C-A-C-G 3´
                              ↓ 2-MeImpG
   3´G-[acridine]-C-G-T-G-C-C-C-C-C 5´
                  ‖ ‖ ‖ ‖ ‖
                  G-C-A-C-G  G* G* G* G*
                             ↶
                              ↓ elongation
   3´G-[acridine]-C-G-T-G-C-C-C-C-C 5´
                  ‖ ‖ ‖ ‖ ‖ ‖ ‖ ‖ ‖
                  G-C-A-C-G-G-G-G-G 3´
```

Scheme 1. RNA is symbolized by bold, DNA by italic letters. The acridine dye is represented by a black rectangle.

A third technique to analyze nonenzymatic primer extension was reported recently *(32,33)*. It makes use of RNA primers, labeled with the fluorescent dye Cy5 at the 5´-end. The primer and the chain-extended products are separated and detected by gel electrophoresis applying a DNA sequencer. This method combines excellent band separation with high sample throughput and the sensitivity of radioisotope labeling (*see* **Scheme 2** and **Fig. 2**). The reaction of two or more primers can be analyzed simultaneously, if they sufficiently differ in length *(33)* (*see* **Scheme 3** and **Fig. 3**).

Nielsen and Orgel studied the potential of peptide nucleic acids (PNA) to act as templates for the oligomerization of ribonucleotides. For instance, they incubated a PNA-C_{10} oligomer with 2-MeImpG. The reaction forms the corresponding 3´,5´-linked RNA-G_{10} oligomer *(34)*. However, yield and rates are inferior to those observed in experiments using RNA or DNA templates. It is

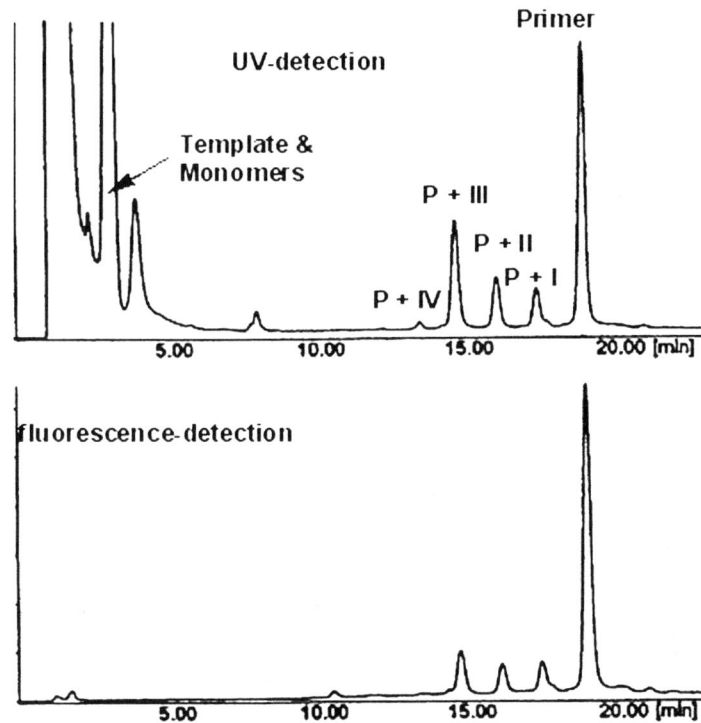

Fig. 1. HPLC analysis of the primer extension experiment depicted in **Scheme 1**.

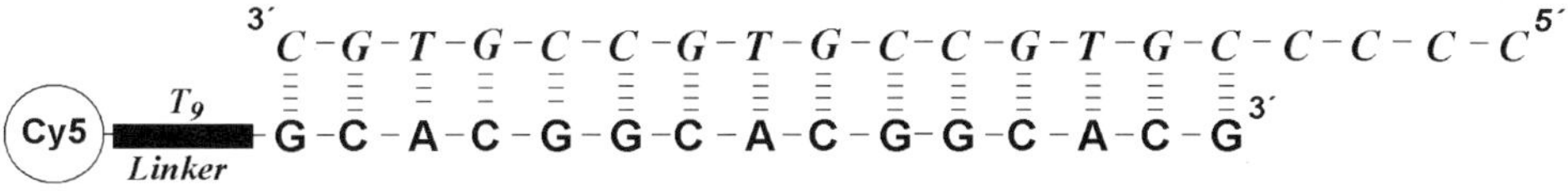

Scheme 2. RNA is symbolized by bold, DNA by italic letters.

also possible to assemble PNA oligomers from PNA–dimers on a DNA template with N-(3-dimethylaminopropyl)-N'-ethyl-carbodimide serving as condensing agent (35).

Herdewijn et al. have introduced two novel classes of nucleic acid analogs, the hexitol nucleic acids (HNAs) (36,37) and altritol nucleic acids (ANAs) (38). Both hybridize with RNA and DNA forming A-type helices. The duplex stabilities decrease in the order HNA/HNA > HNA/RNA > HNA/DNA and ANA/ANA > ANA/RNA > ANA/DNA (38,39). Herdewijn and Orgel and colleagues observed efficient oligomerization of 2-MeImpG in the presence of HNA-C_{10} (40,41) or ANA-C_{10} (42). HNA and ANA are superior templates

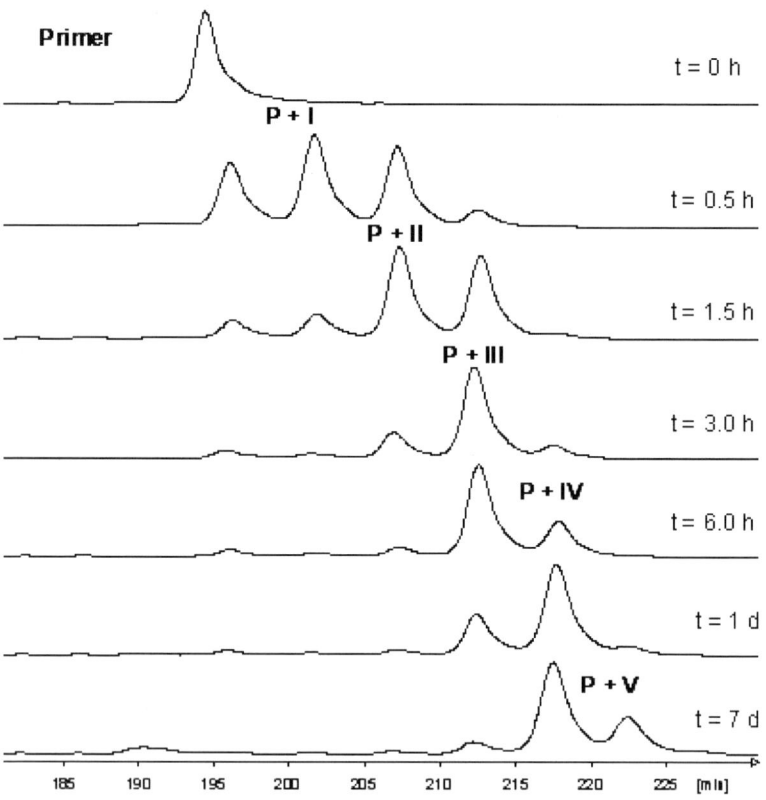

Fig. 2. Analysis of the primer extension experiment depicted in **Scheme 2** applying a DNA sequencer.

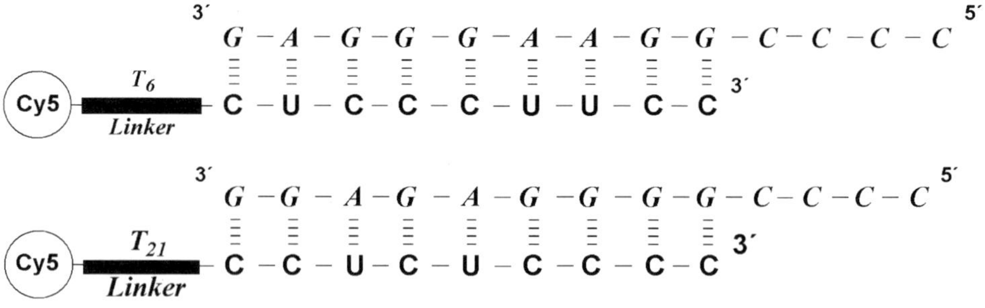

Scheme 3. RNA is symbolized by **bold**, DNA by *italic* letters.

compared to RNA. On the other hand, it is difficult to polymerize activated HNA or ANA monomers on any kind of template. Tetramers are the longest detectable oligomers *(42)*.

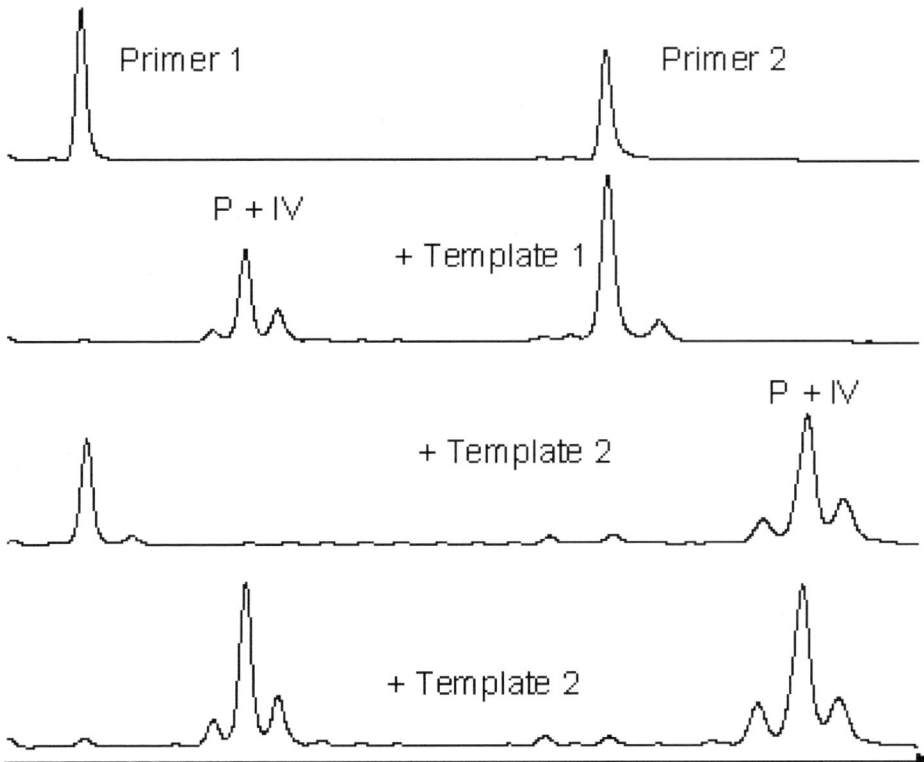

Fig. 3. Simultaneous detection of two independent template controlled primer elongations. *Line 1*: Mixture of primers 1 and 2. *Line 2*: Primer extension in the presence of template 1. *Line 3*: Primer exension in the presence of template 2. *Line 4*: Primer extension in the presence of both templates.

1.1. Practical Examples

Scheme 1 demonstrates the extension of an acridine-labeled RNA primer. When hybridized with a DNA template, the single-stranded part forms a double helical aggregate with the monomer 2-MeImpG. Covalent chain elongation is then monitored by HPLC and UV- or fluorescence detection (**Fig. 1**). The latter method allows for improvement of the sensitivity and the selectivity of the analyis.

Primer extension by four guanosine residues again is shown in **Scheme 2**. However, here the primer is labeled with Cy5. This cyanine dye allows product analysis by gel electrophoresis in a DNA sequencer with superior sensitivity (*see* **Fig. 2**). After several days, the incorporation of a noncoded fifth guanosine unit becomes visible (P + V). Reaction rates do not depend on the analytical method applied (HPLC vs electrophoresis).

Scheme 3 shows how the extension of two primers, differing in sequence and linker length, can be observed simultaneously. When the mixture of primer 1 and 2 is incubated with template 1 and 2-MeImpG, only primer 1 reacts. Conversely, in the presence of template 2, selective extension of primer 2 is observed. Finally, both primers are elongated after incubation with templates 1 and 2, and 2-MeImpG (*see* **Fig. 3**). The band resolution of the sequencer would easily allow a monitor of a third primer as well.

2. Materials
2.1. General
1. Water bath: Lauda RC 6 (±0.05°C).
2. Oligomerization-buffers: A, 0.5M Tris-HCl, pH 7.7, stored at −18°C; B, 0.5M HEPES NaOH, pH 7.7, stored at −18°C (*see* **Note 1**).
3. $MgCl_2$ stock solution: 2.0M (or 2.0M NaCl, 2.0M LiCl, 2.0M KCl, if required) (*see* **Note 2**).
4. Template stock solution: 1 mM in diethyl pyrocarbonate (DEPC)-treated water, stored at −18°C.
5. Primer stock solution: 1 mM in DEPC-treated water, stored at −18°C (*see* **Notes 3** and **4**).
6. Phosphorimidazolide stock solution: 0.5M in DEPC-treated water, freshly prepared (*see* **Note 5**).

2.2. HPLC
1. Stopping buffer A: 8.0M urea in DEPC-treated water.
2. HPLC system: Jasco HPLC pump PU 980, UV-detector Jasco UV 975, fluorescence detector Jasco FP 920, column Merck LiChrospher 100 RP-18 end-capped, 125 × 4 mm; 5 µm running buffers. Buffer A: acetonitrile (ACN)/triethyammonium acetate 0.05M, pH 6.5, 60:40; Buffer B: ACN/triethyammonium acetate (0.05M, pH 6.5) 10:90; software Borwin (JMBS development. v 1.22) (*see* **Note 6**).

2.3. DNA Sequencer
1. Stopping buffer B: formamide.
2. Stock solution TBE buffer (10X): 1M Tris, 0.8M boric acid, 10 mM ethylenediaminetetraacetic acid, pH 8.5 (*see* **Note 7**).
3. Acrylamide/*bis*-acrylamide (19:1) stock solution: 40% in water (Roth) stored at 4°C.
4. Acrylamide solution: 16 % acrylamide/*bis*-acrylamide stock solution, 7M urea, 1X TBE stock solution.
5. Ammonium persulfate (APS): 10% (m/v), stored at −18°C.
6. N,N,N',N'-tetramethylethylenediamine (TEMED).
7. Loading buffer: 5.0 mg/mL dextran blue in formamide, stored at −18°C.

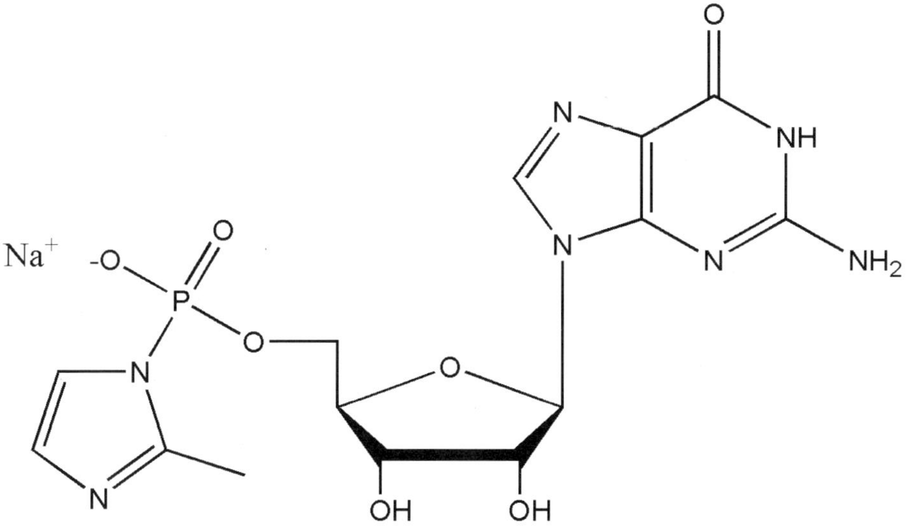

Fig. 4. Structure of guanosine phosphor imidazolide.

8. Running buffer: 0.5X TBE (50 mL 10X TBE in 950 mL Millipore water).
9. Sequencer: ALFexpress (AP Biotech) with Alf Win Instrument Control, (AP Biotech, v 2.00.15a) and AlleleLinks (AP Biotech, v 1.0).

3. Methods

Templates and primers were prepared by standard phosphoramidite chemistry on solid support. Acridine–phosphoramidite was prepared as described *(30)*; Cy5-phosphoramidite was obtained from Glen Research.

3.1. Preparation of Activated Monomers

1.5 mmol nucleotide-5'-dihydrogen phosphate and 15 mmol 2-methyl-1*H*-imidazole (10 equivalents) were dissolved under argon in 10 mL dry dimethyl sulfoxide, 10 mL dry dimethyl formamide, and 4.5 mmol NEt_3 (3 equivalents); 3.3 mmol PPh_3 (2.2 equivalents) were added, and the solution was stirred until all components were completely dissolved. After addition of 4.5 mmol 2,2'-bipyridine 1,1' disulfide (3 equivalents) the yellow solution was stirred for 2 h at room temperature.

This mixture was poured into a solution of 9 mmol $NaClO_4$ (6 equivalents; or $LiClO_4$, respectively; *see* **Note 5**) in 450 mL acetone, 300 mL Et_2O, and 35 mL NEt_3. The precipitated phosphorimidazolide salt was isolated by centrifugation, washed with acetone and Et_2O, and dried *in vacuo*. (*Caution:* Explosive solvent fumes. Use tightly closed centrifugation tubes.) (*See* **Fig. 4**.)

**Table 1
Setup of Oligemerization Experiments Analyzed by HPLC**

	Conc.	Vol (μL)	End conc.
Buffer	0.5M	15.0	0.25M
MgCl$_2$	2.0M	3.0	0.20M
Primer	1.0 mM	3.0	100 μM
Template	1.0 mM	6.0	200 μM
2-MeImp	0.5M	3	50 mM

3.2. Oligomerization Experiments Using HPLC Analysis of Acridine-Labeled Primers (vol 30 μL)

(*See* **Table 1** and **Notes 1–3**.)

1. Mix oligomerization buffer, water, primer, and template together in 1.5-mL Eppendorf tubes, vortex, and centrifuge.
2. Heat for 1 min at 90°C, and store at 10°C for 15 min. Add freshly prepared phosphorimidazolide solution (start of reaction time), vortex and centrifuge again, and store at 10°C. After 1, 2, 3, 7, and 14 d, dilute 3 μL of this solution in 10 μL stopping buffer A.
3. Heat for 1 min at 90°C, and store at −18°C.
4. HPLC running conditions: Linear gradient from 5–36% in running buffer A for 22 min. Flow rate 1 mL/min. Inject 10 μL of the stopped reaction mixture.
5. UV-detection at 260 nm, fluorescence excitation λ_{Ex} at 355 nm, and detection λ_{Em} at 450 nm (*see* **Note 6**).
6. The determination of primer and product concentration from the integrated chromatograms is complicated by the fact that the extinction coefficients of the products increase with the chain length. Therefore the integrals were multiplied by correction factors obtained from division of the primer coefficient (ε_P) by the corresponding coefficient (ε_{P+N}) of the extended products. To estimate ε_{260} of the primers (ε_p) and the elongation products (ε_{p+1}, ε_{p+2}, ε_{p+3}, and ε_{p+4}), the sum of the corresponding extinction coefficients of the monomers was multiplied by the usual factor of 0.9 for single-stranded DNA (acridine 97400, G 11300, C 7400, A 15300, T 9000). To obtain the product's ratio in percentage, each corrected integral was divided by the sum of all corrected areas (primer + products).

3.3. Oligomerization Experiments Using a DNA Sequencer and Cy5-Labeled Primers (vol 30 μL)

(*See* **Table 2** and **Notes 1, 2,** and **4**.)

1. Mix oligomerization buffer, water, primer, and template together in a 1.5-mL Eppendorf tube, vortex, and centrifuge.

**Table 2
Setup of Oligeomerization**

	Conc.	Vol (μL)	End Conc.
Buffer	0.5M	15.0	0.25M
MgCl$_2$	2.0M	3.0	0.20M
Water		6.0	
Primer	1.0 mM	0.9	30 μM
Template	1.0 mM	2.1	70 μM
2-MeImp	0.5M	3	50 mM

2. Heat 1 min at 90°C, and store at 10°C for 15 min. Later, add freshly prepared phosphorimidazolide solution (start of reaction time), vortex, and centrifuge again; store at 10°C. After 1, 2, 3, 7, and 14 d dilute 1 μL of this solution in 99 μL stopping buffer B.
3. Store at –18°C. Dilute 0.5 μL of stopped reaction mixture in 10 μL loading buffer.
4. For gel preparation mix 60 mL acrylamide solution (filtrated through membrane filter) with 400 μL APS and 40 μL TEMED. Pour the solution into an ALFexpress gel cassette with spacer thickness of 0.5 mm.
5. Use a comb with 40 pockets. Allow the gel to polymerize for at least 2 h. Into each pocket pipet 6 μL loading solution (*see* **Notes 8** and **9**).
6. ALFexpress running conditions: 1500 V, 60 mA, 25 W, 55°C; data was collected every 2 s by ALF Win Instrument Control (AP Biotech, v 2.00.15a).
7. For quantification, AlleleLinks 1.0 (AP Biotech) was used followed for numerical analysis. To obtain each product's ratio in percentage, the corresponding integral was divided by the sum of all integrals.

4. Notes

1. Using HEPES or Tris buffer does not change the results.
2. The use of Mg^{2+} is essential.
3. The resolution in HPLC decreases with increased primer length. So the primer must be as short as possible.
4. Cy5-labeled primers: Use a linker of at least 4 Ts between the primer sequence and the dye because very short oligomers are not fully resolved.
5. Li$^+$-imidazolides are less soluble than Na$^+$-imidazolides, but both yielded similar results. Li$^+$-imidazolides are advantageous with templates containing several Gs.
6. When gradients are used for HPLC analysis, fluorescence integrals cannot be used directly to calculate the relative concentration of the elongation products. Emission strongly depends on the changing water content of the solvent. In contrast, product quantification via fluorescence is advantageous under isocratic HPLC conditions.
7. The pH of the running buffer was not adjusted.

8. Remove the comb after 30 min of polymerization.
9. Before loading your probes, clean the pockets thoroughly with a syringe.

References

1. Gestland, R. F. and Atkins, J. F. (1993) *The RNA World.* Cold Spring Harbor Laboratory Press, Cold Spring Harbor, NY.
2. Bartel, D. P. and Unrau, P. J. (1999) Constructing an RNA world. *Trends Cell Biol.* **9,** M9–M13.
3. Joyce, G. F. (2002) The antiquity of RNA-based evolution. *Nature* **418,** 214–221.
4. Cech, T. R. (1990) Self-cleaving and enzymic activity of an intervening RNA sequence from Tetrahymena. *Angew. Chem. Int. Ed. Engl.* **102,** 745–755.
5. Altmann, S. (1990) Enzymic cleavage of RNA by RNA. *Angew. Chem. Int. Ed. Engl.* **102,** 735–744.
6. Bartel, D. P., Doudna, J. A., Usman, N., and Szostak, J. W. (1991) Template-directed primer extension catalyzed by the Tetrahymena ribozyme. *Mol. Cell. Biol.* **11,** 3390–3394.
7. Doudna, J. A., Usman, N., and Szostak, J. W. (1993) Ribozyme-catalyzed primer extension by trinucleotides: a model for the RNA-catalyzed replication of RNA. *Biochemistry* **32,** 2111–2115.
8. Hager, A. J., Pollard, J. D. J., and Szostak, J. W. (1996) Ribozymes: aiming at RNA replication and protein synthesis. *Chem. Biol.* **3,** 717–725.
9. Unrau, P. J. and Bartel, D. P. (1998) RNA-catalysed nucleotide synthesis. *Nature* **395,** 260–263.
10. Johnston, W. K., Unrau, P. J., Lawrence, M. S., Glasner, M. E., and Bartel, D. P. (2001) RNA-catalyzed RNA polymerization: accurate and general RNA-templated primer extension. *Science* **292,** 1319–1324.
11. McGinness, K. E. and Joyce, G. F. (2003) In search of an RNA replicase ribozyme. *Chem. Biol.* **10,** 5–14.
12. Reader, J. S. and Joyce, G. F. (2002) A ribozyme composed of only two different nucleotides. *Nature* **420,** 841–844.
13. Rohategi, R., Bartel, D. P., and Szostak, J. W. (1996) Kinetic and mechanistic analysis of nonenzymatic, template-directed oligoribonucleotide ligation. *J. Am. Chem. Soc.* **118,** 3332–3329.
14. Rohategi, R., Bartel, D. P., and Szostak, J. W. (1996) Nonenzymatic, template directed ligation of oligoribonucleotides is highly regioselective for the formation of 3'-5' phosphodiester bonds. *J. Am. Chem. Soc.* **118,** 3340–3344.
15. Eschenmoser, A. and Dobler, M. (1992) Chemistry of α-amino nitriles: 5. why pentose and not hexose nucleic acids? part I, introduction to the problem, conformational analysis of oligonucleotide single strands containing 2',3'-dideoxyglucopyranosyl building blocks (homo-DNA), and reflections on the conformation of A- and B-DNA. *Helv. Chim. Acta* **75,** 218–259.
16. Eschenmoser, A. (1999) Chemical etiology of nucleic acid structure. *Science* **284,** 2118–2124.

17. Pitsch, S., Wendeborn, S., Jaun, B., and Eschenmoser, A. (1993) Why pentose- and not hexose-nucleic acids? part VII, pyranosyl-RNA('p-RNA'). *Helv. Chim. Acta* **76**, 2161–2183.
18. Pitsch, S., Krishnamurthy, R., Bolli, M., et al. (1995) Pyranosyl-RNA (p-RNA): base pairing selectivity and potential to replicate. *Helv. Chim. Acta* **78**, 1621–1635.
19. Bolli, M., Micura, R., Pitsch, S., and Eschenmoser, A. (1997) Pyranosyl-RNA: further observations on replication: part 5. *Helv. Chim. Acta* **80**, 1901–1951.
20. Bolli, M., Micura, R., and Eschenmoser, A. (1997) Pyranosyl-RNA: chiroselective self-assembly of base sequences by ligative oligomerization of tetranucleotide 2',3'-cyclophosphate. *Chem. Biol.* **4**, 309–321.
21. Schöning, K.-U., Scholz, P., Wu, X., et al. (2002) The α-L-threofuranosyl-(3'-2')-oligonucleotide system (TNA'): synthesis and pairing properties. *Helv. Chim. Acta* **85**, 4111–4153.
22. Wu, X., Delgado, G., Krishnamurthy, R., and Eschenmoser, A. (2002) 2,6-Diaminopurine in TNA: effect on duplex stabilities and on the efficiency of template-controlled ligations. *Org. Lett.* **4**, 1283–1286.
23. von Kiedrowski, G. (1986) A self-replicating hexadeoxynucleotide. *Angew. Chem. Int. Ed. Engl.* **25**, 932–934.
24. von Kiedrowski, G., Helbing, J., Wlotzka, B., et al. (1992) Parabolic reproduction and the origin of replication. *Nachr. Chem. Tech. Lab.* **40**, 478–588.
25. Schöneborn, H., Bülle, J., and von Kiedrowski, G. (2001) Kinetic monitoring of self-replicating systems through measurement of fluorescence resonance energy transfer. *Chembiochem* **12**, 922–927.
26. Müller, D., Pitsch, S., Kittaka, A., Wagner, E., and Eschenmoser, A. (1990) Chemistry of α-aminonitriles: Aldomerization of glycolaldehyde phosphate to rac-hexose 2,4,6-triphosphates and (in presence of formaldehyde) rac-pentose 2,4-diphosphates: rac-allose 2,4,6-triphosphate and rac-ribose 2,4-diphosphate are the main reaction products. *Helv. Chim. Acta* **73**, 1410–1468.
27. Kozlov, I. A. and Orgel, L. E. (2000) Nonenzymatic template directed synthesis of RNA from monomers. *Mol. Biol.* **34**, 781–789.
28. Wu, T. and Orgel, L. E. (1992) Nonenzymatic template-directed synthesis on oligodeoxycytidylate sequences in hairpin oligonucleotides. *J. Am. Chem. Soc.* **114**, 317–322.
29. Kurz, M., Göbel, K., Hartel, C., and Göbel, M. W. (1997) Nonenzymatic oligomerization of ribonucleotides on guanosine-rich templates: supression of the self pairing of guanosine. *Angew. Chem Int. Ed. Engl.* **36**, 842–845
30. Kurz, M., Göbel, K., Hartel, C., and Göbel, M. W. (1998) Acridine-labeled primers as tools for the study of nonenzymatic RNA oligomerization. *Helv. Chim. Acta* **81**, 1156–1180.
31. Kozlov, I. A. and Orgel, L. E. (1999) Nonenzymatic oligomerization reactions on templates containing inosinic acid or diaminopurine nucleotide residues. *Helv. Chim. Acta* **82**, 1799–1805.

32. Hartel, C. and Göbel, M. W. (2000) Substitution of adenine by purine-2,6-diamin improves the nonenzymatic oligomerization of ribonucleotides on templates containing thymidine. *Helv. Chim. Acta* **83**, 2541–2549.
33. Hey, M., Hartel, C., and Göbel, M. W. (2003) Nonenzymatic oligomerization of ribonucleotides: toward in vitro selection experiments. *Helv. Chim. Acta* **86**, 844–854.
34. Schmidt, J. G., Nielsen, P. E., and Orgel, L. E. (1997) Information transfer from peptide nucleic acids to RNA by template directed syntheses. *Nucl. Acids Res.* **25**, 4797–4802.
35. Schmidt, J. G., Christensen, L., Nielsen, P. E., and Orgel, L. E. (1997) Information transfer from DNA to peptide nucleic acids by template-directed syntheses. *Nucl. Acids Res.* **25**, 4792–4796.
36. Van Aerschot, A., Verheggen, I., Hendrix, C., and Herdewijn, P. (1995) 1,5-Anhydrohexitol nucleic acids, a new promising antisense construct. *Angew. Chem. Int. Ed. Engl.* **34**, 1338–1339.
37. Hendrix, C., Rosenmeyer, H., de Bouvere, B., van Aerschot, A., Seela, F., and Herdewijn, P. (1997) 1',5'-Anhydrohexitol oligonucleotides: hybridization and strand displacement with oligoribonucleotides, interaction with RNase H and HIV reverse transcriptase. *Chem. Eur. J.* **3**, 1513–1520.
38. Allart, B., Khan, K., Rosemeyer, H., et al. (1999) D-altritol nucleic acids (ANA): hybridisation properties, stability, and initial structural Analysis. *Chem. Eur. J.* **5**, 2424–2431.
39. Winter, H. D., Lescrinier, E., Aerschot, A. V., and Herdewijn, P. (1998) Molecular dynamics simulation to investigate differences in minor groove hydration of HNA/RNA hybrids as compared to HNA/DNA complexes. *J. Am. Chem. Soc.* **120**, 5381–5394.
40. Kozlov, I. A., Politis, P. K., van Aerschot, A., Busson, R., Herdewijn, P., and Orgel, L. E. (1999) Nonenzymatic synthesis of RNA oligomers on hexitol nucleic acids templates: the importance of the A structure. *J. Am. Chem. Soc.* **121**, 2653–2655.
41. Kozlov, I. A., de Bouvere, B., van Aerschot, A., Herdewijn, P., and Orgel, L. E. (1999) Efficient transfer of information from hexitol nucleic acids to RNA during nonenzymatic oligomerization. *J. Am. Chem. Soc.* **121**, 5856–5859.
42. Kozlov, I. A., Zielinski, M., Allart, B., et al. (2000) Nonenzymatic template-directed reactions on altritol oligomers, preorganized analogs of oligonucleotides. *Chem. Eur. J.* **6**, 151–155.

20

DNase I Footprinting of Small Molecule Binding Sites on DNA

Christian Bailly, Jérôme Kluza, Christopher Martin, Thomas Ellis, and Michael J. Waring

Summary

Nuclease footprinting techniques were initially developed to investigate protein–deoxyribonucleic acid (DNA) interactions but these tools of molecular biology have also become instrumental for probing sequence-selective binding of small molecules to DNA. Here, the method is described and technical details are given for performing deoxyribonuclease (DNase) I footprinting with DNA-binding drugs. An example is presented where DNase I is used (as well as DNase II and micrococcal nuclease) to probe the patterns of sequence-selective recognition of DNA by the anticancer antibiotic actinomycin D. DNase I is a convenient endonuclease for detecting and locating the position of actinomycin-binding sites within GC-rich sequences.

Key Words: Nuclease; DNase I footprinting; DNA-binding drugs; DNase I; DNase II; micrococcal nuclease; anticancer antibiotic; actinomycin D; endonuclease; electrophoresis; polyacrylamide gels; echinomycin; cooperativity; drug–DNA recognition; reverse transcriptase; DNA polymerase I; diaminopurine, inosine; electroelution; dimethylsulfate; densitometric analysis; capillary electrophoresis; sequence selectivity; DNA recognition.

1. Introduction

The technique of deoxyribonucleic acid (DNA) cleavage-protection mapping, commonly referred to as footprinting, was introduced in the late 1970s to study the sequence-specific binding of proteins to DNA *(1)*. The original method relies on the ability of DNA-bound protein molecules to inhibit cleavage of the DNA by deoxyribonuclease (DNase) I at their binding sites. The technique has expanded rapidly to become a major tool in modern molecular biology. It provides a simple means for identifying and differentiating the sites

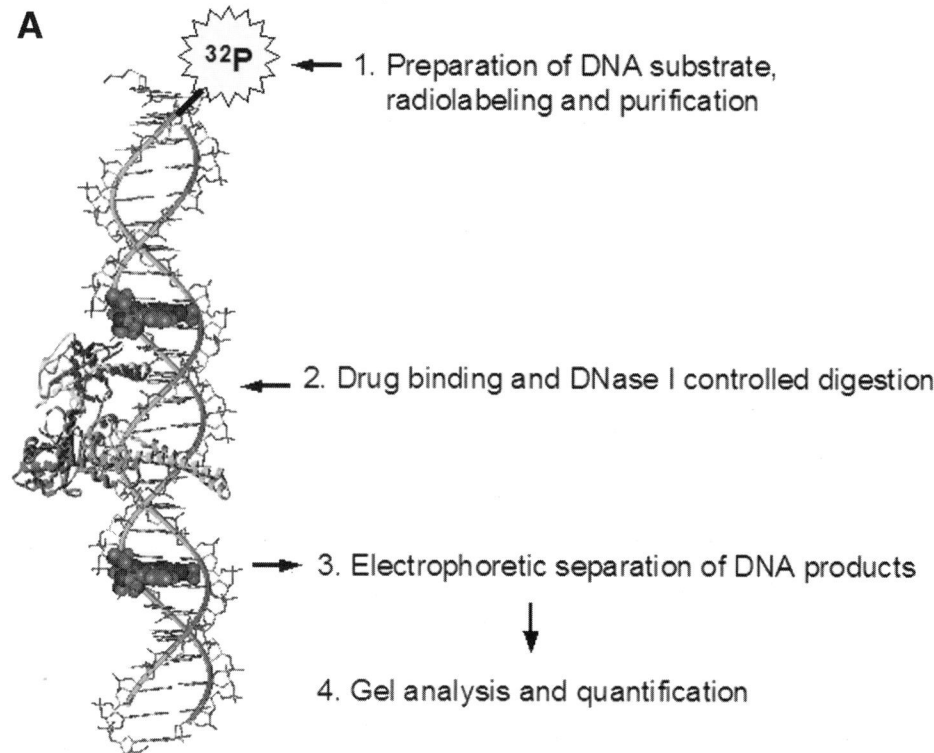

1. Preparation of DNA substrate, radiolabeling and purification
2. Drug binding and DNase I controlled digestion
3. Electrophoretic separation of DNA products
4. Gel analysis and quantification

of reversible (equilibrium) binding of proteins to DNA molecules. Moreover, footprinting titrations permit quantitative characterization of ligand–DNA interactions with the potential to yield thermodynamic and kinetic information *(2–4)*. Since the pioneering study on the *lac* operator–repressor system, the technique has been considerably refined so that nowadays a wide range of enzymic and chemical cleaving agents can be routinely used to detect ligand-binding sites as well as ligand-induced structural changes in DNA *(5)*.

Although footprinting was first applied to study DNA-binding sites for proteins, it rapidly attracted the attention of pharmacologists. Since the early 1980s, footprinting techniques have provided unique tools for studying drug–DNA recognition problems. A variety of nucleases can be used to probe small molecule binding sites but the most simple method relies on the use of the enzyme DNase I (EC 3.1.21.1) *(6)*.

1.1. DNase I Footprinting

The assay depends on cutting the desired piece of DNA and the drug-bound complexes with the nuclease DNase I and separating the resulting fragments

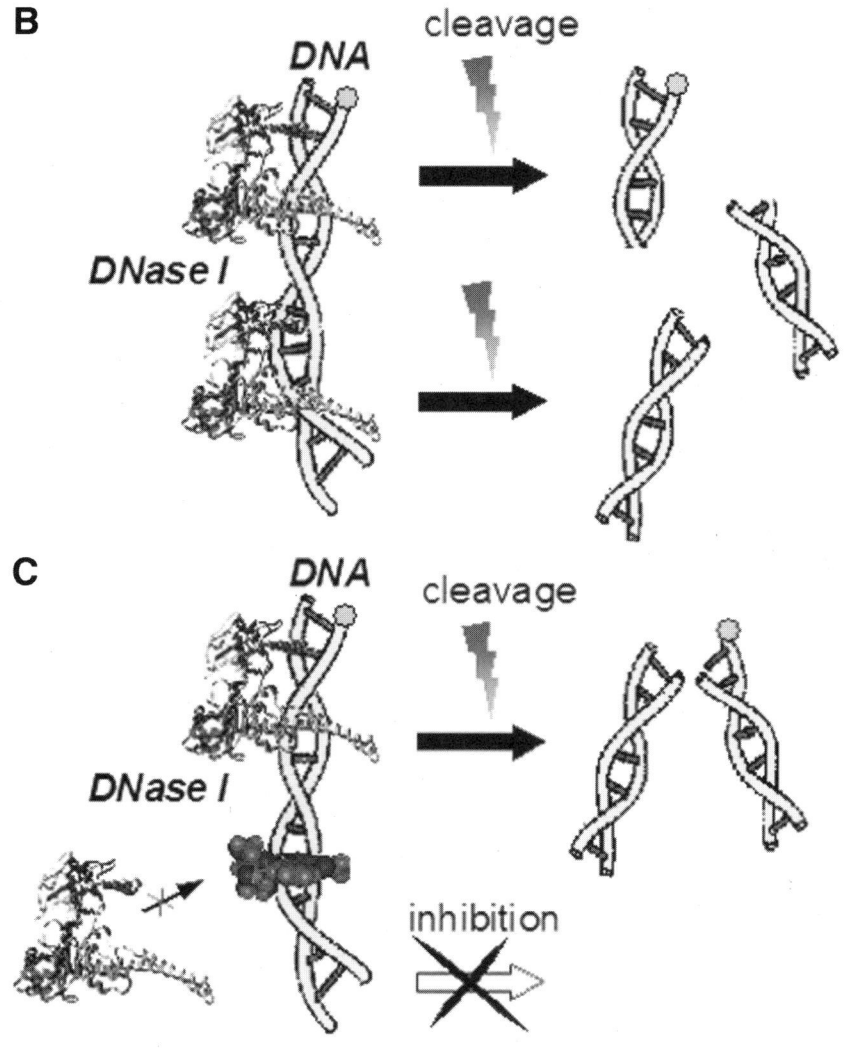

Fig. 1. Overview of the DNase I footprinting assay with (**A**) the four major steps and a schematic illustration of the controlled digestion reaction in the absence (**B**) or in the presence (**C**) of the test compound that inhibits DNA cleavage by the nuclease.

according to size by polyacrylamide gel electrophoresis (PAGE). A typical DNase I footprinting experiment can be decomposed into four steps (**Fig. 1**), and for each step specific criteria must be taken into account, as described in **Subheadings 1.1.–1.4.**

1.1.1. Preparation of the DNA Substrate, Radiolabeling, and Purification

A variety of DNA fragments can be used in footprinting experiments. Almost any DNA sequence can serve as a substrate for DNase I, but in practice it is preferable to use relatively short fragments, 80–300 bp in length, and sequences that are not excessively rich in GC basepairs or contain long homogeneous runs of consecutive AT or GC pairs. Small fragments (<50 bp) require the use of highly reticulated polyacrylamide gels that can be difficult to handle, but long fragments (>300 bp) do not provide more information than shorter pieces of DNA because of the limited resolution (and size) afforded by polyacrylamide gels. Homopolymeric stretches of bases, $(A)_n \cdot (T)_n$ or $(G)_n \cdot (C)_n$, are usually poorly cleaved by DNase I and therefore drug binding to such sequences is less easy to detect. Moreover, GC-rich DNA fragments are often structurally polymorphic, and may give rise to poorly resolved bands during gel electrophoresis.

Many types of DNA sequences have been used to investigate drug binding. A classic example is the 160-bp *Eco*RI-*Ava*I restriction fragment of the plasmid pKMΔ-98 expressed in *Escherichia coli*, usually referred to as the *tyr* T fragment, which directs the synthesis of a major species of tyrosine transfer ribonucleic acid (tRNA) *(7)*. This particular piece of DNA has been used extensively to address the properties of a large variety of drugs that bind to DNA in a sequence-selective manner (*see*, for example, **refs. 8–20**). But more sophisticated substrates, in particular, cloned DNA sequences containing known sites of specific interest, are also frequently used. Among the many examples reported in the literature, two particular cases are worth citing. The first refers to a set of designed DNA fragments each containing two pairs of strong binding sites for the CpG-specific antitumor antibiotic echinomycin: a pair of ACGT sites together with an ACGT site and a TCGA site, either directly adjacent or separated by 2 or 4 A · T basepairs. Quantitative footprinting experiments using these designed substrates have demonstrated that the binding of the antibiotic to CpG steps is highly cooperative. The extent of cooperativity depends on the nature of the sequences clamped by the antibiotic, and it diminishes as the distance between the binding sites is increased. This model study illustrated the utility of the DNase I footprinting methodology to investigate cooperativity in drug–DNA recognition *(21)*. The second example refers to a more recent study with another bidentate quinoxaline derivative, the synthetic compound [N-MeCys(3), N-MeCys(7)]TANDEM, which is structurally close to echinomycin but displays an opposite sequence selectivity with a marked preference for TpA-containing sites. A sequence containing the 136 possible tetranucleotides $[(4^4)/2 + (4^{4/2})/2 = 136]$ was constructed and cloned into the pUC18 vector at the *Bam*HI site. Two *Hin*dIII-*Sac*I restriction fragments containing the insert sequence in the two opposite orientations, 5'→3' or 3'→5', were used to investigate the preferred binding sites for [N-MeCys(3), N-MeCys(7)]TANDEM *(22)*. These two fragments, referred to as universal DNA substrates, provide potentially useful tools for determining the sequence selectivity of a variety of small molecules, minor groove binders in particular *(23)*.

To illustrate the technique, in the next section we refer to two other DNA species that have been used over the years to probe drug binding to DNA. In our laboratories we have found it particularly useful to employ two DNA fragments of 117 and 265 bp obtained simultaneously on digestion of the plasmid pBlueScribe (pBS) (Stratagene) with the restriction enzymes *Pvu*II and *Eco*RI. The plasmid pBS (commercially available from Stratagene, La Jolla, CA) is amplified in *E. coli* and purified by a conventional method based on centrifugation in a cesium chloride gradient or by using specific Qiagen-type columns. The procedure to prepare the two fragments is illustrated in **Fig. 2**. As shown, once the plasmid has been fully codigested with the two restriction enzymes, the resulting 117-mer and 265-mer fragments are rendered radioactive by 3'-[^{32}P]-end labeling at the 3' recessed end of the *Eco*RI site using [α-^{32}P]-deoxyadenosine 5'-triphosphate (dATP) and avian myeloblastosis virus reverse transcriptase or the large (Klenow) fragment of DNA polymerase I. Both enzymes work well for this type of end-labeling. The fragment can also be labeled on the complementary strand at the 5'-end with polynucleotide kinase. In this case, it is necessary first to cut the plasmid with *Eco*RI only, remove the native (unlabeled) phosphate with alkaline phosphatase, label the DNA at the 5'-end by reaction with γ–[^{32}P]-ATP and T4 polynucleotide kinase, and then cut with the second enzyme *Pvu*II (*see* **Note 1**). Indirect labeling methods, such as primer extension or ligation of a [^{32}P]-labeled oligonucleotide, can also be used in footprinting assays, but in most cases end-labeling is the method of choice for preparing a DNA fragment for footprinting experiments by enzymatic addition of a radioactive nucleotide to one end of the molecule. [^{32}P] remains by far the most widely used radioisotope because it can easily be detected by exposure to X-ray film or by a phosphorimager (autoradiography). Alternatively [^{33}P] can be used to get sharper bands, and recently nonisotopic labels have been developed. The use of fluorescent tags is possible, but this method remains relatively unpopular because it is generally less sensitive. The use of a radioisotope remains by far the most favorite method. The target DNA can be a restriction fragment derived from a chosen plasmid, but it can also be amplified by polymerase chain reaction (PCR) using one 5'-end labeled primer. PCR offers several advantages, such as (1) easily changing the length of the DNA fragment by using different primers, in particular when a given fragment

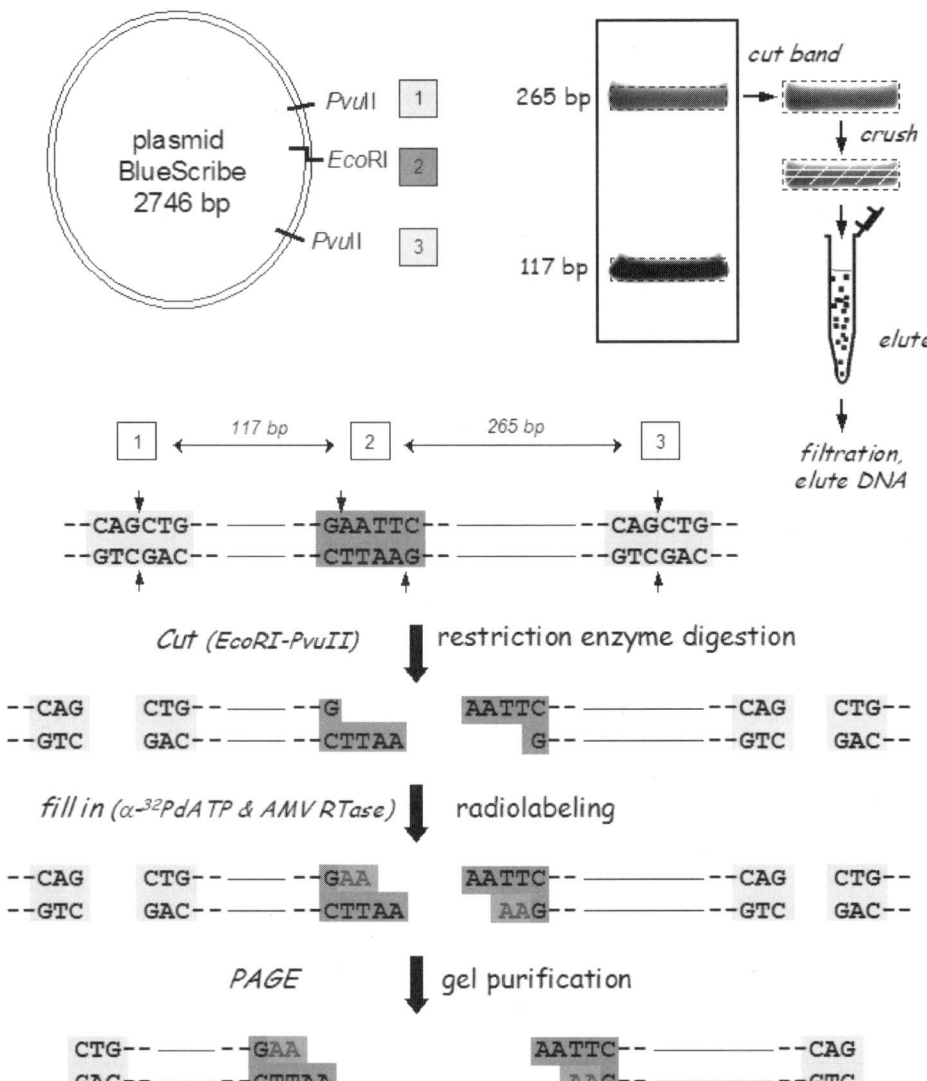

Fig. 2. Preparation of the DNA substrate. The plasmid pBS is digested with *Eco*RI and *Pvu*II to generate two fragments of 177 bp and 265 bp. Treatment with [α-^{32}P]-dATP plus AMV reverse transcriptase provides the radiolabeled materials. Gel purification permits separation of the DNA substrates so that subsequent elution of the DNA from the polyacrylamide gel yields the purified DNA ready for footprinting experiments.

does not present suitable restriction sites; and (2) the possibility to incorporate modified nucleotides. The methodology to introduce a variety of unnatural bases (e.g., 2,6-diaminopurine, inosine, 7-deaza-guanosine, 5-fluorocytosine) has recently been described *(24)*. The DNase I footprinting technique applied to DNA molecules engineered by PCR from appropriately chosen nucleoside triphosphates has been extremely informative for comprehending how small molecules recognize structural elements of the DNA double helix (*see,* for example, **refs. 25–27**).

After completion of the end-labeling reaction, it is necessary to purify the fragment prior to using it in the footprinting reaction. This is usually done by electrophoresis to separate the desired piece of DNA from other unwanted fragments. The position of the labeled fragment in the gel is visualized by autoradiography, as illustrated in **Fig. 2**. The chief technical difficulty that arises is how to recover the labeled fragment. This is often a crucial step where it is possible to lose a large fraction, if not all, of the material if the recovery process is not optimized. Extracting DNA from a polyacrylamide gel, even with a low percentage of acrylamide (5–6%) is not easy. Several procedures have been described, such as freeze–squeeze and electroelution. In the former case, the desired slice of the polyacrylamide gel is cut and frozen in liquid nitrogen or dry ice to break up the structure of the gel. The mixture is then centrifuged through a glass wool plug (or filter-equipped tubes) and the liquid containing the DNA is recovered. In practice this method has proved not very efficient, at least in our hands. The latter method requires sealing the gel slice within a dialysis tube containing buffer and then eluting the DNA from the gel by electrophoresis. Alternatively, one can use electrophoresis to stick DNA to a cellulose membrane placed adjacent to the gel slice, or a specific apparatus that permits electrotransfer of the DNA. But the method that we have found most simple for routine use is based on ordinary diffusion. The gel slice containing the desired DNA is cut from the gel, crushed, and soaked in buffer with gentle agitation for several hours (usually overnight at 4°C). The DNA diffuses from the gel and then can be recovered by filtration or centrifugation. This inexpensive method works relatively well for short DNA duplexes, around 100–150 bp, although the recovery is generally not so efficient for long fragments >200 bp. It is, however, the method that we have used successfully over the past 20 yr.

1.2. Drug Binding and DNase I-Controlled Digestion

DNase I footprinting titrations are conducted in solution by equilibrating a series of drug concentrations with a fixed amount of radiolabeled DNA. In contrast to proteins, binding reactions with small molecules are generally fast, and a 10–30 min incubation period is usually sufficient to form drug–DNA equilibrium complexes. Water-soluble compounds may be easier to handle, but it is not a problem to use an organic solvent such as dimethylsulfoxide (DMSO) to dissolve the drug and perform the footprinting reaction. Up to 3% DMSO can be used in a DNase I footprinting reaction without affecting the nuclease activity. Higher percentages of DMSO can be tolerated (even up to 40% v/v) but the DNase I cleavage pattern becomes altered owing to changes in DNA structure. DNase I is a very stable enzyme, active over a wide range of experimental conditions, from 4 to 70°C, from pH 4.5 to 9.0, as well as at high salt concentrations, up to $1M$ NaCl, for example. In most cases variations in the temperature of the binding reaction by a few degrees do not significantly affect the stability of drug–DNA complexes. It can, however, be preferable to maintain the drug–DNA mixture on ice (around 4°C) to avoid rapid dissociation of the complexes, which would result in a lower extent of inhibition of DNase I cleavage. Addition of the enzyme to the drug–DNA solution in the desired buffer should not unduly perturb the equilibrium. However, the volume of the DNase I stock solution added should be as small as possible to avoid excessive changes in the concentration of the ligand.

Heterocyclic compounds, natural products, peptides, neutral or charged compounds can all be used in footprinting assays providing that the test molecule does not by itself induce DNA cleavages. Even alkylating agents (e.g., the antitumor drug ecteinascidin ET 743) can be employed. The appropriate concentration range for a given drug must be adapted to cover the full range of saturation of the binding sites. It is always advisable to conduct DNase I footprinting experiments at several drug concentrations to make sure that the observed footprints do effectively represent drug-binding sites and not artefactual events attributable to the buffer for example. The goal is to prepare binding reactions over a range of drug concentrations that produce from 0 to 100% fractional saturation of the DNA binding sites. This is especially crucial for quantitative footprinting investigations.

The most important consideration in these experiments is to adjust the DNase I concentration very carefully so that statistically each DNA molecule is cut only once by the nuclease. The level of DNA cleavage must not exceed "single-hit" conditions. This is usually accomplished by conducting preliminary experiments with DNA alone digested with increasing amounts of the enzyme for a fixed exposure time. Then the DNase I concentration producing roughly 30% cleavage of the starting material can be chosen. The exact amount of DNase I will depend on the batch of enzyme and the length of the DNA fragment being used. The nuclease reaction is best performed for a brief period, usually around 3–5 min, and the controlled digestion is stopped either by freezing or by adding a solution containing ethylenediaminetetraacetic acid (EDTA), which quenches DNase I activity by chelating the essential Mg^{2+} and Ca^{2+}

ions. "Stop solutions" containing glycogen, calf thymus DNA, and a high concentration of NaCl, in addition to EDTA, are frequently used *(28)*. In such a case, the DNA samples must be ethanol-precipitated, washed with 70% ethanol, and then resuspended in the denaturing formamide solution prior to loading onto the polyacrylamide gel. A much simpler alternative procedure that we have found convenient with small molecules (but not with proteins) consists of performing the DNase I reaction in a small volume (8 µL) of TN buffer (10 mM Tris, pH 7.0, 10 mM NaCl), stopping the DNase I reaction simply by transferring the tubes to dry ice, lyophilized, and then adding the loading solution containing formamide, which serves to denature both the DNA and the enzyme. In this case the DNA samples can be loaded directly onto a polyacrylamide gel, without any tedious step of ethanol precipitation. The amount of drug and enzyme is very small and does not perturb the electrophoresis. This minimized procedure usually works well with small molecules but not with peptides and proteins for which a purification step is necessary.

1.3. Electrophoresis

The next step consists of loading the DNase I reaction products in the wells of a denaturing polyacrylamide gel to size-fractionate the pool of oligonucleotides by electrophoresis (**Fig. 3**). On a conventional 8% gel, 0.3–0.4-mm thick, 30-cm long, about 150–200 DNA bands can be resolved, but it is generally difficult to assign more than 120 nucleotide positions because the bands at the top of the gel are too close, one to another, to permit assigning them to the cleavage of a particular internucleotide bond unambiguously. However, for the first 100 bands it is easy to determine the linear order of the bases of DNA by reference to markers. Here again there are several chemicals that can be used to generate markers, such as formic acid (for purines, the so-called GA track) and hydrazine (for pyrimidines, the TC track), but the most convenient is probably dimethylsulfate (DMS), which is ideal for identifying the positions of guanine residues in the sequence.

A polyacrylamide gel is obtained by copolymerization of acrylamide and N,N'-methylene-*bis*-acrylamide, which crosslinks the chain in the presence of ammonium persulfate and tetramethylethylenediamine (TEMED). The TEMED catalyzes the formation of free radicals from the persulfate ions, and these initiate polymerization of the acrylamide. Gels are usually cast horizontally between two glass plates separated by spacers 0.3–0.4-mm thick. The concentration of polyacrylamide must be adapted according to the length of the DNA fragments that are to be separated. In most cases 8% polyacrylamide provides the optimal pore sizes in the gel matrix to permit the separation of DNA molecules up to 300 nucleotides long. Of course, every investigator will

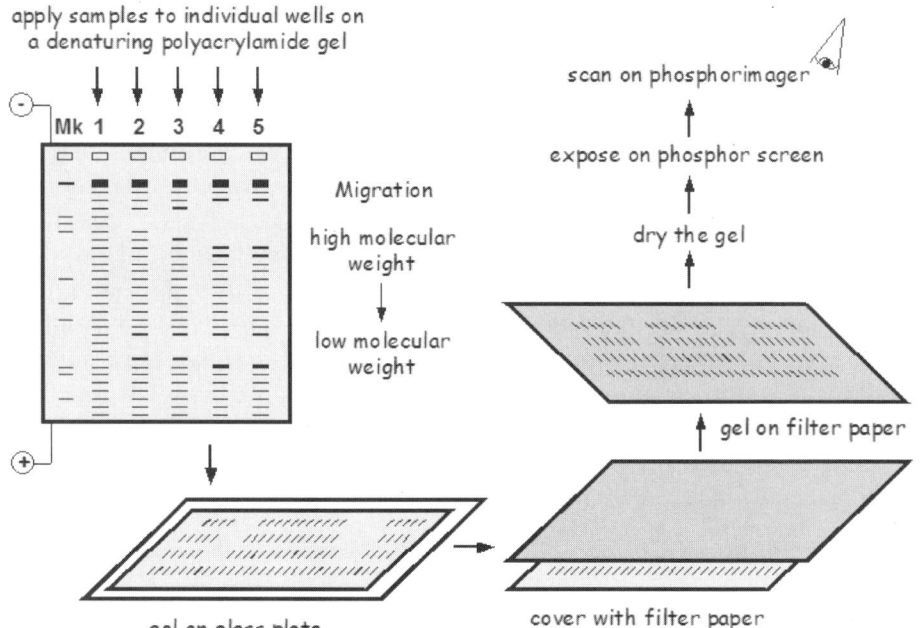

Fig. 3. Schematic illustration of the electrophoresis of DNase I footprinting cleavage products and the transfer of the gel from the glass plate to filter paper for the drying and analysis.

adapt the concentration of the polyacrylamide to suit the requirements of a given experiment. As an example, highly reticulated gels containing up to 15% acrylamide will be necessary to fractionate short oligonucleotides 20–40 nucleotides in length.

Denaturing gels are necessary to stop complementary strands from reannealing, and intramolecular basepairing from forming secondary structures, which would affect the mobility. Three factors contribute to ensure the formation of single-stranded DNA: (1) the addition of formamide to the DNA samples interferes directly with the formation of secondary structures, (2) heating of the samples to 90°C and rapid cooling in ice prior to loading the materials in the wells of the gel will maintain and/or amplify the denaturation, (3) the presence of a high concentration of urea ($8M$) in the acrylamide matrix also helps to disrupt secondary structure. Moreover, the electrophoresis of the single-stranded DNA species is performed at elevated temperature, around 60°C, as a consequence of the heating caused by the electric current in the gel (high voltage, around 65 mA). After electrophoresis, it is preferable to fix the gel in a bath of 10% acetic acid for 5 min to remove much of the urea prior to drying it onto a filter paper using a conventional gel dryer. The dried gel can then be exposed to a storage phosphor screen for a few hours (usually overnight), and the scanning of the plate produces a digitized autoradiogram.

1.4. Gel Analysis and Quantification

The densitometric analysis of the gel is always essential to accurately locate drug-binding sites and to measure relative binding at specific sequences. However, the utility of the densitometric analysis depends crucially on the quality of the gel. Visual inspection of the autoradiogram can reveal immediately whether the test compound recognizes preferred sites. If no DNase I footprint can be seen by eye, it is generally useless to try quantifying the bands to detect weakly binding sequences. Similarly, if the extent of DNase I digestion in each lane is far from constant, it is not appropriate to go further with the data analysis. Attempted comparison using a control (drug-free) lane that is over- or underdigested will give unsatisfactory results. It is always better to spend time repeating the gel experiments rather than trying to extract information from a nonoptimized gel. In difficult cases the success rate may be as low as one good gel out of ten experiments.

There are essentially two levels of gel analysis: scanning to locate the binding sites of a compound and quantitative analysis of binding affinity. Relative affinities can be determined from footprinting plots where band intensities are plotted as a function of the ligand concentration. Analytical procedures used to extract binding constants from the footprinting gels have been detailed recently *(4,28)*. Here we limit the analysis to the so-called differential cleavage plots in which the extent of DNase I cleavage is presented as a function of the nucleotide sequence of the DNA fragment. Such plots are obtained by measuring band intensities in two lanes of the gel, the control lane where the DNA has been digested in the absence of ligand and a lane from a drug-containing DNA sample digested with the enzyme under strictly identical conditions. It is important that the two analyzed lanes come from the same gel, as the extent of digestion may vary from one gel to another. Comparison of the fractional cleavage at successive bonds along the DNA sequence in the presence and absence of the test ligand is commonly represented in the form of a logarithmic plot (*see* the case study in **Subheading 1.5.**) so as to generate negative and positive data points corresponding to sites of DNase I cleavage protection or enhancement.

1.5. A Case Study: DNase I Footprinting of Actinomycin Binding to DNA

The following section describes a typical DNase I footprinting experiment performed with the previously mentioned 265-bp *Eco*RI-*Pvu*II fragment from plasmid pBS to identify the binding sites for the antibiotic actinomycin D

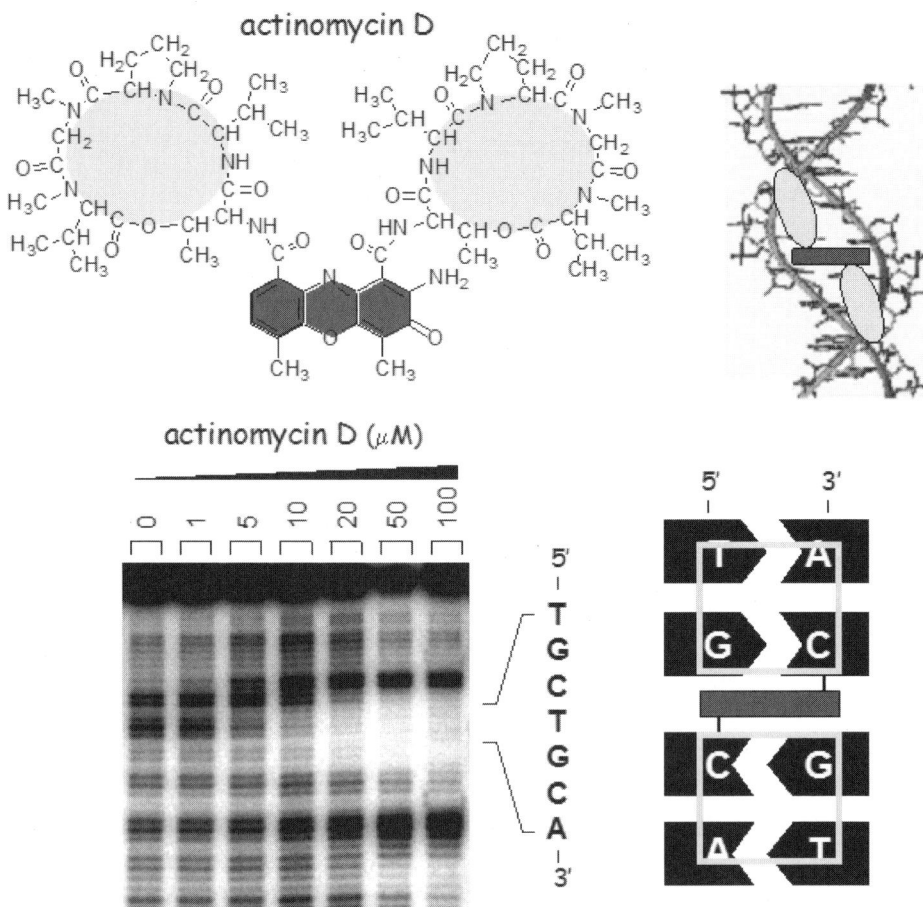

Fig. 4. Structure of actinomycin D and schematic representation of the drug–DNA complexes. The two cyclic peptides fit into the minor groove of DNA, and the planar phenoxazinone chromophore intercalates at the GpC sites. The gel shows a marked footprint that develops on interaction of actinomycin D with a GpC-containing sequence.

(Fig. 4). This drug has valuable antitumor properties related to its propensity for binding to DNA. It is chiefly used in the treatment of Wilms' tumor in children. The ability of actinomycin to intercalate into DNA has been long established *(29–31)* and its binding to DNA is characterized by a general requirement for GC residues *(32–36)*. Previous footprinting studies have shown negligible binding to AT sequences and specific binding to GpC sites *(11,37–39)*. Sequence recognition is provided by the two cyclic peptides that fit into the minor groove of DNA while the phenoxazinone chromophore of actinomycin intercalates at GpC sites, as illustrated in **Fig. 4**. A well-resolved footprint develops at a sequence containing two adjacent GpC steps. The tetranucleotide TGCA represents a high-affinity binding site for actinomycin D *(40)*.

2. Materials

1. General TN buffer to dilute the DNA and the drug solutions: 10 m*M* Tris, 10 m*M* NaCl buffer, adjusted to pH 7.0.
2. TBE electrophoresis buffer: 89 m*M* Tris base, 89 m*M* boric acid, 2.5 m*M* Na$_2$ EDTA, pH 8.3.
3. DNase I: Type IV bovine pancreatic deoxyribonuclease I (DNase I, EC 3.1.21.1, Sigma) is prepared as a 7200 Kunitz U/mL solution in 0.15*M* NaCl containing 1 m*M* MgCl$_2$ (*see* **Note 2**).
4. Drug: The initial stock solution is made by shaking 1 mg of actinomycin D (Sigma) in 2 mL H$_2$O at 4°C for about 2 h. Subsequent dilutions of the drug are made with the TN buffer (*see* **Note 3**).
5. BRL sequencer model S2 and CBS Scientific model DDH-402-33.
6. *Pvu*II buffer: 10 m*M* Tris-HCl, 10 m*M* MgCl$_2$, 50 m*M* NaCl, 1 m*M* dithioerythritol, pH 7.5; as supplied by the manufacturer.
7. DNase I dilution buffer: 2 m*M* MgCl$_2$, 2 m*M* MnCl$_2$, 20 m*M* NaCl, pH 8.0.

3. Methods

3.1. DNA Preparation

1. Digest the pBS plasmid (40 µg) simultaneously with restriction enzymes *Eco*RI and *Pvu*II (50 U each) for 3 h at 37°C in a volume of 100 µL using the *Pvu*II buffer.
2. Add 500 µL of cold ethanol, centrifuge at 13,800*g* for 15 min, remove the supernatant, and wash the pellet with 50 µL of 70% ethanol (optional wash).
3. Redissolve the pellet in 80 µL H$_2$O, and add 20 µL of 5X reverse transcriptase buffer (as supplied by the manufacturer). Gently mix prior to adding 5 µL [α-^{32}P]dATP (3000 Ci/mmol) together with 2 µL avanmyeloblastosis virus (AMV) reverse transcriptase (20 U), and incubate at 37°C for 2 h.
4. Add 12 µL of cold dATP (100 m*M* stock solution stored at –20°C), 1 µL AMV reverse transcriptase, and incubate for a further 2–3 h (*see* **Note 4**).
5. Add 10 µL of loading dye (bromophenol blue, xylene cyanol loading solution from Sigma), and load the solution directly into the well (15-mm wide, 10-mm deep) of a 6% nondenaturing polyacrylamide gel (1-mm thick, 20-cm long), in TBE buffer. Run the electrophoresis at 180V for about 90 min until the bromophenol blue has reached the bottom of the gel.
6. Separate the two plates, cover the gel with plastic film (Saran wrap), and then determine the position of the two DNA bands by autoradiography. A brief exposure (2–4 min) is sufficient to reveal strong black bands corresponding to the positions of the two products. At this stage, scanning of the gel with a hand-held

Geiger counter should give a reading off scale (>3000 cps) over the radioactive bands. Care must be taken while manipulating these radioactive materials. Whenever possible, all manipulations should be performed behind a protective screen.

7. Place the autoradiograph under the glass plate in exactly the same position where it was when the autoradiography was performed, and use it to locate the two bands of interest, as illustrated in **Fig. 3**. Cut the desired band into small pieces with a razor blade, place in a new Eppendorf tube, and crush the gel with a small spatula prior to covering the band with the elution buffer (500 mM ammonium acetate, 10 mM magnesium acetate) or with water (about 600 µL to almost fill up the 1.5-mL tube).
8. Place the tube on a rotating wheel under mild agitation for about 15 h (overnight) at 4°C. It is safe to protect the tube from opening with Parafilm, and avoid exposure to strong light (cover with aluminum foil).
9. Transfer the contents of the tube into a 2-mL syringe equipped with a filter disk (Sartorius, 0.45 µM), and press very slowly to recover the DNA-containing solution through the filter while retaining the residual gel in the syringe. This step must be performed with care to avoid blockage of the filter (or even breakage). Rinse the gel with five times 100 µL of H_2O. This procedure recovers more than 80% of the radiolabel in the gel slice. It works better for the short fragment (117 bp) than for the longer one (265 bp).
10. Precipitate the radiolabeled DNA by adding 3 vol of cold ethanol (stored at −20°C) and centrifuging at 13,800g for 15 min. Wash the pellet with 100 µL 70% ethanol, and finally resuspend the DNA in the working TN buffer. The volume of buffer is adjusted so as to generate about 100 cps per µL on a hand-held counter (*see* **Note 5**).
11. The DNA solution is kept at −20°C and used within 10 d. The exact DNA concentration is not known (estimated in the pM range), but this is not important for routine experiments aimed at identifying drug-binding sites.

We will not describe here the standard protocols used to generate marker lanes. These are fully described in practical molecular biology textbooks *(41)*. The preparation of a G track or a G + A track (obtained by reacting the DNA with dimethylsulfate or formic acid, respectively) is necessary not only to establish the identity of the DNA digestion products but also to monitor the quality of the DNA preparation. It happens from time to time that the radiolabeling at the *Eco*RI site does not proceed correctly, and the presence of a doubly labeled DNA substrate can immediately be spotted from the G track.

3.2. DNase I Footprinting Reaction

The protocol described below has been optimized for actinomycin D, but the same procedure can be applied to a variety of water-soluble compounds (e.g., netropsin, Hoechst 33258, daunomycin). For drugs dissolved in DMSO, it is often (but not always) necessary to use slightly higher amounts of DNase I

DNase I Footprinting

for the controlled cleavage reaction. The basic protocol, simple and efficient, is the following:

1. Combine 2 µL of the 265-bp [^{32}P]-labeled DNA with 4 µL of actinomycin solution. Both the DNA and the drug are dissolved in TN buffer. Perform individual reactions in 1.5 mL Eppendorf tubes with increasing concentrations of actinomycin D: 2, 10, 20, 40, 100, and 200 µM (so as to obtain a final concentration of 1–100 µM).
2. Incubate the mixtures to equilibrium at room temperature for about 2 h. Binding of small molecules to DNA is generally fast, and equilibrium can be reached within a few minutes, but in some cases, such as that of actinomycin with its two bulky cyclic peptides, the binding kinetics are slow and it takes at least 1 h to reach equilibrium. For this reason, it is wise to incubate the drug–DNA mixtures for about 1–2 h when working with a new compound.
3. Initiate the cleavage reaction by adding 2 µL of DNase I (0.005 U/mL final concentration) and mixing gently without vortexing. The enzyme is stored at 200 U/mL in small aliquots (10 µL) at –20°C, in a storing solution (*see* **Note 6**).
4. Incubate for 8 min at room temperature. Stop the reaction by transferring the tubes to dry ice and lyophilize the samples with a SpeedVac for about 30 min until the tubes are just dried (avoid excessive lyophilization as the DNA can be difficult to redissolve fully). Then add 5 µL of the formamide–dyes solution (formamide containing 10 mM EDTA and 0.1% bromophenol blue, 0.1% xylene cyanol). Leave the samples on the bench for at least 10 min to dissolve the DNA.

3.3. Electrophoresis and Autoradiography

1. Heat the tubes at 90°C for 4 min (use of a dry sand bath is convenient), and transfer the tubes to ice prior to loading the products in neighboring wells of an 8% denaturing polyacrylamide gel. Avoid heating the tubes for more than 5 min at 90°C as this can result in partial depurination of the DNA fragments.
2. Separate the products of digestion by PAGE under denaturing conditions (0.3-mm thick, 8% acrylamide gel containing 8M urea) capable of resolving DNA fragments differing in length by one nucleotide.

3.4. Results: Binding of Actinomycin D to GC-Rich Sequences

3.4.1. DNase I Footprinting Data

Figure 5 illustrates the results of a typical DNase I footprinting experiment with the 265-bp fragment from pBS 3'-radiolabeled at the *Eco*RI site. The cleavage reaction was performed in the absence and presence of increasing concentrations of the antibiotic. The products of a Maxam–Gilbert sequencing reaction on the same radiolabeled DNA molecules were also separated by electrophoresis in parallel to allow for direct identification of the sequences protected from cleavage by the nuclease. Several regions that are protected from

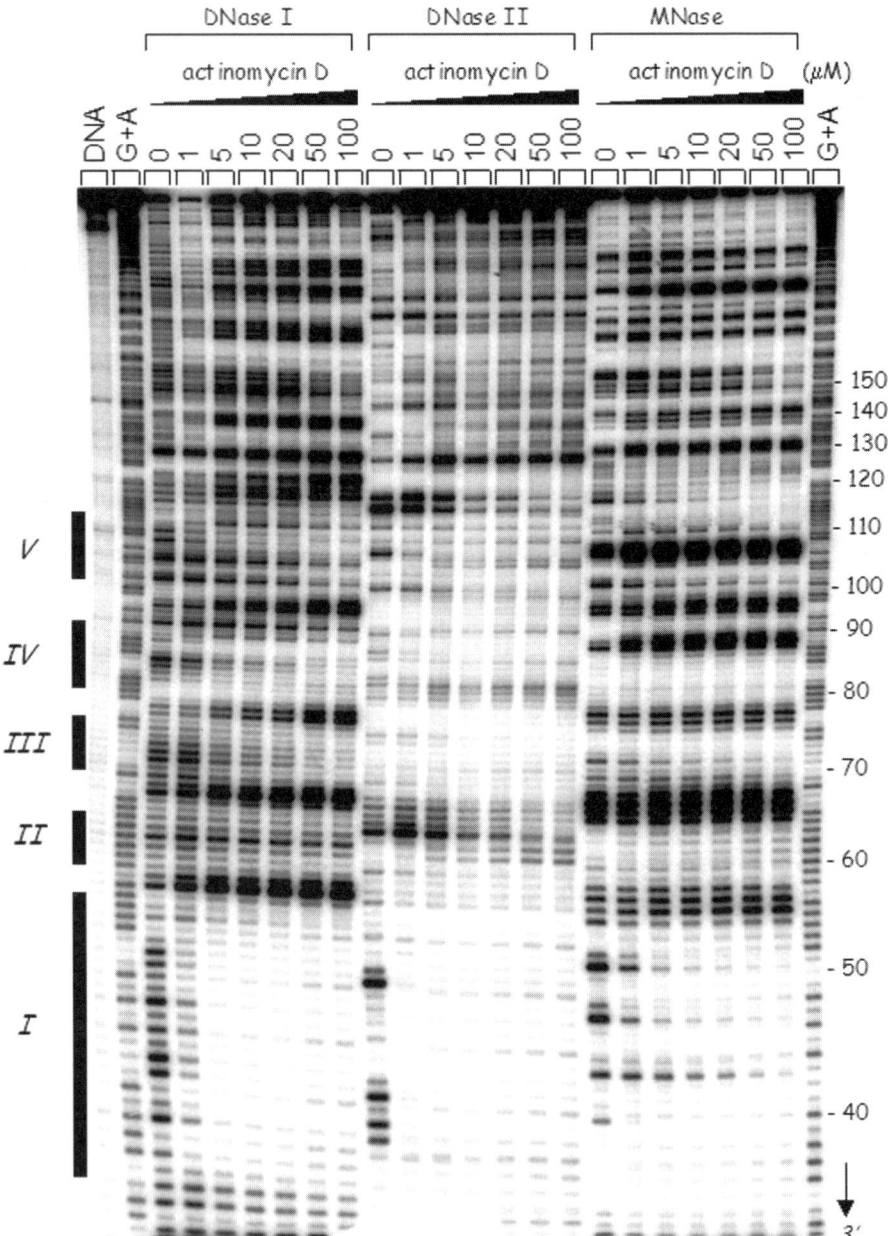

Fig. 5. Footprinting of actinomycin D on the 265 basepair *Eco*RI/*Pvu*II restriction fragment cut out of plasmid pBS. The duplex DNA was 3'-end labeled at the *Eco*RI site with [α-^{32}P]dATP in the presence of AMV reverse transcriptase. Controlled digestion of the DNA–drug complexes was performed with DNase I, DNase II or

DNase I Footprinting

digestion by the enzyme are clearly visible on this gel. Dose-dependent enhancements of cleavage rates relative to those in drug-free control lanes occur at sequences flanking the binding sites. The inhibition of cleavage is particularly strong at the lower site between nucleotide positions 38 and 52 corresponding to the sequence 5'-A*G*CTT*GC*AT*GC*CT*GC*A. As expected, this high-affinity site contains several GpC steps to which the drug is known to bind tightly. The footprint extends up to 12 basepairs, suggesting the presence of overlapping binding sites.

The intensities of the bands in **Fig. 5** were measured by densitometry and converted to numerical probability of cleavage. Differential cleavage plots derived from the DNase I cleavage patterns of the 3'-labeled fragment are presented in **Fig. 6**. In these plots, the regions where the cleavage of DNA by the enzyme has been reduced and enhanced in the presence of the antibiotic appear as negative and positive changes in probability of cleavage, respectively. Five regions of cleavage inhibition were detected by densitometry. Sites IV and V both contain the aforementioned high-affinity sequence 5'-AGCT, and site I presents four consecutive GpC steps. Sites II and III have no 5'-GC dinucleotide step but nevertheless contain the same 5'-CCCT · 5'AGGG site to which the drug can also bind. The data are totally consistent with the known selectivity of the antibiotic.

3.4.2. Footprinting of Actinomycin D With DNase II and Micrococcal Nuclease

In footprinting experiments, the nature of the DNA-cleaving reagent is critical; on its chemical reactivity will depend the sensitivity and the precision with which the binding site can be detected and located. DNase I was the first enzyme used in footprinting assays and remains the most commonly used probe, but other enzymes, such as DNase II and micrococcal nuclease (MNase) can be employed *(42)*. Whenever possible complementary probes should be used to depict fully the molecular recognition process. For comparison, the sequence-selective binding of actinomycin D to the 265-bp fragment was probed further with MNase and DNase II (**Fig. 5**).

One-dimensional scans of a control lane and an actinomycin-containing lane from each gel are shown in **Fig. 6**. The five footprints observed with DNase I

Fig. 5. (*continued*) MNase. The cleavage products of the digestion reactions were resolved on an 8% polyacrylamide gel containing 8*M* urea. The concentration (μM) of the antibiotic is shown at the top of the appropriate gel lanes. Control tracks labeled "0" contained no drug. Tracks labeled G + A represent a formic acid–piperidine marker specific for purines. Numbers at the side of the gels refer to the numbering scheme used in **Fig. 6**. The position of the binding sites (I–V) is indicated by vertical bars.

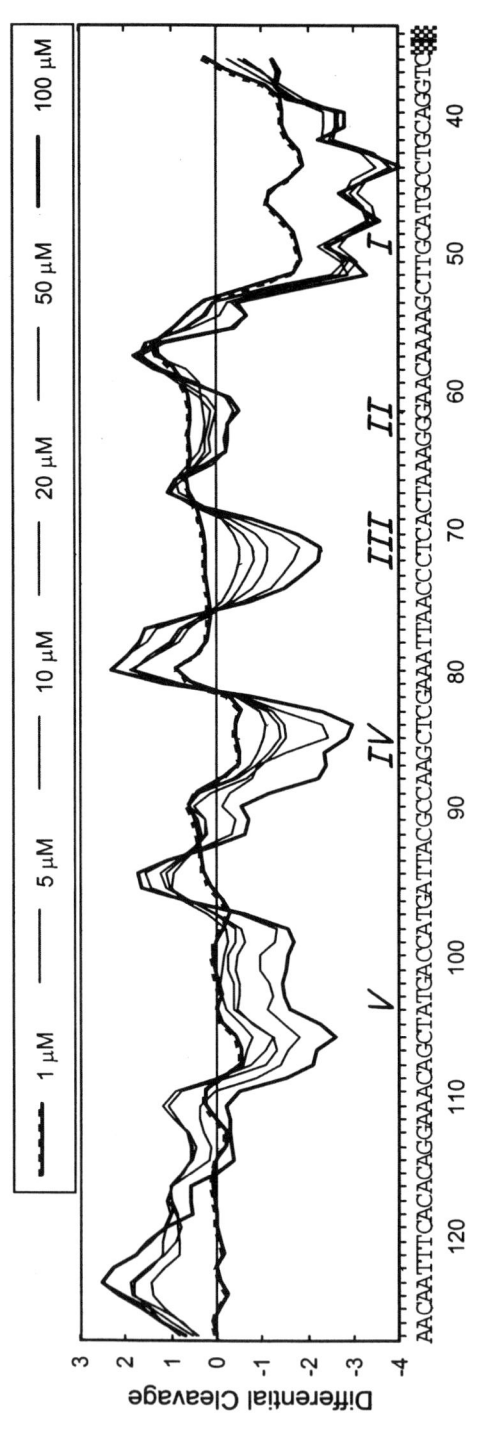

Fig. 6. Differential cleavage plots comparing the susceptibility of the 265-mer DNA fragment to DNase I cutting in the presence of actinomycin. Vertical scales are in units of ln(fa)−ln(fc), where fa is the fractional cleavage at any bond in the presence of the drug and fc is the fractional cleavage of the same bond in the control, given closely similar extents of overall digestion. Each line drawn represents a 3-bond running average of individual data points, calculated by averaging the value of ln(fa)−ln(fc) at any bond with those of its two nearest neighbors. Negative values correspond to a ligand-protected site, and positive values represent enhanced cleavage.

are also detected with MNase and DNase II. However, the MNase-mediated footprints are generally larger than those observed with DNase I so that it is often more difficult to locate the exact position of the binding sites for actinomycin. The larger footprints might reflect a greater sensitivity on the part of MNase to long-range distortions in the DNA helix induced by the bound drug. MNase is notably better at cutting pA and pT bonds than pG and pC, which makes it a particularly useful tool for assessing the binding of AT-selective ligands (e.g., the minor groove binder netropsin *(12)*). Consequently, it is not so well adapted for probing a GC-selective ligand such as actinomycin, even if binding of the antibiotic to the GpC-containing site I is easily detected with this nuclease. DNase II is also useful; it confirms binding of the drug to sites I–V as expected, but the general utility of this nuclease is diminished by the limitation that it often proves more difficult to handle.

3.5. Conclusion

The technique of DNA cleavage-protection mapping, better known as the footprinting assay, has gained widespread popularity in pharmacology as a simple and reliable method for identifying binding sites for small molecules that interact with nucleic acids *(43,44)*. The assay is conceptually simple. Binding of a given drug to particular sites protects DNA locally from cleavage by a nuclease, whereas unbound sequences are accessible for digestion producing fragments, which can be separated by size using PAGE. Typically gels are run so as to compare control (drug-naïve) samples with experimental (drug-treated) samples, each producing ladders of DNA bands from which the sequences of binding sites can be disencrypted. Here we have concentrated on qualitative aspects of the method, but as mentioned above, the use of [^{32}P]-radiolabeled DNA and phosphorimaging affords high-resolution images suitable for quantitative analysis to estimate binding affinities. The method is primarily used to detect drug-binding sites, but it has also proved useful as a means of investigating other aspects of drug–DNA recognition, including kinetics *(40,45,46)* and shuffling between binding sites *(47,48)*. A variety of enzymic and chemical nucleases can be used. DNase I is by far the most convenient and easy to use enzyme, but as briefly illustrated here, DNase II and MNase can furnish additional information to localize drug-binding sites better and/or to detect drug-induced structural perturbations of the DNA double helix. Among the available chemical nucleases, the EDTA–Fe^{II} complex is undoubtedly the most frequently used because it efficiently generates DNA-damaging hydroxyl radicals ideally suited to fine (albeit less sensitive) mapping of binding sites and for DNA structure analysis *(49,50)*. However, a variety of other metal complexes have been described and tested for footprinting analysis *(51,52)*. The method is also well suited to study drug binding to RNA using RNase I (low specificity) or RNase T1 (preferential cleavage at unpaired guanosines) as cleaving probes *(53)*. The various footprinting techniques are therefore extremely useful and widely applicable in pharmacology, particularly in oncology where it is immensely valuable to study DNA sequence recognition by antitumor drugs. It must, however, be admitted that the method can be time-consuming, and it requires high technical skills. Compared with modern machine-aided methods of DNA sequencing and gene expression analysis, footprinting looks conspicuously hand-driven and perhaps old-fashioned, but despite these criticisms it has an established place as one of the most important tools in molecular biology. The future of footprinting most probably lies in the development of instrument-based approaches aimed at avoiding traditional slab-gel electrophoresis. In this regard, capillary zone electrophoresis with laser-assisted fluorescence detection has started to yield interesting results, combining rapid analysis and high sensitivity of detection *(54,55)*. For reasons of both cost and ease of automation, in the foreseeable future a shift toward capillary electrophoresis or another nanotechnology may develop, but for laboratories not so well endowed, it is certain that conventional gel electrophoresis will remain the method of choice, not least because it represents the gold standard against which other technologies must ultimately be assessed.

4. Notes

1. When possible, we recommend performing 3'-end labeling because the DNA fragments resulting from DNase I cleavage terminate with a 3'-phosphate residue and therefore comigrate with Maxam–Gilbert sequence markers, whereas a 5'-end labeled DNA yields fragments bearing a 3'-hydroxy terminus that do not comigrate with the markers, thereby rendering the analysis a little more complicated.
2. The DNase I solution is very stable and can be used for a few years if correctly stored. Freezing and thawing do not reduce the nuclease activity. This stock solution is further diluted in 20 mM NaCl, 2 mM MgCl$_2$, 2 mM MnCl$_2$, pH 8.0, to prepare aliquots (10 µL) at 200 U/mL stored at –20°C. One tube is thawed for each experiment, and the enzyme is freshly diluted to the desired concentration immediately prior to use. The remainder of the diluted solution should not be reused.
3. Concentration of actinomycin D solutions are determined spectrophotometrically applying a molar extinction coefficient of 25,300M^{-1} cm^{-1} at 425 nm.
4. This second step is optional but recommended to avoid double labeling, that is, the production of a mixture of DNA molecules containing either one or two [^{32}P]-dA residues at the *Eco*RI site (G/AATTC). The supplementary reverse transcriptase incubation ensures that all labeled molecules are the same length, regardless of the efficiency of the radioactive labeling step.
5. It is always preferable to work with highly radioactive DNA, even if only a small volume of the working DNA solution is obtained. Typically, this procedure pro-

vides 100–300-μL DNA at 100 cps/μL, at least for the short fragment, and such a solution is sufficient to produce a good phosphorimage from a typical gel within 1 d of exposure.

6. The DNase I concentration needs to be adjusted in preliminary experiments to calibrate the enzyme activity with the test molecule. Actinomycin D does not significantly affect the enzyme activity but other heterocyclic compounds can profoundly inhibit the cutting of DNA by the nuclease, and in such cases the DNase I concentration may have to be substantially increased. This happens with ellipticines, for example *(16)*.

Acknowledgments

Research in the authors' laboratories is supported by the Ligue Nationale Contre le Cancer (Equipe labellisée LA LIGUE), the INSERM and IRCL (to C.B., Lille, France), and Cancer Research UK and the Wellcome Trust (to M.J.W, Cambridge, UK).

References

1. Galas, D. J. and Schmitz, A. (1978) DNase footprinting: a simple method for the detection of protein–DNA binding specificity. *Nucleic Acids Res.* **5,** 3157–3170.
2. Brenowitz, M., Senear, D. F., Shea, M. A., and Ackers, G. K. (1986) Quantitative DNase footprint titration: a method for studying protein–DNA interactions. *Methods Enzymol.* **130,** 132–181.
3. Dabrowiak, J. C., Stankus, A. A., and Goodisman, J. (1992) Sequence-specificity of drug–DNA interactions. In *Nucleic Acid Targeted Drug Design* (Propst, C. L., Perun, T. J., eds.), Dekker, New York, pp. 93–149.
4. Dhavan, G. M., Mollah, A. K. M. M., and Brenowitz, M. (2002) Equilibrium and kinetic quantitative DNase I footprinting. In *Advances in DNA Sequence-Specific Agents,* vol. 4, (Jones, G. B., ed.), Elsevier, New York, pp. 139–155.
5. Nielsen, P. E. (1990) Chemical and photochemical probing of DNA complexes. *J. Mol. Recognit* **3,** 1–25.
6. Bailly, C. and Waring, M. J. (1995) Comparison of different footprinting methodologies for detecting binding sites for a small ligand on DNA. *J. Biomol. Struct. Dyn.* **12,** 869–898.
7. Travers, A. A., Lamond, A. I., Mace, H. A. F., and Berman, M. L. (1983) RNA polymerase interactions with the upstream region of *E. coli tyr*T promoter. *Cell* **35,** 265–273.
8. Low, C. M. L., Olsen, R. K., and Waring, M. J. (1984) Sequence preferences in the binding to DNA of triostin A and TANDEM as reported by DNase I footprinting. *FEBS Lett.* **176,** 414–420.
9. Low, C. M. L., Drew, H. R., and Waring, M. J. (1984) Sequence-specific binding of echinomycin to DNA: evidence for conformational changes affecting flanking sequences. *Nucleic Acids Res.* **12,** 4865–4877.
10. Fox, K. R. and Waring, M. J. (1986) Nucleotide sequence binding preference of nogalamycin investigated by DNase I footprinting. *Biochemistry* **25,** 4349–4356.

11. Fox, K. R. and Waring, M. J. (1987) Footprinting at low temperature: evidence that ethidium and other simple intercalators can discriminate between different nucleotide sequences. *Nucleic Acids Res.* **15,** 491–507.
12. Fox, K. R. and Waring, M. J. (1987) The use of micrococcal nuclease as a probe for drug-binding sites on DNA. *Biochim. Biophys. Acta* **909,** 145–155.
13. Chaires, J. B., Fox, K. R., Herrera, J. E., Britt, M., and Waring, M. J. (1987) Site and sequence specificity of the daunomycin–DNA interaction. *Biochemistry* **26,** 8227–8236.
14. Portugal, J. and Waring M. J. (1987) Comparison of binding sites in DNA for berenil, netropsin and distamycin: a footprinting study. *Eur. J. Biochem.* **167,** 281–289.
15. Portugal, J. and Waring, M. J. (1987) Assignment of DNA binding sites for 4',6-diamidine-2-phenylindole and bisbenzimide (Hoechst 33258): a comparative footprinting study. *Biochim. Biophys. Acta* **949,** 158–168.
16. Bailly, C., Ohuigin, C., Rivalle, C., Bisagni, E., Hénichart, J. P., and Waring, M. J. (1990) Sequence-selective binding of an ellipticine derivative to DNA. *Nucleic Acids Res.* **18,** 6283–6291.
17. Bailly, C., Denny, W. A., Mellor, L., Wakelin, L. P. G., and Waring, M. J. (1992) Sequence-specificity of the binding of 9-aminoacridine- and amsacrine-4-carboxamides to DNA studied by DNase I footprinting. *Biochemistry* **31,** 3514–3524.
18. Bailly, C., Donkor, I. O., Gentle, D., Thornalley, M., and Waring, M. J. (1994) Sequence-selective binding of *cis* and *trans*-butamidine analogues of the anti-*Pneumocystis carinii* drug pentamidine. *Mol. Pharmacol.* **46,** 313–322.
19. Bailly, C., Perrine, D., Lancelot, J. C., Saturnino, C., Robba, M., and Waring, M. J. (1997) Sequence-selective binding to DNA of *bis*(amidinophenoxy)-alkanes related to propamidine and pentamidine. *Biochem. J.* **323,** 23–31.
20. Bailly, C. and Waring, M. J. (1993) Preferential intercalation at AT sequences in DNA by lucanthone, hycanthone, and indazole analogs: a footprinting study. *Biochemistry* **32,** 5985–5993.
21. Bailly, C., Hamy, F., and Waring, M. J. (1995) Cooperativity in the binding of echinomycin to DNA fragments containing closely spaced CpG sites. *Biochemistry* **35,** 1150–1161.
22. Lavesa, M. and Fox, K. R. (2001) Preferred binding sites for [N-MeCys(3), N-MeCys(7)]TANDEM determined using a universal footprinting substrate. *Anal. Biochem.* **293,** 246–250.
23. Joubert, A., Sun, X.-W., Johansson, E., Bailly, C., Mann, J., and Neidle, S. (2003) Sequence selective targeting of long stretches of the DNA minor groove by a novel dimeric *bis*-benzimidazole. *Biochemistry* **42,** 5984–5992.
24. Bailly, C. and Waring, M. J. (2001) Use of DNA molecules substituted with unnatural nucleotides to probe specific drug–DNA interactions. *Methods Enzymol.* **340,** 485–502.
25. Bailly, C., Payet, D., Travers, A. A., and Waring, M. J. (1996) PCR-based development of DNA substrates containing modified bases: an efficient system for

investigating the role of the exocyclic groups in chemical and structural recognition by minor groove binding drugs and proteins. *Proc. Natl. Acad. Sci. USA* **93**, 13,623–13,628.
26. Buttinelli, M., Minnock, A., Panetta, G., Waring, M. J., and Travers, A. A. (1998) The exocyclic groups of DNA modulate the affinity and positioning of the histone octamer. *Proc. Natl. Acad. Sci. USA* **95**, 8544–8549.
27. Crow, S. D. G., Bailly, C., Garbay-Jaureguiberry, C., Roques, B., Ramsay Shaw, B., and Waring, M. J. (2002) DNA sequence recognition by the antitumor drug ditercalinium. *Biochemistry* **41**, 8672–8682.
28. Trauger, J. W. and Dervan, P. B. (2001) Footprinting methods for analysis of pyrrole–imidazole polyamide/DNA complexes. *Methods Enzymol.* **340**, 450–466.
29. Müller, W. and Crothers, D. M. (1968) Studies of the binding of actinomycin and related compounds to DNA. *J. Mol. Biol.* **35**, 251–290.
30. Waring, M. J. (1970) Variation of the supercoils in closed circular DNA by binding of antibiotics and drugs. Evidence for molecular models involving intercalation. *J. Mol. Biol.* **54**, 247–279.
31. Takusagawa, F., Dabrow, M., Neidle, S., and Berman, H. M. (1982) The structure of a pseudo intercalated complex between actinomycin and the DNA binding sequence d(GpC). *Nature* **296**, 466–469.
32. Zhou, N., James, T. L., and Shafer, R. H. (1989) Binding of actinomycin D to [d(ATCGAT)]$_2$: NMR evidence of multiple complexes. *Biochemistry* **28**, 5231–5239.
33. Kamitori, S. and Takusagawa, F. (1994) Multiple binding modes of anticancer drug actinomycin D: X-ray, molecular modeling, and spectroscopic studies of d(GAAGCTTC)2-actinomycin D complexes and its host DNA. *J. Am. Chem. Soc.* **116**, 4154–4165.
34. Gallego, J., Ortiz, A. R., de Pascual-Teresa, B., and Gago, F. (1997) Structure-affinity relationships for the binding of actinomycin D to DNA. *J. Comput. Aided Mol. Des.* **11**, 114–128.
35. Sha, F. and Chen, F.-M. (2000) Actinomycin D binds strongly to d(CGAGACG) and d(CGTCGTCG). *Biophys. J.* **79**, 2095–2104.
36. Robinson, H., Gao, Y.-G., Yang, X., Sanishvili, R., Joachimiak, A., and Wang, A. H.-J. (2001) Crystallographic analysis of a novel complex of acinomycin D bound to the DNA decamer CGATCGATCG. *Biochemistry* **40**, 5587–5592.
37. Lane, M. J., Dabrowiak, J. C., and Vournakis, J. N. (1983) Sequence specificity of actinomycin D and netropsin binding to pBR322 DNA analyzed by protection from DNase I. *Proc. Natl. Acad. Sci. USA* **80**, 3260–3264.
38. Fox, K. R. and Waring, M. J. (1984) DNA structural variations produced by actinomycin and distamycin as revealed by DNAase I footprinting. *Nucleic Acids Res.* **12**, 9271–9285.
39. Goodisman, J., Rehfuss, R., Ward, B., and Dabrowiak, J. C. (1992) Site-specific binding constants for actinomycin D on DNA determined from footprinting studies. *Biochemistry* **31**, 1046–1058.
40. Fletcher, M. and Fox, K. R. (1996) Dissociation kinetics of actinomycin D from individual GpC sites in DNA. *Eur. J. Biochem.* **237**, 164–170.

41. Maniatis, T., Fritsch, E. F., and Sambrook, J. (1982) *Molecular Cloning: A Laboratory Manual.* Cold Spring Harbor Laboratory Press, Cold Spring Harbor, NY.
42. Drew, H. R. (1984) Structural specificities of five commonly used DNA nucleases. *J. Mol. Biol.* **176,** 535–557.
43. Fox, K. R. (1997) DNase I footprinting. In *Drug–DNA Interaction Protocols: Methods in Molecular Biology,* vol. 90 (Fox, K. R., ed.), Humana Press, Totowa, NJ, pp. 1–22.
44. Fox, K. R. and Waring, M. J. (2001) High-resolution footprinting studies of drug–DNA complexes using chemical and enzymic probes. *Methods Enzymol.* **340,** 412–430.
45. Fletcher, M. and Fox, K. R. (1993) Visualising the kinetics of dissociation of actinomycin from individual sites in mixed sequence DNA by DNase I footprinting. *Nucleic Acids Res.* **21,** 1339–1344.
46. Fletcher, M., Olsen, R. K., and Fox, K. R. (1995) Dissociation of the AT-specific bifunctional intercalator [N-MeCys^3,N-MeCys^7]TANDEM from TpA sites in DNA. *Biochem. J.* **306,** 15–19.
47. Bailly, C., Ridge, G., Graves, D. E., and Waring, M. J. (1994) Use of a photoactive derivative of actinomycin D to investigate shuffling between binding sites on DNA. *Biochemistry* **33,** 8736–8745.
48. Fox, K. R. and Waring, M. J. (1986) Footprinting reveals that nogalamycin and actinomycin shuffle between DNA binding sites. *Nucleic Acids Res.* **14,** 2001–2014.
49. Shafer, G. E., Price, M. A., and Tullius, T. D. (1989) Use of hydroxyl radical and gel electrophoresis to study DNA structure. *Electrophoresis* **10,** 397–404.
50. Price, M. A. and Tullius, T. D. (1992) Using hydroxyl radical to probe DNA structure. *Methods Enzymol.* **212,** 194–219.
51. Chow, C. S. and Barton, J. K. (1992) Transition metal complexes as probes of nucleic acids. *Methods Enzymol.* **212,** 219–242.
52. Routier, S., Vezin, H., Lamour, E., Bernier, J. L., Catteau, J. P., and Bailly, C. (1999) DNA cleavage by hydroxy-salen–iron complexes. *Nucleic Acids Res.* **27,** 4160–4166.
53. McPike, M. P., Goodisman, J., and Dabrowiak, J. C. (2001) Drug–RNA footprinting. *Methods Enzymol.* **340,** 431–449.
54. Yindeeyoungyeon, W. and Schell, M. A. (2000) Footprinting with an automated capillary DNA sequencer. *Biotechniques* **29,** 1034–1036.
55. Wilson, D. O., Johnson, P., and McCord, B. R. (2001) Nonradiochemical DNase I footprinting by capillary electrophoresis. *Electrophoresis* **22,** 1979–1986.

21

Gene Targeting Using Peptide Nucleic Acid

Peter E. Nielsen

Summary
A brief overview of the properties of the deoxyribonucleic acid mimic peptide nucleic acid (PNA) is given, and the recent progress in cellular delivery of PNA and of using PNA oligomers for antisense and antigene gene targeting is presented.

Key Words: Gene targeting; antisense; antigene; DNA recognition; peptide nucleic acid; PNA.

1. Introduction

Peptide nucleic acids (PNA) (**Fig. 1**) were originally conceived and developed as agents that sequence specifically target duplex deoxyribonucleic acid (DNA) by triple helix formation *(1,2)*. However, reality showed that although (homopyrimidine) PNAs bind very tightly and with excellent sequence discrimination to duplex DNA targets, they preferably do so via a helix invasion mechanism and not through conventional triple helix formation *(1–4)*. Nonetheless, the PNA structure does to a surprising extent mimic the structure of nucleic acids despite the fact that the chemical resemblance is minimal. In fact PNAs are essentially pseudopeptides (polyamides) with pendant nucleobases.

Over the past decade the properties of PNA have intrigued both chemists and biologists. Chemists to study, elaborate and modify the rather simple structure of PNA *(5,6)* and biologists to exploit the DNA and ribonucleic acid (RNA)-recognizing properties of PNA within antisense and antigene drug discovery, genetic diagnostics, and molecular biology *(7–11)*.

2. Backbone Modifications

The very straightforward chemistry of the PNA backbone is an open invitation to modification, and numerous analogs of the original aminoethylglycin

From: *Methods in Molecular Biology, vol. 288: Oligonucleotide Synthesis: Methods and Applications*
Edited by: P. Herdewijn © Humana Press Inc., Totowa, NJ

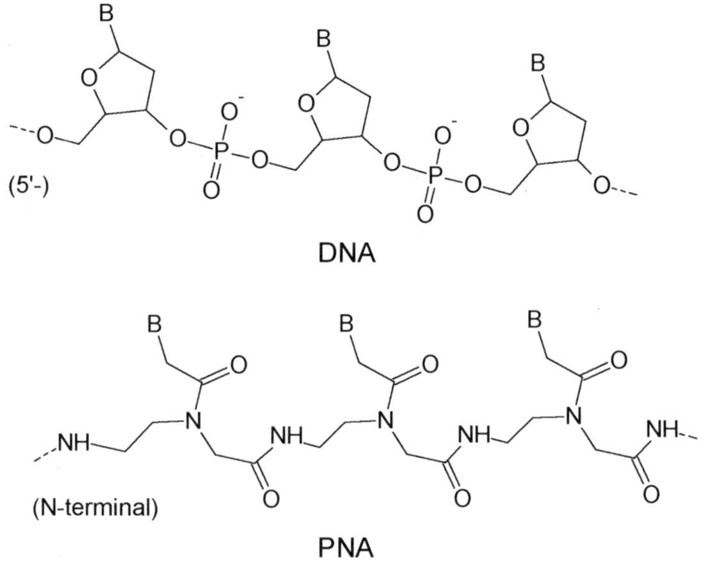

Fig. 1. Chemical structure of PNA compared to DNA.

(aeg) backbone have been designed, synthesized, and studied to understand the DNA recognition properties of peptide nucleic acidsEfforts have also been made to discover molecules with improved DNA/RNA hybridization properties. PNA is unique among high-affinity DNA analogs and mimics in having an acyclic backbone. Acyclic DNA derivatives show very poor hybridization efficacy *(12)*, which is explained by the high flexibility of such backbones leading to a greatly increased loss of entropy on duplex formation as compared to the much more conformationally constrained cyclic backbones, such as the deoxyribose phosphodiester of DNA. Clearly, the two amide bonds in the PNA backbone restrict the conformational flexibility of this backbone considerably. Accordingly, it was found that reduction of the nucleobase side-chain amide to the ethylamino backbone (**Fig. 2**) results in PNA oligomers with significantly inferior hybridization properties ($\Delta Tm - 20°C$ per modification) *(13)*. Along these lines, Leumann et al. have attempted to replace the amide function with an isostructural olefin function (O-PNA). However, such O-PNAs are inferior to aegPNA in terms of DNA/RNA hybridization *(14–16)*. This observation has not been explained, but it may well be connected to differences in hydration of the two backbones *(17)*.

Many attempts have been made to improve on the aegPNA backbone through introduction of a cyclic structure to impose restricted flexibility. The efforts have, however, so far only been met with limited success *(18)*. One

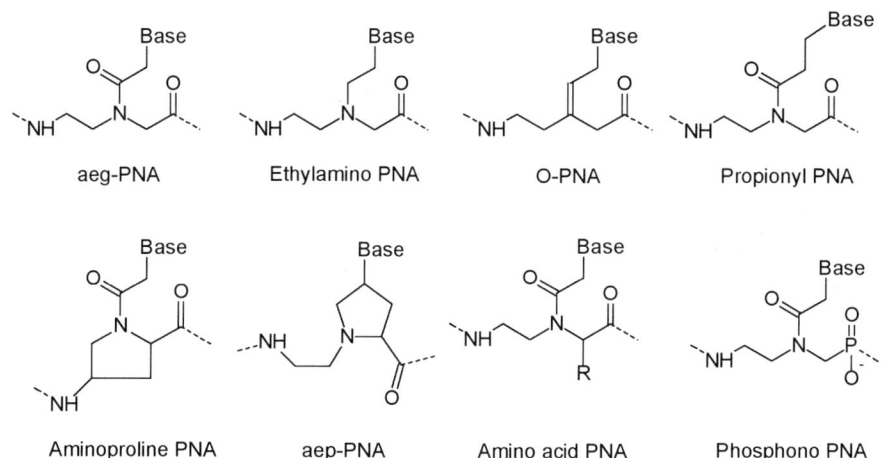

Fig. 2. Modified PNA backbones.

derivative, the aminoethylprolyl (aep) backbone *(19,20)* appears to stabilize the PNA–DNA triplex and at least some duplexes, but the limited data available so far indicate a significant sequence context variation, and more data are required to fully evaluate this backbone modification.

Finally, more nucleotide-like phosphono-PNA has been synthesized, and in particular in combination with prolyl backbone, this alternating structure confers favorable hybridization properties.

3. Non-Natural Nucleobases

A number of non-natural nucleobases have been exploited in a PNA backbone context (**Fig. 3**). Of these the pseudoisocytosine is very useful for pH-independent Hoogsteen recognition of guanine in triplex forming *bis*-PNA clamps *(21)*. The diaminopurine–thiouracil basepair has been exploited in the development of pcPNAs for efficient recognition of duplex DNA by double duplex invasion (*vide supra*) *(22)*. Finally, a range of bi- and tricyclic nucleobase analogs have been tested in efforts to exploit increased stacking interaction to stabilize PNA–DNA duplexes. A significant stabilization was observed with the bicyclic thymine analog (7-chloro-naphthyridinone) *(23)*, although a very dramatic (ΔT_m up to 15°C per modification) stabilization was obtained using the G-clamp *(24,25)*, which is a cytosine analog previously developed in a DNA context, and with similar effects on hybridization potency. It is, however, noteworthy that antisense oligonucleotides containing G-clamps do not appear more potent in cellular antisense assays than unmodified oligonucleotides *(26)*.

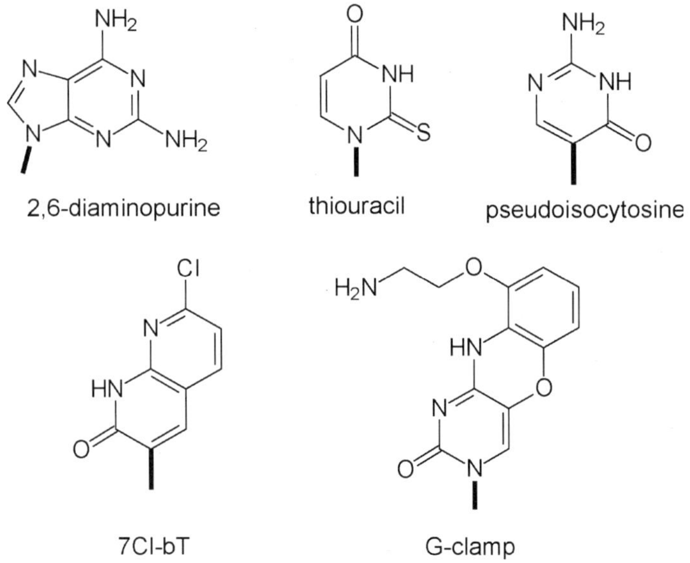

Fig. 3. Examples of nonstandard nucleobases that have been in used PNA oligomers.

4. Genetic Detection

PNA oligomers have been used to improve the performance of a number of "diagnostic" techniques for genetic detection. Two methods in particular are finding widespread applications. PNA clamping of polymerase chain reactions (PCRs) by which a PNA is blocking a PCR primer very effectively suppresses amplification of undesired templates. This is especially useful for point mutation analysis in which two targets differing by only one nucleobase can be selectively amplified *(27–29)*. For instance, it has been shown that single mutant oncogenes in cancer cells from human tissues can be detected in a background of 10^+-fold normal cells *(29)*. Fluorescent *in situ* hybridization(FISH) is another area in which PNA probes perform extremely well *(30–32)*. Finally, PNA beacons *(33)* and other "light-up" probes *(34)* are attracting increasing interest.

5. Antisense mRNA Targeting

PNA–RNA duplexes are not substrates for ribonuclease H. Cellular antisense inhibition of gene expression by PNA must therefore rely on other mechanisms, in particular steric blocking of the translational machinery (ribosomes) or of messenger RNA (mRNA) processing enzymes (e.g., the splicesome). Both principles appear as viable routes to PNA antisense targeting. Although ribosomal translation elongation is not readily arrested by PNAs (or

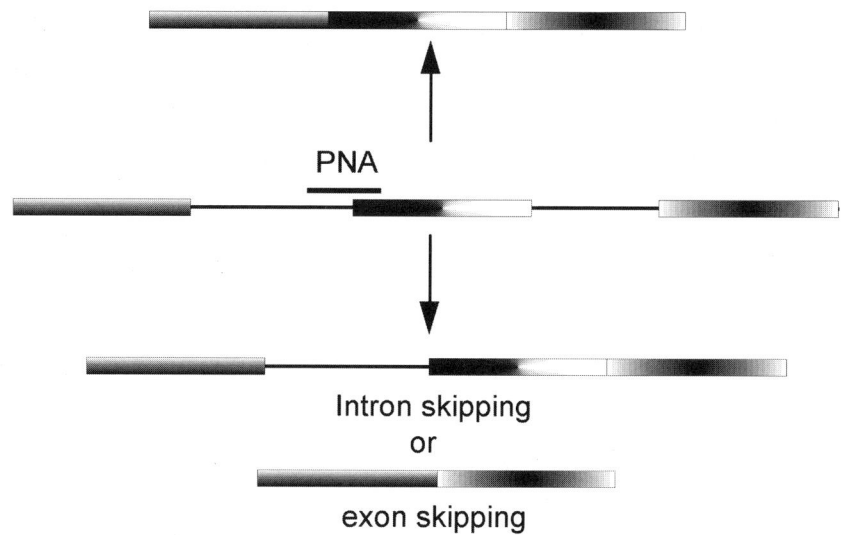

Fig. 4. Effect of PNA targeting of a pre-mRNA intron–exon splice junction. Thick lines signify exons, and thin lines signify introns.

other oligonucleotide analogs or mimics) bound to the coding part of the mRNA *(35–37)*, PNA oligomers targeted to the translation initiation (AUG) region or to the 5-UTR (untranslated region) of the mRNA effectively cause inhibition of protein synthesis, presumably by interfering with ribosome scanning and assembly prior to translation initiation *(38)*.

Likewise, recent results have demonstrated that PNA oligomers targeted to exon–intron splice junctions are potent inhibitors of correct mRNA splicing presumably by steric interference with the spliceosome *(39,40)*. Although no systematic studies have yet been published, at least two outcomes of targeting a 5'-junction (**Fig. 4**) have been found. The spliceosome may either skip the entire exon and thus produce a truncated mRNA missing an exon, or it may skip the intron in the processing and thereby produce a larger mRNA still containing the intron (**Fig. 5**). Therefore, in a drug discovery or molecular biology context, targeting of splice junctions may (in principle) be exploited to correct the splicing of mRNA containing aberrant splice sites (because of mutation), to simply inhibit the synthesis of the correct mRNA (and thus protein product), or perhaps to shift the ratio between biologically functional splice variants. One may also imagine inducing non-natural splice variants with altered (novel) biological function of the resulting protein product. Therefore, splice interference technology may open a range of novel opportunities for gene targeting and drug discovery.

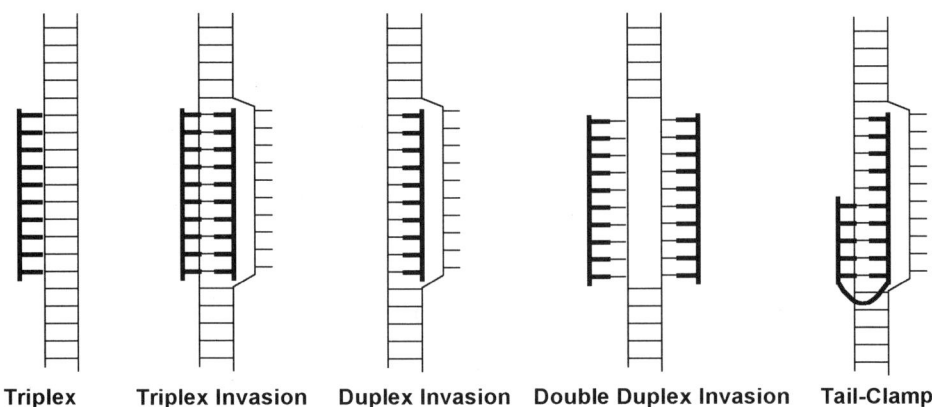

Fig. 5. Five different types of PNA–double-stranded DNA complexes. DNA is schematically drawn as a ladder, and the PNA oligomers are in bold.

6. Duplex DNA Targeting

The concept of PNA was originally aimed at triplex recognition of double-stranded DNA *(1,2)*. However, in contrast to triplex-forming oligonucleotides, homopyrimidine PNAs invade the double helix through the formation of triplex invasion complexes (**Figs. 5** and **6**). Conventional triplexes can be observed with cytosine-rich homopyrimidine PNAs, especially at elevated ionic strength, where triplex invasion is very slow, or as kinetic intermediates (unpublished results and *41,42*). Duplex invasion is seen with mixed purine–pyrimidine PNAs binding to negatively supercoiled DNA, in particular using positively charged PNA–peptide conjugates *(43,44)* or with high-affinity very purine-rich PNAs *(45)*. Mixed purine–pyrimidine sequences can be targeted using sets of pcPNAs, for example, those containing diaminopurine–thiouracil basepairs. Such pcPNAs have greatly reduced affinity for each other owing to steric clashes in the diaminopurine–thiouracil basepair, but bind sequence-complementary DNA with essentially unchanged affinity *(22)*. Finally, triplex and duplex invasion can be combined in two-domain tail-clamp PNAs.

It is well established from in vitro cell-free experiments that triplex invasion complexes are effective inhibitors of transcription initiation as well as of transcription elongation *(35,46)*. Most intriguingly, it was also demonstrated several years ago that these invasion complexes are recognized by RNA polymerase as transcription initiation sites *(47)*. Thus PNA may "mimic" both a transcriptional repressor as well as an activator (transcription factor). Although these effects have also been reported in cell culture experiments *(48,49)*, the results are less clear-cut, and other laboratories are having difficulties reproducing PNA gene activation in cell culture (unpublished results and

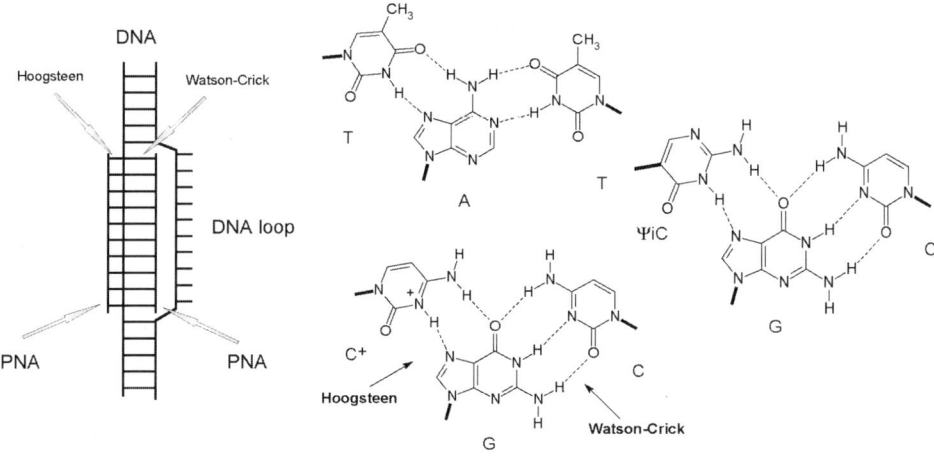

Fig. 6. Triplex invasion by homopyrimidine PNA oligomers. One PNA strand binds via Watson–Crick basepairing (preferably in the antiparallel orientation), whereas the other binds via Hoogsteen basepairing (preferably in the parallel orientation). It is usually advantageous to connect the two PNA strands covalently via a flexible linker into a *bis*-PNA, and to substitute all cytosines in the Hoogsteen strand with pseudoisocytosines (ψiC), which does not require low pH for N3 protonation.

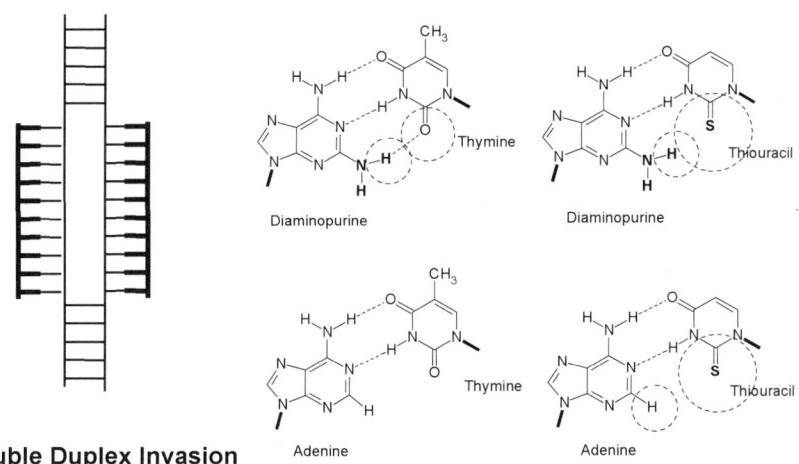

Fig. 7. Double duplex invasion of pseudocomplementary PNAs. To obtain efficient binding the target (and thus the PNAs) should contain at least 50% AT (no other sequence constraints), and in the PNA oligomers all A/T basepairs are substituted with 2,6-diaminopurine/2-thiouracil basepairs. This basepair is very unstable owing to steric hindrance. Therefore the two sequence-complementary PNAs will not be able to bind each other, but they bind their DNA complement very well.

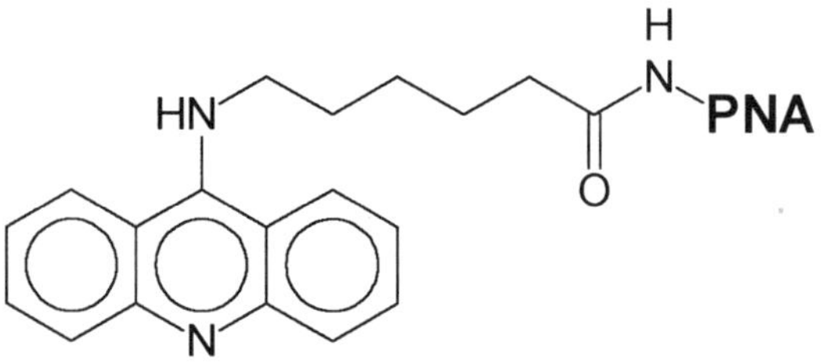

Fig. 8. Acridine–PNA conjugate.

personal communications). Targeting duplex DNA in vivo may also be a challenge because it is known from in vitro studies that physiological ionic conditions (e.g., 140 mM K$^+$, 2 mM Mg^{2+}) are inhibitory to effective helix invasion *(50–52)*. Although natural negative DNA supercoiling may to a large extent alleviate this obstacle *(53)*, and it is possible to construct PNA–peptide in PNA–acridine conjugates (**Fig. 8**) that are able to invade duplex DNA at physiologically relevant ionic conditions *(43,52)*, targeting DNA in a nuclear context is an area where quantitative and well-controlled experiments are highly warranted.

7. Cellular Delivery

Any gene targeting approach requires the PNA oligomer to reach the target in the cell cytoplasm or nucleus. Unfortunately, the largely hydrophilic PNA molecules do not spontaneously cross lipid membranes *(54)*, and they are therefore in general very poorly taken up by living cells *(55,56)*. Within the past few years a number of protocols have been developed that give greatly improved intracellular bioavailability of PNA. Two protocols exploit cationic liposomes, analogous to the standard method for oligonucleotide delivery. However, because PNAs are not inherently anionic and therefore do not spontaneously assemble into PNA–liposome complexes, an assembly auxiliary is required. This may be a fatty acid conjugated to the PNA *(57)* preferably through a cleavable inker *(58)*, but the more general and effective approach is to use a partly complementary oligonucleotide to hybridize to the PNA and thereby "piggyback" the PNA into the cationic liposomes and further into the cell *(59)*.

Alternatively a variety of cell-penetrating peptides have been used via chemical conjugation to PNA oligomers (**Table 1**). It was originally reported that several of these peptides (e.g., pAntp) transverse the membrane by a carrier- and energy-independent mechanism *(60)*, but recent careful reevaluations clearly indicate that the main cellular internalization route is via an endosomal pathway *(56,61,62)*. Nonetheless, PNA–peptide conjugates in particular using transportan *(62,63)* or oligo–arginine *(64)* are quite potent cellular antisense

Table 1
Cell Penetrating Pepides

Peptide	Sequence	Cell type[a]	Reference
penetratin (pAntp)	RQIKIWFQNRRMKWKK	JR8/M14, human melanoma	72
Transportan	GWTLNSAGYLLGKINLAALAKKIL[b]	Bowes cells	63
Retro–inverso penetratin	D-(KKWKMRRNQFWVKVQR)	rat neurones	73
pTat	GRKKRRQRRRPPQ	HeLa	62
SV40 Nuclear Localization Signal (NLS)	PKKKRKV	Burkitt's lymphoma	74
Somatostatin	DKDYKDYDK	IMR32	75
	MSVLTPLLRGLTGSARRLPVPRAKIHSL	Mitochondria in IMR32, HeLa, a.o	76
KFF	KFFKFFKFFK	*Escherichia coli*	77
Polyarginine	R9		78

[a]Demonstrated cell types.
[b]Conjugated to the PNA via a disulfide bridge at the central lysine.

agents. However, it is also clear that better delivery systems for PNA are still very much desired.

8. In Vivo Experiments

Only a handful of in vivo animal studies have so far been published using PNA *(65–68)*. Of these a major part concerns neurological targets in the brain *(65–67)*. Unfortunately, these studies do not comply with good standards for antisense studies in terms of proper mismatch controls and/or molecular biology end points (mRNA and/or protein levels including controls), and independent confirmation of the conclusion should be awaited. In one study effective passage of PNA oligomers over the blood–brain barrier was also reported *(67)*, but this could not be independently confirmed as a general feature of PNA *(69)*.

However, a very convincing study showing systemic antisense effects of PNA in a mouse model was recently published *(68)*. In this study Kole and coworkers used a transgenic mouse with an aberrant splice variant of GFP. On treatment with PNA (or comparable antisense agents such as morpholino) targeted to the exon–intron junction of the aberrant splice site, the pre-mRNA is correctly processed, and the cells and tissues thus express functional GFP. Using this model it was shown that intraperitoneal administration of PNA results in significant antisense activation of green fluorescent protein (GFP) in liver, intestine, heart, and kidneys *(68)*.

In this context it is also worth noting that the pharmacokinetic profile of simple PNAs is not very favorable for in vivo activity as the compound exhibits a very small volume of distribution, low tissue availability, and is excreted very quickly through the kidneys *(70,71)*. Thus it should be possible to obtain much higher in vivo efficacy with PNA derivatives having improved bioavailability.

9. Prospects

The PNA field is steadily moving forward in all areas: chemistry, molecular biology, diagnostics, and cell biology, and also very slowly into drug discovery. The recent results in mouse models as well as the emerging information on the bioavailability and pharmacokinetic behavior of PNA oligomers in animals are extremely encouraging. Thus scientist in all areas are still being inspired by the very simple PNA structure and are still able to bring new and exiting ideas into the field, which may eventually also lead this molecule onto new avenues, such as nanotechnology or artificial life.

References

1. Nielsen, P. E., Egholm, M., Berg, R. H., and Buchardt, O. (1991) Sequence selective recognition of DNA by strand displacement with a thymine-substituted polyamide. *Science* **254,** 1497–1500.
2. Nielsen, P. E., Egholm, M., Berg, R. H., and Buchardt, O. (1993) Peptide nucleic acids (PNA): DNA analogues with a polyamide backbone. In *Antisense Research and Application* (Crook, S. and Lebleu, B., eds.).CRC Press, Boca Raton, FL, pp. 363–373.
3. Cherny, D. Y., Belotserkovskii, B. P., Frank-Kamenetskii, M. D., Egholm, M., Buchardt, O., Berg, R. H. and Nielsen, P. E. (1993) DNA unwinding upon strand displacement of binding of PNA to double stranded DNA. *Proc. Natl. Acad. Sci. USA* **90,** 1667–1670.
4. Nielsen, P. E., Egholm, M., and Buchardt, O. (1994) Evidence for (PNA)$_2$/DNA triplex structure upon binding of PNA to dsDNA by strand displacement. *J. Mol. Recognit.* **7,** 165–170.
5. Ganesh, K. N. and Nielsen, P. E. (2000) Peptide nucleic acids: analogs and derivatives. *Curr. Org. Chem.* **4,** 931–943.
6. Nielsen, P. E. (1999) Peptide nucleic acid: a molecule with two identities. *Acc. Chem. Res.* **32,** 624–630.
7. Nielsen, P. E. (2001) Peptide nucleic acid: a versatile tool in genetic diagnostics and molecular biology. *Curr. Opin. Biotechnol.* **12,** 16–20.
8. Nielsen, P. E. (2001) Peptide nucleic acids as antibacterial agents via the antisense principle. *Expert Opin. Investig. Drugs* **10,** 331–341.
9. Nielsen, P. E. (2001) Targeting double stranded DNA with peptide nucleic acid (PNA). *Curr. Med. Chem.* **8,** 545–550.
10. Ray, A and Nordén, B. (2000) Peptide nucleic acid (PNA): its medical and biotechnical applications and promise for the future. *FASEB J.* **14,** 1041–1060.
11. Braasch, D. A. and Corey, D. R. (2002) Novel antisense and peptide nucleic acid strategies for controlling gene expression. *Biochemistry* **41,** 4503–4510.
12. Nielsen, P. E. (1995) DNA analogs with nonphosphodiester backbones. In *Annual Review of Biophysics and Biomolecular Structure*, Stroud, R. M., ed., pp. 167–183.
13. Hyrup, B., Egholm, M., Buchardt, O., and Nielsen, P. E. (1996) A flexible and positively charged PNA analogue with an ethylene-linker to the nucleobase: synthesis and hybridization properties. *Bioorg. Med. Chem. Lett.* **6,** 1083–1088.
14. Cantin, M., Schütz, R., and Leumann, C. J. (1997) Synthesis of the monomeric building blocks of Z-olefinic PNA (Z-OPA) containing the bases adenine and thymine. *Tetrahedron Lett.* **38,** 4211–4214.
15. Roberts, C. D., Schütz, R., and Leumann, C. J. (1999) The synthesis of a thymine-containing E-olefinic peptide nucleic acid (OPA) monomer. *Synlett* 819–821.
16. Schütz, R., Cantin, M., Roberts, C., Greiner, B., Uhlmann, E., and Leumann, C. (2000) Olefinic peptide nucleic acids (OPAs): new aspects of the molecular recognition of DNA by PNA. *Angew. Chem. Int. Ed. Engl.* **39,** 1250–1253.

17. Hollenstein, M. and Leumann, C. J. (2003) Synthesis and incorporation into PNA of fluorinated olefinic PNA (F-OPA) monomers. *Org. Lett.* **5,** 1987–1990.
18. Kumar, V. A. (2002) Structural preorganization of peptide nucleic acids: chiral cationic analogues with five- or six-membered ring structures. *Eur. J. Org. Chem.* 2021–2032.
19. D'Costa, M., Kumar, V. A., and Ganesh, K. N. (1999) Aminoethylprolyl peptide nucleic acids (aepPNA): chiral PNA analogues that form highly stable DNA:aepPNA2 triplexes. *Org. Lett.* **1,** 1513–1516.
20. D'Costa, M., Kumar, V., and Ganesh, K.N. (2001) Aminoethylprolyl (aep) PNA: mixed purine/pyrimidine oligomers and binding orientation preferences for PNA:DNA duplex formation. *Org. Lett.* **3,** 1281–1284.
21. Egholm, M., Christensen, L., Dueholm, K., Buchardt, O., Coull, J., and Nielsen, P. E. (1995) Efficient pH independent sequence specific DNA binding by pseudoisocytosine-containing *bis*-PNA. *Nucleic Acids Res.* **23,** 217–222.
22. Lohse, J., Dahl, O., and Nielsen, P. E. (1999) Double duplex invasion by peptide nucleic acid: a general principle for sequence-specific targeting of double-stranded DNA. *Proc. Natl. Acad. Sci. USA* **96,** 11,804–11,808.
23. Eldrup, A. B., Christensen, C., Haaima, G., and Nielsen, P. E. (2002) Substituted 1,8-naphthyridin-2(1H)-ones are superior to thymine in the recognition of adenine in duplex as well as triplex structures. *J. Am. Chem. Soc.* **124,** 3254–3262.
24. Rajeev, K. G., Maier, M. A., Lesnik, E. A., and Manoharan, M. (2002) High-affinity peptide nucleic acid oligomers containing tricyclic cytosine analogues. *Org. Lett.* **4,** 4395–4398.
25. Ausín, C., Ortega, J.-A., Robles, J., Grandas, A., and Pedroso, E. (2002) Synthesis of amino- and guanidino-G-clamp PNA monomers. *Org. Lett.* **4,** 4073–4075.
26. Holmes, S. C., Arzumanov, A. A., and Gait, M. J. (2003) Steric inhibition of human immunodeficiency virus type-1 Tat-dependent trans-activation in vitro and in cells by oligonucleotides containing 2'-O-methyl G-clamp ribonucleoside analogues. *Nucleic Acids Res.* **31,** 2759–2768.
27. Ørum, H., Nielsen, P. E., Egholm, M., Berg, R. H., Buchardt, O., and Stanley, C. (1993) Single base pair mutation analysis by PNA directed PCR clamping. *Nucleic Acids Res.* **21,** 5332–5336.
28. Thiede, C., Bayerdörffer, E., Blasczyk, R., Wittig, B., and Neubauer, A. (1996) Simple and sensitive detection of mutations in the ras proto-oncogenes using PNA-mediated PCR clamping. *Nucleic Acids Res.* **24,** 983, 984.
29. Behn, M., Thiede, C., Neubauer, A., Pankow, W., and Schuermann, M. (2000) Facilitated detection of oncogene mutations from exfoliated tissue material by a PNA-mediated "enriched PCR" protocol. *J. Pathol.* **190,** 69–75.
30. Lansdorp, P. M., Verwoerd, N. P., van de Rijke, F. M., et al. (1996) Heterogeneity in telomer length of human chromosomes. *Hum. Mol. Genet.* **5,** 685–691.
31. Chen, C., Wu, B. L., Wei, T., Egholm, M., and Strauss, W. M. (2000) Unique chromosome identification and sequence-specific structural analysis with short PNA oligomers. *Mamm. Genome* **11,** 384–391.

32. Perry-O'Keefe, H., Broomer, A. J., Oliveira, K., Coull, J., and Hyldig-Nielsen, J. J. (2001) Filter based PNA *in situ* hybridization for rapid detection, identification and enumeration of specific microorganisms. *J. Appl. Microbiol.* **90,** 180–189.
33. Kuhn, H., Demidov, V. V., Gildea, B. D., Fiandaca, M. J., Coull, J. C., and Frank-Kamenetskii, M. D. (2001) PNA beacons for duplex DNA. *Antisense Nucleic Acid Drug Dev.* **11,** 265–270.
34. Svanvik, N., Westman, G., Wang, D., and Kubista, M. (2000) Light-up probes: thiazole orange-conjugated peptide nucleic acid for detection of target nucleic acid in homogeneous solution. *Anal. Biochem.* **281,** 26–35.
35. Hanvey, J. C., Peffer, N. C., Bisi, J. E., et al. (1992) Antisense and antigene properties of peptide nucleic acids. *Science* **258,** 1481–1485.
36. Knudsen, H. and Nielsen, P. E. (1996) Antisense properties of duplex and triplex forming PNA. *Nucleic Acids Res.* **24,** 494–500.
37. Summerton, J. (1999) Morpholino antisense oligomers: the case for an RNase H-independent structural type. *Biochim. Biophys. Acta* **1489,** 141–158.
38. Doyle, D. F., Braasch, D. A., Simmons, C. G., Janowski, B. A., and Corey, D. R. (2001) Inhibition of gene expression inside cells by peptide nucleic acids: effect of mRNA target sequence, mismatched bases, and PNA length. *Biochemistry* **40,** 53–64.
39. Karras, J. G., Maier, M. A., Lu, T., Watt, A., and Manoharan, M. (2001) Peptide nucleic acids are potent modulators of endogenous pre–mRNA splicing of the murine interleukin–5 receptor–alpha chain. *Biochemistry* **40,** 7853–7859.
40. Sazani, P., Kang, S. H., Maier, M. A., et al. (2001) Nuclear antisense effects of neutral, anionic and cationic oligonucleotide analogs. *Nucleic Acids Res.* **29,** 3965–3974.
41. Praseuth, D., Grigoriev, M., Guieysse, A. L., et al. (1997) Peptide nucleic acids directed to the promoter of the α-chain of the interleukin-2 receptor. *Biochim. Biophys. Acta* **1309,** 226–238.
42. Wittung, P., Nielsen, P. E., and Norden, B. (1997) Extended DNA-recognition repertoire of PNA. *Biochemistry* **36,** 7973–7979.
43. Kaihatsu, K., Braasch, D. A., Cansizoglu, A., and Corey, D. R. (2002) Enhanced strand invasion by peptide nucleic acid–peptide conjugates. *Biochemistry* **41,** 11,118–11,125.
44. Zhang, X., Ishihara, T., and Corey, D. R. (2000) Strand invasion by mixed base PNAs and a PNA-peptide chimera. *Nucleic Acids Res.* **28,** 3332–3338.
45. Nielsen, P. E. and Christensen, L. (1996) Strand displacement binding of a duplex-forming homopurine PNA to a homopyrimidine duplex DNA target. *J. Am. Chem. Soc.* **118,** 2287, 2288.
46. Nielsen, P. E., Egholm, M., and Buchardt, O. (1994) Sequence specific transcription arrest by PNA bound to the template strand. *Gene* **149,** 139–145.
47. Møllegaard, N. E., Buchardt, O., Egholm, M., and Nielsen, P. E. (1994) PNA–DNA strand displacement loops as artificial transcription promoters. *Proc. Natl. Acad. Sci. USA* **91,** 3892–3895.

48. Wang, G., Xu, X., Pace, B., et al. (1999) Peptide nucleic acid (PNA) binding-mediated induction of human γ-globin gene expression. *Nucleic Acids Res.* **27,** 2806–2813.
49. Wang, G., Jing, K., Balczon, R., and Xu, X. (2001) Defining the peptide nucleic acids (PNA) length requirement for PNA binding-induced transcription and gene expression. *J. Mol. Biol.* **313,** 933–940.
50. Peffer, N. J., Hanvey, J. C., Bisi, J. E., et al. (1993) Strand-invasion of duplex DNA by peptide nucleic acid oligomers. Proc. Natl. Acad. Sci. USA **90,** 10,648–10,652.
51. Kurakin, A., Larsen, H. J., and Nielsen, P. E. (1998) Coorperative duplex invasion of peptide nucleic acids (PNA) *Chem. Biol.* **5,** 81–89.
52. Bentin, T. and Nielsen, P. E. (2003) Superior duplex DNA strand invasion by acridine conjugated peptide nucleic acids. *J. Am. Chem. Soc.* **125,** 6378, 6379.
53. Bentin, T. and Nielsen, P. E. (1996) Enhanced peptide nucleic acid (PNA) binding to supercoiled DNA: possible implications for DNA "breathing" dynamics. *Biochemistry* **35,** 8863–8869.
54. Wittung, P., Kajanus, J., Edwards, K., Nielsen, P. E., Norden, B., and Malmström, B.G. (1995) Phospholipid membrane permeability of peptide nucleic acid. *FEBS Lett.* **365,** 27–29.
55. Bonham, M. A., Brown, S., Boyd, A. L., et al. (1995) An assessment of the antisense properties of RNAse H-competent and steric-blocking oligomers. *Nucleic Acids Res.* **23,** 1197–1203.
56. Koppelhus, U., Awasthi, S. K., Zachar, V., Holst, H. U., Ebbesen, P., and Nielsen, P. E. (2002) Cell-dependent differential cellular uptake of PNA, peptides, and PNA–peptide conjugates. *Antisense Nucleic Acid Drug Dev.* **12,** 51–63.
57. Ljungstrøm, T., Knudsen, H., and Nielsen, P. E. (1999) Cellular uptake of adamantyl conjugated peptide nucleic acids. *Bioconjug. Chem.* **10,** 965–972.
58. Bendifallah, N., Kristensen, E., Dahl, O., and Nielsen, P. E. (2003) Synthesis and properties of ester-linked peptide nucleic acid (PNA) prodrug conjugates. *Bioconjug. Chem.* **14,** 588–592.
59. Hamilton, S. E., Simmons, C. G., Kathiriya, I. S., and Corey, D. R. (1999) Cellular delivery of peptide nucleic acids and inhibition of human telomerase. *Chem. Biol.* **6,** 343–351.
60. Derossi, D., Chassaing, G., and Prochiantz, A. (1998) Trojan peptides: the penetratin system for intracellular delivery. *Trends Cell Biol.* **8,** 84–87.
61. Richard, J. P., Melikov, K., Vives, E., et al. (2003) Cell-penetrating peptides: a reevaluation of the mechanism of cellular uptake. *J. Biol. Chem.* **278,** 585–590.
62. Thierry, A. R., Vives, E., Richard, J. P., et al. (2003) Cellular uptake and intracellular fate of antisense oligonucleotides. *Curr. Opin. Mol. Ther.* **5,** 133–138.
63. Pooga, M., Soomets, U., Hällbrink, M., et al. (1998) Cell penetrating PNA constructs regulate galanin receptor levels and modify pain transmission in vivo. *Nat. Biotechnol.* **16,** 857–861.
64. Wender, P. A., Mitchell, D. J., Pattabiraman, K., Pelkey, E. T., Steinman, L., and Rothbard, J. B. (2000) The design, synthesis, and evaluation of molecules that

enable or enhance cellular uptake: peptoid molecular transporters. *Proc. Natl. Acad. Sci. USA* **97,** 13,003–13,008.
65. Fraser, G. L., Holmgren, J., Clarke, P. B. S., and Wahlestedt, C. (2000). Antisense inhibition of delta-opioid receptor gene function in vivo by peptide nucleic acids. *Mol. Pharm.* **57,** 725–731.
66. McMahon, B. M., Stewart, J. A., Bitner, M. D., Fauq, A., McCormick, D. J., and Richelson, E. (2002) Peptide nucleic acids specifically cause antigene effects in vivo by systemic injection. *Life Sci.* **71,** 325–337.
67. Tyler, B. M., Jansen, K., McCormick, D. J., et al. (1999) Peptide nucleic acids targeted to the neurotensin receptor and administered i.p. cross the blood–brain barrier and specifically reduce gene expression. *Proc. Natl. Acad. Sci. USA* **96,** 7053–7058.
68. Sazani, P., Gemignani, F., Kang S.-H., et al. (2002) Systemically delivered antisense oligomers upregulate gene expression in mouse tissues. *Nat. Biotechnol.* **20,** 1228–1233.
69. McMahon, B. M., Mays, D., Lipsky, J., Stewart, J. A., Fauq, A., and Richelson, E. (2002) Pharmacokinetics and tissue distribution of a peptide nucleic acid after intravenous administration. *Antisense Nucleic Acid Drug Dev.* **12,** 65–70.
70. Kristensen, E. (2002) In vitro and in vivo studies on pharmacokinetics and metabolism of PNA constructs in rodents. In *Peptide Nucleic Acids: Methods and Protocols* (Nielsen, P. E., ed.). Humana Press, Totowa, NJ, pp. 259–269.
71. Hamzawi, R., Dolle, F., Tavitian, R., Dahl, O., and Nielsen, P. E. (2004) Modulation of the pharmacokinetic properties of PNA: preparation of galactosyl, mannosyl, fucosyl, N-acetyl-galactosaminyl and N-acetyl-glucosaminyl derivatives of aminoethylglycin peptide nucleic acid monomers and their incorporation into PNA oligomers. **14,** 941–954.
72. Villa, R., Folini, M., Lauldi, S., Veronese, S., Daidone, M. G., and Zaffaroni, N. (2001) Inhibition of telomerase activity by a cell-penetrating peptide nucleic acid construct in human melanoma cells. *FEBS Lett.* **473,** 241–248.
73. Aldrian-Herrada, G., Desarm nien, M. G., Orcel, H., et al. (1998) A peptide nucleic acid (PNA) is more rapidly internalized in cultured neurons when coupled to a retro-inverso delivery peptide. The antisense activity depresses the target mRNA and protein in magnocellular oxytocin neurons. *Nucl. Acids Res.* **26,** 4910–4916.
74. Cutrona, G., Carpaneto, E. M., Ulivi, M., et al. (2000) Effects in live cells of a c-*myc* anti-gene PNA linked to a nuclear localization signal. *Nat. biotechnol.* **18,** 300–303.
75. Sun, L., Fuselier, J. A., Murphy W. A., and Coy, D. A. (2002) Antisense peptide nucleic acids conjugated to somatostatin analogs and targeted at the n-myc oncogene display enhanced cytotoxicity to human neuroblastoma IMR32 cells expressing somatostatin receptors. *Peptides* **23,** 1557–1565.
76. Chinnery, P. F., Taylor, R. W., Diekert, K., Lill, R., Turnbull, D. M., and Lightowlers, R. N. (1999) Peptide nucleic acid delivery to human mitochondria. *Gene Ther.* **19,** 1919–1928.

77. Good, L., Awasthi, S. K., Dryselius, R., Larsson, O., and Nielsen, P. E. (2001) Bactericidal antisense effects of peptide-PNA conjugates. *Nat. Biotechnol.* **19,** 360–364.
78. Wender, P. A., Mitchell, D. J., Pattabiraman, K., Pelkey, E. T., Steinman, L., and Rothbard, J. B. (2000) The design, synthesis, and evaluation of molecules that enable or enhance cellular uptake: peptoid molecular transporters. *Proc. Natl. Acad. Sci. USA* **97,** 13,003–13,008.

22

Nucleic Acid Library Construction Using Synthetic DNA Constructs

Hani S. Zaher and Peter J. Unrau

Summary

This chapter outlines seven synthetic and molecular biology techniques that allow the controlled synthesis of nucleic acid libraries. Specifically: (1) The high-diversity chemical synthesis of point mutations; (2) the high-diversity chemical synthesis of point deletions; (3) the split-bead approach for constructing point mutation or deletion libraries with limited sequence diversity; (4) pool deprotection, gel purification, and quality-control techniques; (5) large-scale polymerase chain reaction amplification for the generation of high-diversity double-stranded deoxyribonucleic acid libraries; (6) type II restriction enzyme digestion techniques for the construction of long-sequence libraries containing minimal fixed sequence; and (7) extension techniques for the rapid synthesis of long, low-diversity oligonucleotide sequences.

Key Words: Diversity; nucleicacid; libraries; DNA synthesis; point mutation or deletions; equiactive phosphoramidite; split-bead synthesis; codon libraries; large-scale pool amplification; PCR reaction; linking pool segments with nonpalindromic spacer.

1. Introduction

The diversity, or the number of distinct sequences, in a nucleic acid library can span an enormous range. High-diversity libraries containing 10^{12}–10^{16} different sequences have been used to screen for nucleic acid aptamers, ribozymes, and functional proteins *(1–5)*. Pools with lower diversity, in the range of 10^8–10^{10}, are commonly used in phage display libraries *(6,7)*. Finally, pools with very low sequence variation are often highly desirable, as in the recent *de novo* synthesis of a 7.5-kb viral genome *(8)*. This chapter discusses techniques that allow the synthesis of nucleic acid libraries spanning this range of diversity.

From: *Methods in Molecular Biology, vol. 288: Oligonucleotide Synthesis: Methods and Applications*
Edited by: P. Herdewijn © Humana Press Inc., Totowa, NJ

When planning the construction of a nucleic acid pool it is important to consider how different synthesis strategies can affect the overall diversity of the pool. Incomplete sampling may result if the desired diversity cannot be achieved by a given method. This limitation may be unavoidable (for example, a 50-nt-long random sequence pool can only be completely represented using 1.3×10^{30} different sequences—in other words, 2 million moles of nucleic acid!), but the possible consequences of a given synthesis technique should be carefully considered before building a library.

A typical 0.2-μmol solid-phase deoxyribonucleic acid (DNA) synthesis cartridge contains controlled pore glass (CPG) (*see* **Fig. 1**) beads chemically derivatized to allow the simultaneous synthesis of up to 10^{17} different DNA oligonucleotides. This large number of reactive sites makes possible the construction of high-diversity pools containing extensive point mutations or deletions using chemical synthesis techniques (*see* **Subheadings 3.1. and 3.2.**). The synthesis of pools containing point mutations or deletions in limited areas of the pool is often useful and can easily be performed without changing the chemistry on a DNA synthesizer using the split-bead strategy (*see* **Subheading 3.3.**). During the split-bead synthesis, fractions of the CPG beads are removed and subjected to independent DNA synthesis. Remixing these fractions and continuing synthesis creates sequence diversity that is limited ultimately by the number of beads (10^6–10^7 for a 0.2-μmol scale column) available in the synthesis.

Understanding the limitations of DNA phosphoramidite chemistry *(9)* is also critical for constructing optimal nucleic acid pools. The yield for long oligonucleotides can suffer from steric problems that result from the gradual filling of the CPG pores during synthesis. To maximize yield, 1000 Å CPG should be used for oligonucleotides longer than about 50 nt and 2000 Å CPG for oligonucleotides 110 nt or longer. Smaller pore sizes can greatly inhibit final yield and should be avoided. Phosphoramidite coupling efficiency (typically 98–99%) also limits oligonucleotide length and yield. Chemical capping of the small amount of uncoupled oligonucleotide produced in each synthesis cycle is not completely efficient, and consequently low levels of point deletions (typically approx 0.2% per residue) are also introduced during the synthesis. Together, incomplete coupling and capping constrain the design of nucleic acid libraries by limiting both sequence diversity and oligonucleotide length.

After the chemical synthesis of the single-stranded DNA required to construct a pool, it must be deprotected, purified, and the quality of the synthesis determined (*see* **Subheading 3.4.**). Long-sequence, high-diversity pools can be constructed by first polymerase chain reaction (PCR) amplifying (*see* **Sub-

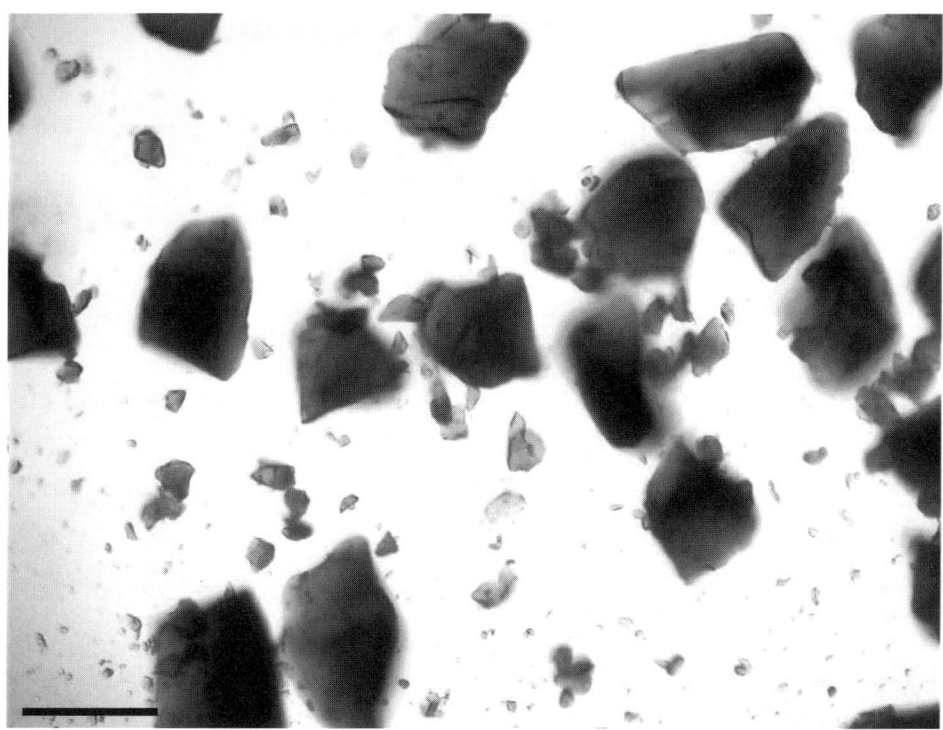

Fig. 1. 1000 Å CPG (dA) beads under the microscope. Scale bar is 20 μm.

heading 3.5.) short pool segments (100- to 140-nt long) that contain restriction endonuclease sites and ligating the digested fragments together using T4 DNA ligase (*see* **Subheading 3.6.**). Finally, a simple method allows the synthesis of long-sequence, low-diversity pools by extending partially overlapping sequences in a PCR-type reaction (*see* **Subheading 3.7.**).

2. Materials

Procedures in **Subheadings 3.1.**, **3.2.**, and **3.3.** require access to a DNA synthesizer. The synthesis of point deletions (*see* **Subheading 3.3.**) requires a programmable DNA synthesizer with a spare reagent port.

2.1. Point Mutation

1. 0.5 g septum-sealed bottles of N^6–benzoyl-dA, N^4–benzoyl-dC, N^2–isobutyryl-dG, and T phosphoramidites (5'-dimethoxytrityl nucleoside 3'-[β-cyanoethyl]) (Applied Biosystems, CA, cat. no. 400330–400333).

2. Acetonitrile (ACN) diluent (<50 ppm H_2O), (Applied Biosystems, cat. no. GEN902005).
3. DNA synthesis columns 1000 Å or 2000 Å CPG (Glen Research, VA, cat. no. 20-2142-42).
4. Dry inert gas source (Argon).
5. Syringes with 21-gage needles.
6. Silanized glass DNA phosphoramidite synthesis bottles and rubber septums.

2.2. Point Deletion

1. 4,4'-Dimethoxytrityl chloride (DMT-Cl) (Fluka, MO, part no. 38827).
2. Anhydrous pyridine and dichloromethane (DCM).
3. Deblock (10% trichloroacetic acid (TCA) in DCM, Applied Biosystems, part no. 400236).

2.3. Split-Bead

1. Glass plate, razor blades, vacuum line, and empty DNA synthesis columns.

2.4. Pool Deprotection, Purification, and Quality Control

1. Saturated aqueous ammonia and butanol.
2. 55°C heating block.
3. 1.5-mL screw-capped Eppendorf tubes (Sarstedt, Germany).
4. 2X formamide loading dye (95% formamide, 5 mM ethylenediaminetetraacetic acid [EDTA], 0.025% xylene cyanol, and 0.025% bromophenol blue).
5. Polyacrylamide gel electrophoresis reagents.
6. TA cloning kit (Invitrogen, CA) or other cloning kits.

2.5. PCR

1. 10X PCR buffer stock: 100 mM Tris-HCl, 500 mM KCl, 15 mM $MgCl_2$, 0.1% gelatin, pH 8.3.
2. 10X deoxynucleotide-triphosphate stock: 2 mM each of deoxyadenosine 5'-triphosphate, deoxycytidine 5'-triphosphate, deoxyguanosine 5'-triphosphate, deoxythymidine 5'-triphosphate, each at pH 7.5 (Amersham Pharmacia, NJ).
3. *Taq* enzyme (*see* **Note 1**).
4. Thermocycler (PCR machine).
5. Agarose gel electrophoresis reagents and equipment.
6. Three temperature-controlled water baths at 94°C, 45°C, and 72°C.
7. 15-mL polypropylene Falcon tubes (Becton Dickinson, NJ).
8. EDTA, anhydrous ethanol, equilibrated phenol, and chloroform.
9. Resuspension solution: 50 mM Tris-HCl, 300 mM NaCl, pH 7.6.
10. 10X TE buffer: 100 mM Tris-HCl, 10 mM EDTA, pH 7.6.
11. Preparative centrifuge and large centrifuge bottles.

Nucleic Acid Library Construction 363

2.6. Linking Pool Segments Using Type II Restriction Enzymes

1. PCR reagents (*see* **Subheading 2.5.**).
2. *Ear*I or *Bbs*I type II restriction enzymes (New England Biolabs, MA).
3. T4 DNA ligase (Invitrogen, CA) and 20 mM adenosine triphosphate (ATP).

2.7. Synthetic Strand Extension

1. PCR reagents (*see* **Subheading 2.5.**).

3. Methods

The methods described as follows outline (1) the chemical synthesis of point mutations; (2) the chemical synthesis of point deletions; (3) the split-bead approach for constructing point mutation/deletion libraries with limited sequence diversity; (4) pool deprotection, gel purification, and quality control; (5) large-scale PCR for the generation of high-diversity libraries; (6) type II restriction enzyme digestion techniques for the construction of long-sequence libraries containing minimal fixed sequence; and (7) extension techniques for the rapid synthesis of long-sequence, low-diversity oligonucleotides.

3.1. Point Mutations

The art of making point mutations with a DNA synthesizer depends on the controlled mixing of phosphoramidite stocks in ratios that precisely control the mutagenic frequency of the nucleotide being coupled *(3,10)*. The most common form of mutagenesis introduces mutations in an unbiased manner by mixing "N" mix (an "equiactive" stock containing the four phosphoramidites at concentrations such that each phosphoramidite is equally likely to couple) with a particular phosphoramidite. By mixing equiactive phosphoramidite with N mix in different ratios it is possible to create a continuous spectrum of mutagenic frequencies from zero (coupling only a pure phosphoramidite) to completely random (coupling using only N mix).

1. After designing the pool, make a table summarizing the number of couplings required to synthesize the pool broken down by mutagenic phosphoramidite type.
2. Calculate the total volume of each mutagenic phosphoramidite required to perform the synthesis. (Check with the operator of your DNA synthesizer to determine phosphoramidite consumption per coupling at the scale you will be performing the synthesis.) Allow for losses that will occur during reagent line priming on the DNA synthesizer.
3. For each type of mutagenic phosphoramidite, calculate the amount of equiactive and N mix required, and enter it into the table started in part 1. A phosphoramidite stock that produces mutations x% of the time is generated by mixing N mix and the appropriate equiactive stock in the ratio 4X/(300-4X) to one (valid up to a mutational frequency of 75%, which corresponds to pure N mix). Where possible

increase the amount of equiactive and N-mix to facilitate measurement with a syringe (*see* **Note 2**).
4. Calculate the total amount of N mix required for the synthesis. As the N mix will be generated by mixing equal amounts of the four equiactive phosphoramidite stocks together, again increase the total volume to facilitate accurate measurement.
5. Calculate the total amount of each of the four equiactive stocks that are required for whole synthesis. Each stock will be used to generate N mix (**step 4**) and mutagenic phosphoramidite mixes (**step 3**). Again increase volumes to facilitate measurement (*see* **Note 3**).
6. Prepare clean and dry phosphoramidite bottles suitable for your DNA synthesizer that will contain all the mutagenic phosphoramidite solutions required for the synthesis. Include one extra bottle to hold the random N mix required to generate the mutagenic phosphoramidite solutions and that can be mounted on the DNA synthesizer if pure N-type couplings are required during the synthesis. Wash previously used phosphoramidite bottles (which should be silanized) carefully with ACN, followed by hot water. Rinse with ethanol. Dry the glassware in a drying oven (50°C overnight, ideally in vacuum) and apply rubber septums.
7. Generation of equiactive phosphoramidite stocks: Fill 10-mL syringes (having 21-gage needles) with dry argon (this is easily achieved using an argon tank and regulator connected to a length of plastic tubing, insert the needle into the end of the tubing, and fill the syringe with argon) and exchange this gas with anhydrous ACN from a septum-sealed bottle. Carefully adjust the volume in the syringes to the following empirically determined volumes: dA 7.0 mL, dC 7.3 mL, dG 8.5 mL, and T 10 mL. Inject through the septum of sealed 0.5-g phosphoramidite bottles by slowly exchanging the ACN in the syringe with the inert gas in the phosphoramidite bottle. Leave the syringe needle in the septum of the bottle. After letting the phosphoramidite dissolve for at least 5 min (making sure to dissolve the powder on the septum of the bottle) exchange the fluid in the bottle with any residual ACN in the syringe to ensure a uniform concentration of equiactive phosphoramidite in each phosphoramidite bottle. If a 10-mL syringe is not required in further steps discard the syringe at this point (*see* **Note 4**).
8. If the total volume of a particular equiactive phosphoramidite stock is larger than the volumes specified in **step 7**, resuspend the appropriate number of phosphoramidite bottles. To ensure uniformity, mix the resuspended phosphoramidite solutions together in a single septum-sealed container before proceeding.
9. Generation of N mix: In the septum-sealed anhydrous container generated in **step 6**, carefully mix equal volumes of the four equiactive phosphoramidite stocks together to make the desired amount of N mix. Use a convenient gradation on an appropriately sized syringe to ensure equal volumes of the four equiactive phosphoramidites are added.
10. Mutagenic phosphoramidite mixes: According to the ratios calculated in **step 3**, mix N mix and appropriate equiactive phosphoramidite into a labeled, septum-covered bottle, being careful not to cross-contaminate the syringes.

11. Follow the appropriate machine-specific procedures to mount phosphoramidite bottles on the DNA synthesizer. General concerns to be respected are that (1) reagents should be kept as anhydrous as possible and (2) any previous reagent should be thoroughly rinsed away by appropriate wash steps before mounting the mutagenic phosphoramidite mix on the synthesizer. Phosphoramidite solutions not mounted on the machine should be stored under argon in septum-covered bottles.
12. Synthesize the pool, respecting the bottle-numbering scheme used by the DNA synthesizer. If the number of mutagenic phosphoramidite solutions required exceeds the number of phosphoramidite ports available on the machine (remembering that four positions are required for regular unmutagenized DNA synthesis), carefully plan the minimum number of bottle changes required to perform the synthesis, and change phosphoramidite bottles during the synthesis. This will require synthesizing the pool in segments (*see* comments in **Subheading 3.3.6.**). If the pool requires mutagenesis of the 3'-most residue, use either a universal support (Glen Research) or mix CPG beads from unused columns using the methods outlined in **Subheading 3.3.** to generate the desired level of mutation.
13. Follow the pool deprotection, gel purification, and quality-control steps outlined in **Subheading 3.4.**

3.2. Point Deletions

This method is completely compatible with directed chemical mutagenesis (*see* **Subheading 3.1.**) and is simply implemented and optimized on a DNA synthesizer having a spare reagent port. The protocol is an extension of a method originally developed by Treiber and Williamson *(11)* and allows the synthesis of high-diversity libraries containing controlled point deletions with variable deletion frequencies *(12)*. While optimized using an 8909 Expedite DNA synthesizer (Applied Biosystems) with software controller, this method should be easily adaptable to other machines (*see* **Subheading 3.2.3.**).

1. Make up 100 m*M* DMT-Cl (mol wt 338.8) in 95% DCM and 5% pyridine. Weigh out DMT-Cl using an analytical balance into clean dry glassware that can be mounted on the spare reagent port of the DNA synthesizer. Add solvent using a clean, dry, graduated cylinder under a blanket of argon gas. Mount the dissolved solution promptly on the DNA synthesizer after first washing (with ACN) and drying the spare reagent delivery lines.
2. A deletion coupling can be performed at any point during a synthesis by using the following modified coupling cycle (The Expedite 8909 synthesizer used for this application delivered an approx 16-µL pulse of reagent in 0.36 s. The column volume was approx 70 µL, *see* **Note 5**.)
   a. Deblock the column, using the standard protocol of 10 rapid pulses of 3% TCA in DCM (deblock) followed by 50 pulses of deblock over 49 s. Wash with 40 pulses of ACN.

b. Deliver 100 m*M* DMT-Cl to the deblocked column (10 rapid pulses from the spare reagent port, followed by 16 pulses spread over either a 2- or 4-min coupling interval). Immediately wash the column (40 pulses of ACN).
   c. Deliver deblock (7 rapid pulses) followed immediately by an extensive wash of the column (40 pulses of ACN) (*see* **Note 6**).
   d. Standard coupling, capping, and oxidation steps followed as usual.
   The 2-min reblocking step gave products that when cloned and sequenced (*see* **Subheading 3.4.**), were found to have an average deletion frequency of 17%. The 4-min deletion protocol yielded an average deletion frequency of 25%. Deletion frequencies were observed to be well balanced, having less than a twofold bias with respect to sequence *(12)*.
3. Overall deletion frequencies can be easily estimated after the synthesis of a library by cloning and sequencing members of the pool population (*see* **Subheading 3.4.**). Deletion rates can be measured directly by synthesizing four pentanucleotide homopolymeric DNA sequences, where the third coupling cycle uses the deletion protocol being optimized. Deprotected oligonucleotides are butanol-precipitated (*see* **Subheading 3.4.**) and radiolabeled with an excess of (^{32}P)-γ-ATP using polynucleotide kinase. Deletion products are resolved on a 20% polyacrylamide sequencing gel and quantified by phosphorImager.

3.3. Split-Bead Synthesis

This procedure is used to make pools of limited sequence diversity containing point mutations, deletions, or insertions. Common uses include the synthesis of degenerate codon libraries and the synthesis of sequences containing limited diversity in localized regions of sequence. Although the relatively small number of CPG beads in a DNA synthesis column limits maximum sequence diversity, this disadvantage is offset by the relative ease with which simple pools can be constructed.

1. Pause DNA synthesis immediately before the residue requiring modification, but after the DNA synthesizer has completed washing the column with ACN in preparation for the next coupling cycle (*see* **Note 7**).
2. Take the column off the machine and connect it to a vacuum line using appropriate connectors (usually a LUER syringe-type fitting), and remove any residual ACN resulting from the final column wash by applying vacuum for several minutes.
3. While still applying vacuum, ensure that all the CPG beads in the column are pulled against the frit on the vacuum side of the column by tapping the column gently. Carefully dismantle the column while keeping the portion that contains the CPG beads attached to the vacuum line.
4. Detach the open column from the vacuum line, and spread the dry CPG beads onto a clean glass plate (*see* **Note 8**).
5. Divide the beads, using a clean razor blade, into fractions of the desired size. The size of each fraction depends on the ratio of the desired modifications within the

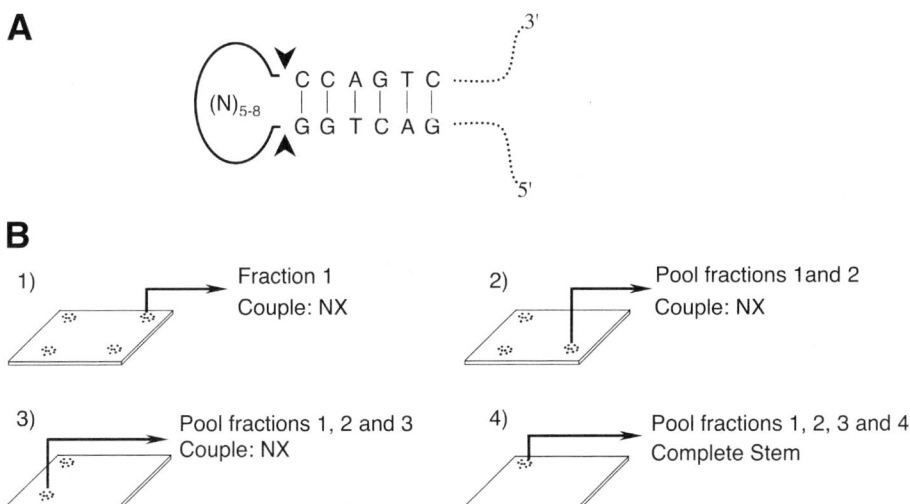

Fig. 2. (**A**) Split-bead synthesis of a 6-bp stem loop possessing a variable 5- to 8-nt-long random sequence loop delimited by the arrows. (**B**) After synthesizing the first constant part of the construct and five of the eight random nucleotides using the sequence 5'-NNN NNC CAG TC . . . (where N is a random position synthesized by coupling N mix equiactive phosphoramidite solution). The column is opened and the beads divided into four fractions. Fractions 1–3 are sequentially added back to the column and the sequence 5'-NX (X is an arbitrary nucleotide representing the synthesis currently on the column) is coupled after each addition. Finally the fourth fraction is added and the remainder of the sequence, 5'- . . . GAC TGG X synthesized.

   final pool. Manual division can be reasonably accurate and is much easier than weighing out bead fractions using a microbalance (which in our hands incurs significant losses owing to static electricity).
6. Point deletions or insertions are constructed by placing the bead fraction destined to have the longer sequence length (i.e., the wild-type sequence for a deletion or the mutated sequence for an insertion) into a column and coupling the desired sequence. Afterward the column is opened as in **step 3** and the leftover beads on the glass plate are sucked back into the column using the vacuum line (**Fig. 2**). Synthesis is resumed until the next modification is required. It is important to realize when entering sequence data into a DNA synthesizer that most synthesizer software assumes that the first nucleotide in the synthesis is attached to the CPG column; thus, all oligonucleotide sequences used part way through a synthesis must include an extra residue at their 3' ends to yield the correct final sequence.
7. Point mutation: Proceed up to **step 5** as described previously. Place each fraction requiring a distinct coupling type into a clearly labeled DNA synthesis column. Couple the desired sequence onto each fraction using the DNA synthesizer

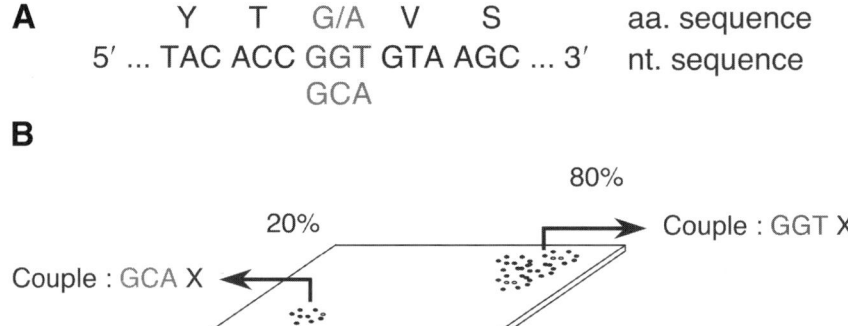

Fig. 3. Split-bead procedure for making a degenerate amino acid codon, having glycine 80% of the time, and alanine the remainder. (**A**) The first unmodified region, 5'-GTA AGC..., is synthesized normally. The column is opened, and beads are divided into two piles in the ratio of 4:1. The larger fraction is coupled with a glycine codon, 5'-GGT X, whereas the smaller is coupled in a second column with an alanine codon, 5'-GCA X. The beads are then pooled together into a single column, and the remainder of the sequence, 5'-... TAC ACC X, is synthesized (X is an arbitrary nucleotide representing the synthesis currently on the column).

(**Fig. 3**). Random or degenerate sequences can easily be synthesized using mutagenized phosphoramidite mixes as described in **Subheading 3.1.** Once the mutation types for each fraction have been synthesized, open all of the columns and mix their beads together on a clean glass plate. Leave one column attached to the vacuum line as in **step 3** and vacuum all of the beads into it. Resume synthesis until the next modification is encountered (*see* **Note 9**).

3.4. Pool Deprotection, Purification, and Quality Control

After completing DNA synthesis, the resulting single-stranded DNA population must be deprotected, purified, and its quality accessed.

1. Deprotect the synthesis by transferring the CPG beads from the synthesis column into a 1.5-mL screw-capped Eppendorf tube. Fill nearly to the top with saturated aqueous ammonia and leave at 55°C for at least 12 h. Ensure that the tube is well sealed—this is easily tested by smell. Transfer the ammonia solution into a 7.5-fold excess of butanol after first cooling it to prevent bumping from the hot ammonia gas. Vortex well and centrifuge at 12,000*g* for 10 min. Decant the supernatant and resuspend the dried pellet in 1 mL of TE. Store the resulting crude, single-stranded DNA at –20°C.
2. The pool should be purified using preparative polyacrylamide gels to remove failed synthesis products and to confirm its general quality. This step is particu-

larly important for long pools, as the capped material that results during a normal synthesis can inhibit the efficiency of PCR. Mix an aliquot of resuspended pool (**step 1**) with an equal volume of 2× formamide loading dye. Load samples onto a preparative polyacrylamide gel of the appropriate percentage. Gel loading should not exceed 1 nmol of DNA per square millimeter of gel-well surface area.
3. Visualization by UV shadowing: Wrap the gel on both sides with Saran wrap and place the resulting sandwich on a thin-layer chromatography plate containing 254 nm absorbing fluorescent indicated. Shine a handheld 254 nm UV light from above to produce shadows corresponding to the location of synthesis products (*see* **Note 10**). A uniform band of the expected pool size with few or no discernable bands under this product should be observed. A long synthesis will produce a noticeable smear below the desired product resulting from the capping of incompletely coupled oligonucleotide sequences. The point deletion protocol (**Subheading 3.2.**), if used extensively during the synthesis, will produce a noticeable broadening of the product band.
4. Excise the single-stranded DNA using a clean razor blade, and place the gel fragments into a fivefold (by volume) excess of 300 mM NaCl. Elute on a rotator overnight at room temperature. Precipitate the nucleic acid by adding 2.5 vol of ethanol and holding the resulting mixture at –20°C for at least one h. Centrifuge at 15,000g for 30 min at 4°C. Resuspend the resulting pellet in TE buffer and estimate concentration by optical density (*see* **Note 11**).
5. Using a small fraction of the purified DNA pool, generate double-stranded material by PCR (**Subheading 3.5.1.**). Clone and sequence at least 10 individual pool isolates (or more depending on the level of mutagenesis used to create the pool). Confirm that the desired pool properties are represented in these sequences before proceeding.

3.5. Large-Scale Pool Amplification

The synthesis of a high-diversity double-stranded DNA pool can require hundreds of milliliters of PCR volume *(3,10)*. Large volumes are required because the maximum concentration of single-stranded synthetic DNA template that can be effectively amplified in a PCR reaction is only in the 25–75-nM range. Moreover, synthetic single-stranded DNA does not extend as efficiently as enzymatically produced DNA, further reducing the effective concentration of template in the PCR reaction. Taken together, these two effects imply that a pool containing 10^{15} different double-stranded sequences may require up to 200 mL of PCR volume to produce.

Reagent quality is critical when performing PCR reactions on such a scale. Pool fragments and PCR primers should be carefully designed, synthesized (*see* **Subheadings 3.1.–3.3.**), and gel-purified (*see* **Subheading 3.4.**) before use.

3.5.1. Pilot PCR

It is essential that analytical-scale PCR reactions be performed before attempting amplification on a large scale. Pilots varying template, primer, magnesium, and enzyme concentrations should all be performed to minimize the number of PCR cycles and guarantee reagent quality. An obvious but important consideration is that pilot PCR reactions must use the same reagents to be employed in the large-scale PCR reaction.

1. Program a thermocycler (PCR machine) for the following cycles: 94°C for 45 s (denaturing step), 50°C for 105 s (annealing step), and 72°C for 115 s (extension step).
2. Using the buffers and reagents prepared for the large-scale PCR reaction, screen an array of conditions varying primer concentration (0.5–1.5 μM) and template concentration (10 to 75 nm). Add *Taq* (0.05–0.1 U/μL) enzyme to each reaction once it has reached 94°C.
3. Collect aliquots from each cycle at the end of each 72°C-extension step. Run aliquots on an agarose gel to see by which cycle PCR product forms (*see* **Note 12**). PCR conditions that produce optimal PCR product in as few cycles as possible should be considered for large-scale PCR.

3.5.2. Large-Scale PCR

1. Prepare three water baths set at 95°C, 72°C, and 45°C (*see* **Note 13**).
2. Mix the reagents (with the exception of *Taq* polymerase) in the ratios determined during the optimization process (**Subheading 3.5.1.**) in a sterile plastic container.
3. Aliquot 10-mL fractions of the reaction mix into 15-mL polypropylene Falcon tubes and place in the 95°C water bath. Add aliquots of *Taq* to each tube after letting the fractions warm up for 10 min. Make sure to cap each tube tightly. Practice this step a few times using water to confirm that the caps seal well at high temperature (*see* **Note 14**).
4. Carry out the following thermal cycles: 95°C for 4 min, 45°C for 5 min, 72°C for 7 min. Times have been lengthened from the pilot PCR to allow for the increased thermal mass of the large-scale reaction tubes. Invert tubes once every minute to ensure even temperatures are reached throughout the reaction volume. If different cycle temperatures are desired, modify the recommended cycle temperatures in the appropriate manner, and monitor temperature changes throughout the PCR cycle using a low-thermal-mass thermometer (*see* **Note 15**).
5. Carry out the number of cycles determined by the pilot PCR. After finishing the amplification, place the large-scale reaction at 4°C, and check the quality of the large-scale amplification on an agarose gel. The large-scale PCR should compare well with the pilot-scale reactions performed previously. If not, another cycle of PCR amplification may be required. Put aside a small aliquot of the large-scale PCR for comparison with the precipitated pool generated later (**step 8**).
6. If amplification is satisfactory, pool the individual amplification aliquots together in an appropriately sized high-speed centrifuge bottle. Chelate free magnesium

by the addition of one molar equivalent of EDTA and bring the NaCl concentration to 300 mM. After mixing well, add 2.5 vol of ethanol, balancing centrifuge bottles pairwise. Shake well and hold the bottles at –20°C, ideally overnight. Spin at 12,000g for 30 min at 4°C. DNA will be pelleted down the outside wall of the centrifuge bottle, and a solid white pellet should be observed at the bottom of the centrifuge bottle. While still chilled, decant the supernatant, and remove with a pipet any excess ethanol solution. Without drying the pellet, dissolve the DNA using resuspension solution making sure to wash the sides of the container. Pipet out the resuspended DNA into a clean container, and repeat the resuspension process a second time to ensure full recovery. Minimize the recovery volume; concentration of the pool by at least 10-fold should easily be attainable (i.e., a total of 10 mL of resuspension solution should be used to dissolve a pellet resulting from 100 mL PCR) (*see* **Note 16**).
7. Perform a phenol–chloroform extraction to remove the large amounts of *Taq* enzyme coprecipitated with the DNA. Add an equal volume of equilibrated phenol to the pool and shake vigorously. Centrifuge at 12,000g for 20 min, and carefully remove the upper aqueous layer. The phenol–aqueous interface is likely to contain visible amounts of protein (*see* **Note 17**). Add an equal volume of chloroform to the extracted material, shake vigorously, and centrifuge (*see* **Note 18**). Again, extract the aqueous layer and add 2.5 vol of ethanol and leave overnight at –20°C. The solution should become flocculent on addition of ethanol. Centrifuge at 12,000g to pellet the DNA, and resuspend in TE buffer supplemented with 50 mM NaCl.
8. Carefully dilute a small amount of the final pool by the ratio of the initial large-scale PCR volume to the final resuspension volume. Compare with an aliquot taken from the initial large-scale PCR (**step 5**) on an agarose gel; recovery should be nearly quantitative.

3.6. Linking Pool Segments Using Type II Restriction Enzymes

This technique can be used to make long, high-diversity double-stranded DNA libraries that cannot be manufactured in one piece using conventional DNA synthesis. The technique uses the ability of some type II restriction enzymes to cut outside of their recognition sequence. The "sticky" ends resulting from digestion with enzymes such as *Ear*I and *Bbs*I make possible the specific ligation of two high-diversity pool segments through a short region of defined sequence the length of the sticky ends. This approach extends a technique developed to construct long, high-diversity libraries using enzymes such as *Ban*I and *Sty*I that leave the restriction recognition sequence within the final pool (**Fig. 4A**) *(3,10)*.

1. Design the pool in a modular fashion, taking into account that DNA synthesis yield decreases exponentially with sequence length (100- to 140-nt segments are practical). Each module of the pool should be linked to its neighbor through a 3-bp (*Ear*I) or 4-bp (*Bbs*I) nonpalindromic spacer. This spacer region allows the specific ligation of adjacent pool fragments having complementary sticky ends

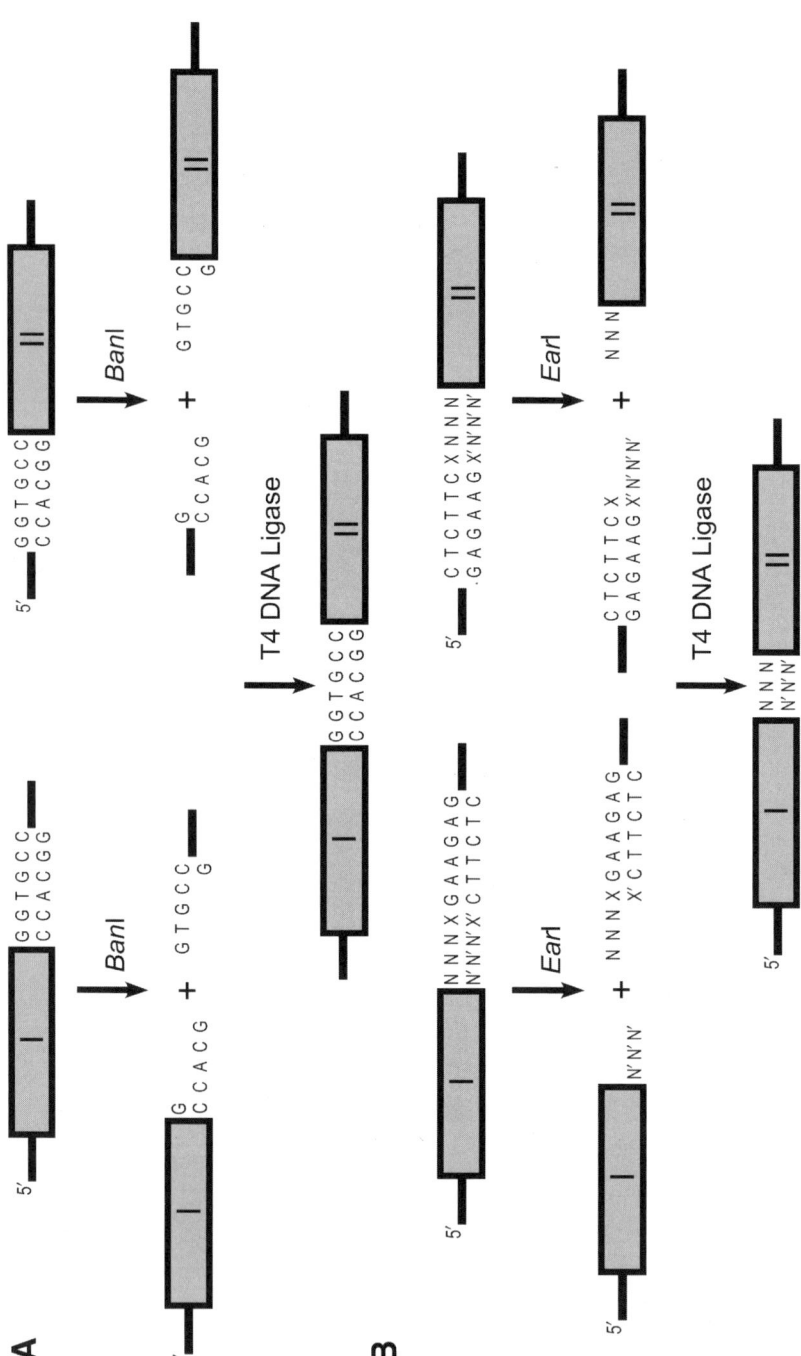

Fig. 4.

generated by restriction digestion of the individual fragments.
2. The sticky end of each fragment is generated by appending to the end of each pool module a primer-binding sequence containing 10 nt of arbitrarily defined sequence (for efficient PCR and restriction digestion), the appropriate restriction enzyme recognition sequence, and the desired spacer sequence. The following sequence elements are required within the primer-binding sequence: 5'-$(X)_{10}$CTCTTCXNNN (*Ear*I) or 5'-$(X)_{10}$GAAGACXXNNNN (*Bbs*I). The nucleotides indicated by an X are arbitrary but defined sequence elements, whereas the nucleotides defined by N specify the spacer sequence required for ligation of the two pool pieces. Each primer-binding sequence requires a partner that is used to amplify the adjacent pool fragment. To generate sticky ends, the partner sequence must be synthesized with the reverse complement of the original N sequence (**Fig. 4B**).
3. Synthesize the segments and PCR primers required for amplification. Gel-purify the resulting oligonucleotides (**Subheading 3.4.**) and use large-scale PCR (**Subheading 3.5.**) to make double-stranded pool segments.
4. Independently digest small aliquots of the appropriate upstream and downstream fragments at high concentration (approx 1 μM) in the appropriate reaction buffer. New England Biolabs buffer no. 2 (10 mM Tris-HCl, 10 mM MgCl$_2$, 50 mM NaCl, 1 mM dithiothreitol, pH 7.9) works well for both *Ear*I and *Bbs*I. When the segments are completely digested, mix the two fragments together, add 0.05 Weiss U/μL T4 DNA ligase, and ATP to a final concentration of 1 mM. Incubate samples overnight at 16°C. Systematically vary the relative fragment concentrations to maximize ligation yield (*see* **Note 19**).
5. Once ligation yield is acceptable, scale up the digestion/ligation reaction by the desired amount. Phenol–chloroform-extract and precipitate the pool as described in **Subheading 3.5.2.**, **step 7**, resuspend the ligated pool in TE buffer supplemented with 50 mM NaCl, and store at –20 °C.

Fig. 4. (*opposite page*) Type II restriction digestion followed by ligation to generate long high-diversity pools using *Ban*I or *Ear*I. Solid lines represent defined sequence elements required to anneal primers during PCR. Boxes represents regions of arbitrary sequence. (**A**) Two PCR-amplified DNA segments containing a *Ban*I site are digested to completion. The enzyme is heat-inactivated (which denatures the cut primer sequences allowing them to reanneal with the PCR primers used in the PCR amplification), and the pool fragments are precipitated after phenol–chloroform extraction. The resulting segments are ligated together using T4 DNA ligase. The restriction site for *Ban*I is still found within the pool sequence. (**B**) PCR-amplified DNA segments containing the *Ear*I restriction site are digested to completion. T4 DNA ligase and ATP are then added to the reaction, allowing the ligation between pool segments having complementary 3-nt long sticky ends (sequence NNN and N'N'N', respectively). The final pool construct cannot be cleaved by *Ear*I as it lacks a restriction site. Undesirable religation of cut primers onto pool segments is minimized by the presence of functional restriction enzyme during the ligation.

3.7. Synthetic Oligonucleotide Strand Extension

This robust procedure can easily be used to construct long DNA strands from small, partially overlapping, synthetic DNA oligonucleotides. If mutations and deletions are particularly undesirable, the synthetic strands must be carefully gel-purified before use.

1. Design a sequence, which should be no longer than approx 450 nucleotides in length. We have found that carefully gel-purified DNA synthesis products contain deletions at a frequency of about 0.2%. This implies that a construct 450 nucleotides long is likely to contain a deletion mutation 60% of the time. Screening five or six isolates by cloning and sequencing should be sufficient to identify a perfect construct.
2. Break the sequence into segments that overlap their neighbors by at least 16 nucleotides of sequence, such that no one segment is longer than 140 nucleotides. The two exterior sequences should be synthesized as though they were long PCR primers (5' primer sense and 3' primer antisense). Inner fragments should be synthesized with overlapping 3' ends allowing each oligo to prime the other for extension. Take care that no sequence can form extensive secondary structure, either with itself or with other oligonucleotides, except in the designed hybridization regions. An important consideration is the well-known ability of *Taq* polymerase to add untemplated dA residues *(13)*. The inner fragments therefore should be designed such that any untemplated incorporation of dA is compatible with the outer flanking sequence. **Figure 5** summarizes these basic design principles.
3. Synthesize the desired oligonucleotide segments using 2000 Å CPG if their length is 110 residues or more. After deprotection, carefully gel-purify a small amount (**Subheading 3.4.**) of each oligo (*see* **Note 20**).
4. Add the two innermost oligonucleotides in equimolar amounts, at approx 50 nm concentration each, and perform one cycle of PCR (**Subheading 3.5.1.**) adding enzyme after incubating the mixture at 94°C for 1 min.
5. Check extension on a polyacrylamide or agarose gel. For longer oligonucleotides, only a small amount of double-stranded material would be expected owing to synthetic lesions in the starting DNA.
6. Dilute the extended product 1000-fold into fresh PCR mix containing the external primer-like sequences at 0.5 μmol concentration. Perform at least 10 PCR amplification cycles (*see* **Subheading 3.5.1.**), stopping when a PCR product of the desired length emerges (as determined by agarose gel).
7. Depending on application, cloning and subsequent sequencing may be required to isolate a construct free of mutation.

4. Notes

1. Large-scale PCR consumes a considerable quantity of this enzyme. Overexpression plasmids for *Taq* are available from a number of laboratories for research purposes.

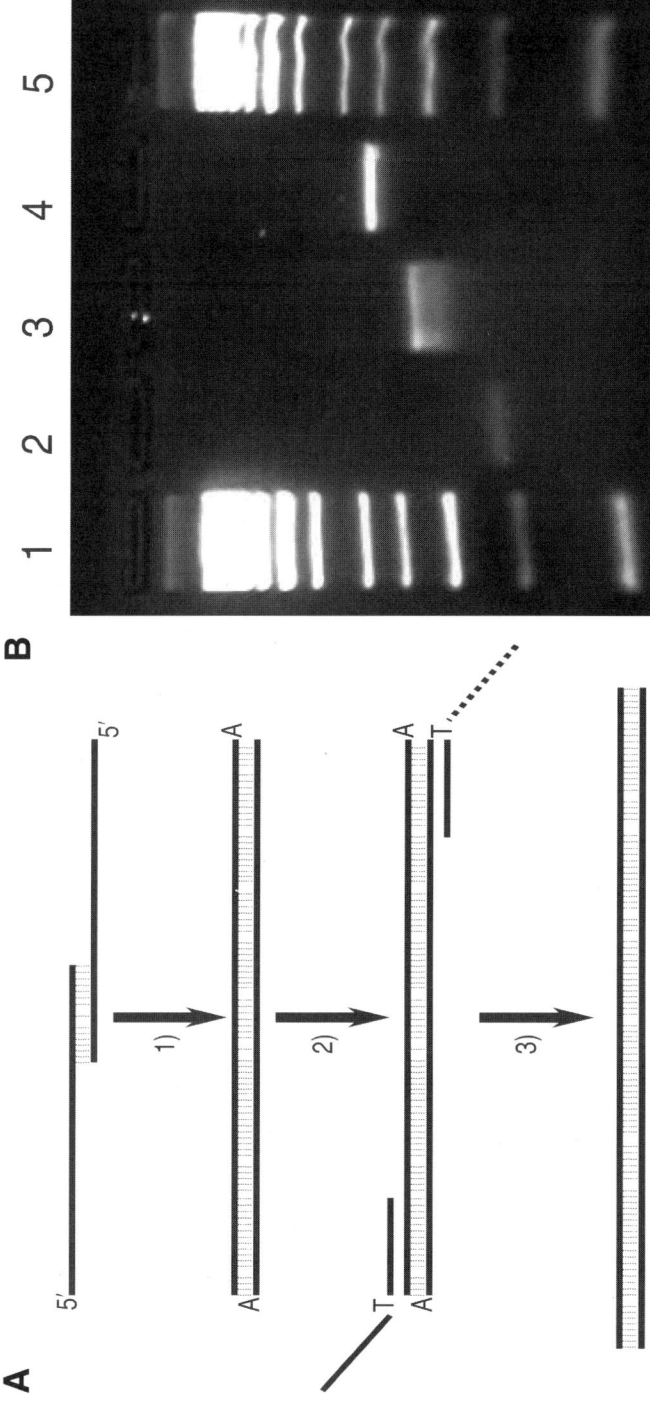

Fig. 5. Synthetic strand extension procedure. (**A**) Schematic representation of the overall procedure: (1) Two long oligonucleotides with complementary sequence at their 3' ends are extended using *Taq* (possibly incorporating untemplated dA residue at their ends). (2) The DNA is diluted 1000 times and long flanking primers added. (3) PCR amplification is carried out to introduce the flanking sequence. (**B**) An agarose gel showing the extension of two 118-nt oligos to make a 220-bp product, followed by extension with a single flanking PCR primer to make a 268-bp product. *Lanes 1* and *5*, 50-bp ladder (starting at 100 bp). *Lane 2*, oligonucleotides before extension. *Lane 3*, extended products; notice the smear of incomplete extension products beneath a definite band. *Lane 4*, PCR product with 5' flanking primer 65-nt long, increasing the final sequence length to 268 bp.

375

2. For example, 7 mL of a 10% mutagenized dA phosphoramidite stock, which couples a dA residue 90% of the time and equal percentages of dC, dG, and T otherwise, could be produced by mixing 6.07 mL of equiactive dA with 0.93 mL of N mix. Increasing these amounts to 6.5 and 1 mL, respectively, permits easier measurement and allows for minor losses.
3. Phosphoramidite reagents are cheap enough that making twice the desired volume of each phosphoramidite is often feasible, making it possible to synthesize the pool twice if required.
4. The different ACN volumes used to make each equiactive mix reflect the differences in molecular weight and reactivity of each phosphoramidite and assume that packing and reagent quality are uniform. We have found that the phosphoramidites supplied by ABI give reproducible results within a batch, but it is highly recommended that pools be cloned and sequenced after synthesis to confirm mutagenic frequencies (*see* **Subheading 3.4.**). Liquid volumes are most accurately determined by drawing a slight excess of ACN into a syringe, tapping the syringe with the needle pointed upward to remove bubbles and then slowly setting the plunger to the correct volume. Work should be performed in a fume hood.
5. Using this protocol on another DNA synthesizer will be sensitive to the length of time reagents, in particular DMT-Cl and deblock, are in the DNA synthesis column.
6. Owing to the relatively short time in which it takes to deblock an oligonucleotide, this step will be quite sensitive to the total time the deblock is on the column. The reactivity of dG phosphoramidite with respect to deblocking and reblocking with DMT-Cl is about sixfold higher than the other three protected phosphoramidites. This factor would lead to significant deletion-sequence bias if DMT-Cl reblocking and deblocking are used independently.
7. Pausing DNA synthesis with trityl "on" rather than "off" is preferable as it allows observation of trityl release after the split-bead procedure. Neither choice will affect the ultimate quality of the synthesis.
8. CPG beads are easily scattered by static, which can be generated by latex gloves.
9. Empty DNA synthesis columns can be purchased from companies such as Glen Research. Alternatively, a used column can be washed thoroughly with ammonium hydroxide to remove traces of the previous synthesis and rinsed with ethanol. Dry thoroughly before use.
10. Make sure eyes and skin are well shielded from UV light. Minimize exposure of the nucleic acid to UV light.
11. There are other elution procedures, notably electroelution, that can give higher recovery of nucleic acid fragments. The soaking procedure outlined here is technically simple and allows the routine recovery of 50–60% of pool material. Ethanol precipitation will not work well for oligonucleotides shorter than about 30 nt, in which case recovery by C18 cartridge can be efficient.
12. Samples taken each cycle can be used to determine if an exponential amplification is taking place by loading one-half the amount of the n + 1 cycle adjacent to the n cycle on an agarose gel.

13. *Caution*: Take steps to prevent burns from the nearly boiling water in the baths. At high-temperature settings, a water bath is likely to change temperature quickly on removal of its cover.
14. Do not use polystyrene tubes as they will melt in the 95°C water bath. For large volumes of PCR, the construction of a tube holder greatly facilitates the ease with which the large-scale PCR is accomplished and can help prevent scalding.
15. Our experience shows little difference in yield when comparing pilot-scale amplification with larger scale conditions. This is likely due in part to the short length of the synthetic strands being amplified.
16. When working with random pools, make sure that the DNA is always resuspended in a buffered solution containing monovalent cations to prevent denaturation. Once denatured (by resuspension in water or by application of heat) a high-diversity pool will not spontaneously renature.
17. Reextracting the aqueous/phenol region with resuspension buffer can increase final pool yield.
18. Avoid the use of polystyrene or polycarbonate plasticware during chloroform extraction as they will dissolve.
19. The restriction enzymes are deliberately kept active in the ligation reaction to recleave any cut ends that are inappropriately ligated back onto pool segments. Ligated segments will not be recognized, as correctly ligated fragments lack a restriction site. Low ligation yields can result if, during the overnight ligation, the restriction enzymes become inactive. This can be prevented by the addition of more enzyme. Alternatively, DNA digests can be heated to 60°C for 20 min to denature the short, cleaved DNA fragments and encourage them to reanneal with the excess PCR primer present in the reaction before addition of ligase and ATP.
20. Increasing coupling and capping times during DNA synthesis can help to ensure high-quality synthesis. Radiolabeling the oligonucleotides using polynucleotide kinase and ^{32}P γ-ATP greatly facilitates the ease in which analytical amounts of DNA can be gel-purified in high purity.

Acknowledgments

The authors wish to thank M. Leroux for helpful comments. This work was supported by grants from NSERC, MSFHR, and CIHR.

References

1. Ellington, A. D. and Szostak, J.W. (1990) In-vitro selection of RNA molecules that bind specific ligands. *Nature* **346**, 818–822.
2. Tuerk, C. and Gold, L. (1990) Systematic evolution of ligands by exponential enrichment: RNA ligands to bacteriophage T4 DNA polymerase. *Science* **249**, 505–510.
3. Bartel, D. P. and Szostak, J. W. (1993) Isolation of new ribozymes from a large pool of random sequences. *Science* **261**, 1411–1418.
4. Roberts, R. W. and Szostak, J. W. (1997) RNA–peptide fusions for the in vitro selection of peptides and proteins. *Proc. Nat. Acad. Sci. USA* **94**, 12,297–12,302.

5. Keefe, A. D. and Szostak, J. W. (2001) Functional proteins from a random-sequence library. *Nature* **410,** 715–718.
6. Scott, J. K. and Smith, G. P. (1990) Searching for peptide ligands with an epitope library. *Science* **249,** 386–390.
7. Barbas, C. F., III., Scott, J. K., Silverman, G., and Burton, D. R. (2001) *Phage Display: A Laboratory Manual.* Cold Spring Harbor Laboratory Press., Plainview, NY.
8. Cello, J., Paul, A. V., and Wimmer, E. (2002) Chemical synthesis of poliovirus cDNA: generation of infectious virus in the absence of natural template. *Science* **297,** 1016–1018.
9. Eckstein, F. (1991) Oligonucleotides and analogues: a practical approach. IRL Press, Oxford, UK.
10. Unrau, P. J. and Bartel, D. P. (1998) RNA-catalysed nucleotide synthesis. *Nature* **395,** 260–263.
11. Treiber, D. K. and Williamson, J. R. (1995) A simple method for preparing pools of synthetic oligonucleotides with random point deletions. *Nucleic Acids Res.* **23,** 3603–3604.
12. Chapple, K., Bartel, D. P., and Unrau, P. J. (2003) Combinatorial minimization and secondary structure determination of a nucleotide synthase ribozyme. *RNA* **9,** 1208–1220.
13. Clark, J. M. (1988) Novel non-templated nucleotide addition reactions catalyzed by procaryotic and eucaryotic DNA polymerases. *Nucleic Acids Res.* **16,** 9677–9686.

23

In Vitro Selection From Combinatorial Nucleic Acid Libraries

Andres Jäschke

Summary

Since the early 1990s, combinatorial deoxyribonucleic acid and ribonucleic acid libraries have been used to isolate specific ligands for a variety of target molecules, as well as nucleic acid-based catalysts for different reactions. These iterative procedures are based on the fact that nucleic acids can be enzymatically amplified. In this chapter, we describe the synthesis of such combinatorial libraries, their analysis, and basic procedures of in vitro selection.

Key Words: In vitro selection; ribozymes; aptamers; solid-phase synthesis; transcription; affinity chromatography.

1. Introduction

Similar to proteins, nucleic acids can fold into intricate three-dimensional structures that are able to specifically bind to other molecules or to catalyze chemical reactions. Although in biological systems only a fraction of these capabilities are used (at least according to current knowledge), combinatorial chemistry allows us to more completely investigate the potential of nucleic acids. Synthetic combinatorial libraries are used to select molecules with specific binding properties (so-called aptamers) or catalytic properties (ribozymes or deoxyribozymes). Compared to other substance classes, nucleic acids have one tremendous advantage—that is, they can be enzymatically amplified. (In biological terms, nucleic acids carry both phenotype [function information] and genotype [genetic information]. This property makes it possible to partition huge libraries into "winners" and "losers," then amplify the winning fraction and repeat the partitioning/amplification procedure until the enriched sublibrary has the desired properties (**Fig. 1**).

From: *Methods in Molecular Biology, vol. 288: Oligonucleotide Synthesis: Methods and Applications*
Edited by: P. Herdewijn © Humana Press Inc., Totowa, NJ

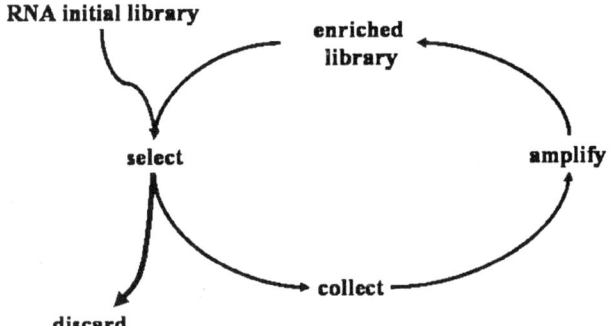

Fig. 1. General scheme of in vitro selection.

In vitro selection *(1–3)*, also termed systematic evolution of ligands by exponential enrichment, can be carried out with either deoxyribonucleic acid (DNA) or ribonucleic acid (RNA) libraries, and modified DNA/RNA libraries can be used as long as the chemical modifications do not interfere with the enzymatic amplifications. Technical details of the amplification procedures vary, depending on the type of nucleic acid library used.

1.1. In Vitro Selection of Aptamers

The most common method for the isolation of nucleic acids with an affinity to a certain target molecule uses affinity chromatography (**Fig. 1**) with a solid matrix to which the target molecule of interest is immobilized (selection step in **Fig. 1**) *(4)*. The combinatorial nucleic acid library is created by automated solid-phase synthesis and then applied to the affinity matrix in a suitable buffer. Unbound nucleic acid molecules are washed away, retaining the desired binding molecules, which are later competitively eluted with a buffer containing the free target molecule. The eluted molecules are collected, a template-dependent polymerase and nucleotides are added, and several hundred copies of each selected nucleic acid molecule are generated. This results in a library in which molecules with affinity to the immobilized target are enriched. Because molecules with high affinity to a given target are very rare, and enrichment of active species in one such round of selection and amplification is normally in the order of 10^2 to 10^3, this scheme has to be repeated several times, until molecules with the desired properties dominate the enriched library. The members of this library are then identified and their nucleotide sequences compared to find common structural motifs.

The published selection experiments were concerned with oligonucleotide binding to all kinds of molecules, from proteins to small monomers such as amino acids or nucleotides, cofactors such as nicotinamide derivatives or biotin, natural RNA binders such as glycoside antibiotics, transition state analogs (TSAs), substituted purines, or organic dyes *(5,6)*. These oligonucleotide binders are generally called aptamers (*lat.* aptus: to fit).

1.2. In Vitro Selection of Ribozymes

Compared to selecting ligands for a given target, the identification of new catalysts is conceptually more ambitious. A catalyst is—per definition—required to leave the reaction unchanged. To preparatively separate the few active molecules in a library from the excess of inactive species, however, one needs a difference in the physical or chemical properties. This is a serious contradiction, but two major detours have been worked out, namely selection against TSAs and direct selection. In the former approach, a stable analog of the transition state of the reaction under investigation is immobilized, and aptamers are generated that bind these molecules as described in **Subheading 1.1.** The binding RNAs are then screened for catalysis of the respective reaction. Although the first part works quite often, the isolated RNAs only occasionally display catalytic activity. Only a few examples are known to date *(5,6)*.

The most successful method to identify nucleic acid catalysts is direct selection *(7–9)*. Those members of a combinatorial DNA or RNA library are isolated that show an accelerated reaction with a substrate X (**Fig. 2**), assuming that these members might combine substrate and catalyst properties in one molecule. For selection purposes, X carries an anchor group—a functional group that is normally not present in RNA, like a thiol or biotinyl group. RNA molecules that react with the substrate automatically acquire the anchor group and can subsequently be isolated by affinity chromatography on a suitably derivatized matrix, for example, activated thiopropyl agarose (disulfide bond formation with the thiol anchor group) or streptavidin agarose (specific interaction with the biotinyl group). Unreacted RNA does not bind to the matrix because it does not contain the anchor group and is removed by washing, whereas bound RNA is isolated and enzymatically amplified. This cycle is repeated until active molecules dominate the library.

The isolated molecules are self-modifiers and not catalysts in the sense of the definition; they do not leave the reaction unchanged, and they perform the reaction only once. Quite often it is possible, however, to rationally dissect the identified self-modifiers into a substrate part and a catalyst part after elucidation of the primary and secondary structure.

To apply this approach to reactions that do not involve RNA as a substrate, the Eaton group and our laboratory developed a variant involving tethered reactants *(10–13)*. To select RNA catalysts for a general reaction A + B → A–B, the potential reactant A is attached to each molecule of a combinatorial RNA

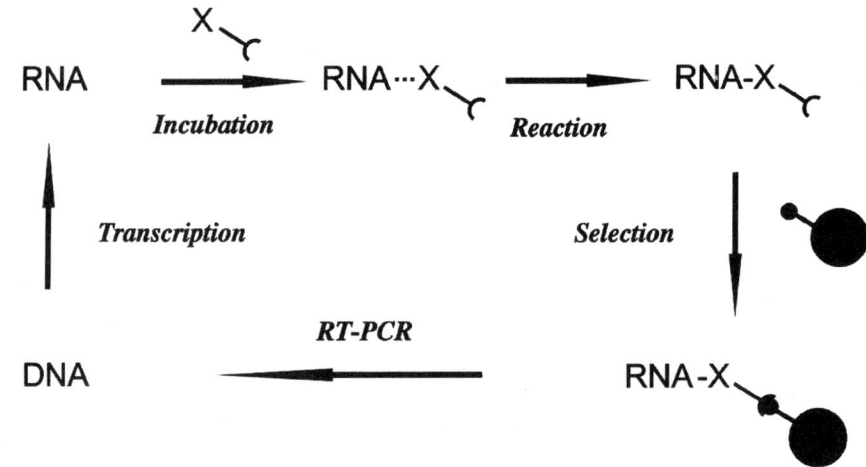

Fig. 2. Direct selection of RNA-modifying ribozymes—general scheme.

library via a long flexible polymeric tether (preferably polyethylene glycol [PEG]). This library of RNA–linker–substrate conjugates is then incubated with reactant B, which carries the anchor group. If an RNA molecule catalyzes the reaction of the attached substrate A with B, it becomes linked to the anchor group and can be selected and amplified (**Fig. 3**).

Our laboratory used this strategy to isolate ribozymes that catalyze a Diels–Alder reaction of tethered anthracene and biotinylated maleimide (**Fig. 4**) *(13)*. The isolated catalysts were later found to efficiently catalyze the reaction between two free reactants, with multiple turnover and a high degree of stereoselectivity *(14,15)*.

1.3. Pool Design and Synthesis—General Considerations

Although it is in principle also possible to chemically synthesize RNA pools, this is normally not a reasonable choice, and pools (or libraries) are synthesized at the DNA level by automated solid-phase synthesis using standard phosphoramidite chemistry. For the enzymatic amplification steps, regions with a defined, constant sequence are needed at both ends of the library to allow for proper binding of the amplification primers (primer binding sites). Between these constant sites, variable regions are located that are either completely random, or biased. Completely random positions are synthesized using mixtures of the phosphoramidites instead of individual monomers. If such a mixture of the four standard phosphoramidites is incorporated at N positions, a library with 4^N different members is generated. At $N = 25$, one gets 4^{25} or 10^{15} different species. With longer randomized stretches, theoretical library size

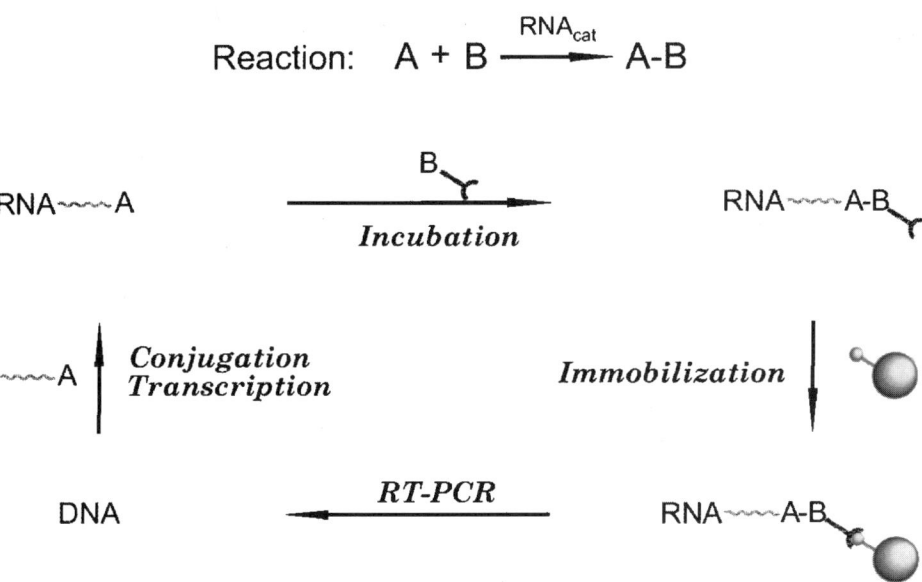

Fig. 3. Direct selection with linker-coupled reactants.

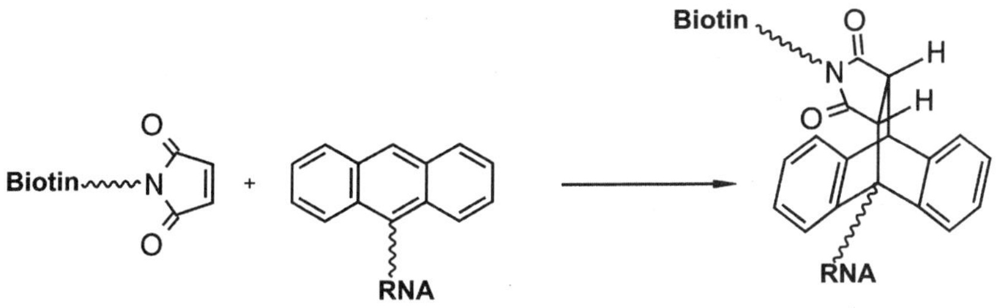

Fig. 4. RNA-catalyzed Diels–Alder reaction.

grows exponentially, although the actual library size is limited by the synthesis scale. Therefore, library coverage becomes incomplete, that is, not all theoretically possible sequences are actually present in the library. The typical yield of full-length, amplifiable DNA molecules with a length >100 that results from a 0.2 µM synthesis is in the order of 1 nmol, that is, 10^{15} molecules.

Sometimes it is desirable not to use completely random sequences, but statistically mutated naturally occurring or designed structural motifs (biased pools). This can be accomplished using "doped" phosphoramidites, that is, monomers that were intentionally contaminated with small amounts of the other three phosphoramidites. The amount of contamination thereby determines the average number of mutations that each pool molecule carries. *Note:* For the synthesis of doped pools, the use of a synthesizer with at least eight monomer ports is strongly recommended.

Primer binding sites must, of course, be synthesized using pure, uncontaminated phosphoramidites.

In addition, it is not uncommon to alternate between constant, random, and biased regions in pool design.

Three more issues should be noted:

1. The T7 RNA polymerase requires a constant "promoter" sequence for recognition. This promoter sequence is not copied and therefore not present in the transcripts. It must be reintroduced into the pool using an overhanging primer.
2. In the synthesis of long oligonucleotides (over 100 nt), acidic depurination during the detritylation steps becomes a serious problem. Apurinic sites are generated, which give rise to molecules that cannot be enzymatically copied. The number of intact, amplifiable molecules needs to be determined by a suitable assay.
3. With completely random pools, it normally does not matter whether the individual molecules contain exactly the specified number of randomized positions. To increase total synthesis yield and obtain a higher pool complexity, the capping steps are often omitted during the synthesis of the random (and sometimes also the biased) parts of the molecules. This can be done by altering the synthesis program, or by attaching bottles filled with dry acetonitrile (ACN) to the Cap A and Cap B ports of the synthesizer. The disadvantage of this method is the length heterogeneity of the synthesized pools, leading to broader bands or peaks in electrophoresis and chromatography.

The protocols described in the **Subheading 3.** relate to our published selection of ribozymes accelerating a Diels–Alder reaction of anthracene and maleimide *(13)*. Readers interested in different selections are therefore required to adjust the protocols to their needs.

2. Materials

1. Oligonucleotide syntheses were performed on an expedite 8900 DNA synthesizer (PerkinElmer).
2. Standard β-cyanoethyl phosphoramidites with the usual protecting groups (benzoyl, isobutyryl, dimethoxytrityl) were used from a variety of manufacturers (Glen Research, Proligo, Cruachem, Roth).
3. For phosphoramidite dilution, dry ACN (<10 ppm H_2O, Proligo) was used.
4. Polymerase chain reaction (PCR) was performed using a PTC-100 thermocycler (MJ Research).
5. T7 RNA polymerase was purchased from Stratagene, [α-^{32}P]-CTP from Amersham, DNase I (RNase free) and SuperScript II RNase H⁻ reverse transcriptase from GIBCO, *Taq* polymerase (including buffer), and bovine serum albumin (BSA) from Boehringer Mannheim.
6. Unlabeled ribonucleoside triphosphates were obtained from Roche and deoxyribonucleoside triphosphates (dNTPs) from MBI.

7. Anthracene-poly(ethylene glycol)-guanosine conjugates (initiator nucleotides) were synthesized as described *(11)*.
8. Primers and synthetic oligonucleotides were either synthesized on an expedite 8900 DNA synthesizer or purchased from IBA Göttingen.
9. Biotin maleimide and streptavidin agarose were purchased from Sigma.
10. Spin filters (Ultrafree-MC 0.45 µM) were from Millipore.
11. Cloning was done with TOPO TA cloning kit (Invitrogen).
12. Sequencing was carried out by the dideoxy method with Thermo Sequenase (Amersham) on a LI-COR DNA Sequencer 4000 (LI-COR Biotech).
13. A: 0.3M sodium acetate buffer, pH 5.5.
14. PCR buffer: 20 mM Tris-HCl, pH 8.4, 50 mM KCl, 4 mM MgCl$_2$.
15. Transcription buffer: 80 mM HEPES, pH 7.5, 1 mM spermidine, 22 mM MgCl$_2$, 10 mM dithiothreitol, 240 µg/mL BSA.
16. Selection buffer: 30 mM Tris-HCl, pH 7.4, 200 mM NaCl, 100 mM KCl.
17. Streptavidin binding buffer: 10 mM HEPES, pH 7.2, 1M NaCl, 5 mM ethylenediaminetetraacetic acid.
18. Denaturing buffer: 0.1M Tris-HCl, pH 7.4, 8M urea.
19. Reverse transcription buffer as supplied with the reverse transcriptase enzyme.

3. Methods
3.1. Preparation of the Single-Stranded DNA Pool

1. Phosphoramidites were dissolved in dry ACN to the concentration recommended by the instrument manufacturer.
2. Primers were synthesized using the standard 1-µmol scale protocol.
3. A 157-mer DNA oligonucleotide (5'-GGAGCTCAGCCTTCACTGC-N$_{120}$-GGCACCACGGTCGGATCC-3') with the random insert of 120 nucleotides was synthesized, where N represents a mixture of A,C,G and T. Owing to the slightly different reactivities of the four phosphoramidites, a molar ratio of 3:3:2:2 A:C:G:T was adjusted by withdrawing and mixing (argon) appropriate amounts from the reservoirs of the standard phosphoramidites. This mixture was attached to an additional monomer port (port 5). A 2000 Å support was packed into a standard synthesis column. The following adjustments were made to the standard 0.2-µmol scale cycle: During synthesis of the randomized part, the capping step was omitted (by deleting the respective lines in the cycle protocol), and the duration of the detritylation step was reduced by 30% to lower depurination.
4. The product was deprotected in concentrated aqueous ammonia overnight (55°C), lyophilized, dissolved in water, lyophilized, dissolved in water, and desalted using a Sephadex G50 column (NICK column, Amersham-Pharmacia).
5. The collected product fraction (200 µL) was mixed 1:1 with formamide and loaded onto a 6% denaturing polyacrylamide gel. The product band was visualized by ultraviolet (UV)-shadowing, excised, crushed with a plastic spatula, eluted several times with buffer A, and precipitated with ethanol.

6. Single-strand DNA was quantified after dissolution in water by UV absorbance at 260 nm; 2.53 nmol of DNA resulted, corresponding to a pool complexity of 1.52×10^{15} (*see* **Note 1**).
7. Owing to acidic depurination during the detritylation step, only a fraction of this DNA can be amplified by template-dependent polymerases. It is neither necessary nor easily possible to preparatively separate these from the rest, but the amount of intact (amplifiable), full-length pool molecules should be determined on an analytical scale, Toward this end, 5'-^{32}P-labeled primer B (complementary to the 3'-end of the pool, 0.02 pmol) was added to an excess of pool molecules (1 pmol). *Taq* polymerase, 0.2 m*M* dNTPs, and buffer B were added, and a single PCR cycle was conducted (cycle conditions: 4 min at 94°C, 6 min at 50 °C, and 10 min at 72 °C). An aliquot of the reaction mixture was loaded onto a 6% denaturing polyacrylamide gel electrophoresis gel, and (by radioactivity) the amount of full-length product on the gel was related to the amount of labeled input primer. The autoradiogram showed a broad smear covering the whole range between primer and full-length product, with the full-length product giving a pronounced band. We determined that 15% of the primer molecules had been elongated to full-length product, giving a practical pool complexity of 2.28×10^{14} (380 pmol).

3.2. Preparation of Double-Stranded DNA Pool and Amplification

1. All synthesized product was subjected to preparative PCR amplification. Because the amount and concentration of template differ greatly from a typical PCR reaction, conditions had to be optimized. PCR was performed on a 20-mL scale. Most conveniently, this can be done by mixing all components and pipetting 100-µL aliquots into 96-well plate thermocyclers, but manual transfer of one large-scale reaction mix in a plastic tube between three different water baths works almost as well. Conditions were as follows: buffer B, 0.2 m*M* dNTPs, primer A (5'-TCTAATACGACTCACTATAGGAGCTCAGCCTTCACTGC-3'; 5 µ*M*) and primer B (5'-GTGGATCCGACCGTGGTGCC-3'; 7.5 µ*M*), 8 cycles (temperature cycle conditions: 94°C, 2 min; 50°C, 6 min; 72°C, 10 min).
2. The reaction was analyzed by agarose gel electrophoresis (2% gel) showing a clear product band with the correct size. The reaction mixture was extracted with phenol/chloroform and the DNA precipitated with ethanol. The amount of DNA synthesized was quantified by UV spectroscopy (A_{260}). This DNA can now be used for transcription.
3. The precipitated PCR product still contains dNTPs and primers, a preparative removal is, however, normally not necessary, but it is necessary to establish how many copies of each DNA template have been prepared. Because these components also absorb at 260 nm, an analytical sample (2% of the total PCR product, dissolved in 50 µL of H_2O) was separated by gel permeation chromatography (Superose 12 column, running in buffer A), which allowed a clean separation of PCR product, primers, and NTPs (in this order). The peak areas established that 15% of the UV absorbance in the samples was owing to the PCR product. Thus,

24 nmol of full-length DNA resulted from the preparative PCR, corresponding to about 63 copies of each amplifiable DNA template. Stock solutions of this PCR product were found to be stable for at least 2 yr at –20°C (*see* **Note 2**).

3.3. Preparation of the Single-Stranded RNA Pool

1. For in vitro transcription, 950 pmol of PCR-amplified DNA (equivalent to 2.5 complexities) was used in a 3-mL transcription reaction (buffer C, 4 mM ATP, CTP, GTP, and UTP each, 2 mM anthracene–PEG initiator nucleotide—only necessary when selecting for Diels–Alderase ribozymes *[13]* and can be omitted otherwise—3 µM [α-^{32}P]-CTP (0.2 µCi/µL), 1 U/µL T7 RNA polymerase) and transcribed at 37°C for 2.5 h.
2. 1200 U DNase I was added and incubation was continued for 30 min to hydrolyze the template DNA.
3. Formamide (3 mL) was added, and the mixture was loaded to several large 9% denaturing polyacrylamide gels (no individual slots, just one wide slot over the whole width of the gel) and separated. Transcript bands were detected by autoradiography, excised (using the X-ray film as template), and eluted with buffer A (three times, 1 h each, 70°C). Radioactivity in all fractions was counted to ensure quantitative elution.
4. The combined eluate fractions were precipitated with ethanol (3 vol, 30 min, –20°C), centrifuged (15 min, 4°C), and the supernatant discarded. The amount of transcript was quantified from the radioactivity of the pellet, and it was established as 43.6 nmol RNA, equivalent to 115 pool complexities.
5. For selecting Diels–Alderase ribozymes, the fraction of RNA conjugated to anthracene–PEG at their 5'-ends had to be quantified. Using analytical reversed-phase HPLC, this fraction was found to be about 50% *(16)*, meaning that about 57 copies of any individual sequence, each connected to an anthracene, were on average present in the starting library.

3.4. In Vitro Selection

1. This step is of course different for every selection, depending on whether aptamers or ribozymes should be selected, and on what the precise target to bind or reaction to catalyze is. Establishing the selection step is normally the most time-consuming step because various parameters have to be investigated. Generally, one has to assume that RNA molecules with specific binding or catalytic properties are very rare, and most of the molecules selected in the first few rounds of a selection are not yet those that perform the desired function. This section describes the selection of new Diels–Alderase ribozymes and is therefore specifically adjusted for this purpose. The reaction occurs between the anthracene, covalently tethered to the RNA, and biotinylated maleimide *(13)*. After reaction, the products contain a covalent linkage between RNA and biotin and can be isolated using commercially available biotin-binding affinity matrices (e.g., streptavidin–agarose).

2. For selecting new ribozymes, we found it useful to have a low but measurable background (i.e., uncatalyzed) reaction. Selection conditions were adjusted so that about 0.1–0.2% of the total RNA would react, just owing to the background reaction, and therefore be selected. An arbitrarily chosen (noncatalytic) RNA would be selected with a probability of 0.1–0.2%. If an RNA molecule accelerates the reaction 100-fold, probability increases to 10–20%, and catalysts for a 1000-fold acceleration or better would be isolated with likelihood close to 100%. Thus, although most of the molecules selected in the first round are not catalysts, the few catalytically active species present are selectively enriched among the selected molecules.
3. 38.8 nmol of the RNA prepared in the previous step (19.4 nmol anthracene–RNA = 51 complexities) was dissolved in 38 mL buffer D (final concentration of anthracene–RNA: 0.5 μM) and heated to 90°C for 5 min to denature the RNA. The reaction mixture was cooled down to room temperature over 20 min.
4. Divalent metal ions were then added to give final concentrations of 5 mM $MgCl_2$; 5 mM $CaCl_2$; 5 μM each of $AlCl_3$, $CoCl_2$, $CuCl_2$, and $MnCl_2$. The reaction was initiated by the addition of 437 µg biotin maleimide in 776 µL dimethyl sulfoxide (DMSO) to give a final concentration of 25 μM biotin maleimide and 2% DMSO, and it was incubated at room temperature for 1 h.
5. After incubation, RNA was ethanol precipitated (first add 1/10 vol of a 10X stock solution of buffer A), the pellets obtained after centrifugation (15 min, 4°C) dissolved in water, combined, and ethanol precipitated again. The resulting pellet was resuspended in buffer E, denatured at 90°C for 5 min, and cooled on ice following incubation for 45 min with streptavidin agarose (2 mL suspension, prewashed twice with buffer E, including 2 mg/mL transfer RNA to saturate possible unspecific RNA-binding sites) at room temperature.
6. The agarose-sample mixture was transferred to a spin filter and washed 20 times with denaturing buffer F and twice with water. All fractions were counted for radioactivity. In the last wash fractions, radioactivity had fallen to background level, although about 16 pmol RNA remained linked to the streptavidin resin, corresponding to 0.09% of the selectable RNA, in agreement with our expectations.

3.5. Reverse Transcription of the Selected RNAs

1. The selected RNAs, still attached to the beads, were directly reverse transcribed. The streptavidin agarose (vol about 900 µL) was suspended in 2.7 mL reverse transcription buffer G, 10 μM primer B, and 0.5 mM of each deoxynucleotide, incubated at 55°C, and the reaction started by addition of 30 µL of Superscript II reverse transcriptase enzyme.
2. After 45 min at 55°C, the whole reaction setup (including the agarose particles) was diluted into a PCR buffer B (final volume 8 mL), primers added, and 8 cycles of PCR were performed.
3. The product was analyzed by agarose gel electrophoresis, and a fraction was used for the 2nd round of selection.

3.6. Further Rounds of Selection

1. The steps of the selections scheme are iterated until the fraction of the selected species increases significantly. Selection pressure can be adjusted by reducing the reaction time in the selection step (to enrich the fastest catalysts—selection for k_{cat}), or by reducing the concentration of the reaction partner (here: biotin maleimide—selection for K_M) (*see* **Notes 3** and **4**).
2. The enriched RNA library obtained after 10 cycles of selection and amplification reacted about 8000 times faster with biotin maleimide than with the starting pool. Pool DNA obtained after round 10 was cloned and sequenced using cloning and sequencing kits following the manufacturer's instructions. At least 16 independent sequence families were found, the best of which accelerated the reaction 20,000-fold. Details are described in *(13–15)*.

4. Notes

1. It must be realized that—from the start of the synthesis until the beginning of the PCR reaction—each DNA molecule is unique. If it is lost, it will not be available for the selection anymore. To keep pool complexity as high as possible, losses must be minimized. All vials should be silanized before use to minimize adsorption problems. When samples are transferred from one vial to another, vessels should be rinsed excessively. Samples needed for analytics should be kept to a minimum, and recycled if possible.
2. It should be noted that the template concentration during PCR amplification is several orders of magnitude higher than in standard PCR protocols. Exponential amplification is not guaranteed any longer under these conditions. Long elongation cycles are necessary to ensure the polymerase has sufficient time to process all available templates. Primer concentrations should be high to ensure that enough primer is available for amplification.
3. To avoid the enrichment of undesired species (e.g., RNAs that bind tightly to streptavidin agarose), occasional rounds of counter selections are recommended. In the selection described here, this was done in rounds 3, 5, 7, and 9 by a preincubation where all components were present except for the biotin maleimide. After the preincubation, the RNA was mixed with streptavidin agarose, the effluent precipitated, redissolved, and the real selection step in the presence of biotin maleimide was performed.
4. Although it is desirable to work with as many different molecules as possible in the first round of selection, the large scale causes additional work and is rather cumbersome. It is therefore recommended to gradually reduce the scale from round to round. In this selection, we reduced the scale of the T7 transcription from 3 mL (round 1) to 400 µL (round 2), 200 µL (round 3), 100 µL (rounds 4–10). The reaction step was reduced from 38.8 mL (round 1) to 5.5 mL (round 2), to 500 µL (rounds 3–10); PCR from 8 mL (round 1) to 600 µL (round 2) 500 µL (round 3), and 400 µL (rounds 4–10). In round 4, one has achieved a scale common to molecular biologists.

References

1. Ellington, A. D. and Szostak, J. W. (1990) In vitro selection of RNA molecules that bind specific ligands *Nature* **346**, 818–822.
2. Ellington, A. D. and Szostak, J. W. (1992) Selection in vitro of single-stranded DNA molecules that fold into specific ligand-binding structures *Nature* **355**, 850–852.
3. Tuerk, C. and Gold, L. (1990) Systematic evolution of ligands by exponential enrichment: RNA ligands to bacteriophage T4 DNA polymerase *Science* **249**, 505–510.
4. Famulok, M. and Szostak, J. W. (1992) In vitro-Selektion spezifisch ligandenbindender Nucleinsaeuren *Angew. Chem. Int. Ed. Engl.* **104**, 1001–1011.
5. Gold, L., Polisky, B., Uhlenbeck, O., and Yarus, M. (1995) Diversity of oligonucleotide functions *Annu. Rev. Biochem.* **64**, 763–797.
6. Jäschke, A., Frauendorf, C., and Hausch, F. (1999) In vitro selected oligonucleotides as tools in organic chemistry *Synlett* 825–833.
7. Joyce, G. F. (1989) Amplification, mutation and selection of catalytic RNA *Gene* **82**, 83–87.
8. Bartel, D. P. and Szostak, J. W. (1993) Isolation of new ribozymes from a large pool of random sequences *Science* **261**, 1411–1418.
9. Lorsch, J. R. and Szostak, J. W. (1994) In vitro evolution of new ribozymes with polynucleotide kinase activity *Nature* **371**, 31–36.
10. Tarasow, T. M., Tarasow, S. L., and Eaton, B. E. (1997) RNA-catalysed carbon-carbon bond formation *Nature* **389**, 54–57.
11. Wiegand, T. W., Janssen, R. C., and Eaton, B. E. (1997) Selection of RNA amide synthases *Chem. Biol.* **4**, 675–683.
12. Hausch, F. and Jäschke, A. (1997) Libraries of multifunctional RNA conjugates for the selection of new RNA catalysts *Bioconjug. Chem.* **8**, 885–890.
13. Seelig, B. and Jäschke, A. (1999) A small catalytic RNA motif with Diels–Alderase activity *Chem. Biol.* **6**, 167–176.
14. Seelig, B., Keiper, S., Stuhlmann, F., and Jäschke, A. (2000) Enantioselective ribozyme catalysis of a bimolecular cycloaddition reaction *Angew. Chem. Int. Ed.* **39**, 4576–4579.
15. Stuhlmann, F. and Jäschke, A. (2002) Characterization of an RNA active site: interactions between a Diels–Alderase ribozyme and its substrates and products. *J. Am. Chem. Soc.* **124**, 3238–3244.
16. Seelig, B. and Jäschke, A. (1999) Ternary conjugates of guanosine monophosphate as initiator nucleotides for the enzymatic synthesis of 5'-modified RNAs *Bioconjug. Chem.* **10**, 371–378.

24

In Vitro Selection Procedures for Identifying DNA and RNA Aptamers Targeted to Nucleic Acids and Proteins

Eric Dausse, Christian Cazenave, Bernard Rayner, and Jean-Jacques Toulmé

Summary

In vitro selection or systematic evolution of ligands by exponential enrichment is a combinatorial procedure that allows the identification of oligonucleotides showing properties of interest—so-called aptamers—through iterative selection/amplification rounds. Libraries containing as many as 10^{14} different sequences can be screened against a wide range of molecules. Ribonucleic acid (RNA), deoxyribonucleic acid (DNA), or chemically modified aptamers generally display high affinity and exquisite specificity of interaction with the target. Aptamers show a promising potential for diagnostic and therapeutic purposes. We describe here methods successfully used in our laboratory for the selection of RNA or DNA aptamers against an RNA structure (the transactivation response element of HIV-1) and a protein (the human ribonuclease H1).

Key Words: Aptamers; SELEX; RNA hairpin; RNase H; transactivation response element RNA; HIV-1; kissing complex; loop–loop interaction; 2'-O-methyl oligonucleotide; chemical interference; combinatorial synthesis; oligonucleotide library.

1. Introduction

Combinatorial strategies offer a powerful alternative to the rational design of ligands for any kind of target. Over the last 10 yr randomly synthesized oligonucleotide libraries have been extensively and successfully used for in vitro selection of either high-affinity ligands or catalytic activities *(1,2)*. Owing to the encoded information allowing the duplication of every selected oligomer by routine polymerase chain reaction (PCR), the diversity of the starting pool reaches values as large as 10^{14}–10^{15} molecules. Each candidate sequence comprises the randomized region, n nucleotides long, flanked by two conserved sequences on the 5' and 3' sides, which are identical for every oligonucleotide in the library. The width of the random window drives the richness of the theoretical diversity, which is equal to 4^n different sequences meaning 4^n different shapes, as the primary sequence dictates the shape resulting from secondary and tertiary folds. The selection then proceeds by repeated rounds of selection (physical separation) and amplification (PCR) of appropriate candidates. Each cycle enriches the population with respect to the criterion of interest leading to the shape(s) adapted to the target. At the end of the selection procedure (typically after 7–15 rounds), termed systematic evolution of ligands by exponential enrichment (SELEX) *(3)*, the winning candidates, so-called aptamers *(4)*, are cloned and sequenced.

The selection can be carried out with deoxyribonucleic acid (DNA) or ribonucleic acid (RNA) pools. A limited number of chemically modified nucleic acids can be similarly handled *(5)* as far as the modification is tolerated by the polymerases used during the enzymatic steps (transcription, reverse transcription, PCR). Moreover, the selection conditions can be adjusted at will. Numerous targets were used for raising aptamers, ranging from small molecules (amino acids, nucleic acid bases, antibiotics, etc.) to macromolecules (proteins, nucleic acids) *(1)*. Oligonucleotides have even been selected against intact viruses *(6)* and live cells *(7)*. These molecules have wide potential for diagnostic and therapeutic applications *(8,9)*.

We describe the selection of DNA and RNA aptamers targeted to structured RNA motifs. We successfully used this method against the transactivation response (TAR) element RNA of HIV-1 *(10,11)*, which on recruitment of viral and host-cell proteins, transactivates the transcription of the retroviral genome. We also illustrate the SELEX procedure by describing the identification of DNA aptamers targeted to the human ribonuclease (RNase H1) *(12)*, a protein involved in several key steps of gene expression and in mediating the effect of antisense oligonucleotides. We describe the design of the DNA library and of the primers, the different sieves used for partitioning, as well as the transcription, amplification, and reverse transcription steps. We also provide a description for optimizing the selected sequence, in particular for improving the nuclease resistance of the aptamers. This is achieved by the introduction of chemically modified residues at those positions that have been shown, by modification interference, as neutral with respect to the properties of the original aptamers.

2. Materials
2.1. General
1. DNA and RNA libraries.
2. Oligonucleotide primers.

3. AmpliTaq Gold™ (PE Applied Biosystems).
4. Dynal magnetic particle concentrator.
5. Dynabeads M-280 or streptavidin magnesphere (Promega).
6. Radiolabeling equipment.
7. Centrex MF disposable microfilter 0.2 μm (Schleicher & Schuell).
8. M-MLV reverse transcriptase, RNase H Minus, point mutant (Promega).
9. Ampliscribe T7 high-yield transcription kit (Epicentre Technologie).
10. TOPO TA cloning kit (Invitrogen).
11. dRhodamin terminator cycle sequencing kit (PE Applied Biosystems).
12. Filters HAWP (Millipore).
13. Vacuum manifold.
14. Expedite 8908 (Millipore) nucleic acid synthesizer.
15. Reagents for oligonucleotide synthesis as listed here were from Glen Research. 2'-O-tert-butyldimethylsilyl-ribonucleoside phosphoramidites (iBu-G, Bz-A, Bz-C, and U), 2'-O-methyl-ribonucleoside phosphoramidites (Dmf-G, Bz,-A, Bz-C, and U), 0.2 mmol columns (iBu-G, Bz-A, Ac-C, and U), activator (0.45M tetrazole in acetonitrile [ACN]), anhydrous ACN, cap mix A (Ac_2O/pyridine/tetrahydrofuran [THF]), cap mix B (10% N-methylimidazole in THF), oxydizing solution (0.1M iodine in THF/pyridine/H_2O) and deblocking mix (3% trichloroacetic acid in methylene chloride).
    Triethylamine trihydrofluoride was from Aldrich.
16. Polyacrylamide gel electrophoresis (PAGE) equipment.
17. R buffer: 20 mM HEPES, 20 mM sodium acetate, 140 mM potassium acetate, and 3 mM magnesium acetate, pH 7.4.
18. E buffer: 10 mM Tris-HCl, pH 7.4, 1 mM ethylenediaminetetraacetic acid (EDTA), and 25 mM NaCl.

2.2. Chemical Interference Experiments

1. Folding buffer: 10 mM Tris-HCl, pH 8.0, 1 mM EDTA, 100 mM KCl.
2. Dimethylsulfate solution (DMS): 2% in water, freshly made.
3. Potassium permanganate solution: 0.5 mg/mL $KMnO_4$, freshly made.
4. Piperidinium formate, pH 2.0, freshly made.
5. Allyl alcohol (stored protected from light).
6. Stop buffer: 0.3M sodium acetate, pH 5.2, 1 mM EDTA.
7. 5 mg/mL transfer RNA (tRNA).
8. Denatured salmon sperm DNA.
9. Binding buffer 2X for electrophoretic mobility shift assay (EMSA): composition to be adapted to the particular protein under study (e.g., for RNase H1: 60 mM Tris-HCl, pH 8.5, 80 mM ammonium sulfate, 40 mM magnesium chloride, and 0.02% 2-mercaptoethanol).
10. TB buffer (0.5X): 45 mM Tris and 45 mM, boric acid.
11. Gel elution buffer: 0.5M ammonium acetate, 1 mM EDTA, and 1/6 vol phenol.
12. Pyrrolidine solution: 1.1M, freshly made.

3. Methods
3.1. DNA SELEX Against RNA Targets
3.1.1. Library, Primers, and Targets

The DNA SELEX is carried out with a DNA random library, chemically synthesized by the addition of an equimolar mixture of the four deoxynucleoside phosphoramidites by automated solid phase and purified by high-pressure (performance) liquid chromatography or PAGE. Usually, the window size of the random library is between 10 and 50 nucleotides, allowing a large diversity of sequences and structures. The random sequence is flanked on 5' and 3' sides by two different fixed sequences for PCR amplification. We have successfully used the following 5' and 3' fixed sequences: 5'GCCTGTT GTGAGCCTCCTGTCGAA and 5'TTGAGCGTTTATTCTTGTCTCCC. The sequences of the upstream and downstream primers for amplifications are consequently 5'GCCTGTTGTGAGCCTCCTGTCGAA for the hybridization to the 5' complement of the library and 5'ACTGACTGACTGACTGACTA-6C3-GGGAGACAAGAATAAACGCTCAA for the 3' end (**Fig. 1**).

Selection is carried out in the R buffer (20 mM HEPES, 20 mM sodium acetate, 140 mM potassium acetate, and 3 mM magnesium acetate, pH 7.4 (*see* **Note 1**).

The RNA targets are synthesized in the laboratory on an Expedite 8908 (Millipore) nucleic acid synthesizer. A biotin is added to the 3' end of the RNA targets (*see* **Note 2**). RNA oligomers are purified by 20% denaturing PAGE. Bands are detected by ultraviolet (UV) shadowing. RNases T1 and A digestions are performed on the 5' end-labeled RNA targets in the R selection buffer to check the RNA structure.

3.1.2. Hybridization and SELEX Procedure (First Round of Selection)

1. Heat the library and the targets separately at 100°C for 1 min and 65°C for 3 min, respectively, cool to 4°C for 1 min, and then equilibrate at room temperature for 5 min.
2. Mix 500 pmol of the counterselectioned library (*see* **Note 3**) with 10 pmol of biotinylated RNA target at room temperature in a final volume of 100 µL of the R buffer for 10 min.
3. Add 150 µg of magnetic beads coated with streptavidin. Incubate for 10 min at room temperature. Candidates with some affinity for the target are retained on beads. Magnetically separate the beads and the supernatant. The supernatant is discarded.
4. Wash the pellet with 100 µL of R buffer to eliminate candidates with poor affinity.
5. Elute the target-bound candidates in 50 µL of H$_2$O by heating for 40 s at 75°C.
6. PCR-amplify the recovered candidates and produce single-stranded DNA for the next round.

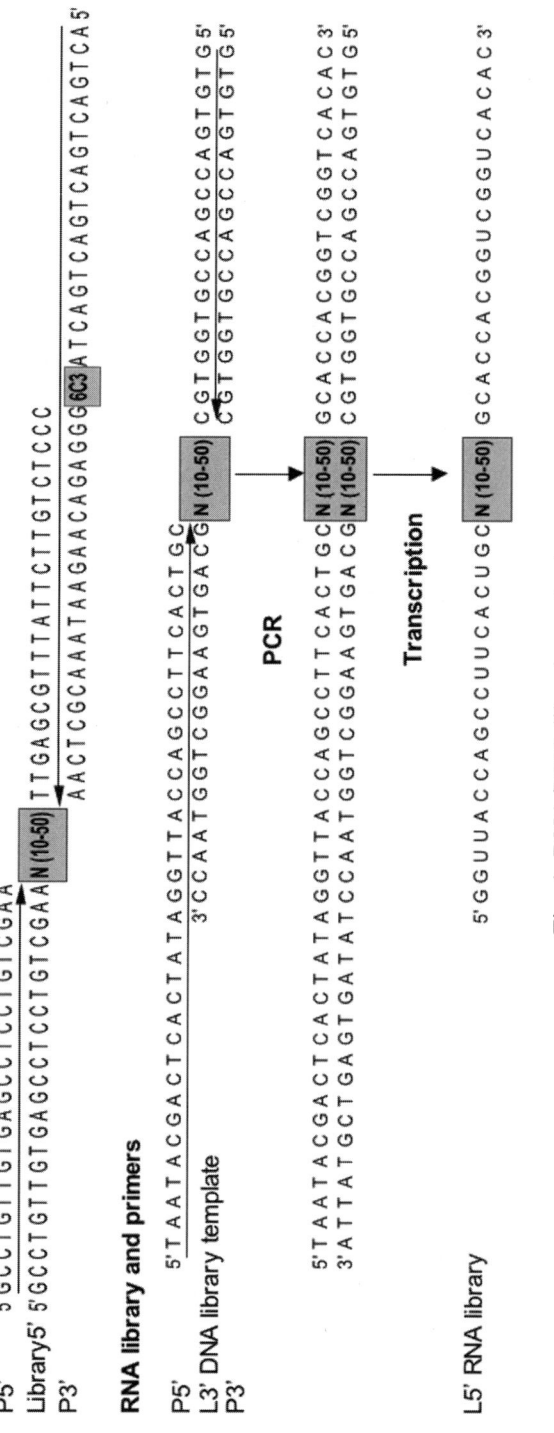

Fig. 1. DNA/RNA libraries and primers.

3.1.3. Single-Stranded DNA Production

To generate single-stranded DNA candidates during each round of the selection procedure, different procedures have been evaluated (*see* **Note 4**). The most suitable in our hands is the method using a terminator primer. The 5' extremity of the primer is made heavier with a succession of six C3 links extended with a DNA stretch of 20 nucleotides (**Fig. 2**) *(13)*. During the PCR, this six C3 region cannot be amplified by the *Taq* DNA polymerase. A PCR product with two strands of unequal length is consequently synthesized. Each strand is then easily purified on a denaturing polyacrylamide gel.

1. Amplify the recovered fragments, using an automatic hot start, with 40 U of AmpliTaq Gold (Applied Biosystems) in 400 μL of the *Taq* buffer containing in addition 200 μM of dNTP, 2 mM of Mg^{2+}, 5% DMSO, 2 μM of each primer labeled with 5 to 10,000 cpm of [γ^{32-}P]ATP (10 mCi/mL) (4500 Ci/mmol) from ICN Pharmaceutical. Then subject the reaction mixture to repeated cycles: (1) 95°C for 10 min, for a preincubation step to activate the AmpliTaq Gold and provide a hot start; (2) 95°C for 1 min, 60°C for 1 min, 72°C for 1 min 30 s, for 30 cycles; (3) 72°C for 2 min, for one final cycle.
2. After precipitation of the PCR product, separate asymmetric single strands by 8% denaturing PAGE (**Fig. 2**). Elute for 1 h at 65°C, in 1 mL of the E buffer, the band corresponding to the library molecular weight and corresponding to selected aptamers. Centrifuge for 10 min at 1500g in a Centrex 0.2 μm MF disposable microfilter, 0.2 μM ethanol precipitate, and quantify the amount of single strand by UV absorbance at 260 nm.

3.1.4. Further Selection Rounds

Single-stranded DNA candidates can then be used for the next round of selection. After several rounds, when the population no longer evolves, (*see* **Note 5**) the candidates can be cloned and sequenced.

Amplify the last round of selection as described above, and add an extra 10-min at 72°C at the end of the PCR to add deoxyadenosine at the 3' end of the PCR product, thus cloning it directly into the vector of the TOPO TA cloning kit from Invitrogen. Transform the *Escherichia coli* TOP10 One Shot™ cells according to the manufacturer's instructions (*see* **Note 6**). Sequence the clones with the dRhodamin terminator cycle sequencing kit from PE Applied Biosystems according to the manufacturer's instructions.

3.2. RNA SELEX Against RNA Targets

RNA SELEX is similar to DNA SELEX except that the library and the RNA amplification steps are modified and reverse-transcription and transcription steps are added.

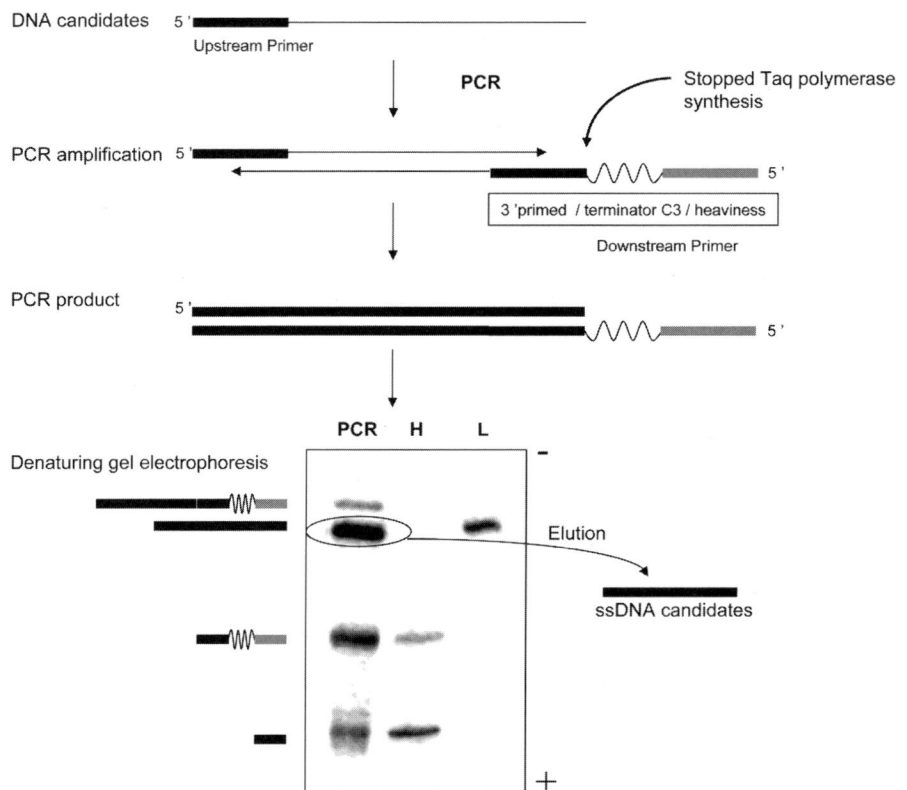

Fig. 2. Single-stranded DNA production by PCR using heaviness-terminator-primer *(13)*. The downstream primer is composed of three parts: (1) the 3' primed sequence, (2) the six C3 terminator, and (3) the heaviness sequence to increase the negative strand molecular weight. This primer allows the production by PCR of a double strand with two strands of inequal length (or molecular weight). After the PCR amplification with the two radiolabeled primers, the two strands are separated on a denaturing polyacrylamide gel (*lane PCR*). From the top to the bottom, negative strand containing the terminator primer, positive single strand representing the candidates with the same migration as the library alone (*lane L*), the terminator primer, and the upstream primer are separated. *Lane H* is a PCR amplification in the absence of candidates. The elution of the positive strand is indicated by an arrow.

3.2.1. Library and Primers

The RNA library is obtained by transcription from a DNA template library (**Fig. 1**). The 5' and 3' invariant sequences of the library are, respectively, 5'GTGTGACCGACCGTGGTGC and 5'GCAGTGAAGGCTGGTAACC.

The sequences of the upstream and downstream primers for amplification are, respectively, 5'TAATACGACTCACTATAGGTTACCAGCCTTCACTGC, including the T7 polymerase transcription promoter complementary to the 3' end of the DNA library, and 5'GTGTGACCGACCGTGGTGC for the hybridization to the 5' complement of the library. After PCR amplification, the transcription of the DNA library template allows the production of the RNA library: 5'GGUUACCAGCCUUCACUGC-$(N)_x$-GCACCACGGUCGGUCACAC. The size of the random region, as for the DNA library, may vary at will (frequently from $x = 10$ to 50 nucleotides).

3.2.1.1. Synthesis of the RNA Library

Amplify the synthetic DNA library template at 0.4 µM with 20 U of AmpliTaq Gold (PE Applied Biosystems) in a final volume of 1 mL of the *Taq* polymerase buffer containing 200 µM of dNTP, 1.5 mM of Mg^{2+}, 5% DMSO, using 2 µM of the upstream primer containing the T7 RNA polymerase promoter and the downstream primer. Then subject the reaction mixture to repeated cycles: (1) 95°C, 10 min, for the preincubation step to activate the AmpliTaq Gold and provide a hot start; (2) 95°C for 30 s, 63°C for 30 s, 72°C for 1 min, for 25 cycles; and (3) 72°C for 5 min, for one final cycle.

3.2.1.2. Transcription Reaction

After phenol/chloroform extraction and precipitation of the PCR product, perform the transcription reaction at 37°C for 2 h in a final volume of 40 µL using the Ampliscribe T7 high-yield transcription kit from Epicentre Technologie. Add 2 µL of RNase-free DNase I at 1 MBU µL for 15 min. Purify the transcription pool by electrophoresis on a 20% denaturing polyacrylamide gel, visualize by UV shadow and extract the RNA as described above for the single-stranded DNA production except that the elution is performed overnight at room temperature. Quantify the amount of RNA by absorbance at 260 nm.

RNA pool can be used for the first round of selection using the selection buffer, the counterselection and SELEX procedures used for DNA SELEX (*see* **Subheadings 3.1.1.** and **3.1.2.**). The following steps must be adapted to or added for the RNA SELEX:

1. Folding of the RNA candidates is achieved by heating the pool of RNA candidates at 65°C for 3 min, keeping them on ice for 1 min, and finally keeping them at room temperature for 5 min.
2. The RNA must then be reverse-transcribed prior to amplification. Anneal the recovered RNA candidates in 10 µL to 3.3 µM of the downstream primer at 70°C for 10 min, and copy the RNA into complementary DNA (cDNA) with 200 U of M-MLV reverse transcriptase, RNase H Minus, point mutant (Promega) at 50°C in a final volume of 20 µL for 50 min. PCR-amplify the cDNA candidates in 1 mL, using the same conditions as described in **Subheading 3.2.1.1.** (*see* **Note 7**).

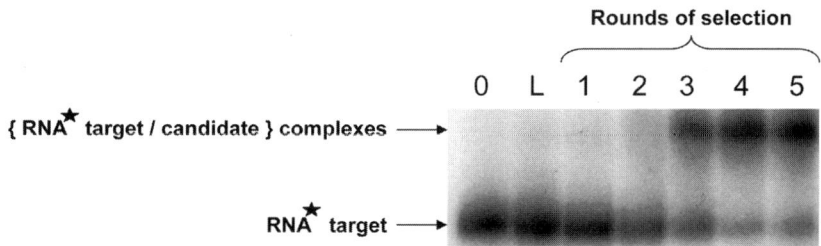

Fig. 3. Analysis of the selection progress. EMSA of [γ^{32-}P]-5' end-labeled RNA target without (*0*) or in presence of 10 n*M* of the RNA library (*L*) or RNA candidates stemming from each round of selection (*lanes 1–5*).

After amplification, transcription, and purification, RNA candidates can be used for the next round of selection. Following reverse transcription, selected candidates are cloned and sequenced as described in **Subheading 3.1.4.**

3.3. Selection of DNA Aptamers to a Protein

The procedure described as follows has been successfully used against the human RNase H1 *(12)*.

3.3.1. Selection

The selection of DNA aptamers (**Fig. 4**) is carried out starting from a library identical to the one described in **Subheading 3.1.1.** The reduction of the background is obtained via a counterselection step for removing nonselective binders (*see* **Note 8**), carried out prior to selection.

3.3.1.1. NEGATIVE SELECTION

Incubate 200 pmol of the ^{32}P-labeled library (or selected sequences in the successive rounds of selection) (*see* **Notes 8**) in 100 µL binding buffer for 15 min at ambient temperature (23°C) with pieces of alkali-treated nitrocellulose filters (HAWP, Millipore) (*see* **Note 9**) for pretreatment of the filters). Collect the pieces of filter with a brief centrifugation, and quantify the amount of radioactive DNA adsorbed by Cerenkōv counting.

3.3.1.2. POSITIVE SELECTION

1. Split the supernatant in two identical parts, one being incubated with 0.1 µ*M* targeted protein, and the other part being incubated in the same conditions, but without protein, for determination of background levels.
2. Incubate 10 min at room temperature (approx 23°C) in 100 µL of selection buffer (example: 30 m*M* Tris-HCl, pH 8.5; 40 m*M* ammonium sulfate; 20 m*M* magnesium chloride, and 0.01% 2-mercaptoethanol for the human RNase H1). The com-

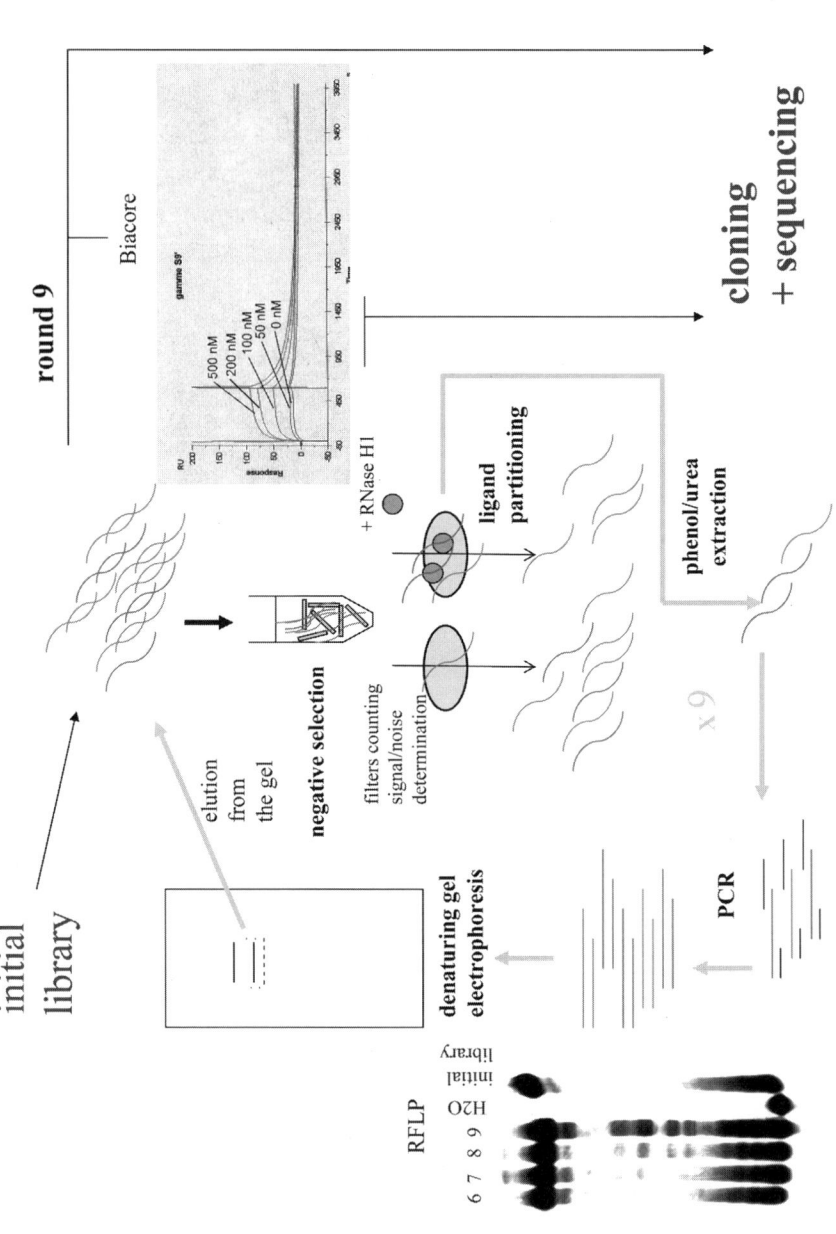

Fig. 4. Schematic description of the SELEX procedure used for the isolation of DNA aptamers binding to the human RNase H1. The 9 rounds of the filter partition amplification are depicted in the central part of the scheme. On the left is presented patterns obtained on the PCR products with a cocktail of restriction enzymes (*see* **Note 8**). On the right are presented BIAcore sensorgrams obtained with aptamers from round 9 interacting with immobilized RNase H1 (*see* **Subheading 3.3.2.** and **ref. 12**).

position of the binding buffer is chosen for the stability and physiological activity of the target protein.
3. Filter each sample independently through alkali-treated filters using a vacuum manifold (5 in. mercury of negative pressure). Wash filters with 3 mL of selection buffer.
4. Extract nucleic acids retained on the filter with $7M$ urea and Tris-buffered phenol: chloroform, pH 7.9, as described by Fitzwater and Polisky *(14)*. Place membranes on a clean glass plate, and cut them into eight pieces with a clean razor blade, then transfer the membranes to a microfuge tube with membrane forceps. Add 200 µL freshly prepared $7M$ urea and 600 µL Tris-buffered phenol:chloroform, pH 7.9. Vortex and hold the tube at room temperature for 30 min. Vortex and spin down briefly and then add 100 µL H_2O. Vortex and spin again. Lightly tamp down the membranes with a pipet tip. Spin the tube at 16,000g for 5 min. Save the aqueous phase and repeat the extraction of the Tris-buffered phenol:chloroform with 100 µL H_2O. Combine and chloroform-extract the aqueous layers, then ethanol-precipitate with 3 µL of 0.25% linear polyacrylamide or 20 µg yeast tRNA carrier.
5. Quantify the amount of nucleic acid recovered by Cerenkōv counting.

3.3.1.3. AMPLIFICATION

PCR-amplify the recovered sequences, precipitate PCR products, separate strands on denaturing electrophoresis, and recover selected aptamers as described in **Subheading 3.1.3.**

Sequences are heated for 2 min at 98°C, chilled on ice for 1 min, then incubated for 5 min at room temperature before being used in another round of selection.

3.3.1.4. SUCCESSIVE ROUNDS OF SELECTION

Perform several rounds of selection as described in **Subheadings 3.3.1.2. and 3.3.1.3.**

For each round quantify the amount of sequences retained by the protein and determine background. Ideally, the background should remain stable and low (less than 1%) all along the selection. If it increases, one can try to modify slightly the procedure (*see* **Note 8**). After a few rounds (3 to 6 rounds, most often) a sharp rise in the amount retained in the presence of the protein is noticed. A few additional rounds should be performed until no further increase is observed. During these last steps the stringency of the selection can be increased (*see* **Note 5**).

When the pool no longer evolves (the ninth round in the case of RNase H1), sequences are cloned, using the TOPO TA cloning kit (InVitrogen, Carlsbad, CA), and sequenced and analyzed on the program sequencing analysis 3.0 (PE Applied Biosystems).

3.3.2. Alternative Selection Step Using BIAcore™

In the case of SELEX carried out against human RNase H1, the population of sequences selected by retention on filters was further selected on BIAcore (**Fig. 4**). Samples were injected on the enzyme immobilized on a BIAcore sensor chip, for example, via primary amino groups with the amine coupling kit from Pharmacia, following the procedure of Haruki et al. *(15)*. Experiments were performed at 23°C in the buffer used for the selection, at a flow rate of 10 μL/min. The volume of nucleic acid samples injected (either the initial library or the products of the last round) was 100 μL; 2 min after the beginning of the dissociation phase, samples were collected for 8 min. DNA from these samples was cloned and sequenced as described in **Subheading 3.1.4.** The BIAcore offered us a last step of selection, alternative to filter retention, and thus potentially eliminated sequences with a high affinity for nitrocellulose.

3.3.3. Analysis of Aptamer and Aptamer–Protein Complexes by Chemical Interference

Labeled DNA fragments are partially modified with a chemical, generating a population of oligonucleotides containing a single modified residue on average. The protein will bind only to oligonucleotides in which the modification does not interfere with site recognition, leading to a retarded band in EMSA: native gel electrophoresis of protein–nucleic acid complexes. The free unshifted band will contain all the modified species with an overrepresentation of species that no longer bind to the target. Extraction and cleavage of DNA from these two bands, followed by denaturing polyacrylamide gel (i.e., sequencing gel), will show missing fragments in the pattern derived from the shifted populations compared to the pattern of the unshifted population. These missing bands identify the sites important for protein binding. We describe, as follows, chemical interference obtained by modification of G residues with DMS (methylation interference), by partial depurination (missing-contact probing of protein DNA interactions), or by permanganate oxidation of thymine residues; all these modifications being used routinely in a simplified chemical sequencing procedure of oligonucleotides *(16)*.

3.3.3.1 PREPARATION OF THE APTAMERS

Modification reactions of the simplified chemical sequencing procedure are carried out in Tris/EDTA buffer containing 80 m*M* KCl so that the aptamers adopt their folded structure (K^+ stabilizes G quartets present in some of the aptamers) (*see* **Note 10**).

Four Eppendorf tubes are prepared as follows: To 1μL radiolabeled aptamer (about 5 10^5 cpm) add 8 μL folding buffer. Heat 2 min at 98°C, cool on ice

2 min, then incubate 2 h at 25°C, and store at 4°C until the modification reaction is initiated.

3.3.3.2. Partial Modification Reactions

1. For modification of G residues, add 1 µL DMS solution, and incubate for 10 min at room temperature. Stop the reaction by addition of 200 µL stop buffer followed by addition of 10 µL tRNA (5mg/mL) and 600 µL ethanol.
2. For depurination (G + A), add 1 µL piperidinium formate, pH 2.0, and incubate 5 min at 65°C. Stop the reaction as in the previous step.
3. For modification of T residues, add 1 µL of permanganate solution, and incubate the resulting solution 3 min at room temperature, followed by addition of 1 µL allyl alcohol, followed by stop buffer, tRNA, and ethanol, as in **step 1**.
4. As a control, add 1 µL H_2O to the fourth tube, which is incubated 10 min at 23°C before adding stop buffer, tRNA, and ethanol.
5. After 1 h at –80°C, centrifuge the tubes for 30 min (14,000g at 4°C). Discard supernatants, dry the pellets a few minutes in a SpeedVac, redissolve in 200 µL stop buffer, and precipitate again by addition of 600 µL ethanol.

3.3.3.3. Gel-Shift Assay and Recovery of Modified Sequences

1. After 1 h at –80°C, centrifuge the tubes, discard supernatants, and air-dry the pellets.
2. Redissolve the pellets in 4 µL binding buffer 2X containing 40 µg/mL denatured salmon sperm DNA (included to prevent nonspecific binding; another currently used polynucleotide for this purpose is poly(dI,dC). After 10 min at room temperature, add 4 µL of the protein in dilution buffer (e.g., for RNase H: 50 mM Tris-HCl, pH 7.9; 40 mM dithiothreitol; 1 mg/mL bovine serum albumin; 50% glycerol). The dilution is chosen to give about 50% retardation in EMSA.
3. After 10 min incubation at room temperature, add 2 µL 5X loading buffer, and load samples, while the current is running, on a native 8% (19:1 acrylamide/*bis*-acrylamide) gel made in 0.5X TB buffer. Electrophoresis is continued at 6W constant power until the bromophenol blue has migrated half the distance of the gel.
4. After autoradiography, locate radioactive bands corresponding to bound (shifted) and free aptamers, by precise superposition of the gel with the autoradiograph, with the help of the radioactive or luminescent markers.
5. Excise and cut the bands in small pieces. Collect in eight different Eppendorf tubes, for bound (upper shifted band) and free aptamers (lower unshifted band) for each modification reaction and control.
6. Elute these overnight with gentle agitation (Eppendorf tubes placed on a rotating wheel or attached to a vortex shaker) in 500 µL of gel elution buffer. Recover the eluates and extract with 1 vol of phenol/chloroform/isoamyl alcohol (50:49:1).
7. Centrifuge, collect the upper aqueous phases, and precipitate nucleic acids by addition of 1 mL ethanol to each tube.
8. After 1 h at –80°C, centrifuge the tubes for 30 min at 14,000g at 4°C, discard supernatants, and dry the pellets in a SpeedVac.

3.3.3.4. Cleavage at Modified Positions: Denaturing PAGE

1. Redissolve the pellets in 10 μL H_2O. Add 100 μL of 1.1M pyrrolidine (freshly made), and incubate the tubes 15 min at 90°C for cleavage at modified sites. Reduce to dryness in a SpeedVac, resuspend in 20 μL H_2O, and redry twice.
2. After final drying, dissolve samples in 10 μL H_2O, and quantify radioactivity by Cerenkôv counting. Mix identical amounts of radioactivity (usually about 2 10^4 cpm) with formamide containing tracking dyes, heat 1 min at 98°C, and load onto a 20% acrylamide 7M urea sequencing gel. For each modification reaction, load bound (**B**) and free (**F**) aptamers side by side. After electrophoresis the gel is wrapped in Saran and autoradiographied (**Fig. 5**).

3.3.4. Synthesis of Chemically Modified Aptamers

2'-O-methyl-oligonucleotides (2'-O-Me) were introduced as nuclease-resistant analogs able to form stable duplexes with complementary RNA strands *(17)* and chimeric 2'-O-Me/oligodeoxynucleotides have been reported to display enhanced antisense potencies *(18)*. Because 2'-O-Me modification is not tolerated by polymerases used during the enzymatic steps, the SELEX methodology cannot be applied for screening 2'-O-Me-modified oligonucleotide libraries. However, such chemical modification can be introduced *a posteriori* in selected DNA or RNA aptamers providing that every modification in the aptamer is probed for maintaining the original affinity and selectivity. This can be achieved by chemical interference (*see* **Subheading 3.3.3.**).

3.3.4.1. Chimeric 2'-O-Me/RNA Oligonucleotide Synthesis

Chimeric 2'-O-Me/RNA oligonucleotides are assembled on a 0.2-mmol scale. Standard cycles (as provided by the manufacturer) are used for the incorporation of regular ribonucleotides, whereas slightly modified cycles are used for 2'-O-Me ribonucleotides. For the coupling steps, 5 pulses (instead of 10 pulses for regular ribonucleotides) of a mixture of phosphoramidite and activator (tetrazole) are sent to the column followed by a 940-s wait period.

Deprotection is carried out according to the procedure reported for oligoribonucleotides (www.Glenres.com). The regular deprotection is applied for the cleavage from the support and the removal of base and phosphate protecting groups. Removal of silyl protecting groups is achieved by treatment with triethylamine trihydrofluoride. Then the oligonucleotide is desalted by butanol precipitation and purified by 7M urea PAGE.

3.3.4.2. Cleavage from the Support and Removal of Base and Phosphate Protecting Groups

1. Open the synthesis column and pour the support into a sealable vial. Add 1 mL of ethanolic ammonium hydroxide to the vial, and incubate at 55°C for 17 h. Cool the sealed vial and open cautiously.

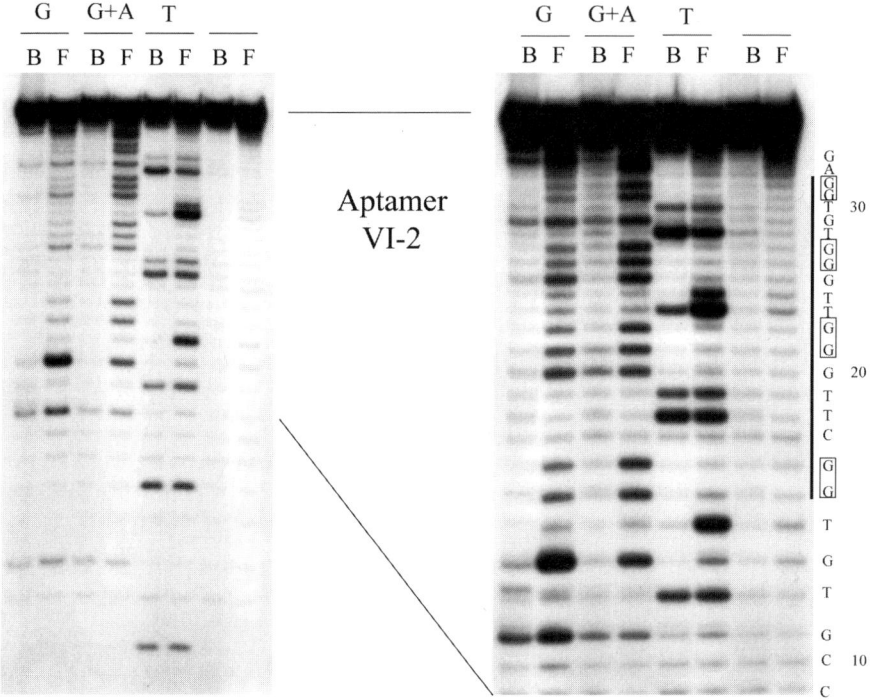

Fig. 5. Autoradiographs of chemical interference experiments carried out on aptamer VI-2 interacting with RNase H1 *(12)*. Identical amounts of free (**F**) or protein-bound aptamers (**B**) were cleaved at the site of modification by treatment with pyrrolidine and loaded side by side on a sequencing gel. The gel on the right is a longer migration time for detailing the bases constituting the central unimolecular quadruplex formed by stacking of two guanine quartets. Nucleotides belonging to the intramolecular quadruplex are located by the black bold line; G residues involved in G quartets are boxed. Following chemical modification of T residues with potassium permanganate, strong interferences are noticed for T14, T23, and T24, whereas T12, T18, and T19 do not interfere. Similar observations can be made for samples modified with DMS (*lane G*) or depurinated (*lane G + A*).

2. Remove the supernatant solution from the support using a sterile pipet, and rinse the support twice with 1 mL of ethanol:H_2O (3:1) mix. Evaporate the combined solutions to dryness.

3.3.4.3. REMOVAL OF SILYL PROTECTING GROUPS

Add 0.3 mL net triethylamine trihydrofluoride to the residue, and leave the tube at room temperature for 24 h. Agitate the solution by frequent shaking. Quench the reaction by adding 0.4 mL of 1*M* triethylammonium acetate and 1 mL of sterile H_2O.

3.3.4.4. Desalting and Purification

Desalt the chimeric 2'-O-Me/RNA oligonucleotide by butanol precipitation, and purify it by 7M urea PAGE.

4. Notes

1. The folding of nucleic acids is strongly dependent on the ionic composition of the medium. Mg^{2+} ions are of particular importance for RNA structures.
2. Alternatively biotinylated anchors—oligo (dT), for example—can be used to capture long RNA molecules in which the complementary sequence—(oligo(dA)— constitutes one end.
3. Different streptavidin aptamers have been identified from random libraries using in vitro selection. To avoid selecting these sequences that have affinity for the beads, the DNA pool is mixed, prior to selection, with 100 µg of magnetic beads coated with streptavidin (alternately with Dynabeads M-280 or streptavidin magnesphere Promega beads), and preequilibrated three times in the R buffer. For monitoring the evolution of the nonspecific DNA pool, candidates retained with the beads are eluted in 50 µL of H_2O and quantified by UV absorption spectrometry. The selection is then carried out with DNA candidates that have not been retained on the beads and that have been carefully collected.
4. Alternative methods include:
    a. *Asymmetric PCR*: The simple and rapid technique described by Gyllensten and Erlich *(19)*consists in amplifying DNA with an excess molar amount of one of the primers. The typical ratio between the two primers can be 50 to 0.5. The PCR amplification initially generates a double-stranded template until the primer at the lowest concentration is consumed. The amplification then produces an excess of full-length single-stranded DNA. (The double-stranded DNA amplification at the first cycles of the PCR, which contaminates the accumulation of the single-stranded DNA, constitutes the major limitation of this technique).
    b. *Exonuclease λ digestion*: The λ exonuclease selectively digests one strand of a DNA duplex from a 5' phosphorylated end leaving the complementary strand intact. Using this rapid enzymatic step, single-stranded DNA can be produced by digesting an intact PCR product. Using a phosphorylated 5' primer, a double-stranded PCR product with a 5' phosphorylated strand can be obtained. Whereas the phosphorylated strand is 5' → 3' degraded by the nuclease, the unphosphorylated strand remains intact and can be purified by PAGE. (The λ exonuclease also degrades nonphosphorylated substrates but at a greatly reduced rate.)
    c. *Biotin–streptavidin purification*:This procedure involves the use of a biotinylated 5' reverse primer. After PCR amplification, separate on streptavidin-coated magnetic beads the double-stranded biotinylated PCR product from the supernatant. Denature the isolated duplex DNA, and elute the nonbiotinylated single strand. This method based on the capture of the PCR product, using a 5' biotinylated primer on streptavidin magnetic beads,

is unsuitable when the selection procedure makes use of biotin/streptavidin to capture the target. In this case, the target could be captured by free streptavidin that is released with the single strand rather than being captured on magnetic beads at the next selection round.
5. In the first rounds of selection, the stringency must be low enough to retain in the selected pool the few sequences that display affinity for the target. The washing of the beads can be modulated. Only one wash step can be performed at the beginning of the selection. In subsequent rounds of selection, to keep only high-affinity candidates, the beads are washed two or more times. The candidate/target ratio, the candidate, and the target concentrations are other parameters that can be changed to increase the stringency. During the selection, the concentration of the candidates is decreased at each round and the candidate/target ratio must be kept high to promote the competition.
   a. Evaluate the evolution of the affinity of the RNA or DNA population for the RNA target by EMSA at each round of selection (**Fig. 3**). Incubate 1 nM of 5' end-labeled target with increasing concentrations of aptamer for 20 min in 10 µL of R buffer at room temperature.
   b. Run the sample on a native gel (10 or 15% [w/v], thickness 1 mm, 19:1 acrylamide:*bis*-acrylamide according to the size of the target) in 50 mM Tris-acetate (pH 7.3 at 20°C) and 3 mM magnesium acetate (TAC buffer) at 100 V and 4°C for 15 h. Quantify by Instant Imager (Packard Instrument).
6. Perform only 15 cycles of PCR to minimize the *Taq* polymerase mutations.
7. The r*Tth* polymerase can alternatively be used instead of the reverse transcriptase and the Taq polymerase. In a single tube, r*Tth* polymerase can perform reverse transcription at elevated temperature and PCR process. cDNA synthesis at higher temperature by r*Tth* increases the denaturation of the secondary structure of RNA.
8. The pool of candidates can be prepared using 5' end-radiolabeled primers added in trace amounts to the bulk of unlabeled primers used for PCR. This facilitates the detection of sense and antisense strands on the denaturing gel and also permits the quantification of the amount of sequences retained on the filter in the presence, or in the absence, of the target protein. Also, radiolabeled PCR products can be used as substrate for a cocktail of restriction enzymes (*Alu*I, *Hae*III, *Hha*I, and *Mbo*I, for example; *see* **Fig. 4**). After several rounds of selection, this should generate a distinct restriction pattern indicating that the population is evolving *(21)*.
9. In general, nucleic acid passes through nitrocellulose filters, whereas nucleic acid bound to protein is trapped on the filter (although generally designed as nitrocellulose filters, they are made in reality of a mixture of cellulose acetate and nitrocellulose, so some authors prefer to design them as "modified cellulose" filters). Filter retention assay is therefore a convenient method for partitioning free and bound candidates. One recurrent problem of this methodology is the selection of sequences possessing a high affinity for nitrocellulose, a phenomenon exacerbated by the use of DNA sequences. These are more hydrophobic than RNA owing to the absence of the 2'-OH group on the sugar moiety and, thus, have a higher tendency to stick to the filters, producing high backgrounds. To avoid the

selection and the carryover of such sequences, two precautions can be adopted:
a. According to McEntee et al. *(20)*, filters are soaked in 0.5M KOH for 20 min at 22°C, extensively washed with H_2O, then washed for 45 min in 100 mM Tris-HCl, pH 7.5, and stored in the same buffer at 4°C. This treatment reduces the background to acceptable levels (less than 1% in our conditions), but the filters become mechanically fragile and have to be handled with care.
b. Prior to filter retention selection, the pool of sequences is preincubated with pieces of nitrocellulose to remove potential "nitrocellulose-binding sequences."
Despite the systematic pretreatment of the filters and the negative selection step, it may happen that the background increases sharply after several rounds. In such a situation, try to add a second preincubation step with nitrocellulose, to extend the volume of washing, or to reduce the number of PCR cycles.
10. The procedure described has been optimized for aptamers binding to the human RNase H1 and can be used as a starting point. For other proteins buffers might be varied depending on the requirements for the binding. This might necessitate adjusting the time and the concentration of the reagents used for the modification reaction.

Acknowledgments

We are grateful to F. Ducongé, C. Boiziau, and F. Pileur for their contribution to the elaboration of the procedures described in this chapter. We also thank S. Da Rocha and E. Ledan-Schuester for careful reading of this manuscript.

References

1. Osborne, S. E. and Ellington, A. E. (1997) Nucleic acid selection and the challenge of combinatorial chemistry. *Chem. Rev.* **97,** 349–370.
2. Wilson, D. S. and Szostak, J. W. (1999) In vitro selection of functional nucleic acids. *Annu. Rev. Biochem.* **68,** 611–647.
3. Tuerk, C. and Gold, L. (1990) Systematic evolution of ligands by exponential enrichment: RNA ligands to bacteriophage T4 DNA polymerase. *Science* **249,** 505–510.
4. Ellington, A. D. and Szostak, J. W. (1990) In vitro selection of RNA molecules that bind specific ligands. *Nature* **346,** 818–822.
5. Kusser, W. (2000) Chemically modified nucleic acid aptamers for in vitro selections: evolving evolution. *J. Biotechnol.* **74,** 27–38.
6. Pan, W., Craven, R. B., Qui, Q., et al. (1995) Isolation of virus-neutralizing RNAs from a large pool of random sequences. *Proc. Natl. Acad. Sci. USA* **92,** 11,509–11,513.
7. Homann, M. and Goringer, H. U. (1999) Combinatorial selection of high affinity RNA ligands to live African trypanosomes. *Nucl. Acid. Res.* **27,** 2006–2014.

8. Brody, E. N., Willis, M. C., Smith, J. D., Jayasena, S., Zichi, D., and Gold, L. (1999) The use of aptamers in large arrays for molecular diagnostics. *Mol. Diagn.* **4,** 381–388.
9. Toulmé, J.-J. (2000) Aptamers: selected oligonucleotides for therapy. *Curr. Opin. Mol. Ther.* **2,** 318–324.
10. Boiziau, C., Dausse, E., Yurchenko, L., and Toulmé, J. J. (1999) DNA aptamers selected against the HIV-1 TAR RNA element form RNA/DNA kissing complexes. *J. Biol. Chem.* **274,** 12,730–12,737.
11. Ducongé, F., Di Primo, C., and Toulmé, J. J. (2000) Is a closing "GA pair" a rule for stable loop–loop RNA complexes? *J. Biol. Chem.* **275,** 21,287–21,294.
12. Pileur, F., Andréola, M. L., Dausse, E., et al. (2003) Selective inhibitory DNA aptamers of the human RNase H1. *Nucl. Acid. Res.* **31,** 5776–5788.
13. Williams, K. P. and Bartel, D. P. (1995) PCR products with strands of unequal length. *Nucl. Acid. Res.* **23,** 4220, 4221.
14. Fitzwater, T. and Polisky, B. (1996) A SELEX primer. In *Combinatorial Chemistry* (Abelson, J. N., ed.), Academic Press, San Diego, CA, pp 275–301.
15. Haruki, M., Noguchi, E., Kanaya, S., and Crouch, R. J. (1997) Kinetic and stoechiometric analysis for the binding of *E. coli* ribonuclease HI to RNA–DNA hybrids using surface plasmon resonance. *J. Biol. Chem.* **275,** 22,015–22,022.
16. Williamson, J. R. and Celander, D. W. (1990) Rapid procedure for chemical sequencing of small oligonucleotides without ethanol precipitation. *Nucl. Acid. Res.* **18,** 379.
17. Iribarren, A. M., Sproat, B. S., Neuner, P., Sulston, I., Ryder, U., and Lamond, A. I. (1990) 2'-O-Alkyl oligoribonucleotides as antisense probes. *Proc. Natl. Acad. Sci. USA* **87,** 7747–7751.
18. Monia, B. P., Lesnik, E. A., Gonzalez, C., et al. (1993) Evaluation of 2'-modified oligonucleotides containing 2'-deoxy gaps as antisense inhibitors of gene expression. *J. Biol. Chem.* **268,** 14,514–14,522.
19. Gyllensten, U. B. and Erlichl, H. A. (1988) Generation of single-stranded DNA by the polymerase chain reaction and its application to direct sequencing of the HLA–DQA locus. *Proc. Natl. Acad. Sci. USA* **85,** 7652–7656.
20. McEntee, K., Weinstock, G. M., and Lehman, I. R. (1980) RecA protein-catalyzed strand assimilation: stimulation by *Escherichia coli* single-stranded DNA-binding protein. *Proc. Natl. Acad. Sci. USA* **77,** 857–861.
21. Bartel, D. P. and Szostak, J. W. (1993) Isolation of new ribozymes from a large pool of random sequences. *Science* **261,** 1411–1418.

25

Short Double-Stranded Ribonucleic Acid as Inhibitor of Gene Expression by the Interference Mechanism

Jean-Rémi Bertrand, Andrei Maksimenko, and Claude Malvy

Summary

The rapid development of the small interfering ribonucleic acid (siRNA)-induced inhibition of the gene expression at the RNA level offers to research groups a new strategy for the understanding of gene functions. The siRNA approach is close to antisense oligonucleotide technology and takes advantage of the progress of chemically synthesized oligoribonucleotides. This approach for the mammalian cells was described by Elbashir et al. at the beginning of 2001, and in this chapter we describe methods for the design of siRNA molecules, solutions for efficiently transfecting cells, and methods for analyzing the inhibition of targeted genes. Methods for in vivo approach are also proposed.

Key Words: siRNA; interference; short double-stranded RNA; inhibition of gene expression; oligonucleotide.

1. Introduction

The specific inhibition of gene expression by short nucleic acids has been an important challenge for many laboratories. The reason resides in fundamental approaches for the understanding of gene functions and for therapeutic applications. The development of the automatic oligonucleotides synthesis was among the first innovations that have allowed these strategies. Most of the antisense oligonucleotides used are from the deoxyribonucleic acid (DNA) series in single-stranded structure. Because of their low nuclease resistance, ribonucleic acid (RNA) oligonucleotides are believed not to be efficient. The mechanism of action of phosphodiester and phosphorothioate antisense oligonucleotides is owing to the cellular ribonuclease (RNase) H activity, which hydrolyses RNA in the RNA· DNA structure formed between the target RNA and the antisense oligonucleotide. Because RNA hybridized oligoribonucleotides do not form a substrate for this enzyme they are not efficient in such an approach. In addition, the messenger RNA (mRNA) · RNA duplex is not able to block ribosomes as peptide nucleic acid and morpholino oligonucleotides are able to do. More recently, a new strategy resulting in silencing of the targeted RNA has been described, first in plant, and then in nematode, *Drosophila,* and embryonic mammalian cells. This strategy involves double-stranded RNA molecules and presents a very high efficiency for gene inhibition (for reviews, *see* **refs.** *1–3*). This strategy originates from a system present naturally in many eukaryotic cells with the aim of protecting the cell against viruses. The mechanism of interference relative to the target gene involves many steps. First, after the introduction of a long double-stranded RNA (for instance, a viral RNA) into the cell, it is hydrolyzed by a cellular endonuclease: the DICER. This results in the accumulation in the cell of short double-stranded RNAs, which are responsible for the interference effect and called short interfering RNAs (siRNAs). Then the siRNAs interact with proteins to form an activated double-stranded ribonuclease (RNase), which is responsible for the specific recognition of the target RNA and then induces a strand break. This key riboprotein complex is called RNA-induced silencing complex (RISC). Then the cleaved mRNA is totally destroyed by unspecific cellular RNases (**Fig. 1**). However, this whole mechanism of interference is not efficient in mammalian cells. In fact, the introduction of cells of long double-stranded RNAs results in the cell death by induction of a double-stranded RNA-dependent protein kinase. This indeed results in the activation of the interferon response and in the phosphorylation of the eukaryotic initiation factor 2 (eIF2) resulting in a translation inhibition *(4)*. It was shown in 2001 that in mammalian cells the use of 21–23 oligoribonucleotide basepairs in place of long double-stranded RNA results in a very efficient inhibition of the targeted gene without unspecific cell death *(5,6)*. This result opens the way for many applications of siRNAs for academic research as well as for therapeutic applications.

The use of siRNAs as inhibitors of gene expression presents many advantages in comparison with the antisense oligonucleotide strategy. Nevertheless, this assertion has to be moderated by the imperfect knowledge of this new strategy. Both these strategies are close in their action mechanism (**Fig. 2**), but they differ in the RNase involved in the cleavage of targeted mRNA. For antisense oligonucleotides, the natural role of RNase H seems to be essentially involved in DNA replication, but for siRNA, the RISC complex is naturally devoted to this activity. The siRNA strategy does not necessitate specially modified ribo-oligonucleotides because the introduction of modified bases induces a loss in their efficiency except when this occurs at the two bases of the 3' extremity of siRNA *(7)*. This finding simplifies greatly the choice of

siRNA as Inhibitor of Gene Expression

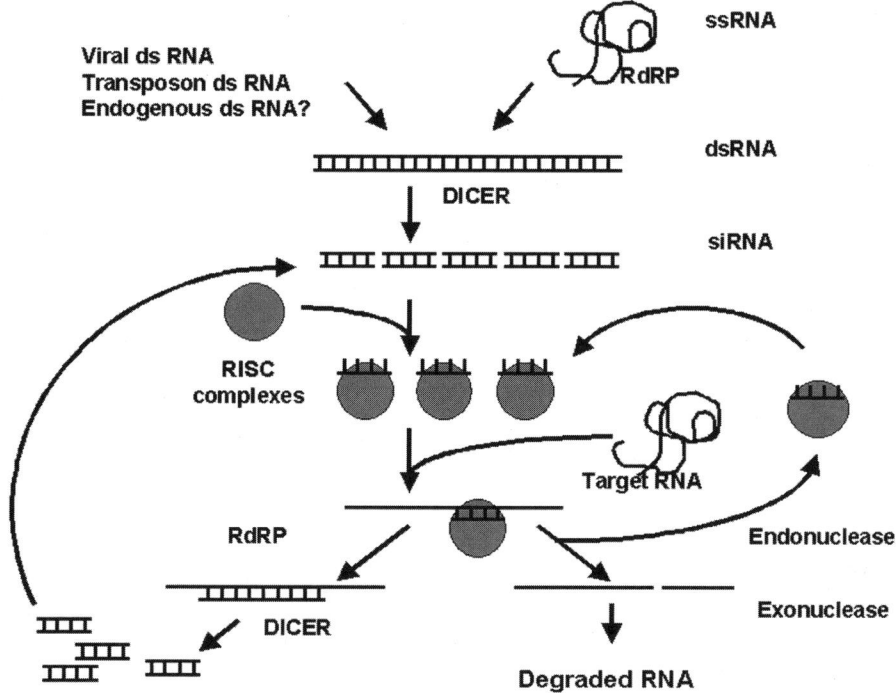

Fig. 1. The interference mechanism. The introduction into a cell of a long double-stranded RNA or its synthesis by the RNA-dependent RNA polymerase induces a cascade mechanism resulting in the disappearance of the corresponding RNA. First, the double-stranded RNA is cleaved in short siRNAs by the DICER enzyme (similar to an RNase III). Then these siRNAs interact in the RISC complex with the target RNA to induce its degradation. The antisense siRNA fragment may also act as a primer for the RdRP polymerase. This leads to the generation of a new long double-stranded RNA, which is then cut by the DICER enzyme in new siRNAs resulting in an amplification of the phenomenon *(8)*.

chemistry for such molecules. This is not true for antisense oligonucleotides where the use of modified bases is necessary to render them resistant toward nucleases.

At this date, the laboratories using this new strategy have described many successes with a very good efficiency and specificity, and one sequence out of five (Jayasena personal communication) selected by the method described by Tuschl et al. *(9)* is efficient. In spite of the high cost for siRNA synthesis, the confirmation coming from this ratio renders this strategy very attractive for academic research in which exploring RNA to find an efficient sequence is expensive.

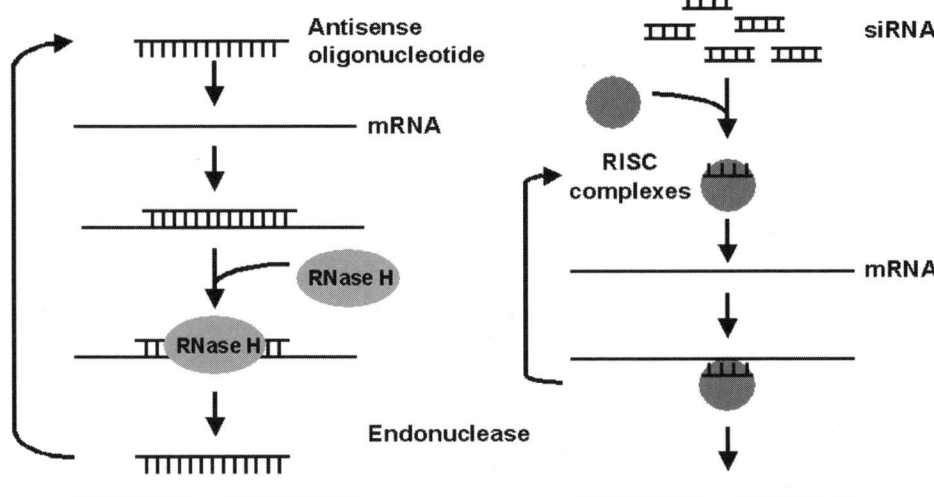

Fig. 2. Comparative mode of action of antisense oligonucleotide and siRNA. In both strategies, antisense oligonucleotide and activated RISC complexes after the targeted mRNA cleavage are able to be used again for the destruction of another complementary mRNA molecule.

2. Materials

1. Hybridization buffer: 100 mM potassium acetate, 30 mM HEPES–KOH, pH 7.4, 2 mM magnesium acetate.
2. Thermostatic bath for the incubations.
3. Centrifuge for Eppendorf tubes.
4. T7 RNA polymerase transcription buffer: 40 mM Tris-HCl, pH 7.9, 6 mM MgCl$_2$, 10 mM NaCl, and 2 mM spermidine.
5. 10 mM dXTP mix solution.
6. Yeast pyrophosphatase (cat. no. I 1643, Sigma).
7. 40 U/µL RNasin (Promega).
8. T7 RNA polymerase (New England Biolabs).
9. RQ1 RNase-free deoxyribonuclease (DNase) (Promega).
10. 2M Sodium acetate, pH 4.
11. Ethanol.
12. Miniprotean II electrophoresis apparatus (Bio-Rad) and all the material necessary for polyacrylamide gel preparation.
13. Stains-All (cat. no. E9379, Sigma).
14. Dimethyl formamide (DMF) (Sigma).
15. T4 polynucleotide kinase buffer 1X: 70 mM Tris-HCl, pH 7.6, 10 mM MgCl$_2$, 1 mM dithiothreitol.
16. 3000 Ci/mM [γ-^{32}P]ATP (ICN).

17. T4 polynucleotide kinase (Promega).
18. Sephadex G50 (Pharmacia).
19. 1M N-methylmorpholine buffer, pH 7.5 (Sigma).
20. 1-ethyl-3 (3'-dimethylaminopropyl)-carbodiimide (Sigma).
21. LiClO$_4$ (Sigma).
22. Acetone (Merck).
23. Fetal calf serum (FCS) (GIBCO Invitrogen).
24. Phenol:chloroform:*iso*-amyl alcohol (25:24:1).
25. Formamide (Sigma).
26. TE buffer: 40 mM Tris-HCl, pH 8.0, 2 mM ethylenediaminetetraacetic acid (EDTA).
27. S2 sequencing electrophoresis apparatus (Gibco-BRL).
28. PhosphorImager (Storm 840, Molecular Dynamics) or equivalent.
29. Cytofectine (Gene Therapy Systems).
30. Polystyrene tube.
31. Complexes formation buffer: 10 mM HEPES buffer, pH 7.4, 100 mM NaCl.
32. 6-well plates for cell culture (Falcon).
33. Phosphate-buffered saline (PBS) (GIBCO Invitrogen).
34. Solution D: 4M guanidium thiocyanate, 25 mM sodium citrate, pH 7.0, 0.5% sarcosyl, 0.1M β-mercaptoethanol.
35. 2M Sodium acetate, pH 4.
36. Saturated water phenol (Qbiogen).
37. Chloroform: *iso*-amylalcool (49:1).
38. Isopropanol (Prolabo).
39. 70% Ethanol solution.
40. Spectrophotometer and adapted ultraviolet cells.
41. 37% Formaldehyde (Sigma).
42. Horizon 58 agarose minigel (GIBCO BRL).
43. Major oocyst proteins (MOPS) (Sigma).
44. Solution 1: MOPS 1X, formaldehyde 6.4%, formamide 50%.
45. Solution 2: 1 mM EDTA, 0.25% bromophenol blue, 0.25% xylene cyanol, 100 µg/mL ethidium bromide.
46. 10 mg/mL ethidium bromide (Sigma).
47. BA-S 85 nitrocellulose (Schleicher & Schull).
48. Prime-a-gene labeling system (Promega)
49. 3000 Ci/mM [α-32P] deoxycytidine 5'-triphosphate (dCTP) (ICN).
50. 20X Sodium saline citrate (SSC) (Sigma): 3M NaCl, 0.3M sodium acetate, pH 7.0.
51. 50X Denhart solution: 5 g Ficol type 400 (Pharmacia), 5 g polyvinylpyrolidone (Sigma), 5 g bovine serum albumin fraction V (Sigma), water for 500 mL final volume.
52. Salmon sperm DNA (Roche).
53. Yeast transfer RNA (tRNA) (Roche).
54. Sodium dodecyl sulfate (SDS) (Sigma).

55. Bioprofil gel analyzer (Vilber Lourmat, France).
56. Reporter lysis buffer (RLB) (Promega).
57. Bincinchoninic acid (BCA) protein assay reagent (Pierce).
58. Glyo-Fixx solution (Shandon).
59. Loading buffer 6X: 10 mM Tris-HCl, pH 7.4, 30% glycerol, 0.01% bromophenol blue.

3. Methods
3.1. Design of a siRNA Sequence

The siRNA molecule consists of a 21–23-mer double-stranded oligoribonucleotide with two bases overhanging at their 3' extremities *(6)*. Hairpins with 19–29 basepair stems and 4–9 base loops are efficient too *(10)*. The design of such hairpin structure is dependent on the chosen synthesis strategy.

The target sequence is selected far from the protein interaction area on the mRNA. Then the 5' and 3' untranslated sequences have in principle to be eliminated owing to the presence of sequences involved in the transcription regulation and the ribosome binding. Nevertheless, different groups have described some efficient siRNAs targeted to such sequences *(11,12)*.

Tuschl et al. proposed a process for the selection of a target sequence on the desired mRNA (*[9]*; http://www.mpibcp.gwdg.de/abteilungen/100/105/sirna_u.html).

1. Selection of the open reading frame sequence of the gene of interest and elimination from it of the first and last 50 bases to suppress the sequences possibly involved with protein interaction.
2. Search for the motif sequence AA(N19)TT and selection of those with a 50% G/C content. A range between 30 and 70% is considered acceptable. Nevertheless, recently a correlation between a high-efficiency and a low G/C content was described *(13)*. The RNA complementary to AA(N19) will constitute the antisense strand of the siRNA. If no suitable sequences are found, an AA(N21) motif can be researched. The antisense sequence must be complementary to the target sequence. For the sense fragment it is not important that the two bases at the 3' end be complementary to the target sequence. They may be replaced by two deoxytymidines. Therefore the sense fragment becomes (N19)TT. A symmetrical siRNA molecule with a TT at each 3' extremity will be much more favorable for the formation of the RISC complexes (*see* **Note 1**).
3. Verify by a Basic Local Alignment Search Tool (BLAST) search in the expressed sequence tag library the selected siRNA sequence to ensure that the targeted gene is unique (http://www.ncbi.nlm.nih.gov/BLAST/).
4. If all the required conditions are not met, pass on to the next siRNA potential target sequence (*see* **Note 2**).

For the inhibition of gene expression experiment it is important to select a control siRNA. Because the introduction of a mismatch at the center of the siRNA sequence greatly decreases the inhibitory effect *(14)*, a good control may be obtained by the introduction of either 3- or 4-base substitution on a strand to keep the same base composition. After BLAST control this may constitute a good control molecule.

3.2. Synthesis of the siRNA

3.2.1. Synthesis of siRNA by Chemical Methods

Single-stranded fragments of the siRNA selected sequence may be synthesized with the (β-cyanoethyl) phosporamidite methods as indicated by the oligonucleotide synthesizer manufacturer. Many companies offer to chemically synthesize siRNAs and send them to researchers as single-stranded oligoribonucleotides, or directly as double-stranded ready-to-use RNA, such as Dharmacon Research, US (www.dharmacon.com); Eurogentech, Belgium (http://www.eurogentec.be/hp/hp.htm); MWG Biotech, Germany (http://www.mwg-biotech.com/html/all/index.php); and Qiagen, Germany (http://www.qiagen.com).

They offer assistance in selecting the siRNA sequence on a researcher's personal target mRNA, and they supply it online. This is a very convenient but as yet still expensive solution to obtain a siRNA molecule. We expect the price to decrease over time as a function of the market.

3.2.2. Enzymatic Synthesis of siRNA

Another way to obtain siRNA molecules is to produce them by an enzymatic method in vitro. Two strategies may be used: (1) the synthesis of each strand separately and, later, formation of the double-stranded siRNA by hybridization *(15)* or (2) the design of a hairpin structure *(10)*. For this synthesis strategy, in all cases, the sequence for the siRNA is determined as described in **Subheading 3.1.** Then, in vitro transcription is done by a phage promoter such as the T7 RNA polymerase model. The siRNA sequence is then AAG(N17)CTT. The introduction of a G and a C at the extremities of the target sequence is owing to the fact that T7 RNA polymerase starts the transcription with a G *(16)*.

The single-stranded matrix to be ordered for the synthesis of the RNA strand consists of an 18 mer sequence containing the T7 RNA polymerase promoter sequence at its 3' extremity, followed by the sequence corresponding to the sense or antisense strand of the siRNA sequence. A common complementary 18-mer sequence for the T7 promoter has to be ordered too (**Fig. 3**).

For the transcription reaction, we use the following steps as described early by Donzé and Picard *(15)*:

1. Annealing of the matrix oligonucleotide:
    a. In an Eppendorf-type tube add 1 nmol of the matrix oligonucleotide and 1 nmol of 18-mer T7 promoter oligonucleotide in 50 µL of TE buffer.

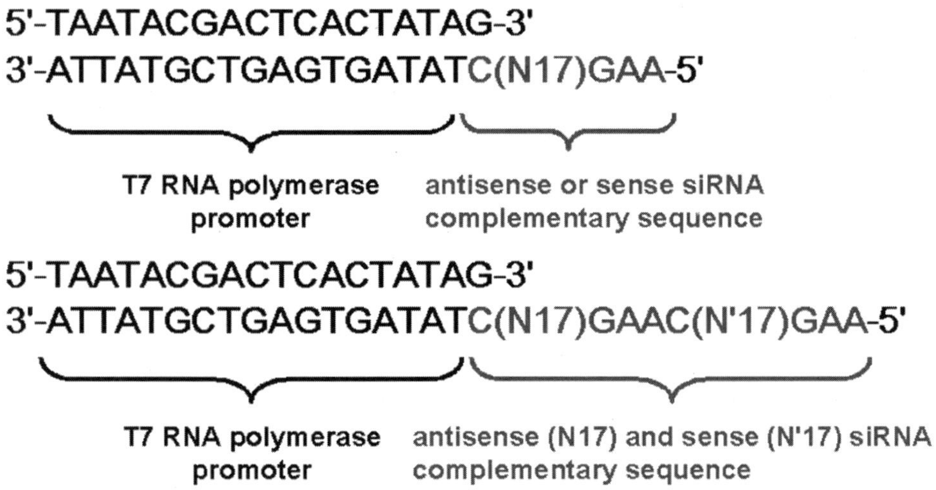

```
5'-TAATACGACTCACTATAG-3'
3'-ATTATGCTGAGTGATATC(N17)GAA-5'
   _____/_____/
            |              |
      T7 RNA polymerase   antisense or sense siRNA
          promoter        complementary sequence

5'-TAATACGACTCACTATAG-3'
3'-ATTATGCTGAGTGATATC(N17)GAAC(N'17)GAA-5'
   _____/_____/
            |                  |
      T7 RNA polymerase   antisense (N17) and sense (N'17) siRNA
          promoter        complementary sequence
```

Fig. 3. Structure of the oligonucleotides used as templates for the enzymatic synthesis of siRNAs. The top is for the synthesis of each single-stranded 21-mer RNA, and the bottom corresponds to the matrix DNA for the synthesis of a hairpin siRNA.

b. Heat for 2 min at 95°C.
c. Cool down slowly to obtain double-stranded DNA.
d. The final concentration is 20 pmol/μL of matrices solution.
2. Transcription: In two separate tubes transcribe the sense and the antisense strands as follows:
   a. In a tube mix 10 μL of the matrices solution (200 pmol).
   b. 5 μL of 10X T7 RNA polymerase transcription buffer.
   c. 1 mM Each ATP, UTP, CTP, and GTP.
   d. 0.1 U Yeast pyrophosphatase (cat. no. I 1643, Sigma).
   e. 40 U RNasin (Promega).
   f. 100 U T7 RNA polymerase (New England Biolabs).
   g. Water for 50 μL final volume.
   h. Incubate for 2 h at 37°C.
   i. Add 1 U of RQ1 RNase-free DNase (Promega) for 15 min at 37°C.
3. SiRNA duplex formation:
   a. The early synthesized sense and antisense strands are mixed without a further purification.
   b. Heat 5 min at 95°C, followed by a 1-h incubation at 37°C.
   c. Double-stranded siRNAs are then ethanol-precipitated by adding 10 μL of 2M sodium acetate, pH 4.0, and 275 μL ethanol.
   d. Mix gently and hold for 1 h at –20°C.
   e. Centrifuge 15 min at 12,000g.
   f. The pellet is washed with 70% ethanol.

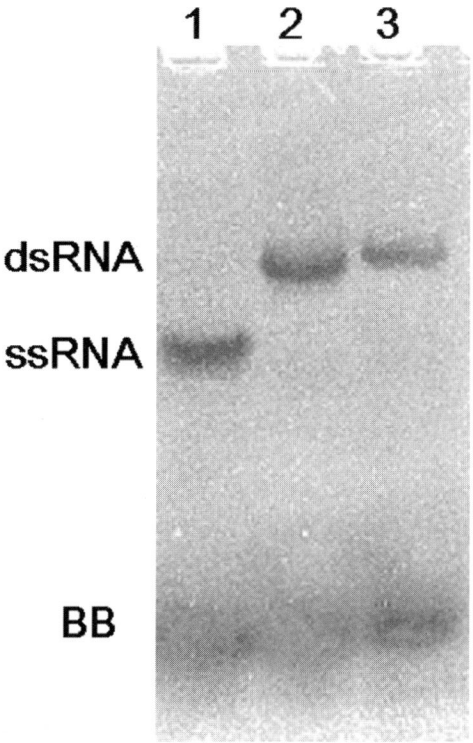

Fig. 4. Control of the double-stranded formation of siRNA after hybridization. A 21-mer single-stranded RNA (ssRNA) (5 µL of a 20 µM solution) (*lane 1*), or siRNA as a double-stranded RNA (dsRNA) solution hybridized in potassium acetate, HEPES, magnesium buffer (5 µL of a 20 µM solution) (*lane 2*), or in 150 mM NaCl solution (5 µL of a 10 µM solution) (*lane 3*), are analyzed in native PAGE and detected after Stains-All coloration. BB is bromophenol blue used, as a visualization control for gel migration.

 g. Dry and dissolve in 50 µL H$_2$O.
 h. The siRNA solution must be kept at –20°C.

3.3. Formation of Double-Stranded siRNA

The formation of duplexes is performed after the concentration of the RNA solution is determined. For a 21-mer solution, 1 optical density (OD) corresponds to a concentration of 30 µg/mL (4.46 µM), according to Sambrook et al. *(17)*.

1. We prepare a 20 µM solution by mixing an equimolar quantity of one sense and one antisense strand in hybridization buffer *(9)*. The use of a hybridization solution such as 150 mM NaCl is also possible (**Fig. 4**).

2. The solution has to be heated for 5 min at 95°C followed by 1 h at 37°C.
3. The siRNA solution must be stored at –20°C.

3.4. Control for Duplex Formation

To check whether the siRNA solution is in a duplex form, a very useful method consists in a native polyacrylamide gel electrophoresis (PAGE). The migration is done in a MiniProtean II (Bio-Rad).

1. Prepare a 20% acrylamide gel in 25 mM Tris-acetate pH 7.0, 50 mM sodium acetate, and 10 mM magnesium acetate buffer.
2. 5 μL of the duplex solution or a single-stranded RNA are mixed with 1 μL of loading buffer 6X and then dropped onto the gel.
3. Migration conditions are 20 V/cm for 1 h.
4. The PAGE gel is immersed in 0.5 g/L Stains-All in a 50% DMF solution for 5 min in darkness.
5. Rinse in water. In light, the staining of the gel disappears quickly, except at the nucleic acid location where it progressively fades.
6. When the band corresponding to the RNA reaches the right intensity the gel may then be photographed or dried (**Fig. 4**).

3.5. Determination of siRNA Resistance Toward Nucleases

The efficiency of nucleic acid as an inhibitor of gene expression is mainly dependent on its ability to stay intact for a long time in a biological medium. This leads us to check on whether siRNA is stable in a culture medium.

3.5.1. siRNA Labeling

The labeling of single-stranded sense or antisense RNA is done at 50 μL final volume:

1. 25 μL of a 20 μM RNA solution is added in an Eppendorf tube.
2. Add 5 μL of T4 polynucleotide kinase buffer 10X.
3. 3000 Ci/mM 4 μL [γ-^{32}P]ATP (ICN).
4. 20 U T4 polynucleotide kinase (Promega).
5. Incubate 1 h at 37°C.
6. Prepare a Sephadex G50 gel filtration column (Pharmacia) in a 1-mL syringe (*see* **Note 3**).
7. Centrifuge for 4 min at 1300g. The labeled oligonucleotide is in the filtrate.

3.5.2. Protection of the 5'-Terminal Phosphate of Oligonucleotides by Ethylation

To protect the terminal [^{32}P]-phosphate against enzymatic digestion the following modified methods are used:

siRNA as Inhibitor of Gene Expression

1. The single-stranded 5'-labeled siRNAs are dissolved in 100 μL of 1M N-methylmorpholine buffer, pH 7.5 (Sigma), 20 mM $MgCl_2$ containing 50% ethanol.
2. The solutions are supplemented by 10 mg of 1-ethyl-3 (3'-dimethylaminopropyl)-carbodiimide (Sigma).
3. The reaction proceeds for 16 h at 4°C.
4. The oligonucleotide ethyl esters are precipitated with 1 mL of 2% solution of $LiClO_4$ in acetone for 15 min at room temperature.
5. Centrifuge for 15 min at 12,000g.
6. Repeat **steps 4** and **5** twice.
7. The pellet is washed with acetone.
8. Mix with the unlabeled oligonucleotide to obtain the desired concentration, and then hybridize with the complementary strand as described in **Subheading 3.3**. The labeled double-stranded siRNA are then used to study their resistance to enzymatic digestion *(18)*.

3.5.3. Measurement of siRNA Stability in Serum

The determination of the siRNA nuclease degradation rate is carried out in FCS (GIBCO Invitrogen). To avoid the enzymatic [^{32}P]-labeled phosphate cleavage, the oligonucleotides with the protected terminal phosphate prepared as described above are used.

1. The [^{32}P]-labeled siRNA at 10 μM concentration is incubated in 120 μL of the corresponding medium at 37°C.
2. At various times, 15 μL aliquots are removed.
3. Supplement with 15 μL of 50 mM EDTA and freeze at −20°C.
4. Extract samples twice with phenol:chloroform: *iso*-amyl alcohol (25:24:1).
5. Precipitate the oligonucleotides from aqueous fractions with 10 vol of acetone containing 2% $LiClO_4$ for 15 min at −20°C.
6. Centrifuge 15 min at 12,000g.
7. Dry the pellet.
8. Dissolve in 5 μL of formamide:water (4:1), 0.01% bromophenol blue, and 0.01% xylene cyanol.
9. The samples are analyzed by electrophoresis in a 20% denaturing polyacrylamide gel TBE buffer (0.089M Tris-borate, pH 8.3, 0.02M EDTA) at 30 V/cm for 2 h.
10. The resulting gel is scanned using the phosphorImager (Storm 840, Molecular Dynamics).

The degradation of oligonucleotides is quantified as a ratio of the signal efficiency of bands corresponding to intact and degraded oligonucleotides. Uncertainty in degradation percentage is estimated at ±5% based on repetitions of experiments. Comparison between siRNA and its single-stranded 21-mer RNA counterpart shows that double-stranded siRNA is very resistant

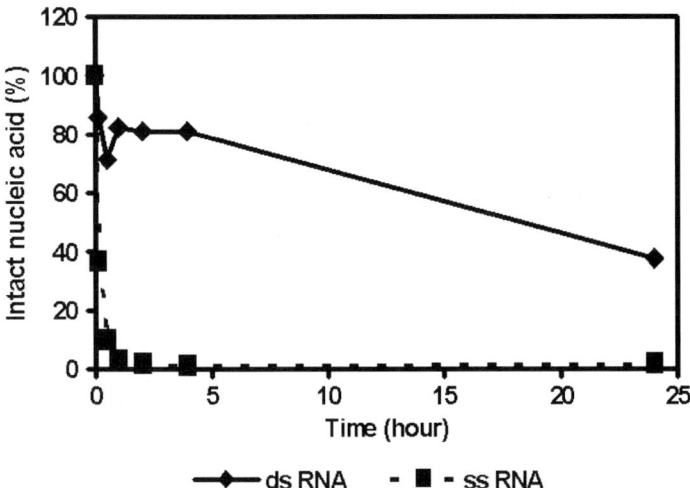

Fig. 5. Stability of siRNA in 100% FCS. Labeled siRNAs are incubated in pure fetal calf serum. At different times, aliquots are protein-extracted and analyzed on 20% PAGE containing 7M urea. The intact siRNA fragment is quantified on the gel after phosporImager analysis. The siRNA is very stable in these conditions with a half-life of 18 h compared to the <5 min obtained for the corresponding single-stranded RNA.

to nuclease degradation (**Fig. 5**). This result explains in part the very good efficiency of such a strategy (*see* **Note 4**).

3.6. Transfection of siRNA into Cell Culture

To be efficient, the siRNA molecule needs to penetrate into the cell before interacting with the target mRNA. This is a steep limiting of the action of siRNA and it depend on the cell type and the conditions of culture. Many companies encourage the use of transfection agents, which they claim are very efficient for such an application. For instance, cationic lipids are very susceptible to the presence of serum *(19)*. It is the reason why protocols for cell transfections are generally proposed in a culture medium (1) without serum at the time of the transfection, (2) with addition of serum to the cells 2–4 hr later, followed by a further incubation time before expression analysis. In these conditions, Lipofectine 2000 (Invitrogen), oligofectamine (Invitrogen), JetSI (Polyplus transfection), TransIT-TKO (Merus), Gensilencer siRNA (GTS) are proposed to be efficient agents for siRNA in absence of serum. Nevertheless, many of them are ineffective if serum is added to the medium at the time of the transfection. To measure the inhibition of many genes, the absence of serum in the medium might be of no consequence. It is different if the aim is to study the effect of suppression of a gene involved in a mechanism where the privation of serum may be an inductor of a phenotype such as apoptosis, differentiation, or signal transduction.

Cell electroporation is an alternative method and is successful for T lymphocytes siRNA transfection *(20)*.

In ours hands cytofectine (GTS) gives a good transfection efficiency in medium containing serum. Nevertheless, we were less successful with the transfection of suspension growing cells.

3.6.1. Preparation of the siRNA-Transfection Agent Complexes for a 100-nm siRNA Concentration

For an experiment in triplicate, we prepare a volume corresponding to 3.5 wells.

1. Mix in a polystyrene tube 157.5 µL of 10 mM HEPES buffer, pH 7.4, 100 mM NaCl, and 17.5 µL of 20 µM siRNA.
2. In an other tube dilute 4 µL of cytofectine transfection agent with 196 µL of 10 mM HEPES buffer, pH 7.4, and 100 mM NaCl.
3. Then 175 µL of the diluted cytofectine is added to the siRNA solution.
4. Mix gently and keep at room temperature for 15 min before adding to the cells.
5. We prepare under the same conditions a complex with a mismatched siRNA as a control.

3.6.2. Cell Transfection

One day before the transfection, the cells are seeded in a 6-well plate with $4\ 10^5$ cells per well.

1. Remove the culture medium and add 900 µL new medium, without selection agent if it is used for gene selection of cell strain.
2. Add 100 µL of the siRNA–cytofectine complex and mix by stirring gently.
3. Incubate for the desired time (*see* **Note 5**).
4. A control with the buffered solution is done (*see* **Note 6**).

3.7. Measurement of Gene Inhibition by siRNA

The determination of gene inhibition after the cell treatment by siRNA takes place at three levels:

1. Measurement by Northern blot or by quantitative reverse transcriptase-polymerase chain reaction (RT-PCR) of the expression of the targeted mRNA.
2. Determination by Western blot or after immunoprecipitation of the expression of the corresponding protein.
3. Measurement of the phenotype resulting from the mRNA inhibition.

3.7.1. Quantification at the RNA Level

3.7.1.1. RNA Purification

After incubation of the cell culture in 6-well plates, RNA extraction is done by the guanidium thiocyanate method *(21)*:

1. Wash one time with PBS buffer (GIBCO Invitrogen).
2. Lyse the cells with 400 µL of solution D.
3. Transfer the solution to an Eppendorf tube.
4. Homogenize by vigorously vortexing and pipetting.
5. Add 45 µL of 2*M* sodium acetate, pH 4.0, and vortex.
6. Add 400 µL of saturated phenol water (Qbiogen) and vortex.
7. Add 120 µL of chloroform:*iso*-amylalcool (49:1) and vortex.
8. Centrifuge 15 min at 12,000*g*.
9. 300 µL of the aqueous fraction (superior) is precipitated with 300 µL of isopropanol for 2 h at –20°C.
10. Centrifuge 15 min at 12,000*g*.
11. The RNA pellet is washed with 70% ethanol.
12. Centrifuge for 15 min at 12,000*g*.
13. The RNA pellet is dried at room temperature and dissolved in 10 µL H$_2$O containing 0.5 U/µL RNasin (Promega).

The RNA concentration is determined by spectrophotometry with the assumption that 1 OD$_{260}$ corresponds to a 30 µg/mL solution *(17)*.

3.7.1.2. Detection of the Targeted RNA by Northern Blot

1. 2 µg of total RNA are diluted in 3 µL H$_2$O.
2. Add 5 µL of solution 1.
3. Add 1.5 µL of solution 2.
4. Heat for 5 min. at 65°C.
5. Drop on a 1% agarose gel containing 6.29% formaldehyde (Sigma) in MOPS 1X buffer *(17)* with a Horizon 58 minigel (GIBCO) for 1.5 h at 80 V.
6. The RNA is blotted on BA-S 85 nitrocellulose (Schleicher & Schull) overnight in 10X SSC buffer *(17)*.
7. Heat the nitrocellulose for 2 h at 64°C.

Specific RNAs are detected by hybridization with a radiolabeled single-stranded probe.

1. 50 ng of probe (500–1000 bp) are labeled with the Prime-a-gene labeling system (Promega) using 3000 Ci/m*M* [α-32P] dCTP (ICN) as indicated by the supplier for 1 h at 37 °C.
2. The labeled probe is purified on a Sephadex G50 (Pharmacia) minicolumn as indicated in **Note 3**.
3. Heat the probe for 5 min at 95°C before addition to the hybridization buffer.

Hybridization is performed using the probe with 0.1 mL of hybridization solution per cm² of nitrocellulose membrane as follows:

1. Hybridization buffer is prepared in a 15-mL Falcon tube by mixing the following substances:
2. 2.5 mL of 20X SSC.
3. 100 µL of 50X Denhart solution.
4. 100 µL of salmon sperm DNA 10mg/mL; heat 5 min at 95°C.
5. 20 µL of yeast tRNA 10 Mg/mL.
6. 100 µL of 10% SDS.
7. 250 µL of $1M$ KPO$_4$, pH 7.0.
8. 5 mL of formamide.
9. The prehybridization step is done with 5 mL of the buffered solution for 2 h at 42°C (for 4.8 cm² nitrocellulose). The last 5 mL is conserved for the hybridization steps.
10. Hybridization occurs overnight at 42°C in the same medium where the heated probe was first added.
11. Wash the nitrocellulose four times at room temperature in 2X SSC and 0.1% SDS for 3 min.
12. Wash the nitrocellulose three times at room temperature in 0.1X SSC and 0.1% SDS for 3 min at 54°C.
13. Quickly dry the nitrocellulose between two Whatman 3MM papers.
14. The nitrocellulose is analyzed on a Storm 840 phosphorImager, and the corresponding RNA is quantified.

The gene expression is expressed as a percentage of untreated cells after the normalization of the detected RNA by a control gene (*see* **Note 7**).

Instead of blotting, quantitative RT-PCR may be done and starts generally with 1 µg total RNA.

3.7.2. Determination of the Protein Expression

For protein expression measurement, we used 6-well plates.

1. The cells on the plates are washed twice with a PBS solution.
2. The cells are scraped in the presence of 400 µL RLB (Promega).
3. Incubate the cells for 10 min at room temperature.
4. Centrifuge the cells for 15 min at 12,000*g*.
5. Proteins are detected in the supernatant. The RLB solution preserves the protein structure and then allows the detection of enzymatic activities (*see* **Note 8**).
6. The total protein concentration is detected by the BCA colorimetric method (Pierce).

The dose effect may be performed by using variable volumes of the siRNA–cytofectine complexes as presented in **Fig. 6**.

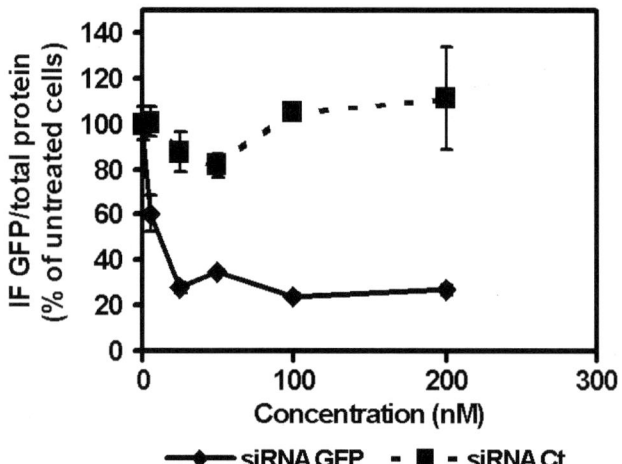

Fig. 6. Inhibition of GFP expression in HeLa cells by siRNAs. Dose effect is shown after a 5-h incubation with siRNA targeted to the GFP mRNA or with a nontargeted siRNA. A plateau response was observed for concentrations up to 25 nm. This effect may be owing to a saturation of the pool of RISC complexes.

3.8. Inhibition of Gene Inhibition In Vivo

Animals experiments are developed with siRNA molecules *(22–24)*. For these experiments, mice are injected in the liver with 40 µg of siRNA *(23)*, or 50 µg siRNA in 1 mL PBS is rapidly injected in the tail vein *(24)*, and injections are repeated after 8 and 24 h. Then gene expression is detected.

For HeLa tumor cells *(21)*:

1. 2×10^6 cells in 100 µL PBS are subcutaneously inoculated into nude mice.
2. Seven days after tumor inoculation, subcutaneous tumors are of visible size.
3. 100 µL of GFP-targeted siRNA (20 µg) or control siRNA vectorized with 30 µg cytofectine is injected in the tumor.
4. Twenty-four hours later, the tumors are discarded and sectioned in two parts.
5. The first one is used for histological analysis after immersion in Glyo-Fixx solution (Shandon).
6. The second one is frozen in liquid nitrogen for Northern blot analysis. Extraction of the total RNA is performed after homogenization in solution D as described in **Subheading 3.7.1.1**.

4. Notes

1. If the target mRNA does not contain a suitable AA(N21) sequence, a CA(N21) may be researched. Then, sense and antisense may be synthesized as a (N19)TT because the introduction of modification at the 3' ends of the siRNA has very few

consequences on their inhibitory efficiency. It is important to notice that the 19 bases at the center of the siRNA molecule must form a perfectly matched duplex, and introduction in this sequence of a modified base might suppress its inhibition potential. Nevertheless, it is possible to introduce different modifications at the 3' extremities without consequences to their inhibitory effect *(7)*. It is why many companies propose to synthesize siRNA with the introduction of two deoxyribothymidines in place of the two ribouridine at the 3' extremity of each strand, to decrease the siRNA synthesis price and to protect them from nuclease degradation *(9)*.
2. Because the siRNA strategy is recent, we have not yet enough statistical information to design a rule. Nevertheless these simple rules are sufficient to design efficient siRNAs for biological applications. As a precaution it may be preferable to select two sequences to be sure to obtain a good gene inhibition.
3. To prepare a Sephadex G50 column, we need first to hydrate some Sephadex G50 powder into sterile water for half an hour. Prepare a column in a 1-mL syringe by introducing two rings of Whatman GF/C filters with the assistance of the syringe plunger. Then introduce full-hydrated Sephadex G50 resin with a pipet. Introduce the syringe in a 15-mL Falcon tube and centrifuge 4 min. at 1300g. Discard the water in the Falcon tube and add a 1.5-mL Eppendorf tube without the cap (cut it properly). Then you may drop on the surface of the resin the labeled oligonucleotide you have to purify. Centrifuge 4 min. at 1300g. The purified oligonucleotide is in the Eppendorf tube.
4. Following the same method, the stability of siRNA may be studied in a cell lysate or in another biological medium.
5. The time of incubation is dependent on two parameters:
   a. The half-life of the targeted protein.
   b. The half-life of the siRNA in the cell. It has been reported that siRNAs after transfection are able to block gene expression for more than 48 h *(7,14)*.
   In these conditions if the gene of interest has a short half-life, an inhibition may be observed after 5 h incubation *(22)*. For long half-life protein, it is possible to perform a new transfection after 24 or 48 h. One has to be careful that the number of cells to be seeded on the plate at the beginning of the experiment is not too large because the confluence might result in cell death and render difficult the result analysis.
6. The incubation of the cells with the transfection agent alone may be an incorrect control because in some cases it is cytotoxic. This cytotoxicity of cytofectine disappears when it forms a complex with nucleic acids.
7. The normalization of the RNA expression is done by 18 S or 28 S RNA measured by ethidium bromide fluorescence with Bioprofil gel analyzer (Vilber Lourmat, France) before nitrocellulose heating. The normalization may also be done on the same nitrocellulose membrane with another Northern blot-detected gene (G3PDH). As far as expression-vector transfected cells are concerned, the gene responsible for the antibiotic selection of the cells might also be used for normalization (neomycin phosphotransferase, puromycin acetyl transferase).

8. For GFP detection, fluorescence is directly detected on the lysate (ex: 473 nm; em: 505 nm).

Acknowledgments

The authors thank M. Pottier for technical assistance and the "service d'experimentation annimale" (P. Ardouin and M. Stanciu). This work was supported by the Association pour la Recherche sur le Cancer (Grant no. 4310).

References

1. Tushl, T. (2001) RNA interference and small interfering RNAs. *Chembiochem.* **2**, 239–245.
2. Hannon, G. J. (2002) RNA interference. *Nature* **418**, 244–251.
3. Shuey, D. J., McCallus, D. E., and Giordano, T. (2002) RNAi: gene-silencing in therapeutic intervention. *Drug Discov. Today* **7**, 1040–1046.
4. Stark, G. R., Kerr, I. M., Williams, B. R., Silverman, R. H., and Schreiber, R. D. (1998) How cells respond to interferon. *Annu. Rev. Biochem.* **67**, 227–264.
5. Elbashir, S. M., Harborth, J., Lendeckel, W., Yalcin, A., Weber, K., and Tuschl, T. (2001) Duplexes of 21-nucleotide RNAs mediate RNA interference in cultured mammalian cells. *Nature* **411**, 494–498.
6. Elbashir, S. M., Lendeckel, W., and Tuschl., T. (2001) RNA interference is mediated by 21- and 22-nucleotide RNAs. *Genes Dev.* **15**, 188–200.
7. Holen, T., Amarzguioui, M., Wiiger, M. T., Babaie, E., and Prydz, H. (2002) Positional effects of short interfering RNAs targeting the human coagulation trigger tissue factor. *Nucleic Acids Res.* **30**, 1757–1766.
8. Sijen, T., Fleenor, J., Simmer, F., et al. (2001) On the role of RNA amplification in dsRNA-triggered gene silencing. *Cell* **107**, 465–476.
9. Elbashir, S. M., Harborth, J., Weber, K., and Tuschl, T. (2002) Analysis of gene function in somatic mammalian cells using small interfering RNAs. *Methods* **26**, 199–213.
10. Yu, J. Y., DeRuiter, S. L., and Turner, D. L. (2002) RNA interference by expression of short-interfering RNAs and hairpin RNAs in mammalian cells. *Proc. Natl. Acad. Sci. USA* **99**, 6047–6052.
11. McManus, M. T., Haines, B. B., Dillon, C. P., et al. (2002) Small interfering RNA-mediated gene silencing in T lymphocytes. *J. Immunol.* **169**, 5754–5760.
12. Vickers, T. A., Koo, S., Bennett, C. F., Crooke, S. T., Dean, N. M., and Baker, B. F. (2003) Efficient reduction of target RNAs by small interfering RNA and RNase H-dependent antisense agents: a comparative analysis. *J. Biol. Chem.* **278**, 7108–7118.
13. Parsons, N., (2003) *Communication to the Antisense and siRNA Technologies:* conference organized by SMI at the Hatton, London, 12–13 February.
14. Amarzguioui, M., Holen, T., Babaie, E., and Prydz, H. (2003) Tolerance for mutations and chemical modifications in siRNA. *Nucleic Acids Res.* **31**, 589–595.
15. Donzé, O., and Picard, D. (2002) RNA interference in mammalian cells using siRNA synthesized with T7 RNA polymerase. *Nucleic Acids Res.* **30**, e46.

16. Milligan, J. F., Groebe, D. R., Witherell, G. W., and Uhlenbeck, O. K. (1987) Oligoribonucleotide synthesis using T7 RNA polymerase and synthetic DNA templates. *Nucleic Acids Res.* **15,** 8783–8798.
17. Sambrook, J., Fritsch, E. F., and Maniatis, T. (1989) *Molecular Cloning: A Laboratory Manual.* Cold Spring Harbor Laboratories, Cold Spring Harbor, NY.
18. Maksimenko, A. V., Gottikh, M. B., Helin, V., Shabarova, Z. A., and Malvy, C. (1999) Physico-chemical and biological properties of antisense phosphodiester oligonucleotides with various secondary structures. *Nucleosides Nucleotides* **18,** 2071–2091.
19. Kang, S. H., Zirbes, E. L., and Kole, R. (1999) Delivery of antisense oligonucleotides and plasmid DNA with various carrier agents. *Antisense Nucleic Acid Drug Dev.* **9,** 497–505.
20. McManus, M. T., Haines, B. B., Dillon, C. P., et al. (2002) Small interfering RNA-mediated gene silencing in T lymphocytes. *J. Immunol.* **169,** 5754–5760.
21. Chomczynski, P. and Sacchi, N. (1987) Single step method of RNA isolation by acid guanidium thiocyanate-phenol-chloroform extraction. *Anal. Biochem.* **162,** 156–159.
22. Bertrand, J.-R., Pottier, M., Vekris, A., Opolon, P., Maksimenko, A., and Malvy, C. (2002) Comparison of antisense oligonucleotides and siRNAs in cell culture and in vivo. *Biochem. Biophys. Res. Commun.* **296,** 1000–1004.
23. McCaffrey, A. P., Meuse, L., Pham, T. T., Conklin, D. S., Hannon, G. J., and Kay, M. A. (2002) RNA interference in adult mice. *Nature* **418,** 38–39.
24. Song, E., Lee, S. K., Wang, J., et al. (2003) RNA interference targeting Fas protects mice from fulminant hepatitis. *Nat. Med.* **9,** 347–351.

Index

A

2'-ACE chemistry, 34, 42
Acridine-labeled oligonucleotide primers, 307, 311, 314
Acridine-PNA conjugate, 350
Actinomycin D, footprinting, 319, 329, 334, 335
Adamantane carbonyl chloride (AdCl), 83
Affinity chromatography, 380, 381
2'-Aldehyde oligonucleotides, 220
Altritol nucleic acids (ANAs), 309
Aminoethylglycin PNA, 343
Aminoethylpropyl (aep) backbone, 345
Analysis of RNA cleavage, 266
Anchor group, 382
Antigene strategy, 252, 343
Antisense oligonucleotide (AON)-RNA hybrid, 66
Aptamer, 17, 379, 380, 381, 391, 402
Arabinonucleic acids, 68
Aryl H-phosphonate monoesters, 85
A-type conformation, 307
Autoradiography, 333
8-aza-7-deazapurines, 165, 166, 168, 172, 177

B

Base-modified oligonucleotides, 165
5-(Benzylmercapto)-1H-tetrazole, 18
Biotin phosphoramidites, 225, 226, 230, 232, 235, 237
Biotin-labeled oligonucleotide, 225, 237, 238
Bis(2-oxo-3-oxazolidinyl) phosphinic chloride, 83
Bis(*N,N*-di-isopropylamino) chlorophosphine, 85
Bromomethyl- or aryldiazomethane reactive group, 241
5-(Bromomethyl)fluorescein (5-BMF), 242, 244, 245, 248

C

Camptothecin (CPT), 251, 252, 256
Carbomethoxy-2-thiouracil, 192
Carbomethoxy-2-thiouridine, 194
5-Carbomethoxymethyl-2-thiouridine, 193
Cationic liposomes, 350
Cell targeting, 205
Cell-penetrating peptides, 350
Cellular penetration, 252
Chemical interference, 402
Chemically modified aptamers, 404
Chimeric 2'-*O*-Me/RNA oligonucleotides, 404
Chimeric monomers, 147
3-Chloro-4-hydroxyphenylacetic acid, 106, 109, 121
Circular oligonucleotides, 101, 102, 121
Cleavage of RNA, 261

Combinatorial nucleic acid libraries, 379
Combinatorial synthesis, 391
Combined sulfurization-capping, 53
Covalent conjugation, 252
2-Cyanoethyl *H*-phosphonate, 85
2-Cyanoethyl (*N*-isopropylamino-chlorophosphine), 85
Cy5-labeled RNA, 261, 262, 270, 308
Cyclohexene nucleic acids, 68
Cytofectine (GTS), 423

D

Delivery of oligonucleotides, 205
Densitometric analysis, 329
Deoxyribonuclease (DNase), 319
Determination of siRNA resistance toward nucleases, 420
Diamine linker, 255
Diaminopurine, 345
DICER, 412
Diels-Alder reaction, 382
Diels-Alderase ribozymes, 387
5,5-Dimethyl-2-oxo-2-chloro-1,3,2-dioxaphosphinane, 83
Dimethylthioureum disulfide, 51, 52, 59
Dinucleoside *H*-phosphonates, 87
Diol groups, 207, 215, 220
2'-Diol, 220
Diphenyl *H*-phosphonate, 85
Direct sequencing of genomic DNA, 294
DNA
  chips, 241
  cleavage, 242
  random library, 394
  SELEX against RNA targets, 394
  -binding drugs, 319
DNase I, 321, 322, 325, 326, 332

Double-stranded RNA-dependent protein kinase, 412
Double-stranded DNA pool, 386
Drug–DNA complexes, 326
Duplex flexibility, 68
Duplex stability, 177, 179

E

Electroelution, 325
Electrophoresis, 327, 333
Enzymatic synthesis of a siRNA, 417
5-Ethylthio-1*H*-tetrazole, 38, 41

F

Fimers, 291, 293
  design, synthesis and purification, 298, 299
Fluorescent *in situ* hybridization (FISH), 346
Fluorescently labeled probes, 261, 275, 323
2'-Fluoro-arabinonucleic acids, 68
Fluorophore/fluorescene, 277, 281, 285, 3111
Footprinting assay, 323, 337
Formation of double-stranded siRNA, 419
Fragment coupling of peptides, 206
Freeze–squeeze, 325

G

G-clamp, 345
Gel analysis and quantification, 329
Gel-shift experiments, 262, 268
Gene inhibition
  by siRNA, measurement of, 423
  in vivo, 426
Gene-specific reagents, 251
Glycosylation reaction, 168
Genomic deoxyribonucleic acids (DNA), sequencing of, 291

Index

H

Halogenated nucleoside residue, 179
Helix invasion, 343
Hexitol nucleic acids, 309
High-density DNA chip, 242
HIV-1 Tat protein, 262
H-Phosphonate chemistry, 81, 82, 84, 87, 88
Hydrazone linkage, 207, 221
Hydrogen peroxide, 10
4-Hydroxyproline, 153
Hypermodified nucleotides, 187, 193
HypNa-pPNA, 147, 148, 149, 151, 155, 156, 159

I

In vitro selection, 379, 380, 387
Inactivation of RNase, 265
Interference mechanism, 411
Intracellular bioavailability of PNA, 350
Iodophosphates, 89

L

Labeled nucleic acids, 242
Labeling during cleavage (LDC) approach, 242, 245, 248
Labeling plus cleavage (LPC) procedure, 244, 245, 248
Large-scale pool amplification, 369
Ligand-DNA interactions, 320
Ligation, 101
Linking pool segments using type II restriction enzymes, 371
LNA oligonucleotide(s), 127, 128, 138
LNA–DNA chimeras, 127, 128, 131, 138
Locked nucleic acid (LNA), 127, 166

Long noncharged tethering arm, 226
L-threonine, 199
Lysine transfer ribonucleic acid, 187

M

Meta-biotinphenyl-methyldiazomethyl(*m*-BioPMDAM), 243, 244, 245, 247, 248
Meta-chloroperbenzoic acid, 10
Methoxyoxalamido (MOX) and succinimido (SUC) precursor strategies, 226, 293
Methylaminomethyl-2-thiouridine, 187, 196
Methylcarboxymethyl-2-thiouridine, 187
2-Methyl-phosphorimidazolides, 307
Methylthiopurin-6-yl-carbamoyl-L-threonine, 199
2-Methylthiopurin-6-yl-carbamoyl, 200
Microbial genomic DNA isolation, 300
Minor groove binders, 323
Molecular beacon, 273, 275
  characterization of, 282
  design, 280
  multiplex detection using, 285
  synthesis, 281

N

N^2-isobutyrylguanin-N^9-acetic acid, 162
N^4-benzoylcytosine-N^1-acetic acid, 160
N^6-benzoyladenin-N^9-acetic acid, 160
Nonenzymatic oligomerization reactions, 305, 307
Nucleases, 148

footprinting techniques, 319
resistance, 101
Nucleic acid
  aptamers, 359
  labeling, 241

O

O-alkyl hydroxylamine, 307
5'-O-carbonate protecting group, 3
5'-O-silyl ether protecting group, 37
Oligonucleotide,
  library, 359, 391
  synthesis, 91
  synthesis in solution, 81
Oligopyrimidine-oligopurine sequences, 252
Optimization using SYBR Green real-time PCR, 278
Oxime, 207
  and hydrazone ligation, 214, 221

P

PCR, 276, 369, 386, 398
Peptide nucleic acid (PNA), 147, 166, 308, 343
Peptide-oligonucleotide conjugate, 205
Peroxy anion deprotection and oxidation, 3
Phage display libraries, 359
Phenoxazinone, 331
Phenylacetyl disulfide, 58
Phosphite triesters, 82
Phosphono peptide nucleic acid (Phosphono-PNA), 147, 345
Phosphorimidazolide, 51, 91, 313
Phosphorothioate, 51, 59, 68, 91, 411
  linkage, 102
  LNA, 142
Phosphotriester method, 101, 102, 119

Pivaloyl chloride, 83
Plasmid pBlueScribe (pBS), 323
PNA clamping of polymase chain reactions, 346
PNA–peptide conjugates, 348, 350
Point deletions, 365
Point mutations, 363
Polymerase chain reaction (PCR), 291, 323, 360, 391
Polymerases, 101
Pool deprotection, purification, and quality control, 368
Prebiotic RNA synthesis, 306
Primer design, 278
Proteases, 148
Protein–nucleic acid interactions, 262, 319
Pseudoisocytosine, 345
Pyranosyl-RNA, 306
Pyrazolo[3,4-d] pyrimidines, 165, 166, 168
Pyridinium H-pyrophosphonate, 85

Q

Quadruplex, 102

R

Real-time
  amplification assays, 274
  polymerase chain reaction (PCR), 273, 284
Reverse transcription of the selected RNAs, 388
Ribonuclease H, 65, 346
Ribonucleic acid (RNA) synthesis, 92
Ribozymes, 17, 381, 382
RNA
  cleavage, 66, 242
  fragmentation patterns, 262
  library, 381
  ligand interactions, 261, 268

SELEX against RNA targets, 396
synthesis, 17, 33, 94, 187, 189
-DNA hybrids, 65, 66
-induced silencing complex (RISC), 412
RNase H, 66, 391, 411, 412
-mediated cleavage, 68, 77, 392, 399

S

Salicylchlorophosphite, 85
Selection of DNA aptamers to a protein, 399
Selection step using BIAcore™, 402
Short double-stranded RNA, 411, 412
Single nucleotide polymorphisms (SNPs), 275
Single-stranded DNA pool, 385
Single-stranded RNA pool, 387
siRNA, 17, 33, 387, 411, 412
  sequence, design of, 416
Solid-phase synthesis, stepwise, 205, 206
Solution-phase coupling, 205, 206
Specific and nonspecific hybridization, 293
Splice interference technology, 347
Split-bead synthesis, 360, 366
Stop solutions, 327
7-Substituted 8-aza-7-deazapurine, 179,
Succinimido (SUC) precursor strategies, 293
Sulfur-transfer reagent, 51
SYBR Green, 274
SYBR Green PCR, 284
Synthetic oligonucleotide strand extension, 374

Systematic evolution of ligands by exponential enrichment (SELEX), 392, 404

T

Target design, 278
Tert-butyl hydroperoxide, 41, 104, 109, 187
Tert-butyldimethylsilyl (TBDMS) protection (+ 2'-*O*-TBDMS), 18, 94
Tert-butyldimethylsilyl chloride, 193, 200
Tert-butyldimethylsilyl triflate, 199
Tetrazole, 9, 53, 104
Thiazolidines, 207, 214, 221
2-Thiomethyl-*N*-6-carbamoylthreonyl-adenosine, 187
Threofuranosyl nucleic acid, 306
Topoisomerases, 252
Trans-4-hydroxy-L-proline, 147
Transactivation response element (TAR), 262, 392
Transcription initiation, 187
Transfection of siRNA into cell culture, 422
Triethylamine trihydrofluoride $Et_3N(3HF)$, 18, 41, 190, 405
Trimethylsilyl chloride, 89
Triple helix, 102, 148, 251, 252, 343
Triplex and duplex invasion, 348
Two-step synthesis cycle, 3

W

Whole-genome direct sequencing protocols, 293, 301